T0204031

Mechanical
Tribology

Mechanical Tribology

Materials, Characterization, and Applications

edited by

George E. Totten

G. E. Totten & Associates, LLC
Seattle, Washington, U.S.A.

Hong Liang

University of Alaska
Fairbanks, Alaska, U.S.A.

CRC Press
Taylor & Francis Group
Boca Raton London New York

CRC Press is an imprint of the
Taylor & Francis Group, an **informa** business

CRC Press
Taylor & Francis Group
6000 Broken Sound Parkway NW, Suite 300
Boca Raton, FL 33487-2742

First issued in paperback 2019

© 2004 by Taylor & Francis Group, LLC
CRC Press is an imprint of Taylor & Francis Group, an Informa business

No claim to original U.S. Government works

ISBN-13: 978-0-8247-4873-9 (hbk)
ISBN-13: 978-0-367-39436-3 (pbk)

This book contains information obtained from authentic and highly regarded sources. Reasonable efforts have been made to publish reliable data and information, but the author and publisher cannot assume responsibility for the validity of all materials or the consequences of their use. The authors and publishers have attempted to trace the copyright holders of all material reproduced in this publication and apologize to copyright holders if permission to publish in this form has not been obtained. If any copyright material has not been acknowledged please write and let us know so we may rectify in any future reprint.

Except as permitted under U.S. Copyright Law, no part of this book may be reprinted, reproduced, transmitted, or utilized in any form by any electronic, mechanical, or other means, now known or hereafter invented, including photocopying, microfilming, and recording, or in any information storage or retrieval system, without written permission from the publishers.

For permission to photocopy or use material electronically from this work, please access www.copyright.com (http://www.copyright.com/) or contact the Copyright Clearance Center, Inc. (CCC), 222 Rosewood Drive, Danvers, MA 01923, 978-750-8400. CCC is a not-for-profit organization that provides licenses and registration for a variety of users. For organizations that have been granted a photocopy license by the CCC, a separate system of payment has been arranged.

Trademark Notice: Product or corporate names may be trademarks or registered trademarks, and are used only for identification and explanation without intent to infringe.

Library of Congress Cataloging-in-Publication Data
A catalog record for this book is available from the Library of Congress.

Visit the Taylor & Francis Web site at
http://www.taylorandfrancis.com

and the CRC Press Web site at
http://www.crcpress.com

Preface

There are many texts and handbooks available describing surface characterization and various aspects of tribology. However, there has been a need for a single text that combines these subjects and integrates them into various application technologies. This book was developed to address this need from a thorough, mechanistic perspective.

This book is divided into two parts. Part One, "Material and Tribological Characterization" describes surface characterization methodologies, concentrating on the chemical structure of surfaces (Chapter 1) and the physical structure of surfaces (Chapter 2). These two chapters provide a thorough discussion of a wide range of surface structural characterization methodologies. This discussion is followed by an extensive, in-depth treatment of the surface properties and tribology of plastics (Chapter 3), a description of a relatively new thermal analysis surface characterization that combines microthermal surface analysis and atomic force microscopy (AFM) for surface structural elucidation (Chapter 4). A rigorous discussion of the macro- and micromechanical properties of ceramics is provided in Chapter 5.

In addition to surface structural analysis, various methodologies for examining lubrication and wear are also discussed in Part One. These chapters include: methodologies for characterizing and scuffing and seizure wear processes (Chapter 6), characterization of wear mechanisms by wear mapping (Chapter 7), and the measurement and characterization of thin film lubrication (Chapter 8).

Part Two, "Tribological Applications" focuses on selected characterization methodologies that include: metal cutting (Chapter 9), metal forming (Chapter 10), textile manufacturing (Chapter 11), and biotribology (Chapter 12). These topics were selected because they are omitted in many texts but are vitally important manufacturing technologies, especially as they relate to a wide range of surface modification technologies. This discussion is followed by a topic of increasing importance in the tribological community, biocompatible metals and alloys (Chapter 13). Finally, Chapter 14 discusses a relatively little known process, epilamization barrier films, which prevent oil from spreading and creeping in lubricating applications.

This book will not only be useful to material scientists and engineers, mechanical engineers, tribologists, and lubrication engineers as an invaluable addition to their collection, but it may also be used as a textbook for advanced undergraduate or graduate courses in tribology and wear.

We are grateful for the vital assistance of the various international experts for their contributions. Special thanks to the staff at Marcel Dekker, Inc., and Richard Johnson for their patience and invaluable assistance.

George E. Totten
Hong Liang

Contents

Part Two   Tribological Applications

Contributors

Viktor P. Astakhov, Ph.D., Dr.Sci Astakhov Tool Service Company, Rochester Hills, Michigan, U.S.A.

Besim Ben-Nissan, Ph.D. Department of Chemistry, University of Technology, Sydney, Sydney, Australia

Narelle Brack, Ph.D. Department of Physics, Centre for Materials and Surface Science, La Trobe University, Bundoora, Victoria, Australia

Tadeusz Burakowski, D.Sc., Ph.D., M.Sc. Eng. Department of Mechanical Engineering, Radom Technical University, Radom, Poland

Namita Roy Choudhury, Ph.D. Ian Wark Research Institute, University of South Australia, Mawson Lakes, South Australia, Australia

Naba Dutta, Ph.D. Ian Wark Research Institute, University of South Australia, Mawson Lakes, South Australia, Australia

Scott Edwards, Ph.D. Ian Wark Research Institute, University of South Australia, Mawson Lakes, South Australia, Australia

Alexia W. E. Hodgson, Ph.D. Department of Materials, Swiss Federal Institute of Technology, Zurich, Zurich, Switzerland

Peter D. Hodgson, Ph.D. School of Engineering and Technology, Deakin University, Geelong, Victoria, Australia

Hong Liang, Ph.D. Department of Mechanical Engineering, University of Alaska, Fairbanks, Fairbanks, Alaska, U.S.A.

Christina Y. H. Lim, Ph.D. Department of Mechanical Engineering, National University of Singapore, Singapore

S. C. Lim, Ph.D. Department of Mechanical Engineering, National University of Singapore, Singapore

Jianbin Luo, Ph.D. State Key Laboratory of Tribology, Tsinghua University, Beijing, China

Stephen Michielsen, Ph.D. Department of Textile and Apparel Technology and Management, College of Textiles, North Carolina State University, Raleigh, North Carolina, U.S.A.

Kazuhisa Miyoshi, Ph.D., Dr. Eng. Department of Materials, National Aeronautics and Space Administration, Glenn Research Center, Cleveland, Ohio, U.S.A.

Wolfgang H. Müller, Dr. rer. nat. habil. Dipl.-Phys. Technische Universität, Berlin, Berlin, Germany

Nikolai K. Myshkin, Ph.D., D.Sc. Department of Tribology, Belarus National Academy of Sciences, Gomel, Belarus

Mark I. Petrokovets, Ph.D., D.Sc. Department of Tribology, Belarus National Academy of Sciences, Gomel, Belarus

Giuseppe Pezzotti, Ph.D. Department of Chemistry and Material Engineering, Kyoto Institute of Technology, Kyoto, Japan

Paul J. Pigram, Ph.D. Department of Physics, Centre for Materials and Surface Science, La Trobe University, Bundoora, Victoria, Australia

Maria Provatas, B.Sc. Ian Wark Research Institute, University of South Australia, Mawson Lakes, South Australia, Australia

Zygmunt Rymuza, Ph.D., D.Sc., Meng. Department of Mechatronics, Institute of Micromechanics and Photonics, Warsaw University of Technology, Warsaw, Poland

Dirk Jan Schipper, Dr. ir. Department of Surface Technology and Tribology, University of Twente, Enschede, The Netherlands

Bing Shi Department of Mechanical Engineering, University of Alaska, Fairbanks, Fairbanks, Alaska, U.S.A.

Marian Szczerek, D.Sc., Ph.D., M.Sc. Eng. Department of Tribology, Institute for Terotechnology, Radom, Poland

Nguyen Duc Tran Ian Wark Research Institute, University of South Australia, Mawson Lakes, South Australia, Australia

Waldemar Tuszynski, Ph.D, M.Sc. Eng. Department of Tribology, Institute for Terotechnology, Radom, Poland

Emile van der Heide, Dr. ir. Department of Tribology, TNO Industrial Technology, Eindhoven, The Netherlands

Sannakaisa Virtanen, Ph.D. Department of Materials, Swiss Federal Institute of Technology, Zurich, Zurich, Switzerland

Heimo Wabusseg, Ph.D. Department of Materials, Swiss Federal Institute of Technology, Zurich, Zurich, Switzerland

Shizhu Wen, B.Sc. State Key Laboratory of Tribology, Tsinghua University, Beijing, China

1
Surface Characterization of Materials

Paul J. Pigram and Narelle Brack
La Trobe University, Bundoora, Victoria, Australia

Peter D. Hodgson
Deakin University, Geelong, Victoria, Australia

I. INTRODUCTION

A. The Nature and Importance of Surfaces

The surface of a material is the interface between the bulk and the external phase in direct contact with the material. The external phase may be solid, liquid, or gas. The surface of a material may be defined in a number of ways, depending principally on the interaction being considered. In fundamental terms, the outermost layer of atoms of the material composes the surface. The physical and chemical behavior of these atoms, however, is strongly influenced by atomic layers in the vicinity, to a depth of the order of several nanometers into the bulk. In practical terms, surface modification and the application of thin films, for example, for lubrication, creates a functional surface region of the order of 100 nm thick [1].

Controlling and characterizing the behavior of surfaces is central to physical and chemical tribology. Friction and wear processes occur at surfaces and interfaces and are a manifestation of the physical and chemical characteristics of the materials in question. Surface modification is the means by which these processes are controlled or mitigated. For example, the inherent wear properties of a material depend on parameters such as hardness, structure at the surface, and chemical reactivity. Erosion may be influenced by preferential surface segregation of species and processes occurring at the surface such as oxidation and degradation. Wear behavior may be controlled by the application of hard surface films, as in plasma nitriding, and appropriate lubrication regimes.

Analytical tribology necessarily involves the determination of the chemical, electronic, and structural characteristics of the surface and wear debris. Modern surface analytical techniques provide a comprehensive understanding of tribological mechanisms via spectroscopy, imaging, and depth profiling.

B. Surface Properties and Processes Occurring at Surfaces

Surface phenomena, which determine surface properties and processes, occur on scales ranging from tenths to hundreds of nanometers. Figure 1 summarizes the characteristic length scales of tribological phenomena and associated materials properties.

1

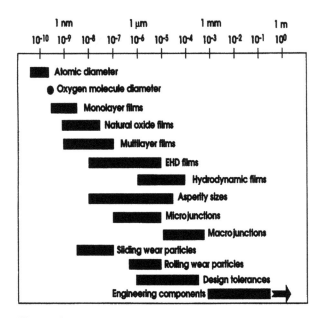

Figure 1 Surface-related phenomena in tribology and associated length scales. (From Ref. 2.)

The characterization of surface processes is also influenced by similar factors. Bonding between atoms and adhesion, for example, involve atomic and molecular interactions on a nanometer or subnanometer length scale. Similarly, the work function of a material and the nature of electrical contact between different materials are determined by the properties of a nanometer-scale interfacial or surface region. Wettability is a function of surface chemistry, cleanliness, and roughness. Surfactants radically alter wettability through the formation of molecular monolayers, bilayers, and aggregated structures at the surface [3]. In catalytic processes, interactions involve the adsorption of individual atoms or molecules at the surface, reaction of adsorbed species, and subsequent desorption of products. Optical absorption by materials is also a function of the characteristics of the surface region. It follows then that tribological phenomena, such as friction, wear, and lubrication, are strongly affected by processes occurring on the nanometer or subnanometer scale.

Length or depth scales of the order of 10 nm are characteristic of important surface chemical processes including preferential segregation of species, the formation of surface oxides, and corrosion. Surface passivation is implemented on a similar scale via surface chemical modification. Thin films applied to surfaces are often up to 100 nm in thickness. In the tribology context, these may include thin lubricating layers applied to the surface to reduce friction or modification of the surface region itself for the purpose of wear mitigation. Surface coatings, functional films, sensors, and optical structures have thicknesses on this scale. The distinction between a modified surface region and a discrete surface coating is arbitrary. However, layer structures with thicknesses exceeding 100 nm may have both surface and bulk properties in their own right, necessitating separate consideration from the underlying substrate material.

C. Methodologies for Characterization

1. Overview of Characterization Strategy

A comprehensive physical and chemical surface characterization of a material requires the selection of an appropriate set of analytical techniques and methodologies. Table 1 presents an overview of the principal areas to be considered. It is important to recognize that no one analytical technique will necessarily supply all the information required. The use of techniques in combination provides a workable solution for characterization.

A robust characterization strategy can be constructed using the following stages:

1. Clear identification and definition of the issue/problem to be investigated
2. Identification of one or more possible hypotheses to be tested
3. Clear definition of the questions to be answered by the characterization and the nature of the information to be obtained; for example, identification of an unknown material via chemical and molecular surface spectroscopy, or determination of a coating layer profile via elemental depth profiling
4. Assignment of one or more analytical techniques to each question to generate objective data by the most reliable and efficient route
5. Choice of experimental methodologies and acquisition of quality data, sufficient to address the analytical questions. Data processing, interpretation, and integration of results from different techniques
6. Evaluation of data and proposal of a solution or answer to the question and testing of the validity of the outcome. Confirmation of the reproducibility of the results and consistency of the findings from different techniques
7. Presentation of the outcome

This strategy has been adapted from a recently published, comprehensive work on surface and interface analysis edited by Rivière and Myhra [4]. The work contains an extended discussion of characterization strategy, properties of surface analytical techniques and case studies and is recommended for readers seeking further information.

Table 1 Physical and Chemical Characteristics of Materials and Analytical Methods

Physical characteristics	Chemical characteristics
Structure	Elemental composition
Morphology, topography	• Spectroscopy
Mechanical properties	• Imaging
Stress	• Depth profiling
	Chemical composition
	• Spectroscopy
	• Imaging
	• Depth profiling
	Molecular composition
	• Spectroscopy
	• Imaging
	• Depth profiling

Table 2 Techniques Frequently Deployed in Characterizing the Elemental Composition
of Surfaces

Technique	Properties
AES	*Description:* An electron beam technique with elemental information obtained by analysis of electron-induced Auger secondary electrons emitted from the sample surface. Can detect elements from lithium in the periodic table. Overlaying secondary electron micrographs and elemental images can be obtained, allowing compositional analysis of topographic features. AES is a key surface analytical technique for conducting and semiconducting samples. *Spatial resolution:* Of the order of 10 nm in a modern instrument with a field emission electron gun. Resolution is influenced by sample features, with rough and particulate samples returning lower-resolution images. *Depth of analysis:* Of the order of 5 nm, determined by inelastic scattering of emerging Auger electrons. *Limitations:* High surface sensitivity rules out the use of conducting overlayers on insulating samples to mitigate surface charging. Insulating samples cannot be analyzed.
XPS	*Description:* A photoelectron technique with elemental information obtained by analysis of x-ray excited photoelectrons emitted from the sample surface. Can detect elements from lithium in the periodic table. Charge neutralization allows high-resolution spectra to be obtained from insulators using a monochromatic x-ray source. XPS is a key surface analytical technique for all vacuum compatible sample types. *Spatial resolution:* Imaging instruments can achieve resolutions of the order of 1 μm. Resolution is influenced by sample features, with rough and particulate samples returning lower-resolution images. *Depth of analysis:* Of the order of 5 nm, determined by inelastic scattering of emerging photoelectrons. *Limitations:* Elemental detection sensitivity is limited to approximately 0.5 at.%
SIMS	*Description:* An ion beam technique with elemental information obtained y the massanalysis of secondary ions sputtered from the sample surface by a primary ion beam or pulse. Mass spectra are commonly generated using quadrupole, magnetic sector, and time-of-flight systems. Can detect all mass fragments; this capability is particularly useful for light elements such as hydrogen and for heavy clusters. Charge neutralization in modern instruments allows high-resolution spectra to be obtained from insulators. *Spatial resolution:* Imaging instruments can achieve resolutions better than 100 nm. Resolution is influenced by sample features, with rough and particulate samples returning lower-resolution images. *Depth of analysis:* Ranges from less than one monolayer in static SIMS analyses to 10 μm and greater in depth profiling tasks. *Limitations:* SIMS produces mass spectra, and the data obtained are often information-rich, requiring substantial analysis and interpretation. While high mass resolution techniques such as TOF-SIMS facilitate identification of elemental species, peak overlaps or interferences are a serious consideration in identifying unknown materials.

2. Elemental Composition

Surface characterization of materials in any field starts with the elemental composition of the surface region. The selection of a technique or techniques depends critically on the nature of the information required. Issues such as depth of analysis or surface sensitivity are most important in the investigation of thin film deposition, adsorption, delamination, and segregation of species in multicomponent systems. Table 2 shows a selection of techniques deployed in the elemental characterization of surfaces. The basic properties of each technique are listed with comments on experimental limitations.

3. Chemical and Molecular Information

Processes occurring at surfaces frequently involve chemical reactions, for example, corrosion, polymerization, bonding and adhesion, and the participation of a range of species with similar elemental constituents but different chemical forms and structures. The ability to distinguish different chemical forms of the same element and to gain information about the structure of complex molecular species is critical in developing a complete understanding of problems and processes. These issues arise routinely in the characterization of polymers, and complex organic systems including lubricants and biomaterials. Distinguishing between the different chemical states of carbon, for example, can be achieved using x-ray photoelectron spectroscopy (XPS). Important complementary structural information is gained by analyzing mass spectra from time-of-flight secondary ion mass spectrometry (TOF-SIMS). Table 3 shows a selection of techniques deployed in the chemical and molecular characterization of surfaces. As in the previous section, the basic properties and limitations of each technique are listed.

4. Further Techniques for Surface and Near-Surface Characterization

The surface characterization techniques tabulated above are in common use and are well suited for the investigation of surface-related phenomena associated with tribology. There are many other surface characterization techniques available that are used for specific sample types, used in studies of the fundamental electronic and structural properties of materials, and emerging as valuable routine analytical tools. A selection of these techniques is summarized below.

Important fundamental surface science techniques include those based around diffraction, vibrational phenomena, and photoelectron emission. Low-energy electron diffraction (LEED) and reflection high-energy electron diffraction (RHEED) are ultrahigh vacuum techniques used for probing the structural properties of the outermost layers of crystalline surfaces. For example, growing semiconductor layers can be monitored using RHEED; LEED provides information on adsorption sites on the surface of model catalyst materials and on the mechanisms of thin film growth at the atomic scale.

Ultraviolet photoelectron spectroscopy (UPS) is a photoelectron technique that probes the valence band typically using 21.2 eV radiation (He I) generated by a helium discharge source. Wavelengths between 8 and 45 eV are possible using different gases and conditions in the discharge source. Figure 2(a) and (b) shows schematic descriptions of the experimental arrangement for UPS and the photoelectron emission process, respectively. UPS is generally used to investigate bonding interactions and the electronic structure of crystalline materials. UPS is not deployed in imaging mode; the analysis area is defined by the spot size of the photon source and electron energy analyzer acceptance area. The analysis

Table 3 Techniques Frequently Deployed in Characterizing the Chemical and Molecular Composition of Surfaces

Technique	Properties
AES	*Description:* As per Sec. I.C.2. Peak positions in Auger electron spectra are influenced by the chemical environment of the atom in question. Analysis of these chemical shifts has not received as extensive treatment as chemical shifts in XPS spectra. Useful information, however, can be obtained about carbon chemical states, metals and their corresponding oxides, and resonance phenomena. Data interpretation may be complicated by the three step Auger electron emission process. Data reduction tools such as principal component analysis are now available to aid the interpretation of more complex spectra [5]. *Spatial resolution:* As per Sec. I.C.2. *Depth of analysis:* As per Sec. I.C.2. *Limitations:* General lack of availability of reference works, which collate and validate results from multiple sources.
XPS	*Description:* As per Sec. I.C.2. XPS is an excellent tool for surface chemical analysis. Chemical shifts in the elemental peaks are often easy to interpret, with many reference works and software packages available to aid the process. *Spatial resolution:* As per Sec. I.C.2. Note that chemical shifts are difficult to image adequately when component peaks are closely spaced and low in intensity in comparison with the overall peak envelope. *Depth of analysis:* As per Sec. I.C.2. Angle-dependent XPS, the acquisition of a series of spectra as a function of electron emission angle, allows higher surface sensitivity and is well suited to the analysis of ultrathin overlayers. *Limitations:* As per Sec. I.C.2. XPS provides information on chemical or oxidation state; analysis of materials, for example, polymers, with identical bond types but different bonding arrangements or structures may require input from other techniques such as SIMS and infrared spectroscopy.
SIMS	*Description:* As per Sec. I.C.2. TOF-SIMS is well established as the principal technique for the molecular characterization of surfaces. Mass spectra can be obtained with high mass resolution and high mass range. The technique is widely applied in the investigation of materials such as polymers, biomaterials, pharmaceuticals, lubrication systems, and semiconductors. *Spatial resolution:* As per Sec. I.C.2. *Depth of analysis:* As per Sec. I.C.2. *Limitations:* As per Sec. I.C.2. While TOF-SIMS has a high mass range, the ionization of larger molecules often results in the production of many smaller fragment ions. Chemical matrices are sometimes employed to enhance the yield of high mass fragments. Alternative ion sources are also employed for similar purposes.

Table 3 Continued

Technique	Properties
Infrared spectroscopy and imaging	*Description:* A widely used vibrational spectroscopy with a number of specific adaptations for surface characterization. The attenuated total reflection (ATR) mode of operation provides enhanced surface sensitivity under ambient conditions, using a prism in intimate contact with the sample to achieve multiple bounces of the infrared beam along the sample surface. Adsorbate studies may be undertaken using reflection-absorption infrared spectroscopy (RAIRS) with the sample housed in a controlled-environment cell or, in some instances, an ultrahigh vacuum chamber. Modern instruments use a Fourier transform signal detection strategy, hence the acronym, FTIR. *Spatial resolution:* FTIR microscopes achieve spatial resolutions of the order of 10–20 μm. The synchrotron-based infrared spectromicroscopy variant of this technique achieves spatial resolutions an order of magnitude better than lab-based instruments due principally to a much brighter photon source. *Depth of analysis:* The sampling depth in an ATR experiment depends on the prism material and is typically in the range of 0.5–2 μm. *Limitations:* Sampling depth is suitable for thin films on surfaces; thinner features such as adsorbed layers are often difficult to analyze adequately due to signal-to-noise constraints.

area is typically several millimeters in diameter. Synchrotron implementations of this technique use probe sizes of the order of micrometers at modern facilities. UPS is extremely surface sensitive with a depth of analysis of the order of 1 nm, determined by inelastic scattering of emerging photoelectrons. Hence, excellent vacuum conditions are required to avoid surface contamination.

Synchrotron-based photoelectron spectroscopy provides additional opportunities for characterizing the electronic properties of crystalline materials via angle-resolved photoelectron spectroscopy and determining surface structure and the adsorption sites of molecules on surfaces via photoelectron diffraction. Both are variants of laboratory-based techniques that take advantage of the unique properties of synchrotron radiation including a high-intensity, focused photon source and the ability to use a high-quality monochromator to sweep continuously through a range of photon energies. The latter allows surface sensitivity to be maximized, greatly enhances spectral resolution, and provides the opportunity for new types of experiments.

Synchrotron radiation XPS has the same general properties as laboratory-based XPS, except use of a synchrotron photon source allows any wavelength to be selected as the probe. Synchrotron beam lines can be designed to supply photons with very narrow energy spread, thus allowing higher-resolution spectra to be acquired. Photon line width is of critical importance when analyzing closely spaced chemical component peaks. The wavelength of the incident radiation may be adjusted to achieve very high surface sensitivity by minimizing the inelastic mean free path of emerging photoelectrons, detection of higher-energy photo-

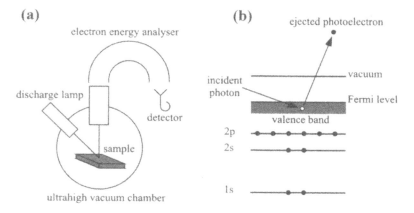

Figure 2 (a) Experimental arrangement for UPS featuring a hemispherical sector analyzer; (b) schematic description of the photoelectron emission process.

electron peaks inaccessible using laboratory x-ray sources, and optimization of detection sensitivity by maximizing photoionization cross sections.

X-ray absorption spectroscopy (XAS) describes a group of synchrotron-based techniques, which provide unique information about the surface structural and chemical properties of materials. Extended x-ray absorption fine structure (EXAFS) is a technique that studies the oscillatory variation in x-ray absorption with photon energy beyond the absorption edge for a given element. Analysis of these oscillations allows structural information to be extracted, in particular, the local order around individual types of atoms [6]. Fine structure is also observed in the x-ray absorption spectrum in the region immediately adjacent to the absorption edge, a narrow zone of approximately 50 eV in width. This phenomenon is known as near edge x-ray absorption fine structure (NEXAFS).* The near-edge structure is a function of the distribution of antibonding orbitals; hence chemical shifts are observed. Chemical states at the surface of materials can be studied in a similar manner to XPS. Linear polarization of the synchrotron light allows studies of molecular orientation on surfaces to be undertaken. NEXAFS theory and applications are comprehensively discussed in a recent book by Stöhr [7].

A synchrotron-based technique known as photoemission electron microscopy (PEEM) is emerging as a most useful analytical tool. PEEM produces images of electrons emitted from surfaces with a spatial resolution in some specialized instruments as small as a few nanometers but generally of the order of a hundred nanometers, far higher spatial resolution than any standard photoelectron spectrometer. The microscope can be operated as a detector for NEXAFS experiments, thus producing surface chemical images with submicrometer spatial resolution. High operating voltages within the microscope result in some sample charging issues and difficulties analyzing insulators.

Ion scattering spectroscopy (ISS) is an ion beam technique with elemental information obtained by analysis of primary ions inelastically scattered from atoms in the sample surface. A useful surface analytical technique in principle, ISS finds most applications in fundamental studies of surface structure and as a complement to Auger electron spectroscopy

_____

* This phenomenon is also described as x-ray absorption near edge structure (XANES).

(AES) [8]. ISS is used in macroscopic analyses only in laboratory-based instruments but is highly surface sensitive, influenced by the submonolayer presence of adsorbates on the surface. ISS is experimentally more complex than competing techniques such as XPS, AES, and SIMS.

The surface wettability of liquids on solids is of central importance in a range of industrial processes including application of coatings, adhesion and release issues, dispersion, and lubrication. Contact angle analyses provide a direct measure of the wetting properties of a surface and hence an indirect measure of the surface chemical composition. Surface energy and its relationship to surface functionalization can be investigated. Contact angle measurements are most useful when used in conjunction with techniques such as XPS and TOF-SIMS. Some instruments have mapping capability with a spatial resolution of the order of a few millimeters. The contact angle is very sensitive to the outermost monolayer on flat surfaces, surface roughness, and chemical heterogeneity. Although it does not provide a direct chemical or molecular characterization, it is very valuable used in combination with other techniques. Figure 3(a) and (b) shows a typical experimental configuration for contact angle measurement using sessile drops and a schematic definition of contact angle and the associated interfacial tensions.

Related techniques include the measurements of dynamic contact angle, surface tension, and interfacial tension using the Du Noüy ring method, pendant drop method, drop volume method, and Wilhelmy plate method. The reader is referred to Ref. 9 for a comprehensive coverage of these areas and others.

Scanning probe microscopies (SPM) are built around the interaction between a fine tip and a surface. Several groups of techniques exist, the dominant groups being electron tunneling techniques and tip–surface force techniques. The scanning tunneling microscope (STM) was the first instrument to be demonstrated, emerging from the IBM Zürich research laboratories in 1982 [10]. STM and a related technique, scanning tunneling spectroscopy (STS), rely on positioning a sharp tip in close proximity to the surface of interest, applying a bias voltage and measuring an electron tunneling current. Scanning the tip produces images related to surface topography and surface electronic properties with atomic resolution. Tips require very small radii of curvature to be effective and are generally prepared using electrochemical etching of tungsten or platinum–iridium wires. Figure 4(a) and (b)

(a)   **(b)**

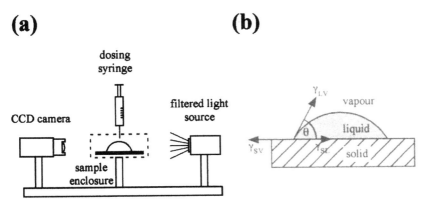

Figure 3 (a) Schematic diagram of a typical contact angle measuring instrument; (b) sessile drop showing solid, liquid, and vapor phases, contact angle, θ, and interfacial tensions, γ_{SV}, γ_{SL}, and γ_{LV}.

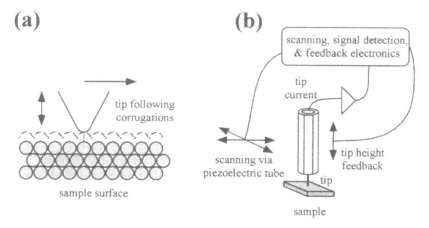

Figure 4   (a) Schematic diagram of STM operating mode; (b) a typical configuration for control systems required for operation and data acquisition.

shows schematic descriptions of the mode of operation and control systems of the STM, respectively.

The atomic force microscope (AFM) followed in 1986 [11]. AFM relies on positioning a micromachined assembly incorporating a lever arm and sharp tip in contact with the surface or in close proximity to the surface. Scanning the tip across the surface produces images related to topography, friction, and a variety of other surface properties depending on tip and instrument configuration. Soft substrates may be analyzed by tapping rather than dragging the tip during the scan.

The ability of STM and AFM to image surfaces with atomic-scale resolution under ambient conditions and under ultrahigh vacuum is a defining feature of this group of techniques. In subsequent years, many variants of these two instruments have been devised allowing spectroscopy and high-resolution studies of surface properties such as friction, magnetic force, electrochemical behavior, and electrical conductance. There is extensive literature available on scanning probe microscopies; the reader is referred to the references for several starting points [12].

II.   PHYSICAL PROPERTIES AND CHARACTERIZATION

A.   Topography and Physical Characterization

1.   Topography of Surfaces

For most applications of surface engineering, it is extremely important to be able to quantify the physical shape of the surface. In metal-forming operations, the roughness of both the stock and tool will determine the friction and wear behavior and will influence the effectiveness of lubrication. For example, in some cold forging operations, the wire is drawn immediately before forging. In this case, the roughness of the incoming rod must be sufficient to allow effective bonding between the surface and the lubricant so that some of the lubricant is still present on the surface after the draw for the subsequent forging operation. In wear situations, however, the topography of the hard tool surface can damage the surface of the product.

Hence, the main aim is to be able to understand and quantify the topography, generally referred to as the roughness, of the surface and to then relate this to the behavior of interest. Historically the topography characterization was mainly performed in only one direction (2-D when considering the depth or height of the surface along that line), but increasingly 3-D techniques are being used.

a. 2-D Roughness Measurements. The surface roughness or surface texture relates to the change in height, or profile, of the surface along a line on the surface, relative to some ordinate. Typically, this is measured using a stylus moved across the surface in a straight line. The stylus is free to move perpendicular to the surface. The accuracy of this reading depends on the stylus run, or cut-off length. The smoother the surface, then the shorter the cut-off length required for an accurate measurement. More recently, laser techniques have also been used to provide this information. It is also possible to have portable devices that can be used in the field to provide in-service measurements of the evolving roughness.

The most common measure of surface roughness is R_a which is the average height of peaks over a given length [13]. The typical measurement length is generally restricted to a few millimeters, and hence it is important to realize that this is only a local measure and a number of measurements at different parts of the surface are required to obtain a meaningful representation of the average surface profile.

While R_a is the most common measure of surface roughness, it has been found that other parameters may be more appropriate measures of the surface topography. These include:

Height parameters: R_{max}, the largest individual roughness depth; R_t, the difference between the highest and lowest measurements; R_q, the geometric mean of all values.

Waviness parameters: W, used to characterize the spacing of the peaks.

Skewness: R_{sk}, measures the symmetry of the profile about the mean value and is used to distinguish between shape profiles with the same R_a.

Slope and curvature: these are hybrid parameters that combine aspects of height and wavelength.

The simple R_a measure for height displacements y over a length of measure l is given by:

$$R_a = \frac{1}{l} \int_0^l |y(x)| dx$$

where R_a can be simply viewed as the centerline average or arithmetical average and represents a line drawn through the profile so that the area filled with material equals that unfilled. A variation on this is the root mean square or R_q:

$$R_{q,rms} = \left(\frac{1}{l} \int_0^l y(x)^2 dx \right)^{0.5}$$

These parameters are the most common in surface engineering studies. An example of the evolving surface roughness for a range of tool coatings is shown in Fig. 5. Here the uncoated tool shows a rapid increase in roughness as the number of parts stamped increases. Experience in the plant had shown that a critical roughness value of 4.5 μm was associated with galling and scoring. The variable coatings altered the rate of increase in R_a. These measurements were performed in the production environment using a portable device.

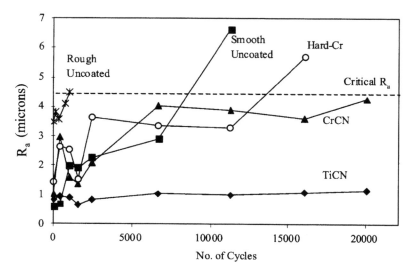

Figure 5 Variation in tool roughness with part numbers during the stamping of heavy channels, for different tool coatings. (From Ref. 14.)

  b. 3-D Roughness Measurements.   The roughness along a line is obviously a limited representation of the surface morphology. Many surfaces are not random and exhibit different roughness characteristics when measurements are normal to each other. For example, in metalworking operations the roughness in the line of metal flow will be quite different to that normal to the flow direction. In other cases, the manufacturing operations used to create the surface will either deliberately, or as a by-product, give quite different surface textures in different directions.

  Early methods used multiple stylus measurements of surfaces, with the data then combined to reconstruct the surface. One issue here is in defining the reference datum for these measurements [15]. However, with increasing use of computers, electronics, lasers, and other technologies, it is now relatively simple to create 3-D surface maps.

  Recent major developments, particularly at nanometer spatial resolution for 3-D measurements, have been the scanning tunneling microscope and the atomic force microscope, discussed in Sec. I.C.4. Between these and the conventional stylus method, there are the focus detection systems and those based on interferometry. The focus system uses a laser beam that is projected onto the surface and then the lens is adjusted in the height direction with the surface as the focal point. The movement of the lens is then the record of the height change. A schematic diagram of such a system is given in Fig. 6.

  These techniques now allow the ready construction of 3-D maps of the surface topography of a range of materials and roughnesses. Figure 7 summarizes the relative use of various techniques in providing roughness information as a function of the spatial distance over which they can measure.

  Stout and Blunt [15–17] have attempted to define a set of appropriate 3-D roughness parameters (Table 4). Many of these parameters are similar to the 2-D parameters above, although some are designed to provide information related to the anisotropy of the surface texture. S_{tr} is specifically related to the texture anisotropy and is based on the ratio of fastest to slowest decay in the autocorrelation function of the profile in all directions. Stout has stated that values less than 0.3 indicate strong anisotropy or texture, whereas values greater

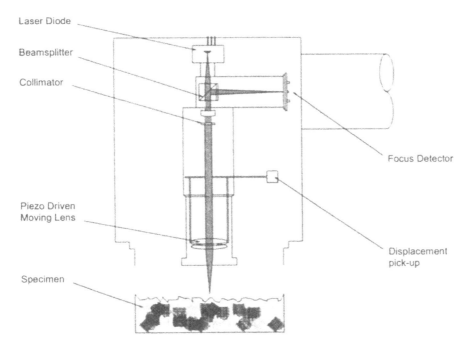

Figure 6 Schematic representation of laser focus instrument for 3-D topography measurements.

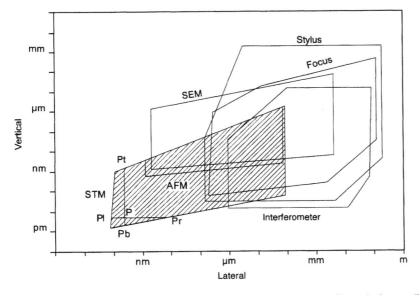

Figure 7 Vertical and lateral resolution of different 3-D profile techniques. (From Ref. 15.)

Table 4 3-D Roughness Parameters

Amplitude parameters	
S_q	Root-mean square deviation of the surface
S_z	Ten point height of the surface
S_{sk}	Skewness of the surface
S_{ku}	Kurtosis of the surface
Spatial parameters	
S_{ds}	Density of summits of the surface
S_{tr}	Texture aspect ratio of the surface
S_{td}	Texture direction of the surface
Hybrid parameters	
$S_{\Delta q}$	Root-mean square slope of the surface
S_{sc}	Arithmetic mean summit of the surface
S_{dr}	Developed surface area ratio
Functional parameters characterizing bearing and oil-retention properties	
S_{bi}	Surface bearing index
S_{ci}	Core oil-retention index
S_{vi}	Valley oil-retention index

Source: Ref. 16.

than 0.5 suggest a high degree of isotropy. The actual direction of the texture is represented by S_{td}; this is important only for S_{tr} values below 0.3.

Nearly all studies of coating technologies utilize topography measures. Most still use the simple measures of R_a and R_q, partly because of a lack of systematic work to correlate the other parameters with behavior. Increasingly, though, there is a need to use more complex parameters for the development of surfaces, with the simple R_a measure more appropriate for quality control or a field measurement of the change in topography.

2. Mechanical Characterization

Understanding of the behavior of the component under load requires knowledge of the modulus of elasticity, the yield stress, and the ultimate tensile strength. Although the thickness of the coating may often be very small compared with the bulk and, hence, is not expected to affect the overall average properties of the component, it will have a major impact on the local deformation behavior. Obtaining accurate and meaningful information about the actual tensile behavior of the coating is extremely difficult. More often, the material is simply characterized by the hardness, although new techniques such as nano-hardness allow more information than just a single average hardness to be obtained. The strength between the coating and the substructure is also an important mechanical characterization of the parameter, while the friction and wear behavior have also led to the development of a range of tests.

a. Equivalent Tensile Properties. The term "equivalent" is used here to show that the aim of these techniques is to obtain information that would be typically measured in a tensile test of the substrate. It is possible to perform simple tensile tests of some coating materials. This is more common for thicker coatings that are not alloyed with the substrate. In this case, a coating can be made separately from the component and then tested. However, even this is extremely difficult and subject to a number of sources of error. The coating has to be free of defects, as this will cause early failure. It must also have uniform thickness and low levels of residual stress [18].

The modulus has been measured by beam bending [19]. This involves using micro-machining techniques to remove the layers around and below a small beam of the coating. By using a sensitive probe, such as a nanoindentor tip, it is possible to measure the deflection of the beam as a function of the applied load and to then use standard equations to determine the modulus. It should also be possible to use this method to measure the yield point, as in bending this is determined from the departure from the linear elastic behavior. However, this is very difficult and prone to a number of sources of error. Again, the uniformity of the beam and removal of stress concentration at the coating/substructure joint are issues.

An alternative beam method has recently been applied to determine the elastic modulus as a function of temperature in surface coatings for hot forging dies [20]. Here the deflection of the beam due to thermal stresses was used. The method involved heating samples 0.25 mm thick, 8 mm wide, and 80 mm long to temperatures from room temperature to 400°C. The strip was assumed to behave as a composite plate with known thermal expansion values for each layer (valid only for a discrete coating on a substrate); the elastic modulus of the substrate was also known. The elastic modulus in the surface layer was obtained by following the deflection during heating. The values of the modulus appeared to be close to those given by other workers for similar coatings, even though the thickness of the coating was relatively small in comparison with the thickness of the beam (e.g., 7 μm anodized layer on 250 μm H13 tool steel).

b. Hardness. The above methods are either extremely difficult or may have limited applicability to coating behavior. By far the most common technique to gain information relating to the likely mechanical behavior of a surface is the hardness test. Increased hardness of a surface will generally give greater wear resistance, as expressed by Archard's law:

$$W = k(Pl/H)$$

where W is the volume of material removed by wear, l is the sliding distance, P is the load, H is the hardness, and k is a constant that is a function of shape and material. This is a simplistic relationship, but it did form the basis for many early developments in wear-resistant surface treatments.

Hardness tests involve pushing a hard indentor into the surface of a sample to measure resistance to plastic deformation. For coatings, the preferred hardness tests are the Vickers and Knoop tests. The Vickers hardness test uses a diamond indentor in the shape of a square pyramid. The hardness, HV_p, is given by:

$$HV_p = 1.854P/d^2$$

where P is the load (kg) and d is the measured indentation diagonal (mm).

The Knoop test uses a blunt, elongated indentor, which can reduce the tendency for cracking of the coating. The test can be used to highlight any anisotropy in the coating and is useful for testing narrow grains or thin layers. In this case, the hardness, HK_p, is given by:

$$HK_p = 14.2P/d^2$$

although in this case d refers to the length of the longer diagonal.

For all hardness tests of coatings or other surface treatments the main issue is whether the hardness measurement reflects the properties of the layer of interest alone or whether it is also affected by subsurface layers or the bulk sample. Figure 8 shows the Vickers hardness of a hard chrome layer on a die as a function of coating thickness for two different loads. Even for the thickest coating, the 2.5 kg load is still increasing in hardness toward the value

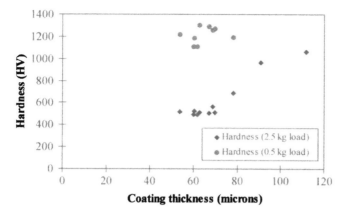

Figure 8   Hardness as a function of coating thickness at 2.5 and 0.5 kg indenter loads.

obtained with the 0.5 kg load. Hence, in the case of the higher load the interaction volume is greater than the coating thickness.

The influence of the substrate on the measured properties of a coating is reduced as the size of the indent relative to the coating is decreased. ASTM standard E 384 requires that the depth of indentation must not exceed 10% of the sample thickness (in this case the coating). This has led to the adoption of micro- and nanohardness tests for thin coatings. These test methods reduce the depth of the indents to just a few micrometers or nanometers. In the case of nanoindentations, the size of the indents is too small to be determined optically so hardness is quantified by measuring force and displacement of the indenter during the test. This means that there is no direct conversion between hardness values determined by nanoindentation and those measured by conventional hardness tests where the permanent indent is measured optically.

As mentioned earlier, one problem with thin coatings is the determination of mechanical properties. This is a result of the samples being much smaller than those used in conventional mechanical tests. The coatings are not always able to be removed from the substrate without great difficulty. Force and displacement can be measured continuously as nanoindentation is carried out. These data can be used to determine material properties such as hardness, elastic modulus, and stress–strain curves [21]. Figure 9 shows a schematic diagram of the load penetration curve under typical test conditions for a pointed indentor. The total displacement is given by the sum of the elastic and plastic components, although allowance also needs to be made for the machine compliance and the initial penetration required before the force exceeds the background noise level. There are various assumptions required regarding the indentor shape when determining the effective elastic modulus, which are covered by Swain [22].

This technique has been used to determine the elastic modulus of laser-deposited composite boride coatings on steel [23].

Nanoindentation is also used for measurement of time-dependent properties such as stress relaxation and creep of thin films. If the deformation of the material is to be measured, it is important to make sure that the time taken for indentation is short enough so that the results are not affected by material flow; that is, the loading rate is much faster than the creep rate.

Another advantage of using smaller indents and lighter loads is that cracking of the coating can be avoided or reduced. Thin, brittle coatings may crack or delaminate during

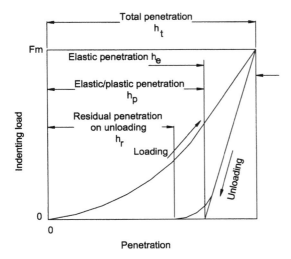

Figure 9 Schematic force–penetration curve obtained in nanohardness or other continuous hardness measurement test. (From Ref. 22.)

indentation, particularly when the substrate is soft. While the adhesion tests described in the next section are more popular, delamination during indentation has been used to assess adhesion of a coating to a substrate, particularly in the case of hard films on metallic substrates. The mechanics of the delamination have recently been modeled and the effective adhesion energy determined for diamond-like carbon films on tool steel [24].

 c. Adherence Tests. In the previous section it was shown how the hardness test often provides some information related to the adhesion of the coating. However, the hardness test is not designed for this purpose and the direction of loading is not that which commonly will cause adhesion problems. Bull [25] has suggested that the experimental adhesion strength (EA) is related to the basic adhesion strength (BA), but that there are also other extrinsic factors related to the test conditions and intrinsic properties, such as the physical properties mentioned above that will affect this measure.

 The scratch test is the most widely used for hard coatings. The concept is simple and relies on a hard indenter being dragged across the surface under an increasing load. For thin coatings (less than approx. 20 µm) the indentor is dragged across the coated surface. The load can be changed during a test or between individual tests to determine the critical load for delamination of the coating. At loads less than the critical load, the coating has damage such as scratching, plastic deformation, and chipping of the coating. At the critical load, delamination will occur and the coating will be removed from the substrate. The scratch test, like most mechanical tests, is most useful for predicting performance when the types of failure and the loading are similar to those occurring during the test, for example, coatings that experience mild abrasive wear in-service.

 A schematic description of the force trace obtained from a scratch test is given in Fig. 10. Here there are two critical changes in the profile at C1T and C2T. These two measurements were used in a round-robin test of equipment and gave consistent results between the different laboratories. C1T represents the first significant change in force, while C2T is the first substantial change in the tangential force gradient.

 For thicker films such as plasma-sprayed coatings, a polished cross section is used. In this case, a fixed load is used and the indentor travels toward the substrate/coating

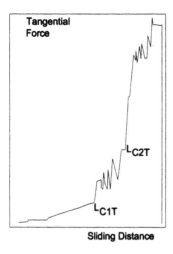

Figure 10 Schematic description of the results from a scratch test. (From Ref. 26.)

interface from the substrate. The load is increased between successive scratches until, at a critical load, the indentor pulls the coating from the substrate, creating an interfacial crack (Fig. 11).

    d. Friction and Wear.   The above sections have dealt largely with properties that have a simple physical meaning. There are other physical characteristics that are not true material properties per se, but which depend on the actual service conditions. Of particular relevance to surface-treated materials are the friction and wear. The friction behavior is a function of a number of parameters such as load, sliding velocity, geometry, and the nature

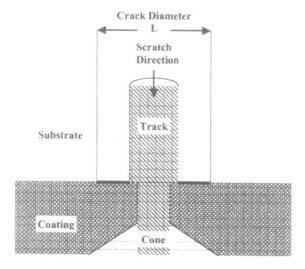

Figure 11 Schematic description of scratch adhesion test for thick coatings. (From Ref. 25.)

of the object the surface is in contact with, including the lubricant. The same is also true for wear. Hence, there are now myriads of tests that are used to quantify these parameters. Even for one simple deformation process there may be many tests used. For example, in sheet metal forming there are tests to handle the different types of deformation conditions that may exist around the die, from the flat faced friction test for the blankholder [27] to the drawbead simulator for the drawbeads [28] to actual draw tests for other regions [27,29].

The objectives of the test need to be clearly understood before choosing one test over another. In the case of friction tests there is an increasing need to be able to provide accurate friction values for numerical modeling. The aim then is to obtain a very accurate and hence representative test where the full range of inputs can be examined and the speed of testing is not an issue because only a few tests will be performed for a given condition. However, wear tests are often designed to provide accelerated information so that materials or other factors can be ranked. Here months or years of service must be simulated in hours or days.

Under deformation conditions the friction coefficient [30] is a function of the following:

- Workpiece
- Tool
- Lubricant
- Process
- Surroundings

One way to group metal forming focused tribology tests is given in Table 5. These cover both friction and wear tests. They have been divided into *process* and *simulative* tests, where the former refers to tests where the basic process conditions are maintained, whereas in the latter case there can be marked deviations from the actual process kinematics. In each case, it is possible to have direct tests where the friction parameter is actually measured, or indirect tests where another parameter is measured and then related back to the friction parameter, often through modeling. An example of this is force in compression, where the measured load is converted into a force and then finite element methods, or analytical modeling, is used to predict the force as a function of the friction coefficient. From this prediction the "average" friction coefficient is obtained, assuming all other parameters in the model are accurately known.

Obviously, the process tests are the most representative of actual service conditions. However, it is also clear that it would be extremely difficult and time-consuming to evaluate

Table 5   Grouping of Tribology Tests

Process tests	Simulative tests
A. Upsetting	G. Global plastic deformation
B. Forging	H. Local plastic deformation
C. Extrusion	I. Elastic deformation
D. Drawing of rod/wire	
E. Rolling	
F. Drawing of sheet/strip. Ironing	

Source: Ref. 30.

the full range of parameters that should be considered when developing a new coating. For example, there are often strong interactions between a number of parameters so that it is not a simple case of testing for an effect in isolation from other changes. One example is that the effectiveness of a lubricant can be a function of the load, the surface roughness, the composition of the surface(s) and the temperature. Hence, it is more common to use simulative tests in the initial development stage and then to use a cut-down number of process tests for final validation and performance ranking. The actual comparison between laboratory and field performance is still quite limited [30], although the general rankings and trends are maintained where appropriate simulative tests have been used.

The most common simulative test for wear is the pin on disc test (Fig. 12), particularly for basic wear information of surface-treated metals [31,32]. This is suitable for simulating sliding wear conditions and the load and environment conditions are easily adjusted. The test is relatively simple and widely used in the surface engineering industry. Other variants that are commonly used to evaluate coatings are the ball on disc and the pin on ring (or plate). There are some issues with this test that limit the application to bulk forming operations. In a forging or rolling operation, for example, the tool is constantly exposed to fresh material that is being plastically deformed during the operation. In the pin on disc test, this is not the case and there is only elastic loading of the tool and workpiece. However, for simple sliding conditions this test and the various modifications can provide a quick ranking. It should be noted that there have been different results depending on which material was the pin or the disc [31]. For abrasive wear, there are also a number of simple tests and associated issues [33].

The measurement of wear can be as simple as following weight loss, through to more precise measurements of the topography, depth of wear, or changes in other features, such as the acoustic emission during testing.

3. Secondary Electron Microscopy

In scanning electron microscopy (SEM), a focused electron beam is scanned across the surface of a sample resulting in the emission and/or reflection of electrons and x-rays. From the collection of these radiations, it is possible to gain information on surface topography, composition, phase constitution, and crystallographic orientation; all of which are of

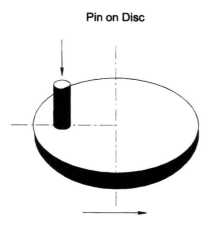

Pin on Disc

Figure 12 Schematic description of the pin on disc tribology test.

considerable interest to the surface scientist. In many instances, the measurements are conveniently plotted as images in gray scale or color contrast that are readily converted into appropriate numerical quantities by either raw data or image analysis.

The radiation emitted from the material can be traced to an interaction volume immediately beneath the point of incidence of the narrow (1–100 nm) electron beam. The size of this volume determines the resolution and is characteristic of the radiation type. Figure 13 shows a schematic diagram of the relative interaction volumes for the various radiations. The smallest volume examinable by standard SEM equipment is that from which secondary electrons emanate while characteristic x-rays are emitted from the largest volume. The absolute volume of interaction is determined by the material (i.e., atomic number), probe diameter, and accelerating voltage.

Secondary electrons are generated when incident electrons ionize atoms in the surface layer. More than one secondary electron may be generated from an incident electron. The low energy of these electrons means that they only escape the sample if they are generated close to the surface. The number of secondary electrons reaching the detector is therefore very sensitive to the topography of the surface.

Backscattered electrons are "reflected" back out of the surface after collision with constituent atoms of the sample. The greater the atomic number, the higher the intensity of backscattered electrons. These electrons therefore carry information on the atomic composition of the sample. Regions of higher average atomic number appear brighter on images generated from backscattered electrons. Importantly, under suitable conditions backscattered electron images can also display crystallographic orientation contrast. This arises from the different density of atoms per unit surface area for different crystallographic orientations.

Characteristic x-rays are generated by atoms following the emission of a secondary electron. The x-ray energy emitted is characteristic of the atom and can be analyzed using energy dispersive spectrometry (EDS) or wavelength dispersive spectrometry (WDS) to yield quantitative elemental composition.

Secondary electron microscopes (see Fig. 14), in common with most advanced surface analytical tools, require a high vacuum or, in some cases, ultrahigh vacuum environment for operation. Vacuum compatibility imposes some restrictions on sample types and sample-handling protocols. However, recently, so-called environmental or variable-pressure SEMs have become readily available. These instruments operate at high pressures and constitute a considerable technical achievement, as it is necessary to segment the electron gun column

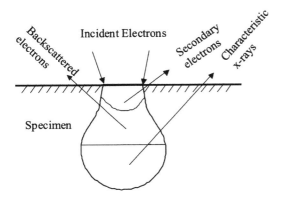

Figure 13   Schematic diagram of the interaction volume of incident electrons.

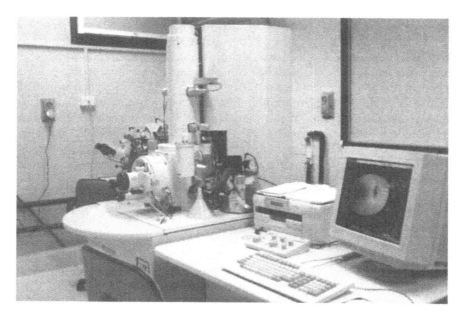

Figure 14 A scanning electron microscope incorporating a cold stage for imaging samples with poor vacuum compatibility (Jeol JSM-6340F FESEM).

into a number of regions of significantly varying pressure. Along with a number of other innovations, these instruments also make use of the ionization of the atmospheric gases to reduce sample charging and thereby permit examination of nonconducting samples. Dirty and wet samples are readily examined in an environmental SEM.

Irrespective of the vacuum in the sample chamber, the electron gun must be in an ultrahigh vacuum environment. Traditional electron sources comprise thermionic tungsten or LaB_6-type filaments. A filament current heats the filament and "boils" off electrons, which are then accelerated and filtered through a negatively charged Whenelt cap, which aids in collimating the beam, and a small hole in the accelerating anode. So-called field emission gun (FEG) sources involve the extraction of electrons from the very sharp tip of a tungsten/zirconium oxide emitter using the phenomenon of quantum tunneling in the presence of a strong electric field. The tip may (hot or Schottky FEG) or may not (cold FEG) be heated. The electron beam generated from these guns is typically smaller and brighter than those generated from thermionic emitters. A comparison of some of the main attributes of the different electron sources is presented in Table 6.

The electron beam is directed onto the sample surface through a series of electromagnetic lenses and often also through one or more apertures. The beam is then scanned over the sample in a raster, or held in a single location, depending on the application. The detection of secondary electrons is often performed by an Everhart–Thornley detector, which comprises a Faraday cage with an adjustable potential positioned in front of a scintillator followed by a light pipe and a photomultiplier tube. The output of the photomultiplier tube is proportional to the total number of electrons collected. These detectors are typically mounted to the side of the sample. There are a number of different types of backscattered electron detectors. Some are similar to the Everhart–Thornley detector but without the Faraday cage. Others are composed of diodes (semiconductors) that are

Table 6 Comparison of Electron Sources

Emitter type	Thermionic	Thermionic	Cold FE	Schottky FE
Cathode material	W	LaB_6	W	ZrO/W
Operating temperature (K)	2800	1900	300	1800
Effective source radius (nm)	15.000	5000	2.5	15
Emission current density (A cm^{-2})	3	30	17,000	5300
Total emission current (μA)	200	80	5	200
Normalized brightness (A/cm^{-2} sr^{-1} kV^{-1})	1×10^4	1×10^5	2×10^7	1×10^7
Maximum probe current (nA)	1000	1000	0.2	10
Beam noise (%)	1	1	5–10	1
Emission current drift (% hr^{-1})	0.1	0.2	5	<0.5
Operating vacuum (hPa)	$\leq 1 \times 10^{-5}$	$\leq 1 \times 10^{-6}$	$\leq 1 \times 10^{-10}$	$\leq 1 \times 10^{-8}$
Cathode regeneration	Not required	Not required	Every 6 to 8 hr	Not required
Sensitivity to external influence	Minimal	Minimal	High	Low

Data courtesy of LEO Electron Microscopes Ltd.

positioned adjacent to the incident beam, immediately above the sample. Energy dispersive spectroscopy employs x-ray detectors, which are also based on the interaction of the emitted radiation with a semiconductor. Only in the case of these detectors, the semiconductor is chilled with liquid nitrogen to reduce the thermionic creation of charge carriers. Wavelength dispersive detectors rely on diffraction and therefore require considerably more space than EDS detectors. WDS, however, is more suitable for analyzing the presence of low-concentration elements. Backscattered electrons from an inclined sample can also be used to generate a projection of Kikuchi lines, which comprise a pattern that in most cases can be used to uniquely identify the crystallographic orientation. These lines are formed by the diffraction of electrons and are detected using a phosphor screen in combination with a charge-coupled device (CCD) camera.

The uses of secondary electron imaging in surface engineering are almost as broad as optical imaging. However, the maximum magnification is much greater (up to 3 million times) as is the depth of field. The latter is particularly useful in the examination of wear surfaces, which are typically quite rough. Figure 15 shows examples of SEM images of a galvanneal-type zinc coating on a steel substrate. The needlelike structure is typical of the zinc-rich zeta phase.

4. Transmission Electron Microscopy

In contrast to scanning electron microscopy, transmission electron microscopy (TEM) hinges on forming images using a magnified electron signal obtained from incident electrons that travel through a thin (~200 nm) specimen. This technique enables very high lateral

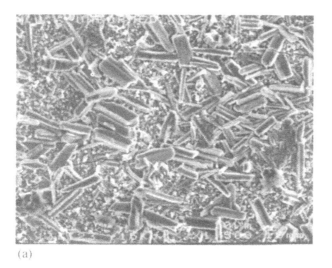

(a)

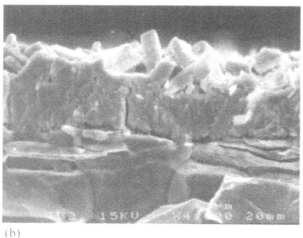

(b)

Figure 15 Secondary electron image of a galvanneal surface coating on steel, (a) view normal to the surface and (b) view perpendicular to the surface normal. These coatings are employed widely on automotive steels. (Images courtesy of D. Haynes and B. Perrot, BHP Billiton Ltd.)

resolution (~ 0.2 nm) but necessitates onerous destructive sample preparation [34]. It is not a tool employed often by the surface scientist and will only be considered here in passing.

Electron guns similar in principle to those employed in SEMs are also incorporated in TEM instruments. In the case of TEM, the beam energy is typically much higher (100–600 keV compared with ~ 20 keV). The beam is finely focused onto the specimen through a series of apertures. In the case of scanning transmission electron microscopy (STEM) the beam is scanned in a raster across the sample. The transmitted signal is magnified through a series of electromagnetic objective and projector lenses and, in some cases, filtered to remove inelastically scattered electrons. The image is typically generated on a phosphor screen, which is viewed through a vacuum-sealed viewing window.

In surface science, TEM is employed in the examination and identification of very fine precipitates/particles that may have formed in surface regions following diffusion heat treatments. Another example is the detection of small amorphous regions and 5 nm crystallites in Ni–P coatings [35].

B. Stresses and Stress Measurement

1. Nature of Stresses

Residual stress refers to the stress that exists in a component after a manufacturing operation or in service when the external forces are removed. These may be induced by thermal stresses such as quenching, or mechanical action such as peening, or chemical reactions such as those involving phase transformation where there is a volume change. The residual stress can act at three levels [36]:

Type 1: macrostresses that operate on a larger scale than the grain size and which can vary over considerable distances
Type 2: microstresses that act on the scale of an individual grain in the microstructure
Type 3: atomic-scale defects, such as those that arise due to lattice mismatch or other highly local factors

By definition, surface treatments are altering the state of the bulk material. Most of these treatments will produce a structure that will have quite different properties from the bulk. In most cases, these differences extend beyond simple chemical/composition differences to differences in the mechanical and physical properties. These can lead to residual stresses in or near the surface, either through the formation of the coating or later through different responses to external factors. From the above discussion, it is clear that the residual stresses in surface engineering are typically Type 1 and operate on the scale of the coating and an interaction volume within the substrate.

Examples of the stress-inducing processes involved in surface treatment include direct processes, such as shot peening, through to indirect processes such as diffusion techniques, which lead to differences in phases, through the coating interaction region. In some processes, thermal treatment will also introduce residual stresses, such as quenching in heat treatment. Even if these are not present or they are eliminated through subsequent heat treatment, the difference in the properties between the surface layers and the bulk of the sample can give rise to stress differences in service. For example, large differences in the elastic limit between the surface and the substrate could mean that one would plastically deform before the other, which on unloading could lead to residual stresses.

The effect of the residual stress on the performance of the component can vary, depending upon the application. Shot peening is a simple surface treatment that introduces compressive residual stresses. These are extremely beneficial to the fatigue life of the component. Tensile stresses increase the likelihood of premature failure and can activate certain chemical processes (e.g., stress corrosion cracking).

2. Stress Measurement

The various techniques can be broadly divided into destructive and nondestructive, and it is clear that no single technique currently provides appropriate coverage and accuracy for the range of situations where the residual stress is to be measured. Tables 7–9 from Ref. 36 summarize the current techniques that are suitable for surface engineering types of

Table 7 Practical Issues for Residual Stress Measurement Methods Used for Coatings

Key techniques	Contact or noncontact	Destructive?	Lab-based or portable	Practical issues					Level of expertise required
				Availability of equipment	Speed	Standards available	Cost of equipment		
Hole drilling	Contact	Semi	Both	Widespread	Fast/Med	ASTM E837-99 [4]	Low		Low/Med
X-ray diffraction	Noncontact	No—unless depth profiling	Both	Generally available	Fast/Med	No [4]	Med		Med
Synchrotron	Noncontact	No	Laboratory	Specialist	Fast	No	Strategic/ government facility		High
Curvature and layer removal	Contact	Yes	Laboratory	Generally available	Med	No	Low		Low/Med
Raman	Noncontact	No	Both	Specialist	Fast	No	Low		Med

Table 8 Materials and Preparation Requirements

Technique	Material type	Surface preparation	Surface condition
Hole drilling	Metals, plastics, ceramics	Light abrasion-strain gauge	Flat—preparation must not affect stresses
X-ray diffraction	Metals, ceramics	Important	Important
Synchrotron	Metals, ceramics	Important	Not critical
Curvature and layer removal	All	Not critical	Not critical
Raman	Metals, ceramics	Not critical	Not critical

problems; this reference also summarizes other techniques that are more suited to bulk residual stress measurements. For coating-related residual stress measurements, by far the most common technique is x-ray, with hole drilling and curvature also commonly used. Other recent overviews can be found in the literature [37–40].

Synchrotron radiation techniques are typically used for fundamental studies of surface properties rather than stress. They àre more generally used for bulk analysis of residual stresses, whereas Raman has only been used in limited application, to date. Most of this work is drawn from a recent study by the National Physical Laboratory in the United Kingdom, which is reviewing residual stress measurement techniques and aiming to establish future guidelines [36].

a. Hole Drilling. The hole drilling method has been extensively used for larger components. The principle is quite simple and relies on measuring the local relaxation of the structure through the introduction of the hole. This local strain change can then be transformed into a stress. The hole drilling method is generally considered as destructive, although in some cases small holes can be introduced and then repaired without affecting the life of the component. Because of the long history of this test, there is an ASTM standard [41].

Figure 16 shows one arrangement for this test. Strain gauges are fixed to the sample surface and then a hole is introduced. The residual strain, ε_r, which can then be converted to a stress, can be calculated from:

$$\varepsilon_r = (\sigma_{max} + \sigma_{min})\overline{A} + (\sigma_{max} - \sigma_{min})\overline{B}\cos 2\beta$$

where σ_{max} and σ_{min} are the maximum and minimum principal stresses, respectively, β is the angle from the x axis to the direction of the maximum principal stress and $\overline{A}$ and $\overline{B}$ are material constants (see the ASTM standard referenced above for further details).

An obvious drawback of this method is that it provides an average for the depth of the hole. To obtain information about the residual stress profile away from the surface, which is of most interest in surface engineering, it is necessary to take measurements at a range of depths. Other limitations are that the technique is only accurate as long as the depth is less than the diameter of the hole. It has a relatively low strain sensitivity compared with other methods and there are potential errors from surface preparation and the dimensions of the hole. The technique has been applied to coatings, but is not suitable for very thin coatings because the stress will change sharply over very small distances and material removal to that level of accuracy is not practical with most methods [38].

Table 9 Physical Characteristics of Residual Stress Test Methods

| Key techniques | Resolution | Penetration | Stress type | Physical characteristics | | | Sampling area/volume |
				Stress state	Stress gradient	Uncertainty	
Hole drilling	50–100 μm	= hole diameter	Macro	Uniaxial, Biaxial	Yes—difficult to interpret	Varies with depth	1–2 mm diameter 1–2 mm deep
X-ray diffraction	20 μm depth 1 mm laterally	5 μm—Ti 50 μm—Al 1 mm—layer removal	Macro Micro	Uniaxial, Biaxial	Yes—with layer removal	Limited by several factors	0.1–1 mm^2 0.05–0.1 mm
Synchrotron	20 μm lateral to incident beam 1 mm parallel	> 500 μm 100 mm—Al	Macro Micro	Uniaxial, Biaxial Triaxial	Yes	Limited by grain sampling	0.1 mm^3
Curvature and layer removal	Depends on material and measurement method	Not applicable	Macro	Uniaxial, Biaxial	Yes		
Raman/ fluorescence	0.5 μm	Surface	Macro	Uniaxial, Biaxial	No		

Figure 16 Schematic diagram of standard (Type A) strain gauge arrangement for residual stress measurement using the hole drilling technique.

The NPL study (referenced above) has identified a number of issues that need to be considered in the hole drilling method, including:

- Drilling method and wear
- Drill speed and feed
- Operator skill
- Strain gauge (factor, quality, installation)
- Location
- Instrumentation

These are particularly important for engineering applications, but are obviously less relevant for laboratory- and small-scale-based studies of coatings.

The main advantages of the hole drilling method are that it is a quick and inexpensive method and will generally identify general trends in the residual stress profile.

b. Curvature Method. This involves measuring the curvature of a strip of material as the coating (or substrate) is removed [38]. This destructive method is well suited to laboratory investigations of the effects of various parameters on the coating. However, there are a number of difficulties with this approach. It is limited to very simple shapes and has mostly been applied to strip or plate samples. Also, there have been problems in interpreting the outcomes from this technique because of the difficulty in relating the shape after etching to the residual stress profile.

The analysis of thin films is based on the Stoney formula [42]:

$$\sigma_{\text{film}} = \frac{Et_0^2(K - K_0)}{6(1 - v)t}$$

where E is Young's modulus, t_0 is the substrate thickness, t is the film thickness, K is the curvature of the substrate, K_0 is the curvature of the substrate without the coating, and v is Poisson's ratio.

The curvature can be measured by a variety of methods depending on the requirements. These include profilometry, strain gauges, laser scanning, and optical microscopy.

c. X-Ray. The x-ray method is probably the most widely used within the surface engineering field. Generally, it is a destructive method, as samples have to be prepared for laboratory examination. However, there are a number of modern instruments that promise in situ measurements, which are highly suited to examination of the field residual stress. It would appear, though, that most of the field-based measurements are really suited only for crude quality control due to the likely errors arising from both instrumentation and operators.

The basic principle of the x-ray methods is that the interplane spacing in the crystal in the stressed state differs from the unstressed state. By taking one measurement with the beam normal to the surface and another at an angle ψ to the normal, the normal stress in the direction ϕ is given by:

$$\sigma_\phi = \left[E/(1 + v) \right] \left[d_\psi - d_\mathrm{n}/d_0 \right] / \sin^2 \psi$$

where d_n is the spacing in the stressed state at the normal angle, d_0 is the spacing in the unstressed state at the normal angle, d_ψ is the spacing in the stressed state from the measurement at angle ψ to the normal, E is Young's modulus, and v is Poisson's ratio.

This is the basis of the two-tilt technique where the measurements are taken relative to some arbitrary angle using two exposures. If this is repeated for two further values of ϕ then the biaxial stress state is obtained. The more common method is the $\sin^2 \psi$ method, which follows from the observation that a plot of lattice spacing against this parameter measured over a number of angles will give the stress. There are several issues that need to be considered in this method and these are covered in more detail in the literature [36–40].

d. Synchrotron. This method uses high-energy x-rays and so gives much higher penetration. It can also provide much greater spatial resolution than laboratory-based x-ray instrumentation through control of the much narrower beams. It is not really suited to the types of problems mentioned above and is very expensive. However, there may be some limited applications with advances in analytical technology.

e. Raman Spectroscopy. Raman spectroscopy relies on the interaction between a laser light source and atoms. The Raman lines are sensitive to changes in hydrostatic stress and so can measure changes in the surface strain. It has not been used extensively in these types of studies but has some application for these types of problems.

REFERENCES

1. Vickerman, J.C. Introduction. In *Surface Analysis—The Principal Techniques*; Vickerman, J.C., Ed.; John Wiley and Sons: Chichester, 1997; 1–7.
2. Holmberg, K.; Mathews, A. Tribology of coating. In *Coatings Tribology*, Tribology Series, 28; Dowson, D., Ed.; Elsevier, 1994; 33–124.
3. Shaw, D.J. *Introduction to Surface and Colloid Chemistry*; 4th Ed.; Butterworth Heinemann, Oxford 1992; 64–114. Chapter 4, Liquid-gas and liquid-liquid interfaces.
4. Myhra, S.; Riviere, J.C. Elements of problem-solving. In *Handbook of Surface and Interface Analysis: Methods for Problem Solving*; Riviere, J.C., Myhra, S., Eds.; Marcel Dekker Inc.: New York, 1998; 9–22.
5. Smith, G.C. Compositional analysis by Auger electron and X-ray photoelectron spectroscopy. In *Handbook of Surface and Interface Analysis: Methods for Problem Solving*; Riviere, J.C., Myhra, S., Eds., Marcel Dekker Inc.: New York, 1998; 159–208.
6. Donnet, C. Problem-solving methods in tribology with surface-specific techniques. In *Handbook of Surface and Interface Analysis: Methods for Problem Solving*; Riviere, J.C., Myhra, S., Eds.; Marcel Dekker Inc.: New York, 1998; 697–746.

7. Stöhr, J. *NEXAFS Spectroscopy*, Springer Series in Surface Science 25, Springer-Verlag: Berlin, 1996.
8. Woodruff, D.P.; Delchar, T.A. Incident ion techniques. In *Modern Techniques of Surface Science*; Cambridge University Press, 1986; 195–278.
9a. *Handbook of Surface and Colloid Chemistry*; Birdi, K.S., Ed.; CRC Press: Boca Raton, New York, 1997.
9b. Shaw, D.J. *Introduction to Surface and Colloid Chemistry*, 4th Ed.; Butterworth Heinemann, Oxford, 1992.
9c. Adamson, A.W.; Gast, A.P. In *Physical Chemistry of Surfaces*, 6th Ed.; John Wiley and Sons: New York, 1997.
9d. *Contact Angle Wettability and Adhesion*; Mittal, K.L., Ed.; VSP: Utrecht, 1993.
10. Binnig, G.; Röhrer, H. Scanning tunnelling microscopy. Helv Phys Acta 1982, *55*, 726.
11. Binnig, G.; Quate, C.F.; Gerber, C. Atomic force microscopy. Phys. Rev. Lett. 1986, *56*, 930.
12a. *Scanning Tunneling Microscopy and Spectroscopy*; Bonnel, D.A., Ed.; VCH: New York, 1993.
12b. General principles and applications to clean and adsorbate-covered surfaces. In *Scanning Tunneling Microscopy I*; 2nd Ed.; Guntherodt, H.-J., Weisendangar, R., Eds.; Springer Verlag: Berlin, 1994; General Principles and Springer Series in Surface Science, Vol. 20.
12c. Theory of STM and related scanning probe methods. In *Scanning Tunneling Microscopy III*; Weisendangar, R., Guntherodt, H.-J., Eds.; Springer Verlag, Berlin, 1993; Springer Series in Surface Science, Vol. 29.
13. Song, J.F.; Vorburger, T.V. Surface texture. In *ASM Handbook, Friction, Lubrication and Wear Technology*; ASM International: Metals Park, Ohio, 1992; Vol.18, 334–335.
14. Fabijanic, D.M.; Barnett, M.R.; Hodgson, P.D.; Cardew-Hall, M. Improving the galling resistance of automotive stamping tools. In *Metal Forming 2000*; Pietrzyk, M., Kusiak, J., Majta, J., Hartley, P., Pillinger, I., Balkema, A.A., Eds.; Rotterdam, 2000; 641–646.
15. Stout, K.; Blunt, L. *Metrology and properties of engineering surfaces*. Proceedings of the 2nd International Australian Conference on Surface Engineering—Coatings and Surface Treatments in Manufacturing; Subramanian, C., Stafford, K.N., Eds.; Modbury Press Pty Ltd.: Pooraka, South Australia, 1994; W66–78.
16. Stout, K.J.; Blunt, L. *Nanometers to microns, three dimensional surface measurements in bioengineering*. Proceedings of the 2nd International Australian Conference on Surface Engineering—Coatings and Surface Treatments in Manufacturing; Subramanian, C., Stafford, K.N., Eds.; Modbury Press Pty Ltd.: Pooraka, South Australia, 1994; K1–15.
17. Stout, K.J.; Blunt, L. *Three Dimensional Surface Topography*, 2nd Ed.; Penton Press: London, 2000.
18. Brotzen, F.R. Evaluation of mechanical properties of thin films. In *ASM Handbook, Surface Engineering*; ASM International: Metals Park, Ohio, 1994; Vol. 5, 642–646.
19. Nix, W.D. Mechanical properties of thin films. Metall Trans 1989, *20A*, 2217–2245.
20. Zheng, X.; Grummett, S.J.; Thomson, P.F.; Lapovok, R.Y. *A study of the physico-mechanical properties of surface treated H13 die steel*. Tooling 99; Institute of Materials Engineering and Tooling Industry Forum of Australia, Melbourne, 1999; 187-192.
21. Oliver, W.C.; Pharr, G.M. An improved technique for determining hardness and elastic modulus using load and displacement sensing indentation experiments. J. Mater. Res. 1992, *7*, 1564–1583.
22. Swain, M.V. In *Indentation based mechanical characterisation of thin films on substrates*. Proceedings of the 2nd International Australian Conference on Surface Engineering—Coatings and Surface Treatments in Manufacturing; Subramanian, C., Stafford, K.N., Eds.; Modbury Press Pty Ltd.: Pooraka, South Australia, 1994; W79–W94.
23. Agarwal, A.; Dahotre, N.B. Mechanical properties of laser-deposited composite boride coating using nanoindentation. Metall. And Mater. Trans. 2000, *31A*, 401.
24. Michler, J.; Tobler, M.; Blank, E. Thermal annealing behaviour of alloyed DLC films on steel: determination and modelling of mechanical properties. Diamond Relat. Mater. 1999, *8*, 510–516.

25. Bull, S.J. In *Scratch adhesion testing and significance for coating performance*. Proceedings of the 2nd International Australian Conference on Surface Engineering—Coating and Surface Treatments in Manufacturing; Subramanian, C., Strafford, N., Eds.; Modbury Press Pty Ltd.: Pooraka, South Australia, 1994; W95–W113.

26. Matthews, A. *Variability in coatings and test results*. Proceedings of the 2nd Australian International Conference on Surface Engineering—Coating and Surface Treatments in Manufacturing; Subramanian, C., Strafford, K.N., Eds.; Modbury Press Pty Ltd.: Pooraka, South Australia, 1994; W38–W48.

27. ASTM D4173-82. Standard Practice for Evaluating Sheet Metal Forming Lubricant. ASTM: West Conshohocken, Pennsylvania, 1982.

28. Nine, H.D. Testing for sheet metal forming. In *Novel Techniques in Metal Deformation Testing*; Wagoner, R.H., Ed.; Metallurgical Society of AIME: Warrendale, Pennsylvania, 1983; 31-46.

29a. Wang, W.; Wagoner, R.H. A Realistic Friction Test for Sheet Forming Operations. Society of Automotive Engineers Paper Number 930807. SAE Trans.–J. Mat. Manuf. 1993, *2*, 915–922.

29b. Duncan, J.L.; Shabel, B.S.; Gerbase-Filho, J. A Tensile Strip Test for Evaluating Friction in Sheet Metal Forming. Society of Automotive Engineers Paper Number 780391. SAE Technical Paper Series: Warrendale, Pennsylvania, 1978; 1–8.

29c. Schey, J.A.; Smith, M.K. In *Report of NADDRG Friction Committee on reproducibility of friction tests within and between laboratories*. Sheet Metal and Stamping Symposium, International Congress and Exposition, SAE Special Publication No. 944; Society of Automotive Engineers: Warrendale, Pennsylvania, 1993; 261–267.

30. Bay, N.; Wibom, O.; Ravn, B.G.; Olsson, D.; Testing of friction and lubrication in metal forming—A review of test methods. *unpublished results*.

31. Subramanian, C. *Wear testing of coatings: methods, selection and limitations*. Proceedings of the 2nd International Australian Conference on Surface Engineering—Coatings and Surface Treatments in Manufacturing; Subramanian, C., Stafford, K.N., Eds.; Modbury Press Pty Ltd.: Pooraka, South Australia, 1994; W122–W136.

32. Czichos, H. Basic tribological parameters. In *ASM Handbook, Friction, Lubrication and Wear Technology*; ASM International: Metals Park, Ohio, 1992; Vol. 18, 474-479.

33. Blickensderf, R.; Laird, G.J., II. J. Testing and Evaluation, 1988, *16*, 516.

34. Cowley, J.M. Imaging. In *High Resolution Transmission Electron Microscopy and Related Techniques*; Buseck, P.R., Cowley, M., Eyring, L., Eds.; Oxford University Press: New York, 1992; 3–37.

35. Burnell-Gray, J.S. Datta, P.K. Electroless nickel coatings—case study. In *Surface Engineering Casebook*; Woodhead Publishing: Cambridge, 1996; Chapter 3, 49–69.

36. Kandil, F.A.; Lord, J.D.; Fry A.T.; Grant, P.V. A Review of Residual Stress Measurement Methods—A Guide to Technique Selection. National Physical Laboratory (NPL) Report MATC(A)04. Middlesex, United Kingdom, February 2001.

37. Borland, D.W. *Residual stress measurement—methods, limitations and significance*. Proceedings of the 2nd International Australian Conference on Surface Engineering—Coatings and Surface Treatments in Manufacturing; Subramanian, C., Stafford, K.N., Eds.; Modbury Press Pty Ltd.: Pooraka, South Australia, 1994; W114–W121.

38. Withers, P.J.; Bhadeshia, H.K.D.H. Residual stress Part 1—measurement techniques. Materials Science and Technology. 2001, *17*, 355–365.

39. Withers, P.J.; Bhadeshia, H.K.D.H. Residual stress Part 2—nature and origins. Materials Science and Technology. 2001, *17*, 366–375.

40. Sue, J.A.; Schajer, G.S. Stress determination for coatings. In *ASM Handbook, Surface Engineering*; ASM International: Metals Park, Ohio, 1994; Vol. 5, 647–653.

41. ASTM Standard E 837-99, Standard Test Method for Determining Residual Stresses by Hole Drilling Strain Gage Method. ASTM: West Conshohocken, Pennsylvania, 1999.

42. Hoffman, R.W. The mechanical properties of thin condensed films. In *Physics of Thin Films*; Hass, G., Thun, R.E., Eds.; Academic Press: New York, 1966; 211–273.

2
Surface Characterization Techniques: An Overview

Kazuhisa Miyoshi
National Aeronautics and Space Administration, Glenn Research Center, Cleveland, Ohio, U.S.A.

I. INTRODUCTION

As soon as one is confronted with a system where surface properties are involved, one is in trouble. The difficulty with surfaces in most practical situations is that conditions above, at, and below the engineering surface are extremely complex.

The properties of modified engineering surfaces in practical applications can be determined by material and surface analytical techniques [1]. Material and surface analyses are evolutionary disciplines. A number of analysis techniques are available for studying modified surfaces, thin films, and coatings from the atomic and electronic levels to macroscopical engineering component levels. Such techniques include a variety of physical, chemical, material, and mechanical characterizations. They can provide information that will allow one to select the materials, surface treatments (surface modification techniques and conditions), thin films and coatings, and environments best suited for a particular technical application.

Most materials used in high-technology applications, ranging from high-temperature oxidation, corrosion, thermal insulation, erosion, and wear to hydrophobicity, low adhesion (no stick), and low friction, have a near-surface region with properties differing greatly from those of the bulk material. In general, surface modifications or protective surface coatings (thermal barrier coatings, environmental barrier coatings, and mechanical barrier coatings) are desirable, or may even be necessary for a variety of reasons, including unique properties, light weight, engineering and design flexibility, materials conversion, or economics. These objectives can be attained by separating the surface properties from the bulk material properties [2,3].

Surface characterization (diagnostic) techniques are now available for measuring the shape; chemical, physical, and micromechanical properties; composition; and chemical states of any solid surface [1,4–10]. Because the surface plays a crucial role in many thermal, chemical, physical, and mechanical processes, such as oxidation, corrosion, adhesion, friction, wear, and erosion, these characterization techniques have established their importance in a number of scientific, industrial, and commercial fields [1,4–10].

This chapter deals with the application of surface characterization techniques to the development of advanced surface modification technology and processes of tribological coatings and films. These techniques can probe complex surfaces and clarify their inter-actions in mechanical systems and processes. The primary emphases are on the use of these techniques as they relate to surface modifications, thin films and coatings, and tribological engineering surfaces, and on the implications rather than the instrumentation. Finally, a case study describes the methodologies used for surface property measurements and diagnostics of chemical vapor-deposited (CVD) diamond films and coatings.

II. SURFACE CHARACTERIZATION TECHNIQUES

Although a wide range of physical and chemical surface analysis techniques is available, certain traits common to many of them can be classified from two viewpoints. Most tech-niques involve electrons, photons (light), x-rays, neutral species, or ions as a probe beam striking the material to be analyzed. The beam interacts with the material in some way. In some techniques, the changes induced by the beam (energy, intensity, and angular distribution) are monitored after the interaction, and analytical information is derived from observing these changes. In other techniques, the information used for analysis comes from electrons, photons, x-rays, neutral species, or ions that are ejected from the specimen under the stimulation of the probe beam. In many situations, several connected processes may be going on more or less simultaneously, with a particular analytical technique picking out only one aspect (e.g., the extent of incident light absorption, or the kinetic energy distribution of ejected electrons).

Furthermore, many mechanical techniques are available for assessing the surface roughness and micromechanical properties of material surfaces. Most techniques in this category involve mechanical contacts between a probe and a material surface.

Table 1 briefly summarizes the popular analytical techniques available today for studying the properties and behaviors of solid surfaces:

1. Profilometry and quantitative measurements of film thickness, plastic deforma-tion, and fracture damage
2. Surface microprobes for hardness and mechanical strength measurements
3. Elemental composition and chemical state measurements
4. Microstructure, crystallography, phase, and defect measurements.

The table allows quick access to what types of information are provided by these analytical techniques. In addition, it provides typical vertical resolution (or depth probed), typical lateral resolution, typical types of solid specimen, and popularity. The reader will find the basic principles and instrumentation details for a wide range of analytical techniques in the literature (e.g., Refs. 4–10). However, the analytical instrumentation field is moving rapidly, and within a year, current spatial resolutions, sensitivities, imaging and mapping capabilities, accuracies, and instrument costs and sizes are likely to be out of date. Therefore, these references should be viewed with caution. This table should be used as a quick reference guide only.

III. ELEMENTAL COMPOSITION AND CHEMICAL STATE OF SOLID
      SURFACES

Elemental composition is perhaps the most basic information about materials, followed by chemical state information, phase identification, determination of structure (bond structure, atomic sites, bond lengths, and angles), and defects (Table 1(c) and (d)). The elemental and chemical states, phase, microstructure, and defects of a solid often vary as a function of depth into the material, or spatially across the material. Many techniques specialize in addressing these variations down to extremely fine dimensions, as small as on the order of angstroms in some cases (refer to Fig. 1). Requests are made for physical and chemical information as a function of depth, to depths of 1 mm or so (materials have about 3 million atomic layers per millimeter of depth). This upper region at the material surface affects a broad spectrum of properties: elemental composition, contamination, adhesion, bonding, corrosion, surface strength and toughness, hardness, chemical activity, friction, wear, and lubrication. Knowing these variations is of great importance when selecting and using coatings, modified surfaces, and materials.

For surfaces, interfaces, and thin films, there is often little material to analyze—hence the need for many microanalytical methods such as those listed in Table 1. Within microanalysis, it is often necessary to identify trace components down to extremely low concentrations (parts per trillion in some cases), and a number of techniques specialize in this aspect. In other cases, a high degree of accuracy in measuring the presence of major components may be of interest. Usually, the techniques that are good for trace identification do not accurately quantify major components: Most complete analyses require the use of multiple techniques, the selection of which depends on the nature of the specimen and the desired information.

It is important to know the sampling (information) depth of the analytical tool to be used. For example, let us compare the analysis of two different specimens. The first has an atomic layer of one or more impurities on the surface; the other has these atoms distributed homogeneously within the specimen. When a conventional x-ray microprobe with a sampling depth of about 1000 nm is used, the specimens produce signals of equal intensity and it is not possible to differentiate between them. One can sometimes get an indication of whether there is a bulk impurity or a surface segregation (or a thin film) by lowering the electron beam voltage or by measuring at grazing incidence. However, this is not possible with only a few atomic layers. On the other hand, because of their insufficient detection limits, Auger electron spectroscopy (AES) or x-ray photoelectron spectroscopy (XPS) will produce two quite different spectra: in the first specimen, a strong signal from the impurity layer caused by the low sampling depth; in the second specimen, only the spectrum of the pure bulk material with no indication of impurities.

Ion scattering spectroscopy (ISS) and secondary ion mass spectrometry (SIMS) are extremely surface-sensitive characterization tools, followed by AES and XPS [or electron spectroscopy for chemical analysis (ESCA)]. Electron probe microanalysis (EPMA) and energy-dispersive x-ray spectroscopy (EDS) analysis are bulk characterization tools. Thus, knowing both the structure of real surfaces and the capabilities of the various characterization techniques is of great importance.

It was during the 1960s that the amazing growth and diversification of surface analytical techniques began and evolved with the development of two types of ultra-high-vacuum electron spectroscopy—AES closely followed by XPS. The combination of the all-encompassing definition of surface engineering and tribology with these surface analytical techniques, including a variety of electronic, photonic, and ionic spectroscopies and microscopies, reflects the trend of surface engineering and tribology today.

A number of techniques are now available for measuring the composition of any solid surface. The most widely used techniques for surface analysis are AES, XPS (or ESCA), and SIMS (Table 1(c)). These techniques are well suited for examining extremely thin layers, including contaminant layers and oxide layers.

Table 1 Popular Analytical Techniques for Surface, Thin Film, Interface, and Bulk Analysis of Coatings, Modified Surface Layers, and Materials

Technique	Main information	Vertical resolution (depth probed, typical)[a]	Lateral resolution (typical)	Types of solid specimen (typical)	Use (popularity)
(a) Profilometry and quantitative measurements of film thickness, plastic deformation, and fracture damage					
Optical profiler and laser interferometry	3D and 2D imaging, morphology, profilometry, topographical mapping, film thickness, wear volume, scar and crater depth, defects	–0.1 nm	A few submicrons to a few tens of microns	All	Medium
Confocal microscopy	3D and 2D imaging, morphology, profilometry, topographical imaging, film thickness, wear volume, scar and crater depth, defects	Variable; from a few nanometers to a few microns	Optical, 0.5–4 μm; SEM, 1–50 μm	Almost all	Medium
Optical scatterometry	Profilometry, topographical imaging/mapping, periodical structure, morphology, defects	$\geq$0.1 nm	A few submicrons to a few tens of microns; $\geq$laser wavelength, $\lambda/2$ for topography	Almost all	Not common
Light microscopy (general)	Imaging, morphology, damages, defects	Variable	Variable	All	Extensive
Stylus profilometry	Profilometry, topographical tracing, film thickness, morphology, scar and crater depth, wear volume	0.5 nm	100 nm	Almost all; flat smooth films	Extensive
Scanning tunneling microscopy (STM)	Topographical imaging, compositional mapping, morphology, profilometry, film thickness, spectroscopy, structure, defects	<0.03–0.05 nm	Atomic	Conductors	Medium
AFM or scanning force microscopy (SFM)	Topographical imaging, friction force mapping, morphology, profilometry, film thickness, wear volume, scar and crater depth, structure, defects	<0.03–0.05 nm	Atomic to 1 nm	All	Medium

Technique	Information	Lateral resolution	Analysis depth	Materials	Use
Variable-angle spectroscopical ellipsometry (VASE)	Film thickness, microstructure, optical properties	Tens of nanometers to microns	Millimeter	Planar surface and interface	Medium
X-ray fluorescence (XRF)	Film thickness (1–10^4 nm), element composition (qualitative mapping)	10 μm	10–150 μm	All but low-Z elements: H, He, and Li	Extensive

(b) Surface microprobes for hardness and mechanical strength measurements

Technique	Information	Lateral resolution	Analysis depth	Materials	Use
Mechanical strength microprobe (microhardness measurements)	Microscale hardness, creep deformation, plastic deformation, fracture toughness, strength, anisotropy	0.3 nm	Variable; atomic to a few tens of microns using STM, SFM, AFM, or optical microscopy	All	Extensive
Mechanical strength microprobe (nanohardness)	Nanoscale hardness, Young's modulus, creep deformation, fracture toughness, strength, anisotropy	0.3 nm	Atomic to 1 nm using STM, SFM, or AFM	All	Medium
Micrometer and nanometer scratch hardness measurements	Adhesion failure of thin films and coatings, abrasion resistance, scratch hardness, deformation, friction, fracture, strength, anisotropy	0.3 nm	Variable; atomic to a few tens of microns using STM, SFM, AFM, or optical microscopy	All	Medium

(c) Elemental composition and chemical state measurements.

Technique	Information	Lateral resolution	Analysis depth	Materials	Use
SEM	Imaging, morphology, elemental composition, damages, defects, crystallography, grain structure, magnetical domains	Variable; from a few nanometers to a few microns	1–50 nm in secondary electron mode	Conductors and coated insulators	Extensive
EPMA	Elemental composition, SEM imaging, compositional mapping	1 μm	0.5–1 μm	All	Medium
EDS or WDS	Elemental composition ($Z \geq 5$; boron to uranium), spectroscopy, imaging and mapping	0.02–1 μm	0.5–1 μm for bulk specimens; as small as 1 nm for thin specimens	All	Medium
High-resolution electron energy loss spectroscopy (EELS)	Elemental composition, chemical state, bonding state, imaging	2 nm	1 mm^2	Ultra-high-vacuum-compatible solids	Not common

(Continued on next page)

Table 1 Continued

Technique	Main information	Vertical resolution (depth probed, typical)[a]	Lateral resolution (typical)	Types of solid specimen (typical)	Use (popularity)
AES	Elemental composition (except H and He), chemical state, depth profiling, imaging and mapping	0.5–10 nm	A few tens of nanometers or less	Ultra-high-vacuum-compatible solids	Extensive
XPS	Elemental composition, chemical state, depth profiling, imaging and mapping	A few to several nanometers	5 µm–5 mm	Ultra-high-vacuum-compatible solids	Extensive
Fourier transform infrared spectroscopy (FTIR)	Chemical species, stress, structural inhomogeneity, defects, imaging and mapping	10 nm to microns	20 µm–5 mm	All (solid, liquid, or gas in all forms)	Extensive
Raman spectroscopy	Identification of unknown compounds, chemical state, bonding state, structural order, phase transitions, image and mapping	Few microns to millimeters	1 µm	Solids, liquids, gases, and thin films	Medium
Solid-state nuclear magnetical resonance (NMR)	Chemical state, phase identification, disordered state	—	—	All; not all elements	Not common
RBS	Elemental composition, structure, defects	2–30 nm	1–4 mm; 1 µm in specialized case	Ultra-high-vacuum-compatible solids	Medium
ERS	Hydrogen concentrations in thin films, depth profiling	Varies with depth; 30–60 nm at a depth of 100 nm in Si	1–4 mm	Ultra-high-vacuum-compatible solids	Medium
Secondary ion mass spectroscopy (SIMS)	Chemical state, imaging, elemental composition	0.3–2 nm	10 nm–2 µm	All; vacuum-compatible solids	Extensive

Technique	Information provided	Vertical resolution[a]	Lateral resolution	Materials	Availability
ISS	Elemental composition (outermost mono-atomic layer), imaging (limited)	0.3 nm	150 μm	All; vacuum-compatible solids	Not common
(d) Microstructure, crystallography, phase, and defects measurements					
XRD	Crystalline phases, strain, crystallite orientation and size, atomic arrangements, defect imaging, concentration depth profiling, film thickness	A few microns	None; ~10 μm with microfocus	All	Extensive
Low-energy electron diffraction (LEED)	Surface crystallography and microstructure, surface cleanliness, surface disorder, imaging	~0.4 nm	0.1 mm (~10 μm available)	Single-crystal conductors and semiconductors, insulators and polycrystalline specimens under special circumstances	Medium
TEM	Atomic structure, microstructure, crystallographical structure, defects, imaging and mapping, morphology, chemical bonding	None	≥0.2 nm	Conductors, semiconductors, and coated insulators	Medium
Reflection high-energy electron diffraction (RHEED)	Surface crystal structure 2D and 3D defects	2–10 nm	200 μm × 4 mm	Single-crystal conductors and semiconductors	Medium
Cathodoluminescence (CL)	Chemical state, defects	10 nm to microns	1 μm	All	Not common
Photoluminescence (PL), or fluorescence spectrometry	Band gaps, defects, impurity structure, chemical state, imaging and mapping	0.1–3 μm	1–2 μm	All (solid or liquid)	Medium

[a] Vertical resolution is a measurement of the technique's ability to clearly distinguish a property as a function of depth.

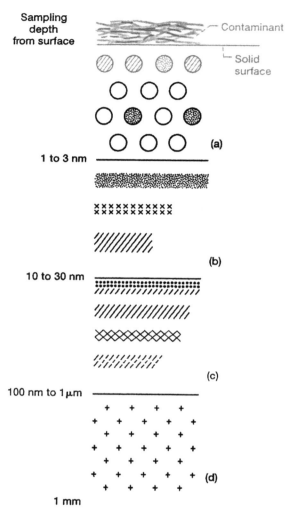

Figure 1 Schematic diagram showing regimes of (a) surface analysis, (b) thin film analysis, (c) interface analysis, and (d) bulk substrate analysis.

A. SEM and EDS

The single most useful tool available today to surface engineers and scientists, tribologists, and lubrication engineers interested in studying the morphology, defects, and wear behavior of material surfaces is, undoubtedly, scanning electron microscopy (SEM). Especially, the combination of the SEM and x-ray analysis using either EDS or wavelength-dispersive x-ray spectroscopy (WDS) provides a powerful tool for local microchemical analysis [7,11]. The use of electron microprobe techniques in SEM is now a well-established procedure. The two techniques, EDS and WDS, differ only in the use of an energy-dispersive, solid-state detector versus a wavelength-dispersive, crystal spectrometer. Successful studies have been carried out to characterize surface-modified materials, thin films and coatings, and surfaces of bulk materials. The scanning electron microscope is often the first analytical instrument used when a quick look at a material is required and the light microscope no longer provides adequate spatial resolution or depth of focus. The scanning electron microscope provides the investigator with a highly magnified image of a material surface. Its resolution can approach a few nanometers, and it can be operated at magnifications from about ×10 to ×300,000. The scanning electron microscope produces not only morphological and topographical information but information concerning the elemental composition in near-surface regions.

In the scanning electron microscope, an electron beam is focused into a fine probe and subsequently raster-scanned over a small rectangular area. As the electron beam interacts with the specimen, it creates various signals, such as secondary electrons (SEs), internal currents, and photon emissions, all of which can be collected by appropriate detectors. The SEM–EDS produces three principal images: secondary electron images, backscattered electron (BE) images, and elemental x-ray maps. Secondary and backscattered electrons are conventionally separated according to their energies. They are produced by different mechanisms. When a high-energy primary electron interacts with an atom, it undergoes either inelastic scattering with the atomic electrons, or elastic scattering with the atomic nucleus. In an inelastic collision with an electron, some amount of energy is transferred to the other electron. If the energy transferred is extremely small, the emitted electron will probably not have enough energy to exit the surface. If the energy transferred exceeds the work function of the material, the emitted electron will exit the solid. When the energy of the emitted electron is less than about 50 eV, by convention, it is referred to as a secondary electron, or simply a secondary. Most of the emitted secondaries are produced within the first few nanometers of the surface. Secondaries produced much deeper in the material suffer additional inelastic collisions, which lower their energy and trap them in the interior of the solid.

Higher-energy electrons are primary electrons that have been scattered without loss of kinetic energy (i.e., elastically) by the nucleus of an atom, although these collisions may occur after the primary electron has already lost some energy to inelastic scattering. Backscattered electrons are, by definition, electrons that leave the specimen with only a small loss of energy relative to the primary electron beam energy, but BSEs are generally considered to be the electrons that exit the specimen with an energy greater than 50 eV, including Auger electrons. The BSE imaging mode can be extremely useful for tribological applications because the energy, spatial distribution, and number of BSEs depend on the effective atomic number of the specimen, its orientation with respect to the primary beam, and the surface condition. The backscatter coefficient, or relative number of electrons leaving the specimen, increases with increasing atomic number [11]. The higher the atomic number Z of a material, the more likely it is that backscattering will occur. Thus, as a beam passes from a low-Z to a high-Z area, the signal due to backscattering and, consequently, the image brightness will increase. There is a built-in contrast caused by elemental differences. BSE images can therefore be used to distinguish different phases, modified surfaces, thin films and coatings, and foreign species of the specimen having different mean atomic numbers (atomic number contrast). For most specimens examined in scanning electron microscopes, except for those that are flat or polished, the specimen both varies in chemistry from area to area and exhibits a varying rough surface. As a result, both atomic number and topographical contrast are present in the BSE signal (as well as in the SE signal). In general, if the high-energy BSEs are collected from the specimen at a relatively high takeoff angle, atomic number information is emphasized. Conversely, if the high-energy electrons leaving the specimen are collected at a relatively low takeoff angle, topographical information is emphasized. For nearly all BSE applications, the investigator is interested in the atomic number contrast and not in the topographical contrast. Note that the backscatter coefficient is defined as the number of BSEs emitted by the specimen for each electron incident on the specimen. Because of the relatively deep penetration of the incident electron beam, combined with the extensive range of BSEs produced, spatial resolution in the BSE mode is generally limited to about 100 nm in bulk specimens under the usual specimen/detector configurations.

Both energy-dispersive and wavelength-dispersive x-ray detectors can be used for element detection in SEM. When the atoms in a material are ionized by high-energy radiation, usually electrons, they emit characteristic x-rays. The detectors produce an output signal that is proportional to the number of x-ray photons in the area under electron bombardment. EDS is a technique of x-ray spectroscopy that is based on the collection and energy dispersion of characteristic x-rays. Most EDS applications are in electron column instruments such as the scanning electron microscope, the electron probe microanalyzer, and the transmission electron microscope. X-rays entering a solid-state detector, usually made from lithium-drifted silicon, in an EDS spectrometer are converted into signals that can be processed by the electronics into an x-ray energy map or an x-ray energy histogram. A common application of the x-ray systems, such as EDS and WDS, involves x-ray mapping, in which the concentration distribution of an element of interest is displayed on a micrograph. The detectors can be adjusted to pass only the pulse range corresponding to a particular element. This output can then be used to produce an x-ray map or an elemental image. Higher concentrations of a particular element yield higher x-ray photon pulse rates, and the agglomeration of these pulses, which appear as dots in the image, generate light and dark areas relating to the element's concentration distribution. In x-ray spectroscopy, the x-ray spectrum consists of a series of peaks that represent the type and relative amount of each element in the specimen. The number of counts in each peak can be further converted into elemental weight concentration either by comparison with standards, or by standardless calculations. Three modes of analysis are commonly used: spectrum acquisition, spatial distribution or dot mapping of the elements, and element line scans.

B. AES and XPS

The surface analytical techniques most commonly used in surface engineering and tribology are AES and XPS (or ESCA). Each can determine the composition of the outermost atomic layers of a clean surface, or of surfaces covered with adsorbed gas films, oxide films, thin films and coatings, reaction film products, surface-modified materials, and frictionally transferred films [1,4–10,12,13].

AES and XPS are generally called "surface analysis" techniques, but this term can be misleading [8,9]. Although these techniques derive their usefulness from their intrinsic surface sensitivity, they can also be used to determine the composition of deeper layers. Such a determination is normally achieved through controlled surface erosion by ion bombardment. AES or XPS analyzes the residual surface left after a certain sputtering time with rare gas ions. In this way, composition depth profiles that provide a powerful means for analyzing thin films, surface coatings, reaction film products, transferred films, and their interfaces can be obtained. Clearly, this capability also makes AES and XPS ideal for studying surface-modified materials and wear-resistant coatings. There are, however, a number of practical differences between the two techniques (e.g., detection speed, background, and spatial resolution) that are generally more advantageous in AES profiling. AES uses a focused electron beam to create secondary electrons near a solid surface. Some of these electrons (the Auger electrons) have energies characteristic of the elements.

As stated above, AES can characterize the specimen in-depth and provide elemental depth profiles when used in combination with sputtering (e.g., argon ion sputter etching) to gradually remove the surface. In addition to energies characteristic of the elements, some of the Auger electrons detected have energies characteristic, in many cases, of the chemical bonding of the atoms from which they are released. Because of their characteristic energies and the shallow depths from which they escape without energy loss, Auger electrons can characterize the elemental composition and, at times, the chemistry of surfaces. The Auger peaks of many elements show significant changes in position or shape in different chemical conditions and environments.

Thus, AES has the attributes of high lateral resolution, relatively high sensitivity, and standardless semiquantitative elemental analysis. It also provides chemical bonding information in some cases. Furthermore, the high spatial resolution of the electron beam and the sputter etching process allow microanalyses of three-dimensional regions of solid specimens.

In XPS, monoenergetic soft x-rays bombard a specimen material, causing electrons to be ejected. The elements present in the specimen can be identified directly from the kinetic energies of these ejected photoelectrons. Electron-binding energies are sensitive to the chemical state of the atom. Although XPS is designed to deal with solids, specimens can be gaseous, liquid, or solid. XPS is applicable to metals, ceramics, semiconductors, and organic, biological, and polymeric materials. Although x-ray beam damage can sometimes be significant, especially in organic materials, XPS is the least destructive of all the electron or ion spectroscopy techniques. The depth of solid material sampled varies from the top two atomic layers to 15–20 layers. This surface sensitivity, combined with quantitative and chemical analysis capabilities, has made XPS the most broadly applicable general surface analysis technique used today, especially in the field of tribology. Like AES, XPS can also characterize the specimen in-depth and provide elemental depth profiles when used in combination with sputtering (e.g., argon ion sputter etching) to gradually remove the surface.

In general, AES provides elemental information only. The AES peaks of many elements, however, show significant changes in position or shape in different chemical environments. On the other hand, the main advantage of XPS is its ability to provide chemical information from the shifts in binding energy. The particular strengths of XPS are quantitative elemental analysis of surfaces (without standards) and chemical state analysis. For a solid, AES and XPS probe 2–10 and 2–20 atomic layers deep, respectively, depending on the material, the energy of the photoelectron concerned, and the angle (with respect to the surface) of the measurement.

The thickness of the outer atomic layers can be determined by studying the attenuation of photoelectrons originating in the bulk material caused by the layers and by studying the variation in intensity of photoelectrons emitted by the layers as a function of thickness.

IV. CASE STUDY: CHARACTERIZATION OF DIAMOND FILMS AND COATINGS

A. Introduction and Background

During the last decade, significant progress has been made in the development of advanced surface films and coatings for engineering and biomedical applications. Some of the most exciting recent developments are superhard coatings and films, such as chemical vapor-deposited diamond, diamond-like carbon (DLC), carbon nitride (CNx), and cubic boron nitride (c-BN) [14].

The commercial potential of CVD diamond films has been established, and a number of applications have been identified through university, industry, and government research studies. CVD diamonds are presently produced in the form of coatings or wafers. CVD diamond film technology offers a broader technological potential than do natural and high-pressure synthetic diamonds because size, geometry, and cost will not be as limiting.

Diamond coatings can improve many of the surface properties of engineering substrate materials, including erosion, corrosion, and wear resistance [14]. Examples of actual and potential applications, such as microelectromechanical systems and environmentally durable barriers, of diamond coatings and related superhard coatings are described in Ref. 15. For example, diamond coatings can be used as a chemical and mechanical barrier for space shuttle check valves, particularly guide pins and seat assemblies [16].

Achieving the quality and distinctive properties of diamond coatings and films requires optimizing deposition parameters through the study of the physical, chemical, and structural changes of coatings and films as a function of deposition parameters. These parameters must not only give the appropriate initial level of quality and properties but must also provide durable coatings and films.

For a material to be recognized as diamond, it must have all of the following characteristics [17,18]:

1. Crystalline diamond morphology and microstructure visible by optical or electron microscopy
2. Single-phase diamond crystalline structure detectable by x-ray or electron diffraction
3. Clear, sharp diamond peak at 1332 cm^{-1} in a Raman spectrum
4. Carbon content (>95 at.%)
5. Low coefficient of friction (0.01–0.05) in air.

This section deals with the application of measurement and characterization techniques required for the technological growth of advanced CVD diamond films and coatings. Each measurement and characterization technique provides unique information. A combination of techniques can provide the technical information required to understand the quality and properties of CVD diamond films, which are important to their application in specific component systems and environments. In this study, the combination of measurement and characterization techniques was successfully applied to correlate deposition parameters and resultant diamond film composition, crystallinity, grain size, surface roughness, and coefficient of friction. An important case study of microwave plasma-assisted CVD diamond films will be highlighted. Some earlier data and experimental details on this research are given in Refs. 19–21.

B. Measurement and Characterization of CVD Diamond

A variety of techniques can be used to characterize CVD diamond films. Measurement and characterization techniques used in this investigation include:

1. SEM and transmission electron microscopy (TEM), to determine surface morphology, microstructure, and grain size
2. Surface profilometry and atomic force microscopy (AFM), to measure surface roughness and to determine surface morphology
3. Rutherford backscattering spectroscopy (RBS) and elastic recoil spectroscopy (ERS), to determine the composition (including hydrogen)
4. Raman spectroscopy, to characterize the atomic bonding state and quality of diamonds
5. X-ray diffraction (XRD), to determine the crystal orientation of diamond
6. Friction measurement, to determine the coefficient of friction and surface properties.

Case studies described in Refs. 19–21 focus attention primarily on microwave plasma-assisted CVD diamond films.

C. Electron Microscopy, Stylus Profilometry, and Atomic Force Microscopy

Transmission electron microscopy offers image and diffraction modes for specimen observation [22]. In the image mode, analysis of transmitted electron images yields information both about atomic structure and about defects present in the material. In the diffraction mode, an electron diffraction pattern is obtained from the specimen area illuminated by the electron beam. The electron diffraction pattern is entirely equivalent to an x-ray diffraction pattern.

Scanning electron microscopy (with energy-dispersive x-ray spectroscopy) is the most useful tool when the researcher needs not only morphological and topographical information about surfaces but also information concerning the composition of near-surface regions of the material. Although diamond is an insulator, it can be studied by using low primary electron beam voltages (5 keV or less) if one is willing to compromise image resolution to some extent. If the diamond is coated with a thin conducting film (10–20 nm thick) of carbon, gold, or some other metal, the coated diamond can be studied with an image resolution of 1–50 nm.

The grain size, surface morphology, and surface roughness of a microwave plasma-assisted CVD diamond film can be controlled by varying the deposition parameters, such as gas-phase chemistry parameters and temperatures (e.g., Table 2). The grain size and surface roughness data were obtained by using TEM and stylus profilometry, respectively. The CVD diamond films referred to in Table 2 can be divided into three groups by grain size: fine (nanocrystalline), medium, and coarse grain. The grain sizes of the fine-grain nanocrystalline diamond films were determined from bright-field and dark-field electron micrographs to

Table 2 Deposition Conditions for Diamond Films of Various Grain Sizes (Nanometer Scale to Micrometer Scale)

Condition	Substrate[a]						
	Si (100)	Si (100)	α-SiC	α-SiC	α-SiC	Si_3N_4	Si_3N_4
Deposition temperature (°C)	860±20	1015±50	1015±50	965±50	860±20	965±50	860±20
Gaseous flow rate (cm³/min)							
CH_4	4	3.5	3.5	3.5	4	3.5	4
H_2	395	500	500	500	395	5800	395
O_2	1	0	0	0	1	0	1
Pressure (Torr)	5	40	40	40	5	40	5
Microwave power (W)	500	100	1000	1000	500	1000	1500
Deposition time (hr)	10.5	140	14	22	21	22	521
Thickness (nm)	1000	4200	5000	8000	1000	7000	800
Grain size (nm)	20–100	1100	3300	1500	22–100	1000	22–100
Surface roughness rms (nm)	15	63	160	92	50	52	35

[a] Scratched with 0.5-mm diamond paste.

be between 20 and 100 nm. The medium-grain and coarse-grain diamond films have grain sizes estimated at 1000–1500 and 3300 nm, respectively. The average surface roughness of the diamond films measured by a surface profilometer increases as the grain size increases, as shown in Fig. 2. Figure 3 shows scanning electron micrographs of fine-grain (nanocrystalline), medium-grain, and coarse-grain diamond films. Triangular crystalline facets typical of diamond are clearly evident on the surfaces of the medium-grain and coarse-grain films.

In an atomic force microscope, a probe tip traverses across a diamond surface and senses the force of interaction between itself and the diamond surface. By monitoring the tip deflection necessary to maintain a constant interacting force, surface topographical data can be obtained on a nanometer scale. Figure 4 shows an AFM image of a chemical vapor-deposited, fine-grain nanocrystalline diamond film on a mirror-polished silicon substrate, along with a histogram and bearing ratio. The CVD diamond surface has a granulated or spherulitic morphology with spherical asperities of different sizes. The surface roughness of the CVD diamond on silicon is 58.8 nm rms.

D. X-ray Diffraction and Electron Diffraction

Although x-ray diffraction is not inherently a surface characterization technique, it offers unparalleled accuracy in the measurement of atomic spacing [22]. XRD was used to determine the structure and crystal orientation of the CVD diamond films [20]. Typical x-ray diffraction patterns for the fine-grain (nanocrystalline) and medium-grain diamond films (Fig. 5) show peaks representing only the diamond film and the silicon substrate. Diffraction peaks corresponding to the {111}, {220}, {311}, and {400} planes, reflective of diamond, are clearly evident. The intensity ratios $I\{220\}/I\{111\}$ were calculated from the x-ray diffraction patterns for these films and were found to be 1.3 and 0.04, respectively. The powder diffraction pattern of diamond with random crystal orientation (ASTM 6-0675)

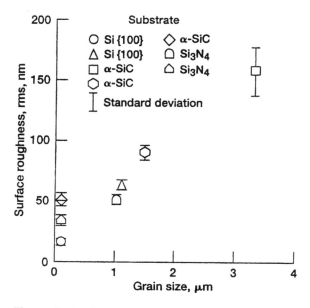

Figure 2   Surface roughness as function of grain size for diamond films.

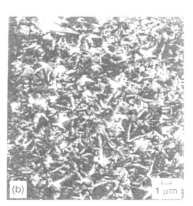

Figure 3 Scanning electron micrographs of diamond films. (a) Fine-grain (20–100 nm) nanocrystalline diamond film on {100} silicon substrate; rms surface roughness, 15 nm. (b) Medium-grain (1100 nm) diamond film on {100} silicon substrate; rms surface roughness, 63 nm. (c) Coarse-grain (3300 nm) diamond film on {100} a-SiC substrate; rms surface roughness, 160 nm.

gives $I\{220\}/I\{111\} = 0.27$. Thus, most of the crystallites in the fine-grain nanocrystalline diamond film are oriented along the $\langle 110 \rangle$ direction, whereas most of the crystallites in the medium-grain diamond films are oriented along the $\langle 111 \rangle$ direction. The well-formed triangular facets observed in SEM micrographs of medium-grain and coarse-grain diamond films confirm the $\langle 111 \rangle$ crystal orientation.

Figure 6 presents a TEM selected area diffraction (SAD) pattern, a TEM bright-field micrograph, and a TEM dark-field micrograph of a free-standing, fine-grain nanocrystalline CVD diamond film [19]. Diffraction rings and dots can be observed in Fig. 6(a). The d-spacings of the diffraction rings were calculated by using an aluminum SAD as a calibration standard and were found to match well with the known diamond d-spacings. No evidence of nondiamond carbon was found in the SAD. This observation indicates that the nondiamond carbon concentration in the nanocrystalline diamond film was extremely small. Careful observation of Fig. 6(b) revealed various nucleilike regions marked N. Diamond grains are distributed radially outward from these nuclei. A grain boundary is formed where the grains from various nuclei meet. As previously mentioned, the grain sizes of the fine-grain nanocrystalline CVD diamond films estimated from the bright-field and dark-field micrographs varied from 20 to 100 nm.

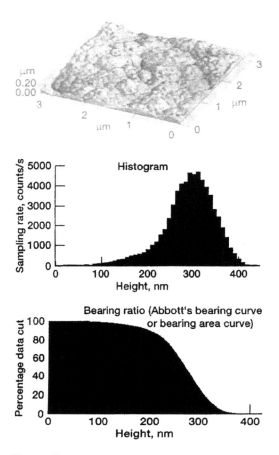

Figure 4   Atomic force microscopy of chemical vapor-deposited, fine-grain nanocrystalline diamond film on mirror-polished silicon substrate.

E. Raman Spectroscopy

Raman spectroscopy is primarily a structural characterization tool [22]. Raman spectra are more sensitive to the lengths, strengths, and arrangements of bonds in a material than to chemical composition. Raman spectra of crystals likewise reflect the details of defects and disorders, rather than trace impurities and related chemical imperfections. The laser optical Raman technique can determine with great confidence the atomic bonding states of the carbon atoms (sp^2 for graphite, or sp^3 for diamond) from their different vibrational modes [23]. Raman spectra result from the inelastic scattering of optical photons by lattice vibration phonons.

   Typical Raman spectra of the fine-grain (nanocrystalline) and medium-grain diamond films (Fig. 7) show one Raman band centered at 1332 cm^{-1} and one centered around 1530 cm^{-1}. The sharp peak at 1332 cm^{-1} is characteristic of the sp^3 bonding of the diamond form of carbon in the film. The very broad peak centered around 1530 cm^{-1} is attributed to the sp^2 bonding of the nondiamond forms of carbon (graphite and other carbon) [24–26].

   More sp^3-bonded (diamond) carbon is produced in larger-grained CVD diamond films (e.g., Fig. 7(b)) than in fine-grain nanocrystalline films, as is evident from the relative

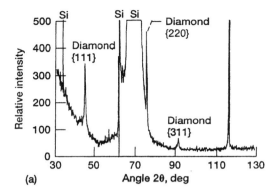

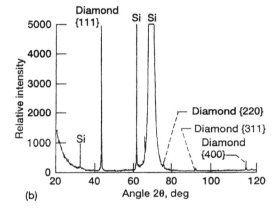

Figure 5 X-ray diffraction patterns of diamond films. (a) Fine-grain (20–100 nm) nanocrystalline diamond film on {100} silicon substrate. (b) Medium-grain (1100 nm) diamond film on {100} silicon substrate.

intensities of the diamond and nondiamond carbon Raman bands [20]. However, the ratio of the intensities of the Raman responses at 1332 cm^{-1} and centered around 1530 cm^{-1} does not indicate the ratio of diamond to nondiamond carbon present in a particular film because the Raman technique is approximately 50 times more sensitive to sp^2-bonded (nondiamond) carbon than to sp^3-bonded (diamond) carbon [25]. Thus, the peak centered around 1530 cm^{-1} for each film represents a much smaller amount of nondiamond carbon in these diamond films than appears at first glance.

F. Rutherford Backscattering Spectroscopy and Elastic Recoil Spectroscopy

Rutherford backscattering spectroscopy is a nondestructive, quantitative depth profiling of thin film compositions and structures, crystallinity, dopants, and impurities [22]. Elastic recoil spectroscopy (hydrogen forward scattering, or proton recoil detection) is the simplest ion beam technique for hydrogen profiling and determining hydrogen concentrations in thin films. In combination with RBS analysis of the same sample, ERS provides concentration profiles and complete compositional analyses of the near-surface regions of the sample material.

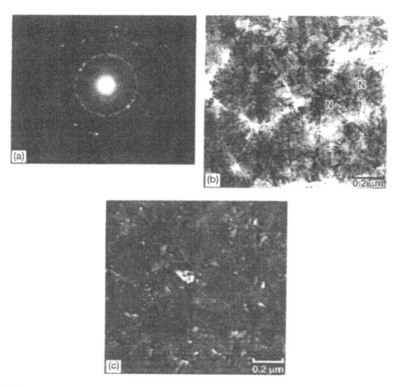

Figure 6   Free-standing nanocrystalline diamond film. (a) Selected area diffraction pattern. (b) Bright-field TEM. (c) Dark-field TEM.

Figures 8 and 9 present RBS and ERS spectra, respectively, of a fine-grain nano-crystalline CVD diamond film [19]. Besides carbon from the diamond film and silicon from the silicon substrate, no other elements were observed in the RBS spectrum. From both spectra, it was estimated that the fine-grain nanocrystalline diamond film consisted of 97.5 at.% carbon and 2.5 at.% hydrogen. (In contrast, the medium-grain diamond films contained less than 1 at.% hydrogen [20].) It was also demonstrated that both carbon and hydrogen are uniformly distributed in the fine-grain nanocrystalline film from the top of the surface to the silicon substrate.

RBS analytical results can also be used to determine diamond film thickness. Figure 8 presents a simulated RBS spectrum of a diamond film with a carbon-to-hydrogen ratio (C/H) of 97.5/2.5 obtained by using the RUMP computer code [27]. In the computer program, the film thickness of the diamond film is taken as a variable. This thickness was obtained from the close match between the observed and simulated RBS, as shown in Fig. 8, and is 1.5 μm at the center of the substrate. The deposition rate was estimated to be 0.14 μm/hr.

G.   Friction Measurement

When the fine-grain (nanocrystalline), medium-grain, and coarse-grain CVD diamond films characterized in previous sections were brought into contact with a natural diamond pin in reciprocating sliding motion in air and in nitrogen, the coefficients of friction varied as the

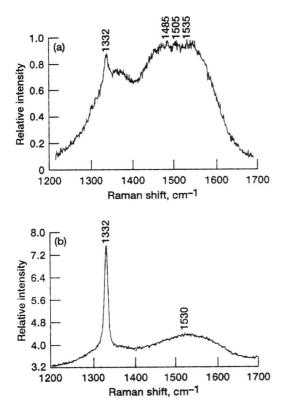

Figure 7 Raman spectra of diamond films. (a) Fine-grain (20–100 nm) nanocrystalline diamond film on {100} silicon substrate. (b) Medium-grain (1100 nm) diamond film on {100} silicon substrate.

pin traveled back and forth (reciprocating motion), retracing its tracks on the diamond films.

Both in humid air (at a relative humidity of 40%) and in dry nitrogen, abrasion occurred and dominated the friction and wear behavior. The bulk natural diamond pin tended to dig into the surface of diamond films during sliding and to produce a wear track (groove). SEM observations of the diamond films indicated that small fragments chipped off the surfaces. When abrasive interactions between the diamond pin surface and the initially sharp tips of asperities on the diamond film surfaces were strong, the friction was high. The surface roughness of diamond films can appreciably influence their initial friction (i.e., the greater the initial surface roughness, the higher the initial coefficient of friction; Fig. 10(a)). Similar frictional results have also been found by other workers on single-crystal diamonds [28] and on diamond coatings [29–31].

As sliding continued and the pin passed repeatedly over the same track, the coefficient of friction was appreciably affected by the wear on the diamond films (i.e., a blunting of the asperity tips). When repeated sliding produced a smooth groove or a groove with blunted asperities on the diamond surface, the coefficient of friction (only due to adhesion) was low, and the initial surface roughness effect became negligible. Therefore, the equilibrium coefficient of friction was independent of the initial surface roughness of the diamond film (Fig. 10(b)).

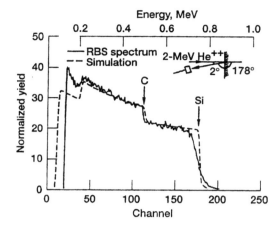

Figure 8 Rutherford backscattering spectrum of fine-grain nanocrystalline diamond film on silicon substrate. (Simulation curve was calculated by using the computer code RUMP.)

H. Summary of Remarks

The technical application and utility of CVD diamond in specific component systems and environments can be achieved if the deposition parameters have been optimized to achieve the desired quality, properties, and durability. To understand the benefits provided by the deposition parameters, and ultimately to provide even better deposition parameters and greater film and coating performance, researchers must use a variety of measurement and diagnostic techniques to investigate the physical, chemical, material, and structural changes in films and coatings produced at different deposition parameters.

The use of measurement and characterization techniques, including friction measurements, was highlighted in the important case study of microwave plasma-assisted, chemical vapor-deposited diamond films. In this study, a combination of measurement and characterization techniques was successfully applied to correlate the coating deposition parameters

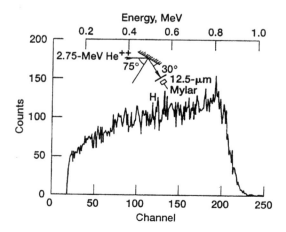

Figure 9 Elastic recoil spectrum of fine-grain nanocrystalline diamond film on silicon substrate.

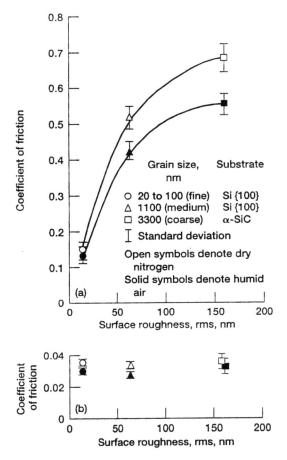

Figure 10 Coefficient of friction as function of initial surface roughness of diamond films in humid air (approximately 40% relative humidity) and in dry nitrogen. (a) Initial coefficients of friction. (b) Equilibrium coefficients of friction.

with the resultant diamond film composition, crystallinity, grain size, surface roughness, and coefficient of friction. These techniques have contributed significantly to the understanding of the quality and properties of diamond films and their surfaces.

V. CONCLUDING REMARKS

To understand the benefits that surface modifications provide, and ultimately to extend these benefits, it is necessary to study the physical, mechanical, and chemical changes they cause. A wide variety of surface characterization techniques is available for assessing the physical, mechanical, and chemical properties of surfaces. Each measurement and characterization technique provides unique information. It should be possible to coordinate the different pieces of information provided by these measurement and diagnostic techniques into a coherent self-consistent description of the surface and bulk properties.

REFERENCES

1. Miyoshi, K.; Chung, Y.W. *Surface Diagnostics in Tribology: Fundamental Principles and Applications*; World Scientific Publishing Co.: River Edge, NJ, 1993.
2. Holmberg, K.; Matthews, A. *Coatings Tribology: Properties, Techniques and Applications in Surface Engineering, Tribology Series 28*; Dowson, D., Ed.; Elsevier: Amsterdam, 1994.
3. Bunshah, B., Ed.; *Deposition Technologies for Films and Coatings: Developments and Applications*; Noyes Publications: Park Ridge, NJ, 1982.
4. Quinn, T.F.J. *Physical Analysis for Tribology*; Cambridge University Press: Cambridge, UK, 1991.
5. Buckley, D.H. *Surface Effects in Adhesion, Friction, Wear, and Lubrication*, Tribology Series #5; Elsevier: Amsterdam, 1981.
6. Glaeser, W.A., Ed.; *Characterization of Tribological Materials*; Butterworth-Heinemann and Manning: Stoneham, MA, 1993.
7. Brundle, C.R., Evans, C.A. Jr., Wilson, S., Eds.; *Encyclopedia of Materials Characterization*; Butterworth-Heinemann and Manning: Stoneham, MA, 1992.
8. Brewis, D.M., Ed. *Surface Analysis and Pretreatment of Plastics and Metals*; Macmillan: New York, 1982.
9. Briggs, D.; Seah, M.P. *Practical Surface Analysis: By Auger and X-ray Photo-Electron Spectroscopy*; Wiley: New York, 1983; Vol. 1.
10. Tsukizoe, T. *Precision Metrology*; Yokkendo Publishing: Tokyo, Japan, 1970; 180–199.
11. Norton, J.T., Cameron, G.T., Sr., Eds.; *Electron Optical and X-ray Instrumentation for Research, Product Assurance and Quality Control, Amray Technical Bulletins*; Amray, Inc.: Bedford, MA, January 1986; Vol. 2(1).
12. Chung, Y.W.; Cheng, H.S. *Advances in Engineering Tribology, STLE SP-31*; Society of Tribologists and Lubrication Engineers: Park Ridge, IL, 1991.
13. Chung, Y.W., Homola, A.M., Street, G.B., Eds.; *Surface Science Investigations in Tribology: Experimental Approaches, ACS Symposium Series 485*; American Chemical Society: Washington, DC, 1992.
14. Molloy, A.P., Dionne, A.M., Eds.; *Wear and Superhard Coatings*; Gorham Advanced Materials, Inc.: Gorham, ME, 1998.
15. Miyoshi, K.; Murakawa, M.; Watanabe, S.; Takeuchi, S.; Wu, R.L.C. *Tribological Characteristics and Applications of Superhard Coatings: CVD Diamond, DLC, and c-BN. NASA/TM-1999-209189*; National Aeronautics and Space Administration: Washington, DC, 1999.
16. Miyoshi, K. *Aerospace Mechanisms and Tribology Technology: Case Studies. NASA/TM-1999-107249*; National Aeronautics and Space Administration: Washington, DC, 1999.
17. Pierson, H.O. *Handbook of Carbon, Graphite, Diamond, and Fullerenes*; Noyes Publications: Park Ridge, NJ, 1993.
18. Miyoshi, K. *Chemical-Vapor-Deposited Diamond Films. NASA/TM-1999-107249*; National Aeronautics and Space Administration: Washington, DC, 1999.
19. Wu, R.L.C.; Rai, A.K.; Garscadden, A.; Kee, P.; Desai, H.D.; Miyoshi, K. Synthesis and characterization of fine grain diamond films. J. Appl. Phys. 1992, 72 (1), 110–116.
20. Miyoshi, K. Friction and wear of plasma-deposited diamond films. J. Appl. Phys. 1993, 74 (7), 4446–4454.
21. Miyoshi, K.; Wu, R.L.C.; Garscadden, A. Friction and wear of diamond and diamond-like carbon coatings. Surf. Coat. Technol. 1992, 54/55, 428–434.
22. Brundle, C.R.; Evans, C.A., Jr.; Wilson, S. *Encyclopedia of Materials Characterization*; Butterworth-Heinemann: Boston, MA, 1992.
23. Nemanich, R.J.; Glass, J.T.; Lucovsky, G.; Shroder, R.E. Raman scattering characterization of carbon bonding in diamond and diamond-like thin films. J. Vac. Sci. Technol. A Vac. Surf. Films 1988, 6 (3), 1783–1787.
24. Cheng, J.J.; Mautei, T.D.; Vuppulahadium, R.; Jackson, H.E. Effects of oxygen and pressure on diamond synthesis in a magnetoactive microwave discharge, J. Appl. Phys. 1992, 71 (6), 2918–2923.

25. Kobashi, K.; Nishinura, K.; Kawate, V.; Horiuchi, T. Synthesis of diamonds by use of microwave plasma chemical vapor deposition: morphology and growth of diamond film. Phys Rev. B 1988, *38* (6), 4067–4084.

26. Wada, N.; Solin, S.A. Raman efficiency measurements of graphite (and Si and Ge), Physica B&C 1981, *105*, 353–356.

27. Doolittle, L.R. Algorithms for the rapid simulation of Rutherford backscattering spectra. Nucl. Instrum. Methods B 1985, *9*, 344–351.

28. Casey, M.; Wilks, J. The friction of diamond sliding on polished cube faces of diamond. J. Phys. D Appl. Phys. 1973, *6* (15), 1772–1781.

29. Hayward, I.P. Friction and wear properties of diamond and diamond coatings. Surf. Coat. Technol. 1991, *49*, 554–559.

30. Hayward, I.P.; Singer, I.L. *Tribological behaviour of diamond coatings, Second International Conference on New Diamond Science and Technology*; Materials Research Society: Pittsburgh, PA, 1991; 785–789.

31. Hayward, I.P.; Singer, I.L.; Seitzman, L.E. Effect of roughness on the friction of diamond on CVD diamond coatings. Wear 1992, *157*, 215–227.

3

Mechanical Behavior of Plastics: Surface Properties and Tribology

Nikolai K. Myshkin and Mark I. Petrokovets
Belarus National Academy of Sciences, Gomel, Belarus

I. INTRODUCTION

Polymers are high-molecular-weight materials made by polymerization of a large number of small units or monomers. The size of a linear macromolecule is a thousand times greater than that of other molecules. The main physical feature of polymer structure lies in the fact that its molecules consists of rigid sections (segments), which have the ability to rotate about one another providing flexibility of chains. Another feature of polymers is a sharp difference between the forces acting along the chains and between the chains. Strong chemical forces link atoms of a polymer chain, whereas the intermolecular forces, which are significantly weaker than the chemical forces, link the chains. The structural features of polymers and the considerable possibility to change their properties provide a wide variety of tribological applications.

II. MECHANICAL BEHAVIOR

A. Viscoelastic Stress–Strain Relations

The mechanical behavior of polymers is governed by their peculiar structure [1–3]. The combination of elasticity and viscosity is typical for their behavior. That is, under condition of small deformation, polymers behave as elastic Hook's body ($\sigma = E\varepsilon$) modeled with spring, and Newton's fluid ($\sigma = \eta \, d\varepsilon/dt$) modeled with dashpot. A combination of these elements gives a simple description of viscoelasticity. Several such models are presented in Table 1. It is unlikely that there exist real materials whose behavior follows the Maxwell or Kelvin bodies. Yet, these models allow researchers to qualitatively estimate how polymers behave in certain situations. In a more general form, the constitutive law is written as an ordinary differential equation with constant coefficients:

$$a_0\sigma + a_1\frac{d\sigma}{dt} + \cdots + a_n\frac{d^m\sigma}{dt^m} = b_0\varepsilon + b_1\frac{d\varepsilon}{dt} + \cdots + b_n\frac{d^n\varepsilon}{dt^n} \tag{1}$$

where σ and ε represent stress and strain depending on time; $a_0,\ldots,a_m,b_0,\ldots,b_n$ are the constant coefficients governing the mechanical behavior of the polymer under study.

Table 1   Simplest Models of Viscoelasticity

Nos.	Model	Equation	Relaxation ($\varepsilon = \varepsilon_0$)	Creep ($\sigma = \sigma_0$)
1	Maxwell	$\dfrac{d\sigma}{dt} + \dfrac{1}{\tau}\sigma = E\dfrac{d\varepsilon}{dt}$	$\sigma = E\varepsilon_0 \exp\left(-\dfrac{E}{\eta}t\right)$	$\varepsilon = \dfrac{\sigma_0}{\eta}t + \varepsilon_0$
2	Kelvin	$\dfrac{d\varepsilon}{dt} + \dfrac{1}{\tau}\varepsilon = \dfrac{1}{\eta}\sigma$	No relaxation	$\varepsilon = \dfrac{\sigma_0}{E}\left(1 - \exp(-\dfrac{E}{\eta}t)\right)$
3	3-element Type a Type b	$\dfrac{d\sigma}{dt} + \dfrac{E_1}{\eta_1}\sigma = (E_0 + E_1)\dfrac{d\varepsilon}{dt} + \dfrac{E_0 E_1}{\eta_1}\varepsilon$	$\sigma = \sigma_\infty + E_1\varepsilon_0 \exp\left(-\dfrac{E_1}{\eta_1}t\right)$	$\varepsilon = \varepsilon_\infty - \dfrac{\sigma_0}{E_1}\exp\left(-\dfrac{E_1}{\eta_1}t\right)$
4	3-element Type a Type b	$\dfrac{d\sigma}{dt} + \dfrac{E_1}{\eta_1}\sigma = \eta_0\dfrac{d^2\varepsilon}{dt^2} + E_1(1 + \dfrac{\eta_0}{\eta_1})\dfrac{d\varepsilon}{dt}$	$\sigma = E_1\varepsilon_0 \exp\left(-\dfrac{E_1}{\eta_1}t\right)$	$\varepsilon = \dfrac{\sigma_0}{\eta_0}t + \dfrac{\sigma_0}{E_1}\left(1 - \exp\left(-\dfrac{E_1}{\eta_1}t\right)\right)$

For the combined stress–strain state described by tensors σ_{ij} and ε_{ij}, the above constitutive law is

$$P_1(D)s_{ij} = Q_1(D)e_{ij}$$
$$P_2(D)\sigma_{ij} = Q_2(D)\varepsilon_{ij}$$

where P_m, Q_m ($m = 1, 2$) are the operator polynomials of the partial derivative with respect to time ($D = \partial/\partial t$), σ_{ii} and ε_{ii} are the stress and strain tensors, s_{ij} and e_{ij} are the deviator components defined as $s_{ij} = \sigma_{ij} - \sigma_{kk}\delta_{ij}/3$, $e_{ij} = \varepsilon_{ij} - \varepsilon_{kk}\delta_{ij}/3$.

Such representation gives us a set of relaxation (retardation) times, which enable the relaxation (creep) curves to be described as a finite series of exponents. What actually happens is that the relaxation (retardation) times are fitted to the measured curves of relaxation (or creep). As a result, a discrete spectrum of relaxation times is found.

However, every so often, the polymer behavior cannot be described by a set of exponents. In this case, the stress–strain relationship is specified in the integral form; given this, it is assumed that there exists a continuous spectrum of relaxation times:

$$s_{ij}(x, t) = \int_{-\infty}^{t} G(t - t')\frac{\partial e_{ij}(x, t')}{\partial t'}dt' \tag{2}$$

where G is the relaxation modulus.

B. Time–Temperature Superposition

Temperature essentially affects the molecular–kinetic processes proceeding in polymers and their mechanical behavior, including tribological properties. As numerous experiments have shown, there exists certain equivalence between the time effect and the temperature effect on the mechanical behavior of polymers. A temperature rise produces such an effect as if the process accelerates. That is, the time scale of a given viscoelastic measurement can be significantly extended, and the experiments can be conducted by the shortcut methods. Table 2 presents some of the first works dealing with the time–temperature superposition (TTS). An example of the master curve is given in Fig. 1. The TTS principle is detailed in Ref. 18. In addition, three main lines of research in this domain can be identified:

1. Validation of the TTS principle.
2. Development and improvement of the experimental techniques for revealing the TTS principle, the criteria of its applicability, and modification of the principle to extend it to various materials in various structural (phase) states.
3. Its application to solving specific problems of the theory of viscoelasticity.

Basic results will be discussed in the same sequence.

The physical approach to TTS validation, even if rather qualitative, is based on the molecular theory [18] advanced in Refs. 19 and 20. According to these two reports, the temperature dependence of the relaxation time τ_r is manifested through density ρ and viscosity η.

Moreover, the expression for τ_r explicitly contains the multiplier T. Hence the relation of the relaxation time at different temperatures T_0 and T can be written as:

$$\tau_r(T)/\tau_r(T_0) = \eta\rho_0 T_0/\eta_0\rho T = a_T$$

Because the relaxation time spectrum H can be represented as [18,21]

$$H = \frac{\rho RT}{M} \sum \tau_r \delta(\tau - \tau_r)$$

Then, as the temperature changes from T_0 to T, the function $H = H(\tau)$ in the log coordinates shifts vertically by $\log \rho T/\rho_0 T_0$ and horizontally by $\log a_T$.

The vertical shift is usually small and can be ignored. Irrespective of the molecular theory underlying the horizontal shift, it corresponds (equivalent) to the assumption that temperature makes the same contribution to all relaxation times: it is a deduction essential for the phenomenological approach.

The phenomenological approach is implemented in Ref. 12, and in consecutive works postulating that the "thermorheologically simple" material is a linear viscoelastic material for which any temperature change is equivalent to a shift in the log time scale. This postulate is based on the structure of viscoelastic functions and the analogy with the Newtonian flow.

Let us consider a polymer satisfying the integral relationships of linear viscoelasticity:

$$e_{ij}(x, t) = \int_{-\infty}^{t} K_1(t - t') \frac{\partial s_{ij}(x, t')}{\partial t'} dt' \tag{3}$$

$$\varepsilon(x, t) = \int_{-\infty}^{t} K_2(t - t') \frac{\partial \sigma(x, t')}{\partial t'} dt' \tag{4}$$

where K_i ($i = 1,2$) is the creep function that, like other viscoelastic functions, bears the following structure

$$f(t) = \int h(\tau) g(t/\tau) d\tau \tag{5}$$

Table 2 Time–Temperature Superposition

Nos.	Author(s) and Reference	Year	Test Condition, Material	Notes
1	Kobeko et al. [4]	1937	Static experiment, rubber-like polymers	Revealing the time–temperature superposition
2	Leaderman [5]	1943	Creep of fibbers, data from Ref. 4	It has been suggested to construct the master curve using the shift of creep data
3	Tobolsky and Andrews [6]	1943	Relaxation of polyisobutylene	Vertical shift of the master curve by T_0/T was proposed
4	Ferry [7]	1950	Dynamic experiment, polyisobutylene	Vertical shift of the master curve by T_0/T was proposed
5	Williams et al. [8]	1950	Fundamental law	WLF equation for the time–temperature superposition was advanced for the temperature range $(T_g, T_g + 100°C)$
6	Andrews and Tobolsky [9]	1951	Relaxation, rubber-like polymers	The TTS was used in calculations
7	Tobolsky et al. [10]	1951	Relaxation, polyisobutylene	The first experimental test of TTS by comparison of static and dynamic experiments
8	Fujino et al. [11]	1963	Polyvinilalcohol derivatives	Relaxation spectrum is divided into portions and the shift function a_T is constructed for each portion
9	Schwarzl and Staverman [12]	1952	Virtual polymer	Class of thermoreologically simple materials described by the shift function is introduced
10	Hilton [13]	1954	Kelvin body	Solution of plane problem on hollow cylinder with radially symmetric temperature field
11	Morland and Lee [14]	1960	Kelvin and Maxwell bodies	Shift hypothesis was extended to the time-dependent temperature field; problem on incompressible cylinder was solved for steady-state temperature
12	Grosch [15]	1963	Rubber	Master curve was plotted for the friction coefficient as a function of speed at different temperatures
13	Maugis and Barquins [16]	1980	Rubber	The WLF equation is applied to description of contact with adhesion
14	Bely et al. [17]	1982	Polyethylene	The TTS with vertical shift was used for describing the real contact area depending on temperature.

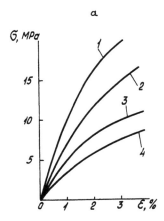

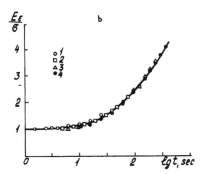

Figure 1 Illustration of the time–temperature superposition: (a) isothermal stress–strain curves for polypropylene; 1—$T = 25°C$, 2—$T = 35°C$, 3—$T = 45°C$, 4—$T = 55°C$; (b) master curve.

Disregarding the temperature dependence of the spectrum $h(\tau)$, relation (5) for two pairs of values of time and temperature (t_0, T_0) and (t, T) is written as

$$f(t_0, T_0) = \int h(\tau) g[t_0/\tau(T_0)] d\tau \tag{6}$$

$$f(t, T) = \int h(\tau) g[t/\tau(T)] d\tau \tag{7}$$

The values $f(t_0, T_0)$ and $f(t, T)$ coincide when $t = t_0/a_T$, where $a_T = \tau(T_0)/\tau(T)$, i.e., the above deduction is achieved.

On the other hand, because the creep component γ depending on viscous flow is determined by the relation

$$\gamma(t, T) = \sigma t/\eta \tag{8}$$

where $\eta \sim \exp(E_\eta/RT)$, then simple transformations yield

$$\gamma = \sigma \exp y, \tag{9}$$

where $y = \ln t + g(T)$; $g(T) = -\ln \eta$

Relation (9) determines the equivalence between time and temperature for the deformation of the Newtonian liquid.

These heuristic considerations have served as a basis for the postulate of the temperature shift function; its analytical formulation for the creep function is as follows:

$$K_T(\ln t) = K_{T_0}[\ln t + f(T)] \tag{10}$$

where $f(T)$ is measured in respect to some arbitrary temperature T_0, and it is a monotonously increasing function of T. Assuming that $a_T \equiv \exp[-f(T)]$, the viscoelastic functions can be expressed through the "reduced" or "local" time. In particular,

$$K_T(\ln t) = K_{T_0}(t/a_T)] = K_{T_0}(\xi) \tag{11}$$

where $\xi = t/a_T$ is the reduced time.

Hence the viscoelastic functions at an arbitrary constant temperature in the new time scale are temperature-independent. This allows writing the constitutive law (3)–(4) as

$$\tilde{e}_{ij}(x, \xi) = \int_{-\infty}^{\xi} \tilde{K}_1(\xi - \xi') \frac{\partial \tilde{s}_{ij}(x, \xi')}{\partial \xi'} \, d\xi' \tag{12}$$

$$\tilde{e}_{ij}(x, \xi) = \int_{-\infty}^{\xi} \tilde{K}_1(\xi - \xi') \frac{\partial \tilde{\sigma}_{ij}(x, \xi')}{\partial \xi'} \, d\xi' \tag{13}$$

The law was first generalized to the nonhomogeneous temperature fields in Refs. 22 and 23.

For practical application of the TTS principle, the temperature relation of the reduction factor a_T should be established. The above procedure of plotting the generalized curve simultaneously allows plotting a_T graphically as a function of T, based on the shift of the relevant curves.

The analytical methods of determination of $a_T(T)$ presume that the temperature dependence of relaxation time (or viscosity) is known [see formula (1)], or they have an empirical nature, such as the known Williams–Landel–Ferry (WLF) equation:

$$\log a_T = -\frac{c_1^0(T - T_0)}{c_2^0 + T - T_0} \tag{14}$$

where c_1^0 and c_2^0 are the universal constants.

To check the validity of the temperature–time superposition principle, it is essential when its modification can be endeavored. Leaving details apart, the following methods can be outlined [12,18,21]:

(a) Concordance of the shapes of the relevant curves describing the viscoelastic function in question; the modification of this method is the conformity between the relaxation (or retardation) time spectra found at different temperatures.
(b) Presence of the linear relation $T-T_0/\log a_T \propto T - T_0$ (the WLF equation is true); in the general case, the curve $\log a_T \sim T$ should be monotonous and continuous.
(c) Coefficients c_1^0 and c_2^0 in Eq. (14) are temperature-independent; hence, in case they coincide when different reference temperatures T_0 are used, the WLF equation is applicable.
(d) Prediction of the results of a long-term static experiment based on the results of a short-term experiment; any modification of this method is an extrapolation of the static experiment within the range of very short time periods (higher frequencies) achievable only with a dynamic experiment.
(e) Comparison of the values of a_T found from the different viscoelastic functions.

According to Ref. 18, the TTS method can be applicable to the data of the transitional and highly elastic deformation and to the finite zone of linear polymers. Nevertheless, application of the method is so promising that intensive effort is underway to modify it to make it applicable to the glassy-like and highly crystalline polymers.

The temperature shift function (the reduction factor) can be useful to formulate the law of deformation, as it is demonstrated above for steady-state homogenous temperature fields.

It can be shown that the principle of conformity in its current definition is applicable only to these fields; meanwhile, the shift function a_T is not determined by x. A number of specific problems are treated in Refs. 13, 14, and 24–27.

III. SURFACE PROPERTIES OF POLYMERS

It is an essential fact that the surface can be treated both as an ideal geometrical object with a highly peculiar topography and a physical object possessing a certain thickness and a specific mechanical behavior. The atoms and molecules belonging to the surface have fewer "neighbors" than those in the bulk. This simple, yet apparent fact has far-reaching consequences for the geometry and physics of a surface: the interactions between its atoms and their neighbors vary, distorting the force field that penetrates to the depth of several interatomic distances (so-called transitional layer). Given this, the excess of energy called surface energy appears. The surface tension is a measure of the surface energy. The solids can be rated in the order of their surface tension into three groups: (1) solids with high surface tension up to several Joules per square meter in vacuum (most of the metals and their oxides); (2) solids with medium surface tension of the order of tenth fractions of Joule per square meter (e.g., ionic compounds); (3) solids with low surface tension (most of the polymers).

Because of the energy, the surface interacts with the environment. This process, known as adsorption, would maintain the elements of neighboring phases on the surface; hence the idea of a "pure surface" is highly conventional. There are physical and chemical types of adsorption. The physical adsorption is characterized by the van der Waals interaction between the adsorbate and the solid surface. As a rule, its energy of the interactions is below 20 kJ/mol of the adsorbate. The polymer films adsorbed on the surface are removed relatively easily, for example, by reducing the ambient pressure.

The chemical adsorption energy is quite high (80–400 kJ/mol), usually producing a monolayer, which is hard to remove even by elevated temperatures, on the surface. Also, chemical reactions between the surface and the active elements in the environment should be considered, such as oxidation. Unlike chemosorption, these reactions result in a bulk phase on the surface.

Hence the surface adsorptivity produces a fine boundary layer with the structure and behavior different from the structure and behavior of surface (transitional) layer of the solid.

Figure 2 shows schematically that the structure of the boundary layer is quite intricate. The boundary layer may be in a diversity of physical states from the nearly gaseous state and up to the solid crystalline state. Both the basic state parameters (temperature, pressure, etc.) and the pattern of interactions with the solid phase determine its state. The mechanical behavior of boundary layers accordingly demonstrates a rich spectrum of properties ranging from viscous and viscoelastic behavior to perfectly elastic one.

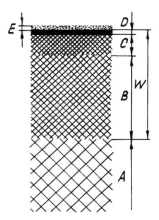

Figure 2 Surface layer structure: A—initial structure; B—region where supermolecular structure is fractured and oriented, as well as where the crystalline phase breaks down partly; C—strongly dispersed layer; D—low-molecular layer; E—gaseous phase; W—working layer.

Incidentally, machining and processing naturally generate normal and tangential stresses, which both alter the behavior of transitional layers and produce an underlying deformed layer.

Therefore the solid surface with region adjacent to bulk can be schematically represented as a laminated system comprising boundary (adsorbed) and deformed layers resting on the solid (bulk) phase of the basic material. Such representation is frequently convenient to analyze and simulate the surface effects in friction and wear. Given this, the molecular field of the solid is some "mixture" of molecular fields of the basic material and the adsorbed layer, the latter frequently shields strongly the molecular field of the basic material.

When two solids approach each other, it is exactly their molecular (surface) fields that come to interact and generate the attractive force responsible for their bonding, or adhesion. Adhesion implies the appearance of molecular bonds between the mated surfaces. The thermodynamic work of adhesion γ between two bodies 1 and 2 equal to the work of reversible adhesive detachment is frequently determined by the Dupre equation [28]:

$$\gamma = \gamma_1 + \gamma_2 - \gamma_{12} \tag{15}$$

where γ_1 and γ_2 are the surface energies of the two surface 1 and 2 before contact (their free energy), and γ_{12} is the interface energy.

The fact that no single theory of adhesion exists so far is an indirect proof of its complexity, although several models of adhesion have been advanced treating its origin from mechanical, adsorptive, electrical, diffusive, or chemical standpoints.

IV. CONTACT PROBLEMS

Peculiar features of the solids contact require a large group of factors to be analyzed, among them contact discreteness, a variety of deformation modes of roughness even within the same contact region, nonhomogeneous properties of materials on their surfaces and in their bulk, variations of these properties, in particular, under the effect of frictional heating, and many others. The modern mechanics of solids endeavors to allow for these factors when formulating and solving contact problems, but not in their integrity, as a rule.

Let us consider some basics of contact mechanics useful for explanation of the polymer contact and tribological behavior. The actual and nominal contact areas are determined based on solutions to the problems of the theory of viscoelasticity and elasticity.

Hertz was the first to solve in 1881 the problem of contact between two elastic bodies with the following assumptions: (1) mated surfaces are homogeneous, isotropic, and their initial contact is concentrated, i.e., it occurs at a point or along the straight line; (2) the deformation is small enough to be able to apply the linear theory of elasticity; (3) the characteristic dimensions of the contact region are small compared with the dimensions of contacting bodies allowing in calculation of local deformation to treat each body as an elastic half-space; (4) bodies have smooth contacting surfaces, i.e., only normal pressures are effective in the contact zone (no friction).

Under these assumptions, the contact problem on penetration of rigid ball into viscoelastic half-space is reduced to solution of the integral equation [29–31]:

$$\frac{1}{\pi} \iint_A \kappa \frac{p(x',y')\mathrm{d}A'}{\rho} = \delta - Ax^2 - By^2 \tag{16}$$

where $\kappa = 3K-4G/6K+2G \ 1/2G$, $\rho = [(x-x')^2 + (y-y')^2]^{1/2}$, K and G are the bulk and shear moduli, and A is the contact area. In the case of circle contact, $A = \pi a^2$.

The contact pressure $p(x,y)$, the contact radius a, and the approach δ are unknown quantities in this equation. For the contact of two balls, the parameters are:

$$p(x,y) = p_{\max}(1 - x^2/a^2 - y^2/a^2)^{1/2}; \ p_{\max} = \left(\frac{6PR^{*2}}{16RE^*}\right)^{1/3}$$

$$a = \left(\frac{3}{4}\frac{PR}{E^*}\right)^{1/3}; \ \delta = \frac{a^2}{R} = \frac{\pi}{2}\frac{p_{\max}a}{E^*} = \left(\frac{9P^2}{16RE^{*2}}\right)^{1/3}$$

Here, E^* is the effective elastic modulus defined as $E^* = E/(1-v^2)$, where E and v are the Young's modulus and the Poisson's ratio of contacting bodies.

In the case of viscoelastic behavior of bodies, the moduli entering Eq. (16) are time-dependent functions, and the contact area A varies in time. Then a viscoelastic analog of Eq. (16) can be written as

$$\frac{1}{\pi} \iint_{A(t)} \kappa^* \frac{p\mathrm{d}A'}{\rho} = \alpha(t) - Ax^2 - By^2 \tag{17}$$

where * stands for convolution and the function κ is determined from the integral equation:

$$2G^*(6K+2G)^*\kappa = 3K+4G \tag{18}$$

Solution of the last two equations for the case of constant load P gives the relation for the contact radius:

$$a^3(t) = \frac{3}{4}PR\kappa(t)$$

and thereby the contact area is

$$A_c = \pi\left(\frac{3}{4}PR\kappa(t)\right)^{2/3}$$

The use of the obtained relations depends on the choice of the constitutive law of the hereditary type. In particular, in case a material has a time-independent Poisson's ratio and kernel of heredity of the exponential type

$$R(t) = \frac{1}{\tau} \exp\left(-\frac{t}{\tau}\right)$$

From Eq. (18), it follows that

$$\kappa(t) = \frac{1 - \mu^2}{E} \frac{1}{1 - \lambda} \left[1 - \lambda \exp\left(-\frac{1 - \lambda}{\tau} t\right)\right] \tag{19}$$

Then

$$A_c = \pi \left[\frac{3}{4} N R \frac{1 - \mu^2}{E} \frac{1}{1 - \lambda} \left(1 - \lambda \exp\left(-\frac{1 - \lambda}{\tau} t\right)\right)\right]^{2/3} \tag{20}$$

Here it should be pointed out that this solution is based on the assumption that the mutual approach of bodies in contact is a nondecreasing function of time.

V. CONTACT WITH ADHESION

A. The Johnson–Kendall–Roberts Model

The Johnson–Kendall–Roberts (JKR) model, sometimes termed as the model of contact mechanics, is based on the assumption on an infinitely small radius of effect of surface forces, i.e., it is assumed that interactions occur only within the contact area [32].

The contact between the elastic sphere and the rigid half-space will be analyzed. Had there been no surface forces, the contact would appear according to Hertz, as the diagram "load P–approach δ" shows it (Fig. 3, branch OC). At a load P_0, point C characterizes the contact with a radius a_0 and the approach δ_0. What happens when surface forces come into action? Following the authors of the model, the contact is mentally loaded additionally up to P_1 (Fig. 3, point A), the contact expands to a_1 and the approaching δ_1, surface forces "switch on" at this stage. After that, when loading is removed, the contact area remains unchanged due to contact adhesion, yet the approaching reduces. The key moment in this situation is that, due to the invariance of the circular contact area, unloading occurs in such a manner as

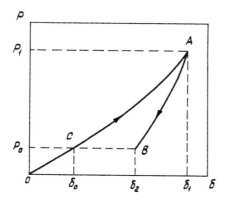

Figure 3 Schematic of loading the adhesion contact.

if an elastic cylindrical die interacted with the circular base of the half-space with a radius a_1, or the Boussinesq problem. In other words, the unloading by the amount $P_1 - P_0$ proceeds along the branch AB (according to Boussinesq), where the state of the contact system at point B is determined by the loading P_0, the contact radius a_1, and the approaching δ_2. The contact system at point B is in the equilibrium state in case of a definite combination of the applied P_0 and apparent P_1 loading and surface energy γ, which is determined further from simple energy considerations.

With such interpretation of the adhesive contact, the distribution of contact pressures is apparently a sum of pressures according to Hertz and Boussinesq (Fig. 4):

$$p(r) = \frac{3P_1}{2\pi a_1^2}\left(1 - \frac{r^2}{a_1^2}\right)^{1/2} - \frac{P_1 - P_0}{2\pi a_1^2}\left(1 - \frac{r^2}{a_1^2}\right)^{-1/2}$$

The minus sign ($-$) indicates that during unloading, the stress according to Boussinesq is tensile. The deformation w at point B is described by the relation:

$$w = \delta_2 - \frac{r^2}{2R}$$

Approach δ_2 here is determined by the relation

$$\delta_2 = \delta_1 - \Delta\delta = \delta_1 - \frac{2(P_1 - P_0)}{3Ka_1} \tag{21}$$

where $1/K = 3/4(1 - v_1^2/E_1 + 1 - v_2^2/E_2)$ is the effective contact modulus.

The contact system is in the equilibrium state when, during fixed approaching, its total energy W_t, comprising the energy of elastic deformation of the sphere W_e and the energy of molecular interactions in the contact zone W_m, reaches the minimum, i.e., $dW_t/da_1|_{\delta = \tilde{n}onst} = 0$. The energy of elastic deformation of the sphere is known to be determined by the relation

$$W_e = \frac{1}{2}\int_0^a p(r)w(r)2\pi r dr$$

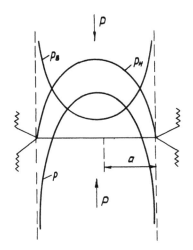

Figure 4 Stress distribution on the adhesion contact.

and the energy of molecular interactions is equal to $W_m = -\pi a^2 \gamma$. Substitution of $p(r)$ and $w(r)$ into the relation for W_e, with the allowance that, in this particular case, the contact radius is equal to a_1, and integration yield

$$W_e = \frac{3}{4} K a_1 \left(\frac{a_1^2}{R} - \delta_2 \right) \left(\delta_2 - \frac{a_1^2}{3R} \right)$$

The total energy of the system is equal to

$$W_t = W_e + W_m = \frac{3}{4} K a_1 \left(\frac{a_1^2}{R} - \delta_2 \right) \left(\delta_2 - \frac{a_1^2}{3R} \right) - \pi a_1^2 \gamma$$

Then the system reaches equilibrium when its parameters relate as follows:

$$\frac{dW_t}{da_1} = \frac{3}{4} K \left(\frac{a_1^2}{R} - \delta_2 \right)^2 - 2\pi a_1 \gamma = \frac{3}{4} K (\Delta \delta)^2 - 2\pi a_1 \gamma = 0$$

Substitution of the value $\Delta \delta$ from Eq. (21) into the latter expression yields a quadratic equation in respect to P_1:

$$(P_1 - P_0) = 6\pi P_1 R \gamma$$

and its solution yields the following expression for the apparent load:

$$P_1 = P_0 + 3\pi R \gamma + \sqrt{6\pi R P_0 \gamma + (3\pi R \gamma)^2} \tag{22}$$

Allowing for the relation between the load P_1 and the contact radius a_1, it is easy to obtain a formula for calculating the radius of adhesive contact (the relevant indexes are omitted):

$$a^3 = \frac{R}{K} \left(P + 3\pi R \gamma + \sqrt{6\pi R P_0 \gamma + (3\pi R \gamma)^2} \right) \tag{23}$$

Hence it is apparent that in a case where there is no adhesion ($\gamma = 0$), the Hertz equation is obtained, while if $\gamma > 0$, the contact area always exceeds that of Hertzian at the same normal load P. It is worthwhile to note that in the case where the load is fully removed ($P = 0$), the contact does not disappear, it remains finite with a radius

$$a = (6\pi R^2 \gamma / K)^{1/3}$$

Only applying a tensile (negative) load can reduce this radius; then, the contacting surfaces would separate at the least load corresponding to the conversion of the radicand into zero in Eq. (23):

$$P_{pull} = -\frac{3}{2} \pi R \gamma$$

and at a nonzero contact area. This circumstance is a typical feature of the JKR model.

The relation between the load and the approach is frequently useful when it is recorded in the following dimensionless form:

$$\tilde{\delta} = \begin{cases} \left(3\sqrt{\tilde{P}+1} - 1 \right) \left[\frac{1}{9} \left(\sqrt{\tilde{P}+1} + 1 \right) \right]^{1/3}, & \tilde{\delta} \leq -3^{-2/3} \\ -\left(3\sqrt{\tilde{P}+1} + 1 \right) \left[\frac{1}{9} \left(1 - \sqrt{\tilde{P}+1} \right) \right]^{1/3}, & -1 \leq \tilde{\delta} \leq -3^{-2/3} \end{cases} \tag{24}$$

where $\tilde{\delta} = \delta / \delta_c$: $\tilde{P} = P/P_c$; $\delta_c = 1/3 (3RP_c/K)^{2/3}$; $P_c = 3/2\pi R \gamma$.

Figure 5 shows the graph of this dependence.

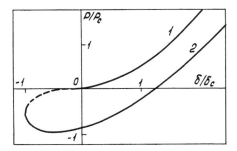

Figure 5 The load on sphere as a function of approach: 1—Hertzian; 2—with consideration for adhesion (JKL).

B. The Deryaguin–Muller–Toporov Model

Deryaguin was the first (1934) to formulate and approximately solve the problem of the effect of elastic contact deformation on adhesion [33]. Then Deryaguin, Muller, and Toporov (DMT) developed it further in their works, resulting in creating a DMT model that describes the contact of an elastic sphere with a rigid half-space. This model is based on two postulates: (1) Surface forces do not change the deformed profile of the sphere and it remains Hertzian; incidentally, further development of the so-called self-coordinated method allowed one to give up this postulate and to determine the shape of the deformed profile in the process of solution of the Hertz integral equation containing also surface forces. (2) The attraction force acts outside the contact circle pressing bodies together, with the contact region being under compression with the Hertz distribution of stresses (Fig. 6). Equilibrium is reached when the deformation is such that the elastic response (the force of elastic restoration of the sphere) F_e counterbalances the joint effect of the applied external load P and the forces of molecular attraction F_s:

$$F_e = P + F_s$$

Let us assume that the attraction forces satisfy the Lennard–Jones law. The profile of the deformed sphere outside the contact circle is known to be described by the equation:

$$z(r, a) = \frac{1}{\pi R} \left[a(r^2 - a^2)^{1/2} - (2a^2 - r^2)\tan^{-1}(r^2/a^2 - 1)^{1/2} \right]$$

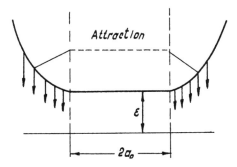

Figure 6 Schematic of contact with adhesion (DMT).

to which the equilibrium state z_0 (the clearance within the contact site) should be added. Then the form of molecular attraction is calculated by direct integration:

$$F_s = 2\pi \int_a^\infty p(z + z_0)r\mathrm{d}r$$

Calculation of this integral is encumbered with certain difficulty, yet a number of approximate formulas have been published that facilitate the use of the model in question. In particular, there is a simple Maugis relation between load and approach obtained for the conditions of the DMT model [34]:

$$\frac{P}{P_c} = \frac{1}{\sqrt{3}} \left(\frac{\delta}{\delta_c}\right)^{3/2} - \frac{4}{3}$$

where

$$P_c = \frac{3}{2}\pi\gamma R, \delta_c \left(\frac{P_c^2}{3K^2R}\right)^{1/3}, K = \frac{4}{3}\frac{1 - v^2}{E}$$

Analysis shows that each of these models is true for certain combinations of physicomechanical and geometric characteristics (Fig. 7). The DMT theory is applied to materials for which the point with coordinates (K,γ) is below the corresponding line plotted with constant radius of asperity tip β. Here, K is the reduced stiffness of contacting materials and γ is interface energy. That is, the JKR theory is valid for soft materials, particularly polymers, whereas the DMT theory is applied to bodies of micrometer sizes having the properties of metals or to bodies of submicrometer sizes having the properties of polymer [35].

C. Contact of Rough Surfaces with Adhesion

The models described above are valid for contact of smooth surfaces. It is common knowledge that real surfaces are rough. Experiments show that roughness drastically affects the adhesion of rough surfaces in contact. As an example, the contact between sphere and rough half-space is considered. The analysis is based on the Greenwood and Tripp model [36], where the load–approach relation (24) is treated as that for a single asperity. The detailed procedure of the analysis is given in Ref. 37. It is easy to show that the total force acting on the sphere is:

$$P = 2\pi P_c D \int_0^{a^*} \int_0^\infty r\overline{P}_i(z - u)\Phi_\alpha(z)\mathrm{d}z\mathrm{d}r$$

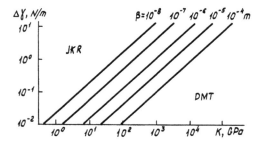

Figure 7   Domains of parameters (K,γ) where either of the two models (JKR or DMT) is correct.

After the change of variables and some manipulations, the force P becomes

$$P = P_c \frac{3\sqrt{2}}{8} \pi \Delta_c \mu \sqrt{\frac{\beta}{R}} I(h, \alpha, \Delta_c) \tag{25}$$

where $P_c = 3/2\pi R \Delta \gamma$ is the adhesion of smooth sphere of radius R; $\mu = 8/3D\sigma\sqrt{2R\beta}$ is a complex parameter of surface topography; D is the asperity density over the nominal area;

$$I(h, \alpha, \Delta_c) = \int_0^{a^*} \int_0^\infty P_1(x) \Phi_\alpha^*(h + a_0 y + b_0 y^2 + \Delta_c x) dx dy$$

$$h = d/\sigma; \quad a_0 = 1 - \zeta/c; \quad b_0 = \frac{3}{2}\left(k - \frac{1}{2}\right)\frac{\zeta}{c^3}$$

$$\zeta = \frac{\sqrt{\pi}}{4} \mu F_{3/2}(h, \alpha) \Gamma(k+1) / \left(k + \frac{1}{2}\right)$$

$$a^* = \frac{3}{4}\sqrt{\pi} c \Gamma\left(k + \frac{3}{2}\right) / \Gamma(k+2); \quad F_{3/2}(h, \alpha) = \int_h^\infty \Phi_\alpha^*(\xi) d\xi$$

Here, $\Phi_\alpha^*(\xi)$ is the Nayak distribution of asperity peaks [38].

Thus, Eq. (25) allows the force compressing two rough surfaces to be estimated, and yet the equation describes the situation where the contact is unloaded. On unloading, the compressive force decreases to zero, whereas the separation remains nonzero and takes a definite value, h_0, which is found from the condition

$$I(h_0, \alpha, \Delta_c) = 0 \tag{26}$$

In order to detach fully the surfaces ($h_0 = 0$), a certain negative (tensile) force should be applied. Then, due to adhesion, the asperities with initial height h_0 are extended by the additional length δ_c, which is found in condition (26), in which the lower limit of integration is replaced with $-L$, where $L = \min(\Delta_c, h-h_0)$. The least value of P is the pull-off force (adhesion force). The calculations have indicated that the roughness initiates a sharp decrease in the adhesion. For example, if roughness σ increases by a factor of 2, adhesion drops by about 2 orders of magnitude. This effect of roughness cannot be explained by the reduction in contact area. It is possible that it is connected with successive breakdowns of adhesive junctions at the elastic recovery of the highest asperities during separation of the contacting surfaces. Some results of the calculations for contact of polymers show satisfactory agreement with the results of Fuller and Tabor experiments [39] (Fig. 8).

VI. FRICTION OF PLASTICS

A. Basic Characteristics and Principles of Friction

The friction theory of polymers, as with other materials, is based on the concept of friction duality [40–44]. This implies an existence of two components of friction caused by deformation and adhesion. Three basic elements involved in friction should be mentioned: (1) real contact area (RCA); (2) interfacial bonds, their type and strength; (3) shearing and rupture of rubbing materials in and about the contact region.

B. Real Contact Area

When two surfaces approach each other, their opposing asperities with maximum height come into contact. As the load is increased, the new pairs of asperities with lesser height

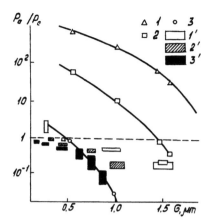

Figure 8 Effect of roughness on adhesion: 1, 1'—$K = 0.29$ MPa, $\Delta\gamma = 0.068$ N/m, $\Delta_c = 1.99$–6.49; 2, 2'—$K = 0.91$ MPa, $\Delta\gamma = 0.040$ N/m, $\Delta_c = 0.66$–2.14; 3, 3'—$K = 3.20$ MPa, $\Delta\gamma = 0.034$ N/m, $\Delta_c = 0.26$–0.84; 1–3—calculation, 1'–3'—experiment. (From Ref. 39.)

make contact, forming individual spots. The overall area of these spots is known as the real contact area (RCA). The surface asperities experience elastic, plastic, or viscoelastic deformation, depending on material behavior. At initial application of load to polymeric bodies, the deformation will be mainly plastic if the polymer is in glassy state, or mainly viscoelastic (or even viscoplastic) if polymer is in a highly elastic state.

A large number of methods have been developed to measure RCA. But there are no universal methods. All the methods have demerits, the major factor of which is that the term "real" is ambiguously determined by experiment. The methods may be conventionally divided into three groups: (1) methods based on the energy transfer (electrical resistance methods, acoustic transmission, optical methods, total internal reflectance, etc.); (2) methods that use mass transfer (radioactive isotopes, coal dust, luminescent paints, etc.); (3) methods based on strain measurements of rough surface layer deformation.

As an example, the apparatus to measure the RCA by total internal reflectance is presented in Fig. 9. Sample 1 fixed in a special mounting is clipped via lever system 17 and ball-bearing 18 to prism 2. To illuminate the contact zone, incandescent lamp 3 is used. In the illuminator housing, lens and field diaphragms 4 are mounted. Light rays from the lamp fall on the internal reflection prism 2, pass through objective 5, and fall on ocular 6. Hand wheels 8 and 9 provide coarse and fine adjustment. The area of contact is photographed onto plate 15 with aid of micrograph head 7 attached to tube 10. Sharp focusing is achieved by prism 11 and lens system 12. Prism 11 can be drawn aside from the path of the light rays by lever 13, which is displaced by bar 14 (when the button of the starting mechanism is depressed).

Figure 10 shows the RCA of polypropylene measured by Mechau's method. These data, as well as an extensive research, give an insight into the factors affecting the RCA formation.

The real contact area is dependent both on the mechanical behavior of the surface layers and their roughness. The latter may vary at compression and friction. The more smooth the surface, the larger is the RCA.

As a rule, an increase in the contact temperature leads to rise in RCA at the expense of variation in the mechanical behavior of the contacting materials, among which polymers

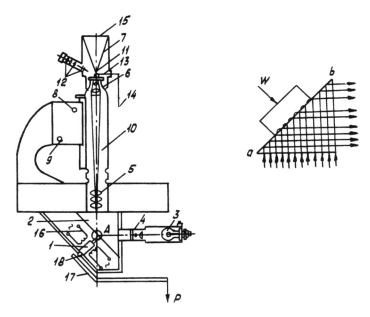

Figure 9 Apparatus for measuring the real contact area by total internal reflection (Mechau's method).

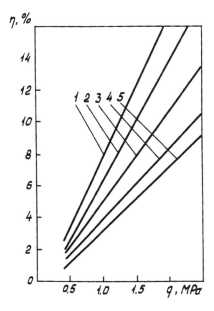

Figure 10 The real contact area of polypropylene-glass pair as a function of load: 1—$R_a = 0.63$ μm, 2—$R_a = 1.25$ μm, 3—$R_a = 2.5$ μm, 4—$R_a = 5$ μm, 5—$R_a = 10$ μm.

and their composites are most sensitive to a change in the temperature conditions and exhibit this property to a large measure. The temperature-dependent RCA may be estimated for the isothermal contact by the conventional procedure [45], with the elastic modulus corresponding to the given temperature. As shown in Ref. 17 for a number of polymers, this approach leads to results that satisfactorily agree with the experimental data.

Of interest is the case where there is a heat flow across the contact zone from a more heated body to a cooler one. When the mating materials are dissimilar, this flow gives rise to an additional ("thermal") contact pressure.

A contact of a rigid rough heated body and smooth elastic half-space is examined [46]. The latter is taken to be homogeneous and isotropic with Young's modulus E, Poisson's ratio v, and thermal expansion coefficient α. The total flow is assumed to penetrate into the smooth half-space.

Following Greenwood–Williamson's approach, we first examine an auxiliary problem for a single heated asperity modeled by a spherical segment of radius R. The problem was solved by Barber [47] with the following assumptions: temperature of the heated sphere is T, while outside of the contact the surface of the half-space is free of the mechanical thermal loads.

A desired solution (i.e., the needed relations between the total load P, the circular contact radius a, and the approach of the sphere δ) is sought as a superposition of the solutions derived for the problems dealing with the mechanical loading and heating. As was shown in Ref. 47, the total load P is the difference of the load P_1 at the isothermal indentation and the load P_2 related to the additional "thermal" pressure, which is required to compensate for a surface distortion (expansion in this case) caused by the heat flow. Formula for calculating the load P_1 was derived elsewhere [48]:

$$P_1 = \frac{ER^2}{1 - v^2} \left[\left(1 + \frac{a^2}{R^2} \right) \tan^{-1} \left[\frac{a}{R} \right] - \frac{a}{R} \right] \tag{27}$$

Because $a/R << 1$, it can be easily shown by expanding $\tan^{-1}(a/R)$ in series that P_1 differs from the Hertzian load

$$P_H = \frac{4}{3} \frac{E}{(1 - v^2)} \frac{a^3}{R}$$

by no more than $8/15(a/R)^5$ resulting in error $\varepsilon < 10^{-5}$ when $a/R = 0.1$. The second component P_2 is [47]

$$P_2 = -\frac{2\alpha T}{\pi} \frac{E}{(1 - v)} a^2$$

Therefore, the total applied load is

$$P = \frac{4}{3} \frac{E}{(1 - v^2)} \frac{a^3}{R} + \frac{2\alpha T}{\pi} \frac{E}{(1 - v)} \tag{28}$$

The relation between the contact radius and the approach was derived in [48]:

$$\delta = a \tan^{-1} \left[\frac{a}{R} \right]$$

From considerations reported in the discussion of Eq. (27), it may be replaced with the familiar relation

$$\delta = a^2/R \tag{29}$$

with error $\varepsilon \leq 1/3(a/R)^3$.

Thus, the contact of a single spherical asperity and a smooth surface is described by Eqs. (28) and (29).

Then, taking the distribution of asperity heights $\varphi(z)$, this solution is extended over a set of asperities. We shall restrict our consideration to the normal distribution which, in a number of cases, does not conflict with the experimental data:

$$\varphi(z) = \frac{1}{\sigma\sqrt{2\pi}} \exp\left[-\frac{1}{2}\frac{z^2}{\sigma^2}\right]$$

Taking into account that a contact spot area is $A_i = \pi R \delta_i$ and the load on a single asperity is

$$P_i = \frac{4}{3}\frac{E}{(1-v^2)}R^{1/2}\delta_i^{3/2} + \frac{2\alpha T}{\pi}\frac{E}{(1-v)}R\delta_i$$

we find the total area of all the contact spots (RCA) and the total load:

$$A_r = \pi R D A_a \int_d^\infty (z-d)\varphi(z)\mathrm{d}z$$

$$P = \frac{4}{3}\frac{E}{(1-v^2)}R^{1/2}DA_a \int_d^\infty (z-d)^{3/2}\varphi(z)\mathrm{d}z + \frac{2\alpha T}{\pi}\frac{E}{(1-v)}RDA_a \int_d^\infty (z-d)\varphi(z)\mathrm{d}z$$

Here, D is the surface density of asperities, A_a is the apparent contact area, d is the separation. Substituting $\varphi(z)$ and going to the nondimensional variables $h = d/\sigma$ and $\xi = z/\sigma$, we obtain

$$A_r/A_a = \pi R D \sigma \frac{1}{\sqrt{2\pi}}\int_h^\infty (\xi - h)\exp\left[-\xi^2/2\right]\mathrm{d}\xi \tag{30}$$

$$\begin{aligned}\frac{P}{A_a}\frac{(1-v^2)}{E} &= \frac{4}{3}R^{1/2}D\sigma^{3/2}\frac{1}{\sqrt{2\pi}}\int_h^\infty (\xi - h)^{3/2}\exp\left[-\xi^2/2\right]\mathrm{d}\xi \\ &+ \frac{2\alpha T}{\pi}(1+v)RD\sigma\frac{1}{\sqrt{2\pi}}\int_h^\infty (\xi - h)\exp\left[-\xi^2/2\right]\mathrm{d}\xi\end{aligned} \tag{31}$$

We introduce the following nondimensional quantities

$$\eta = A_r/A_a; \quad \tilde{P} = \frac{P}{A_a}\frac{(1-v^2)}{E}; \quad \mu = RD\sigma; \quad \tilde{\sigma} = \sigma/R$$

and designations for the integrals

$$F_1(h) = \frac{1}{\sqrt{2\pi}}\int_h^\infty (\xi - h)\exp\left[-\xi^2/2\right]\mathrm{d}\xi, \quad F_{3/2}(h) = \frac{1}{\sqrt{2\pi}}\int_h^\infty (\xi - h)^{3/2}\exp\left[-\xi^2/2\right]\mathrm{d}\xi$$

Then, Eqs. (30) and (31) are rewritten as

$$\eta = \pi\mu F_1(h) \tag{32}$$

$$\tilde{P} = \frac{4}{3}\mu\sqrt{\sigma}\left[F_{3/2}(h) + \frac{3}{2\pi}\frac{\alpha T}{\sqrt{\tilde{\sigma}}}(1+v)F_1(h)\right] \tag{33}$$

These equations describe the dependence of the relative RCA on the nondimensional load in the parametric form. The relative separation h is used as the parameter. For isothermal contact, Eq. (33) is written in the form:

$$\tilde{P}_H = \frac{4}{3}\mu\sqrt{\bar{\sigma}}F_{3/2}(h) \tag{34}$$

The system of Eqs. (32) and (33) makes it possible to find the temperature dependence of RCA at a given load. Given this, the calculation procedure is as follows: fixing the load $\tilde{P}$, we solve the transcendental Eq. (33) for the separation h with each value of temperature T from a given range of its variation and, then, the relative RCA corresponding to the given temperature is determined using Eq. (32).

As an example, we calculated a case of nylon (elastic modulus $E = 1.5$ GPa, Poisson's ratio $v = 0.38$, hardness $H = 0.1$ GPa, and thermal expansion coefficient $\alpha = 1.0 \times 10^{-4}$ 1/K) and some its hypothetical modifications, which are distinguished by the temperature-dependent moduli of elasticity and Poisson's ratios.

The above algorithm is valid for an elastic contact and, therefore, a selection of initial data should be based on one of the familiar criteria of changeover from elastic to plastic contact, for example, on Greenwood–Williamson's criterion (so-called plasticity index):

$$\psi = \frac{E'}{H}\sqrt{\frac{\sigma}{R}}, \text{ where } E' = E/(1 - v^2)$$

According to this criterion, the elastic contact takes place at $\psi < 0.6$, while at $0.6 < \psi < 1.0$, there exists a certain limiting separation and correspondingly critical load whose exceeding leads to plastic contact. As expected, elastic deformation of asperities is observed for very smooth surfaces. Thus, at $\psi < 0.6$, we have $\sigma/R < 0.0012$ for the polymer under consideration. Such characteristics are typical for very smooth surfaces with large curvature radius of asperity peaks and small (down to several hundreds of angstroms) mean square deviation of height [45]. These circumstances were taken into account in our calculation.

As for the nondimensional parameter μ, according to Refs. 36 and 45, it varies within a narrow range (0.03–0.05), although, in general, μ depends on bandwidth parameter when the Gaussian surfaces are examined. In the calculation, $\mu = 0.5$ was taken.

The calculations performed show that, with constant temperature (more exactly, difference of temperatures of bodies in contact), the relative RCA A_r/A_a is near-linearly dependent on the nondimensional load (Fig. 11). The behavior is quite similar to variation in A_r/A_a on P at isothermal contact (Fig. 11, curve 1). One distinction is that the rate of the

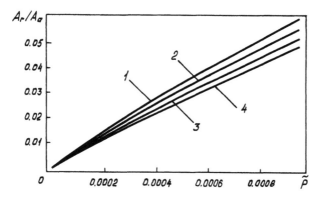

Figure 11 The real contact area A_r/A_a vs. the nondimensional pressure $\tilde{P}$ at different temperatures: 1—$T = 0°C$, 2—$T = 50°C$, 3—$T = 100°C$, 4—$T = 150°C$.

RCA rise (slope of the curves) decreases as the excess temperature elevates. In other words, the same RCA is attained at the higher load, the greater is the excess temperature. It is notable that the load differential depends only on the thermal expansion coefficient, temperature T, and Poisson's ratio. Indeed, as it follows from Eqs. (32–34)

$$\tilde{P} - \tilde{P}_{\mathrm{H}} = \frac{2}{\pi^2} \alpha T (1 + v) \eta$$

The effect of the excess temperature on the relative RCA is shown in Fig. 12. The higher the temperature, the smaller the RCA becomes. In fact, we have the situation that takes place in a sliding contact, where the so-called "thermoelastic instability" appears [30]. The heat flow across a contact spot gives rise to expansion of asperity being in contact and its "buckling" as against the isothermal case. A mean size of contact spot and the RCA becomes smaller. In order to retain the earlier value of RCA, the nominal contact load should be increased.

Thus, it was shown that the RCA of two bodies with different temperatures becomes smaller when the temperature difference increases, and the RCA is always smaller than that in the isothermal case. However, if the mechanical behavior of material is sensitive to change in temperature (for example, the elastic modulus of most polymers drops with increasing temperature), the decrease in RCA obtained above may be "hidden" by the RCA rise due to lowering of the mechanical characteristics. In this connection, our interest is in considering both factors, the thermal expansion of the bodies in contact and the reduction in the mechanical characteristics with increasing the temperature, in their combined action on the RCA.

For simplicity of further analysis, we shall restrict our focus to the polymers with simple thermal–rheological behavior and the applicable time–temperature superposition. If the rheological behavior of such material is governed by relaxation time only, then the simple exponential describes its temperature-dependent modulus:

$$E = E_0 \exp(-\beta T) \tag{35}$$

where β is a constant having a dimensionality of the reciprocal of temperature and conventionally termed as the rheological parameter.

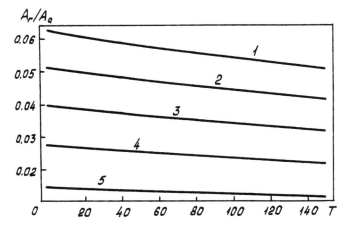

Figure 12 The real contact area $A_{\mathrm{r}}/A_{\mathrm{a}}$ vs. temperature T at different nondimensional pressure $\tilde{P}$, 10^{-4}: 1—$\tilde{P} = 10$, 2—$\tilde{P} = 8$, 3—$\tilde{P} = 6$, 4—$\tilde{P} = 4$, 5—$\tilde{P} = 2$.

Substituting modulus (35) in Eq. (33), we derive a parametric system of equations that describes the temperature dependence of the RCA. This system involves Eq. (33) and the modified Eq. (34). The latter is of the form:

$$\tilde{P}_0 \exp{(\beta T)} = \frac{4}{3}\mu\sqrt{\tilde{\sigma}}\left[F_{3/2}(h) + \frac{3}{2\pi}\frac{\alpha T}{\sqrt{\tilde{\sigma}}}(1+v)F_1(h)\right] \qquad (36)$$

where $\tilde{P}_0 = \tilde{P}/E_0$.

The solution of systems (32) and (36) gives the desired dependence of RCA on temperature. This system is solved numerically. The calculation was performed at several loads with different values of β. Some results are presented in Fig. 13. It is seen that the temperature-dependent RCA may pass the minimum with increasing temperature. The existence of this minimum and its value depend on a combination of the thermal (α) and mechanical (rheological) properties of contacting materials. An increase in the rheological parameter β, i.e., a sharpening of dependence of the modulus on temperature, results in degeneration of the minimum (ascending RCA $\sim E$ curve), all other things being the same (Fig. 13).

A similar situation takes place as parameter β is decreased when the descending dependence of the RCA on temperature is obtained. Nondegenerate minimum is observed within the rather narrow range of β variation. In the case under examination, the range of β is about 0.0014–0.0020. It should also be noted that the load has little effect on a location of this minimum, although its value increases with when the load is increased.

Thus, the temperature-dependent real contact area was analyzed theoretically. It was established that the RCA might pass the minimum at a certain combination of thermal and mechanical properties of mating materials as the contact temperature increases. An existence of the minimum is governed by the competing influence of thermal expansion of material and reduction in mechanical characteristics under the action of frictional heating.

C. The Two-Level Model of Contact

There is no question that the surface microasperities are not smooth, i.e., they have smaller asperities of nanoscale size (of order of tens nanometers), which result from molecular and

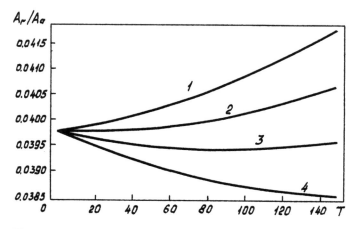

Figure 13   Effect of the rheological parameter β on temperature-dependent real contact area A_r/A_a: 1—$\beta = 0.0020$, 2—$\beta = 0.0018$, 3—$\beta = 0.0016$, 4—$\beta = 0.0014$.

supermolecular structure of polymers. In this case, the RCA should be estimated based on the two-level model of Archard type. The combination of two levels, roughness and subroughness, was examined. To take into account subroughness on large asperities (roughness), the solution of Greenwood and Tripp [36] for contact of rough spheres was used as the governing equation for a single asperity. Analysis of the two-level model showed that the highest asperities of the first level (roughness) come into contact and form the individual contact spots. Yet, contrary to the traditional view, the spots are not continuous but are multiply connected, that is, each of the spots consists of a set of subspots, the total area of which was conditionally named "physical contact area." This area is attributed to contact of submicroasperities (fine or nanoscale roughness) and is less than the real contact area by 1 order of magnitude. The strong interaction between the mating surface may occur within the subspots of the physical contact. The contribution of this physical (and maybe chemical) interaction to total resistance to relative displacement of rubbing surfaces may be very significant.

The two-level model with adhesion is of interest for precision engineering. It can be developed in the same manner as for the above model. A rough sphere modeled an asperity of the first level. The load–contact radius relation for a single subasperity was assumed to result from the JKR or the DMT theory. Without going into detail of calculation procedure, which is described elsewhere [37,49], we shall simply present some results dealing with the physical contact area.

Numerical experiment shows that, for polymeric materials, the physical contact area increases several times because of the surface forces (Fig. 14). This rise is 1.5 times for high-density polyethelene, 3 times for low-density polyethelene, and 2 times for nylon. Under the same conditions, a rise in the real contact area is substantially less. For example, for high-density polyethelene, the RCA increases by 10% when the root-mean-square roughness $\sigma = 0.5$ μm, and by 30% when $\sigma = 0.08$ μm.

D. Adhesion Bonds and Their Formation and Breakdown

When two clean surfaces are brought into contact, the surface forces of attraction responsible for adhesion appear. Because of these forces, the bonds form between the contacting surfaces. The junctions develop on the real contact spots. Formation and rupture

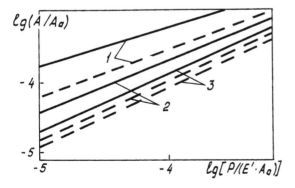

Figure 14 Effect of surface forces on physical contact area: 1—low-density polyethylene, 2—high-density polyethylene, 3—nylon; solid lines—calculation with consideration for surface forces; dotted lines—without consideration for surface forces.

of the junctions are the essence of the adhesion component of friction. The junctions may form at both static contact and friction. The simple model of the adhesive junction formation has been proposed by Bowden and Tabor.

For the majority of polymers, the van der Waals and hydrogen bonding are typical.

The van der Waals bond may occur between any atoms and molecules through dipole–dipole interaction (the atoms are treated as momentary dipoles, because the centers of positive and negative charges are unfair at every time).

The hydrogen bond develops at very short distances in polymers containing the groups OH, COOH, NHCO, and others, in which the hydrogen atom is linked with an electronegative atom. Under favorable conditions, two approaching atoms are linked together by a common proton. This provides a strong and stable compound.

The junctions have to be shared under the applied tangential force. This is the frictional force. That is, the work performed by the frictional force is expanded for damage to the interfacial bonds. The site where fracture occurs is dependent on the relative strength of the junction and the rubbing materials (Fig. 15). If the interfacial bonding is stronger than the cohesive strength of the weaker material, then this material is fractured and the material transfer takes place. Otherwise, fracture occurs at the interface.

In general, the interfacial junctions (their formation, growth, and fracture) are influenced by nature of the surfaces, surface chemistry, and stress state of surface layers (loading conditions). The interfacial junctions, together with the products of their fracture and the highly deformed layers where shear deformation is localized, has been named by Kragelskii as a "third body" [50]. This term implies that the polymer involved in the friction process may possess the properties that drastically differ from its bulk properties.

If the interfacial bonding is stronger then the cohesive strength of the weaker material, then this material is fractured and the polymer transfer takes place. Otherwise, fracture occurs at the interface. As a rule, polymers have the surface forces and the forces between polymer chains, which are nearly equal. Because of this, the fracture often occurs in the bulk of the polymers. This is not always the case. It was observed for metal–polymer contact that metal is transferred to the polymer surface under certain conditions [51].

Electrostatic attraction makes a contribution to the adhesion of polymer contact where an electric double layer is formed owing to transition of electron from one surface to another. Given this, the polymer may be acceptor or donor, depending on the origin of the counterbody. In contact with metal, for example, the metal is the electron donor, and when the contact is broken, the polymer surface goes negative [52].

E. Deformation Component of Polymer Friction

Another source from which the frictional force arises is attributed to deformation taking place when the asperities of two sliding surfaces come into contact with each other. This

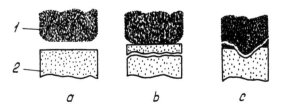

Figure 15 Types of fracture of adhesive junctions formed at friction: a—adhesive fracture, b—cohesive fracture, c—mixed fracture, 1—asperity, 2—counterbody.

deformation is accompanied by dissipation of mechanical energy with different mechanisms depending on deformation mode, sliding conditions, rubbing materials, environment, and other factors. Let us consider some of them.

In the Bowden–Tabor model for sliding friction, the asperities of the harder surface are assumed to plough through the softer one. The ploughing resistance causes a force contributing to the frictional force. This contribution is referred to as the ploughing component of friction, the deformation term. A simplest estimate for conical asperity of semiangle θ gives that the coefficient of friction due to the ploughing term is

$$f_{\rm d} = \frac{2}{\pi} \cos \theta$$

The slope of surface asperities is less than $10°$, that is, the semiangle $\theta > 80°$, and the coefficient $f_{\rm d}$ should be about 0.05 and less. When elastic contact occurs, $f_{\rm d}$ is often assumed to be negligibly small.

It should be noted that, almost without exception, ploughing is accompanied by adhesion and, under certain conditions, the ploughing may result in microcutting, that is, additional work is carried out and the friction is increased.

There are other mechanisms of energy dissipation at deformation. So, when a polymer with viscoelastic behavior slides against a hard rough surface, energy dissipation occurs to high hysteresis losses. This deformation component is known as friction due to elastic hysteresis.

The energy may be also carried away along other channel of dissipation, for example, with elastic waves generated at the interface and outgoing to infinity, owing to nucleation and development of microcracks on sliding surface and in the bulk of material [53–57].

These two processes (adhesion and deformation) are not independent: deformation resulting in a growth of contact area makes for adhesion, and conversely, the stronger adhesion gives rise to deformation. Nevertheless, because the regions where these processes occur vary essentially in size, their additivity is generally accepted as correct. This is a very useful approximation, which allows deeper insight into friction origin.

F. Effect of Load on Friction

An attempt to establish a relation between friction and normal load was first made by Leonardo da Vinci. However, until the present time, there is no common opinion regarding this question. The available experimental data may lead to erratic conclusions. The main difficulty encountered in comparison of experiments conducted by different authors lies in the fact that a unified test technique is absent. Here, the case in point is the shape and dimensions of samples, the values of applied load and sliding velocity, the surface finish, etc.

It is a common knowledge that friction force is proportional to the normal applied load (the first law of friction). Experiments of a number of researchers have shown that this law is valid for some polymers tested under certain conditions (Table 3). Thus the friction coefficient remains practically constant at the load in the range 10–100 N when a steel ball of radius 6.35 mm slides over polytetrafluoroethylene (PTFE), polymethylmethacrylate (PMMA), polyvinylchloride (PVC), polyethylene (PE), and nylon [58]. Other authors have obtained similar results with the same materials, as well as with other components, e.g., with PTFE, polytrifluorochlorethylene (PTFCE), PVC, polyvinylidene chloride (PVDC), PE at loads of 2–15 N [59], with PTFE, PMMA, polystyrene (PS), and PE at loads of 10–40 N [60], and so on.

Table 3 Effect of Load on Friction Coefficient

Nos.	Author(s)	Material	Load, test conditions	Graphical representation
1	Bowers et al. [59]	PTFE, PFCE, PVC, PVDC, PE	2–15 N Steel–polymer	f vs N (flat)
2	Shooter and Thomas [60]	PTFE, PE, PMMA, PC	10–40 N Steel–polymer	f vs N (flat)
3	Shooter and Tabor [58]	PTFE, PE, PMMA, PVC, nylon	10–100 N Steel–polymer	f vs N (rising)
4	Rees [61]	PTFE, PE, nylon	Steel–polymer	f vs N (descending)
5	Bartenev and Lavrentev [54], Schallamach [62]	Rubber	Theory	f vs N (descending)
6	Kragelskii [50]	Rubber	Theory	f vs N (descending)

Outside this range, on the left and right, the proportionality between friction force and applied load breaks down. Thus it was shown that in the range of moderate loads, 0.02–1 N, the friction coefficient decreases as the load increases [61]. Such a behavior may be explained by the elastic deformation of the surface asperities. Of interest is the fact that a similar behavior is characteristic of rubbers for which the elastic deformation is typical [53,62].

On the other side of the proportionality range, the friction coefficient increases with increasing load. This is often explained by plastic deformation of asperities in contact.

Thus friction of polymers as a function of load varies in the manner described by Kragelskii [50]. That is, the friction coefficient passes a minimum, which corresponds to transition from elastic contact (left descending branch of the curve) to a plastic one (right ascending branch of the curve).

It is well to bear in mind that the load can modify the temperature of viscoelastic transitions in polymers and thereby the mechanism of friction.

G. Effect of Sliding Velocity on Friction

It is accepted that friction force is independent of the sliding velocity. This proposal is valid with a good approximation only in the case where the contact temperature varies insignificantly and, as a result, the interface does not change its behavior. But variation in the effect of velocity and of friction temperature presents significant difficulties. Because of this, the results obtained by different researchers should be analyzed with caution. Examples of the great diversity of available results are shown in Table 4.

Table 4 Effect of Sliding Velocity on Friction Coefficient

Nos.	Author(s)	Material	Sliding Velocity, Test Conditions	Graphical representation
1	Shooter and Thomas [60]	PTFE, PE, PMMA, PS	0.01–1.0 cm/sec, Steel–polymer, Limited load	f vs v
2	Milz and Sargent [68]	1—nylon, 2—PS	4–183 cm/sec, Polymer–polymer	f vs v
3	Fort [70]	PETF	10^{-5}–10 cm/sec, Steel–polymer	f vs v
4	White [66]	1—PTFE, 2—nylon	0.1–10 cm/sec, Steel–polymer	f vs v
5	Flom and Porile [64,65]	PTFE	1.1–180 cm/sec, Steel–polymer	f vs v
6	Oloffson and Gralben [63]	Fibers	1.5 cm/sec, Polymer–polymer	f vs v
7	Bartenev and Laverentev [54], Schallamach [67]	Rubber	Theory	f vs v

Speed-independent friction was revealed only within a limited range of velocities (0.01–1.0 cm/sec) for PTFE, PE, PMMA, and PS [60], as well as for fiber–fiber contact [63]. But more complex relationships between friction and sliding velocity are most often observed. Such relationships can be connected with viscoelastic behavior of polymers.

In the range of low velocities, the viscous resistance in the contact zone increases with increasing velocity. When the contact pressure is high, the abnormally viscous flow is observed, which leads to a sharp rise of viscosity due to velocity. Therefore, friction force must increase with velocity (see, e.g., Refs. 64–66). Molecular–kinetic considerations also lead to the same dependence [53,67].

In the range of high velocities, elastic behavior is prevalent in the contact zone and, as a result, the friction force depends only slightly on velocity, or it decreases with velocity (see, e.g., Refs. 68 and 69). In addition, it should be borne in mind that the duration of contact is short at high velocity, leading to a further decrease in the friction force.

In the intermediate range of velocities, all of the above factors are in competition with one another, and a maximum appears in the friction force-sliding velocity curve, the position of which depends on the relaxation properties of polymer (see, e.g., Ref. 70).

It should be recognized that the friction force–sliding velocity relationship essentially depends on the test temperature [71]. When tests are conducted near the glass-transition temperature (high mobility of polymer segments), the sliding velocity has a pronounced effect on the friction, whereas at lower temperature (segments of the main chain are frozen), friction hardly depends on the sliding velocity.

H. Effect of Temperature on Friction

Polymers as viscoelastic materials are very sensitive to frictional heating. It is well known that friction is a typical dissipative process, in which mechanical energy is converted into heat (up to 90–95% according to the available experimental data). The thermal state of friction contact is frequently a decisive factor when evaluating the performance of a friction unit.

It is commonly believed that heat generation at friction is due to the deformation of material in the actual contact spots. In connection with this, some processes can be outlined with their molecular mechanism relating to the transformation of mechanical energy into heat—plastic deformation, hysteresis, dispersion, and viscous flow.

Another source of heat can be attributed to the processes of origination and destruction of adhesive bonds. These processes are most probably nonequivalent energetically, and the energy difference may cause the generation (and in specific situations, absorption) of heat.

It is often believed that the temperature effect on friction can be taken into account by using the mechanical characteristics of polymers measured at proper temperatures. In support of this conjecture, a correlation of friction coefficient with hardness and shear strength was found for some polymers [70–72]. Such correlation is valid only when temperature produces an unessential effect on adhesion. This fact was observed in Ref. 73.

Some characteristic behaviors of friction as a function of temperature are presented in Table 5.

Table 5   Effect of Temperature on Friction Coefficient

Nos.	Author(s)	Material	Temperature, Test Conditions	Graphical representation
1	Shooter and Thomas [60]	1—PS, 2—PTFE	20–80°C Steel–polymer	
2	Ludema and Tabor [72]	1, 2—PCTFE, 3—PP	−50 to +150°C, Steel–polymer $1-v=3.5\times10^{-5}$ cm/sec, $2-v=3.5\times10^{-2}$ cm/sec	
3	King and Tabor [73]	1—PE, 2—PTFE	−40 to +20°C, Steel–polymer	
4	Schallamach [62], Kragelskii [50]	Rubber, AMAH (multicomponent polymer composite)	20–200°C, Steel–polymer	

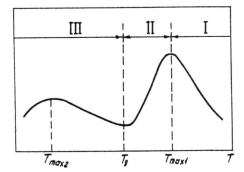

Figure 16 Effect of temperature on polymer friction.

The friction coefficients of thermoplastic amorphous and crystalline polymers as a function of temperature can be expressed in the general form by the Elkin curve shown in Fig. 16 (see Ref. 54).

The curve of the thermoplastic amorphous polymers can be divided into three portions corresponding to the highly elastic state (I), the transition region (II) and the glassy state (III). At elevated temperatures, the force of friction within portion I increases as temperature goes down and reaches the maximum at a temperature $T_{\text{max}1}$. During transition from the highly elastic into the glassy state within portion II, the force of friction reduces, and passes through the maximum again within portion III at a temperature $T_{\text{max}2}$. Hence the curve of the relation between the friction force and temperature has two maxima: the main maximum in the transition from highly elastic to glassy state and the low-temperature maximum in the glassy state. The basic mechanism of friction of polymers in the highly elastic state over smooth surfaces is adhesion. Another friction mechanism appears when a polymer transforms from the highly elastic into glassy state. Mechanical losses contribute more to the bulk redeformation of the surface layers on the polymer, and the volume mechanical component contributes more when the polymer is heated almost to the glass-transition temperature T_{g} until it becomes comparable with the contribution of adhesion.

The above relationship was observed for amorphous polymers and rubbers. A similar dependence was also observed for crystalline polymers [71].

VII. WEAR OF POLYMERS

A. Wear Modes

When solid surfaces are in sliding or rolling relative motion, one can observe a removal of material from one or both of these surfaces. The process is termed as wear. Similar to friction, wear is also a very complicated phenomenon, as illustrated by the fact that there is no single definition for wear in tribology.

Wear is expressed in specified units (length, volume, or mass). The wear process is often described by wear rate. There is no single standard way to express wear rate. The units used depend on the type of wear and the nature of the tribosystem in which wear occurs. For example, wear rate can be expressed as: (1) volume of material removed per unit time, per unit sliding distance, per revolution of a component, or per oscillation of a body; (2) volume loss per unit normal force at unit sliding distance [$mm^3/(N m)$], which is sometimes called the wear factor; (3) mass loss per unit time; (4) change in a certain dimension per unit time; (5) relative change in dimension or volume with respect to the same changes in another (reference) substance.

The reciprocal of wear rate has come to be known as wear resistance, which is a measure of the resistance of a body to removal of material by wear process. Relative wear resistance is sometimes considered using arbitrary standards.

Wear has its origin in atomic diffusion, adhesion, surface film formation, and other phenomena. Three stages of wear may be formulated: interaction of surfaces, changes in surface layer, and damage to surfaces. All the stages are interrelated and occur on individual contact spots.

The surfaces interact in mechanical and molecular ways. The mechanical interaction involves penetration and interlocking of asperities. The molecular interaction is manifested as adhesion or welding in the limiting case.

Changes in surface layer arise from mechanical stresses, temperature, and chemical reactions. Polymers, as a result of their specific structure and mechanical behavior, are more sensitive to these factors.

The normal stress at plastic contact of asperities will be close to the indentation hardness of the softer body. But if the surfaces are very smooth, closely conforming or lightly loaded then the contacts may be elastic. The normal load applied to the system therefore controls the extent of plastic flow at the asperity contacts; if the load is low enough, or the surfaces conform well enough, wear will proceed very slowly, perhaps by an elastic process only (e.g., high-cycle fatigue).

The magnitude and position of the maximum shear stress depend on the coefficient of friction f. For f less than ~0.3, the maximum shear stress and associated plastic flow will lie beneath the surface, and the plastic strain accumulated by each sliding pass is small. This condition is typical for a lubricated system, or for one carrying a protective layer. For f greater than ~0.3, however, the maximum shear stress lies at the surface and large shear strains can be accumulated. Several wear mechanisms dominated by plastic flow have been proposed, involving asperity adhesion and shear, or nucleation and growth of subsurface cracks leading to the formation of lamellar wear particles (delamination wear).

The local temperature at the interface may be substantially higher than that of the environment, and may also be enhanced at the asperity contacts by transient "flashes" or "hot-spots." Temperature exerts an influence on wear of polymers. Thus it was shown that a number of polymers sliding against steel pass a minimum at characteristic temperature [74].

The above-listed mechanisms and many others kept in the shade are the basis for wear process. Yet, the great diversity of the mechanisms and their interrelation make it impossible to carry out rigorous classification of wear processes, although many classification systems have appeared in the literature. It is generally recognized that the most common types of wear of polymers are abrasion, adhesion, and fatigue.

B. Abrasive Wear

Abrasive wear is defined by the ASTM G 40-83 terminology standard as a wear due to hard particles or hard protuberances forced against and moving along a solid surface.

Abrasive wear, also known as abrasion, is one of the most common forms of wear.

The key aspect of abrasive wear is its association with the cutting or plowing of the surface by harder particles or asperities. These cutting points may either be embedded in the counterface, or loose within the contact zone. The former case is commonly called two-body abrasion, and the latter, three-body abrasion.

Abrasion displays scratches, gouges, and scoring marks on the worn surface, and the debris produced by abrasion frequently takes on the appearance of fine cutting chips, similar to those produced during machining, although at a much finer scale. Most of the models associated with abrasive wear incorporate geometric asperity descriptions, so that wear rates turn out to be quite dependent on the shape and apex angles of the abrasive points moving along the surface. The sources of the abrasive solid are numerous, and the nature of the abrasive wear in a given tribosystem will depend to some extent on the manner in which the abrasives enter the tribosystem: whether they are present in the original microstructure as hard phases, enter the system as contaminants from outside, or are generated as debris from the contact surfaces as they wear.

Fundamental mechanisms of abrasive wear are directly related to plastic flow and brittle fracture. Plastic deformation of working surface by harder abrasive particles results in the formation of grooves from which material is removed. Let these abrasive particles be modeled by pyramids or cones indenting the surface at the depth h_i (Fig. 17). When the friction path is rather large, the volume wear per unit sliding distance I is proportional to a sum of the indentation depths:

$$I \propto \sum_i h_i^2$$

The external load W is distributed among all particles so that $W = \Sigma_i W_i$. Then, under condition of plastic deformation, the hardness H is

$$H = \frac{W_i}{B h_i^2}$$

Where it follows that

$$W_i = B H h_i^2$$

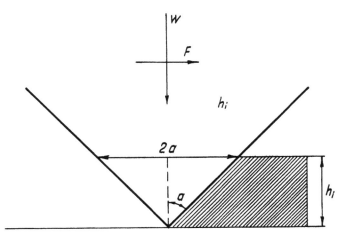

Figure 17 Contact between conical abrasive particle and a surface.

Summing the equality gives

$$W = BH \sum h_i^2 \propto HI$$

or

$$I = K \frac{W}{H} \tag{37}$$

where K is the dimensionless wear coefficient being a measure of the severity of wear. Eq. (37) is a main relation of the abrasive wear. It shows that relative wear resistance (the reciprocal of wear rate) is directly proportional to indentation hardness.

Because of viscoelastic deformation of polymers, their indentation hardness, in contrast with metals, reflects less appropriately the plastic behavior and thereby the above relationship should be considered as a tentative correlation describing the general trend of polymer behavior.

There is another approach to description of abrasive wear. Experiments have shown that the abrasive wear rate is in proportion to $1/\sigma_u \varepsilon_u$, where σ_u is the ultimate tensile stress and ε_u is the corresponding strain. The correlation was introduced by Lancaster and Ratner and is often referred to by their names [74,75].

The abrasion by plastic flow is accompanied by the formation of grooves. There are two distinct modes of deformation when an abrasive particle acts on the plastic material. The first mode is plastic grooving, often referred to as ploughing, in which a prow is pushed ahead of the particle, and the material is continually displaced sideways to form ridges adjacent to the developing groove. No material is removed from the surface. The second mode is named cutting, because it is similar to micromachining and all the material displaced by the particle is removed as a chip.

In two-body abrasion, some asperities produce ploughing, the rest shows cutting, depending on two controlling factors: the attack angle of the particle and the interfacial shear strength expressed as the ratio between the shear stress at the interface and the shear yield stress of plastically deformed material. There is a critical attack angle at which the transition occurs from ploughing to cutting. The value of critical angle depends only on interfacial shear strength for perfectly plastic material; for a real material, it depends also on the work-hardening rate and on its elastic properties, namely on the ratio E/H between the Young's modulus and the surface hardness. The higher is the value of E/H, the larger is the critical attack angle. The ratio can be used as a criterion for selecting plastics exposed to abrasive wear. As the ratio decreases, the wear resistance of plastics increases.

In the case of three-body abrasion, the free abrasive readily penetrates the polymeric surface, which begins to operate as an emery cloth resulting in increasing the wear of countersurface.

C. Adhesive Wear

Adhesive wear results from the shear of friction junctions. The fundamental mechanism of this wear is adhesion, an important component of friction outlined above. This wear process evolves in exactly the same manner as adhesion friction component does: formation of adhesion junction, its growth, and fracture. A distinguishing feature of this wear is that transfer of material from one surface to another occurs because of localized bonding between the contacting solid surfaces. Bely et al. [17] noted that the transfer of polymer is the most important characteristic of adhesive wear in polymers. It is reasonable that the processes associated with other wear types (fatigue, abrasion, and so on) accompany the adhesive wear. Experiment shows that even the static contact of two bodies results in adhesion at the interface. When the bodies are pulled apart, the material transfer can take place from the cohesively weaker material to the other. If the body moves relative to the surface of the other, the transfer film forms on the surface of cohesively stronger material. The amount of the transferred material is dependent on strength of adhesion bonds which, in its turn, is governed by the electronic structure of the mating materials, their compatibility, as well as crystal structure and orientation.

The phenomenon of friction transfer is observed for nearly all materials (metals, ceramics, polymers) and their combinations. The point is whether or not the transfer produces an influence on tribological behavior of the friction pair. In this case, the consequences of material transfer may be significantly distinct [76–78]. If small particles of micrometer size are transferred from one surface to the other, then the wear rate varies only to a small extent. Under certain conditions, the situations take place when thin film of soft material is transferred onto the hard mating surface, for example polymer on metal. Results of the transfer may be as follows. If the transferred polymer film is carried away from steel surface and is newly formed, the wear rate is increasing. In the case that the film is held in place, friction occurs between similar materials that may result in seizure. Spreading of polymer on steel shaft gives rise to abrupt jump of friction force, but the wear changes insignificantly.

It has been known that, under certain conditions, the hard material is transferred on the soft surface. For example, bronze is transferred on polymer. The transferred hard particles are embedded in soft material and serve as abrasive, which scratches the parent's material.

Polymers are most susceptible to friction transfer when rubbing both against metals and polymers. As an illustration, let us consider the friction between polytetrafluoroethylene (PTFE) and polyethylene (PE) [17]. Experiments were carried on wear tester with cylindrical block–shaft (conforming contact) geometry. It has been discovered that PTFE is transferred in the form of flakes of very small size in the initial period of friction. The thickness of the transferred layer increases monotonically and then oscillates about a mean value, whose magnitude and amplitude of oscillations depend on the test conditions, especially on load and sliding velocity (Fig. 18).

The transferred polymer fragment may exhibit a wide variety of forms depending on polymer properties and friction conditions. For example, even near ideally spherical particle

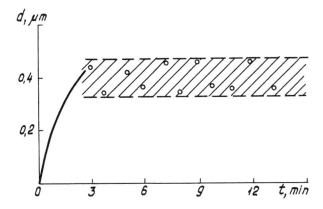

Figure 18 The thickness of the transferred layer of PTFE as a function of friction time (load = 0.05 MPa, sliding velocity = 0.35 m/sec).

is observed during adhesive wear. Such particle is likely to owe its origin to the flexibility of chain molecules of the polymer, thermal condition at the interface, and some other factors, which remain unknown.

Once more, the consequence of polymer transfer is a change in roughness of both surfaces in contact. The roughness of polymer surface undergo large variation during the unsteady wear until the steady wear is reached, while metal surface roughness is modified due to transfer of polymer [79].

D. Fatigue Wear

Fatigue is a change in the material state due to repeated (cyclic) stressing, which results in progressive fracture. Its characteristic feature is the accumulation of irreversible changes, which gives rise to generation and development of cracks. A similar process takes place at friction accompanying nearly all the wear modes. A friction contact undergoes the cyclic stressing at rolling and reciprocal sliding. In addition, each asperity of friction surface experiences sequential loading from the asperities of countersurface. As a consequence, two varying stress fields are brought about in surface and subsurface regions with different scales from the diameter of apparent contact area in the first case to that of local contact spot in the second. These fields are responsible for material fatigue in these regions that leads to the generation and propagation of cracks and the formation of wear particles. This process is named friction fatigue. Unlike the bulk fatigue, it only spans surface and subsurface regions. The loss of material from solid surfaces owing to friction fatigue is referred to as fatigue wear. It has been known that fatigue cracks are initiated at the points where the maximum tangential stress or the tensile strain takes place. The theoretical and experimental studies show that under contact, loading the maximum tangential stress position is dependent on friction coefficient. With low friction coefficient, the point where the shear stress is maximum is located below the surface ($f < 0.3$). When the friction coefficient increases ($f > 0.3$), the point emerges on surface. On the other hand, if a solid is subjected to combine normal and tangential loading, the surface and subsurface regions appear where the tensile strain and thereby frictional heating occur. Therefore cracks may be nucleated in surface and/or below it.

The initiation of the fatigue cracks is aided by defects, which are responsible for stress concentration. These are scratches, dents, marks, and pits on the surface, and impurities, voids, and cavities in subsurface region.

Both surface and subsurface cracks that open due to repeated stressing will gradually grow, join, cross each other, and meet the surface until wear debris, including spalls, are detached after a certain number of stressing.

The simplest model of fatigue wear is based on the Paris equation:

$$\frac{dL}{dN} = A(\Delta K)^n$$

where A and n are the empirical coefficients, dL/dN is the increase in crack length L per stress cycle, and ΔK is difference of stress amplitude in a cycle. The exponent n varies usually from 1.5 to 3.5 for elastomers, and from 3 to 10 for rigid thermoplastics and thermosets.

Fatigue wear rate is dependent on numerous factors: physical, mechanical, and chemical properties of solid surface, lubricant (if it presents), environment, surface quality, temperature, etc. The fatigue wear occurs, even though a direct physical solid contact is absent. The surface film does not eliminate the contact but only smooths it out. Although the friction force is decreased by lubrication and, hence the tensile stress drops, fatigue wear occurs and a number of cycles to surface damage increases insignificantly.

VIII. CONCLUSIONS

A widespread interest in plastics has quickened in the mid-20th century due to the features of their structure, specific mechanical behavior, and the considerable possibility to change polymer properties. Tribological applications of polymers were widely diversified. But creep of polymers, strong dependence of their properties on temperature, low heat conductivity, and sensitivity to the environment posed numerous problems. Extensive studies took years, as shown above, to outline the field of modern engineering, in which the plastics can be applied as tribological materials. It has been established that pure polymers as a rule do not operate efficiently, but can be used as blends or composites. They proved to be of considerable promise as coatings and solid lubricants. The latter are used either in pure form or in composite and laminated structures [80]. Thin polymer films, e.g., self-assembled monolayers formed by chemosorption and physical adsorption of organic molecules (polymers) are prospective boundary lubricants in the fast-growing area of memory storage devices, microelectromechanical systems, and other precision mechanisms [81,82].

It is clear that the progress in engineering will provide a lot of new opportunities in the applications of plastics, so research in their mechanical and tribological behavior will be a challenging and fruitful field of science and technology.

REFERENCES

1. Gross, B. Mathematical Structure of the Theories of Viscoelasticity; Hermann et Cie: Paris, 1953; 74 pp.
2. Alfrey, T. Mechanical Behavior of High Polymers; Interscience Publishers: New York, 1948; 312 pp.
3. Bland, D.R. The Theory of Linear Viscoelasticity; Pergamon Press: New York, 1960; 286 pp.
4. Kobeko, P.; Kuvshinskii, E.; Gurevich, G. Study of amorphous state. Elasticity of amorphous bodies, Proc. USSR Acad. Sci. (Phys.) 1937, 6, (in Russian)329–341.
5. Leaderman, H. Elastic and creep properties of filamentous materials and other high polymers, Text. Res. 1943, 11, 171–193.
6. Tobolsky, A.V.; Andrews, R.D. Systems manifesting superposed elastic and viscous behavior, J. Chem. Phys. 1945, 13, 3–11.
7. Ferry, J.D. Mechanical properties of substances of high molecular weight, J. Am. Chem. Soc. 1950, 72, 3746–3750.
8. Williams, M.L.; Landel, P.E.; Ferry, J.D. The temperature dependence of relaxation mechanisms in amorphous polymers and other glass forming liquids, J. Am. Chem. Soc. 1955, 77, 3701–3707.
9. Andrews, R.D.; Tobolsky, A.V. Elastoviscous properties of polyisobutelene, J. Polym. Sci. 1951, 7, 221–229.
10. Tobolsky, A.V.; Dunell, B.A.; Andrews, R.D. Stress relaxation and dynamic properties of polymers, Text. Res. 1951, 31, 404–421.
11. Fujino, K.; Horino, K.; Kawai, H. Tensile stress relaxation behavior of semicrystalline polymers in terms of validity of time–temperature superposition, J. Colloid Interface Sci. 1963, 18, 119–128.
12. Schwarzl, E.; Staverman, A.J. Time–temperature dependence of linear viscoelastic behavior, J. Appl. Phys. 1952, 23, 838–843.

13. Hilton, H.H. Thermal stresses in thick-walled cylinders exhibiting temperature-dependent viscoelastic properties of the Kelvin type, Proc. 2nd U.S. Nat. Congr. Appl. Mech. 1954, *1*, 547–551.

14. Morland, L.W.; Lee, E.H. Stress analysis for linear viscoelastic materials with temperature variation, Trans. Soc. Rheol. 1960, *4*, 233–241.

15. Grosch, K.A. The relation between the friction and viscoelastic properties of rubber, Proc. R. Soc. Lond. 1963, *A274*, 21–39.

16. Maugis, D.; Barquins, M. Fracture mechanics and adherence of viscoelastic solids, Lee, L.H., Ed.;Adhesion and Adsorption of Polymers; Plenum Publishing Corporation: New York, 1980; (part A), 203–277.

17. Bely, V.A.; Sviridenok, A.I.; Petrokovets, M.I.; Savkin, V.G. Friction and Wear in Polymer-Based Materials; Pergamon Press: Oxford, 1982; 416 pp.

18. Ferry, J.D. Viscoelastic Properties of Polymers; John Wiley & Sons: New York, 1961; 412 pp.

19. Rouse, P.E. A theory of the linear viscoelastic properties of dilute solution of coiling polymers, J. Chem. Phys. 1953, *21*, 1272–1277.

20. Zimm, B.H. Dynamics of polymer molecules in dilute solution. Flow birefringence and dielectric loss, J. Chem. Phys. 1956, *24*, 296–302.

21. Passaglia, E.; Knox, J.R. Viscoelastic behavior and time–temperature relationships, Baer, E., Ed.;Engineering Design for Plastics; Reinhold Publishing Corporation: New York, 1964, 143–198.

22. Muki, R.; Sternberg, E. On transient thermal stresses in viscoelastic materials with temperature dependence, J. Appl. Mech. 1961, *28*, 193–198.

23. Lee, E.H. Stress analysis in viscoelastic materials, J. Appl. Phys. 1956, *27*, 665–672.

24. Hilton, H.H. Viscoelastic analysis, Baer, E., Ed.; Engineering Design for Plastics; Reinhold Publishing Corporation: New York, 1964, 199–276.

25. Freudental, A.M. Effect of rheological behavior on thermal stresses, J. Appl. Phys. 1954, *25*, 1110–1116.

26. Hunter, S.C. Tentative equations for the propagation of stress, strain and temperature fields in viscoelastic solids, J. Mech. Phys. Solids 1961, *9*, 152–164.

27. Muki, R.; Sternberg, E. On transient thermal stresses in viscoelastic materials with temperature-dependent properties, ASME J. Appl. Mech. 1961, *28*, 193–207.

28. Bowers, R.C.; Zisman, W.A. Surface properties, Baer, E.., Ed.;Engineering Design for Plastics; Reinhold Publishing Corporation: New York, 1964, 689–741.

29. Timoshenko, S.; Goodier, J.N. Theory of Elasticity; McGraw-Hill: New York, 1951; 548 pp.

30. Johnson, K.L. Contacts Mechanics; Cambridge University Press: Cambridge, 1987; 452 pp.

31. Goryacheva, I.G. Contact Mechanics in Tribology; Kluwer Academic Publishers: Dordrecht, 1998; 360 pp.

32. Johnson, K.L.; Kendall, K.; Roberts, A.D. Surface energy and the contact of elastic solids, Proc. R. Soc. 1971, *A324*, 301–313.

33. Derjaguin, B.V.; Muller, V.M.; Toporov, Yu.P. Effect of contact deformation on the adhesion of particles, J. Colloid Interface Sci. 1975, *53*, 314–326.

34. Maugis, D. Adhesion of spheres: the JKR–DMT transition using a Dugdale model, J. Colloid Interface Sci. 1992, *150*, 243–269.

35. Myshkin, N.K.; Petrokovets, M.I.; Chizhik, S.A. Simulation of real contact in tribology, Tribol. Int. 1998, *31*, 79–86.

36. Greenwood, J.F.; Tripp, J.H. The elastic contact of rough spheres, ASME J. Appl. Mech. 1967, *34*, 153–159.

37. Sviridenok, A.I.; Chizhik, S.A.; Petrokovets, M.I. Mechanics of Discrete Friction Contact; Nauka i Technika: Minsk, 1990; (in Russian) 272 pp.

38. Nayak, P.R. Random process model of rough surfaces, ASME J. Lubr. Technol. 1971, *93*, 398–407.

39. Fuller, KN.G.; Tabor, D. The effect of surface roughness on the adhesion of elastic solids, Proc. R. Soc. 1975, *A345*, 327–342.

40. Bowden, F.P.; Tabor, D. Friction and Lubrication of Solids; Clarendon Press: Oxford, 1964; 544 pp.
41. Hutchings, I.M. Tribology: Friction and Wear of Engineering Materials; Edward Arnold: London, 1992; 273 pp.
42. Myshkin, N.K.; Kim, C.K.; Petrokovets, M.I. Introduction to Tribology; Cheong Moon Gak: Seoul, 1997; 292 pp.
43. Moore, D.F. The Friction and Lubrication of Elastomers; Pergamon Press: Oxford, 1972; 256 pp.
44. Yamaguchi, Y. Tribology of Plastic Materials; Elsevier: Amsterdam, 1990; 362 pp.
45. Greenwood, J.A.; Williamson, J.B.P. Contact of nominally flat surfaces, Proc. R. Soc. 1966, *A295*, 300–319.
46. Petrokovets, M.I. Effect of temperature on real contact area of rough surfaces, J. Frict. Wear 1999, *20*(2), 1–6.
47. Barber, J.R. Indentation of a semi-infinite solid by a hot sphere, Int. J. Mech. Sci. 1973, *15*, 813–819.
48. Segedin, S.M. The relation between load and penetration for a spherical punch, Mathematica 1957, *4*(1–2), 156–161.
49. Myshkin, N.K.; Petrokovets, M.I.; Chizhik, S.A. Basic problems in contact characterization at nanolevel, Tribol. Int. 1999, *32*, 379–385.
50. Kragelskii, I.V. Friction and Wear; Pergamon Press: Elmsford, 1982; 458 pp.
51. Buckley, D.H. Surface Effects in Adhesion, Friction, Wear, and Lubrication; Elsevier: Amsterdam, 1981; 631 pp.
52. Deryagin, B.V.; Krotova, N.A.; Smilga, V.P. Adhesion of Solids; Consultant Bureau: New York, 1978; 302 pp.
53. Schallamach, A. How does rubber slide? Wear 1971, *17*, 301–312.
54. Bartenev, G.M.; Lavrentev, V.V. Friction and Wear of Polymers; Elsevier: Amsterdam, 1981; 212 pp.
55. Briscoe, B.J. Interfacial friction of polymer composites. General fundamental principles, Klaus, F., Ed.; Friction and Wear of Polymer Composites; Elsevier: Amsterdam, 1986, 25–59.
56. Barquins, M.; Courtel, R. Rubber friction and the rheology of viscoelastic contact, Wear 1975, *32*, 133–150.
57. Briscoe, B.J. Friction of organic polymers, Singer, I.L. Pollok, H.M. Eds.; Fundamental of Friction: Macroscopic and Microscopic Processes; Kluwer Academic Publishers: Dordrecht, 1992, 167–182.
58. Shooter, K.; Tabor, D. The frictional properties of plastics, Proc. R. Soc. 1952, *B65*, 661.
59. Bowers, R.C.; Clinton, W.C.; Zisman, W.A. Frictional behavior of polyethylene, polytetrafluorethylene and halogeneted derivatives, Lubr. Eng. 1953, *9*, 204–209.
60. Shooter, K.; Thomas, R.H. Frictional properties of some plastics, Research 1952, *2*, 533–539.
61. Rees, B.L. Static friction of bulk polymers over a temperature range, Research 1957, *10*, 331–338.
62. Schallamach, A. The load dependence of rubber friction, Proc. Phys. Soc. 1952, *65B*(393), 658–661.
63. Oloffson, B.; Gralben, N. Measurement of friction between single fibers, Text. Res. J. 1947, *17*, 488–496.
64. Flom, D.G.; Porile, N.T. Effect of temperature and high-speed sliding on the friction of teflon on teflon, Nature 1955, *175*, 682–685.
65. Flom, D.G.; Porile, N.T. Friction of teflon sliding on teflon, J. Appl. Phys. 1955, *26*, 1080–1092.
66. White, N.S. Small oil-free bearings, J. Res. Natl. Bur. Stand 1956, *57*, 185–189.
67. Schallamach, A. The velocity and temperature dependence of rubber friction, Proc. Phys. Soc. 1955, *B66*, 1161–1169.
68. Milz, W.C.; Sargent, L.E. Frictional characteristic of plastics, Lubr. Eng. 1955, *11*, 313–317.
69. Tanaka, K. Kinetic friction and dynamic elastic contact behavior of polymers, Wear 1984, *100*, 243–262.

70. Fort, T. Adsorption and boundary friction of polymer surfaces, J. Phys. Chem. 1962, 66, 1136–1143.
71. Vinogradov, G.V.; Bartenev, G.M.; Elkin, A.I.; Mikhailov, V.K. Effect of temperature on friction and adhesion of crystalline polymers, Wear 1970, 16, 358–368.
72. Ludema, K.C.; Tabor, D. The friction and viscoelastic properties of polymeric solids, Wear 1966, 9, 329–348.
73. King, R.T.; Tabor, D. The effect of temperature on the mechanical properties and the friction of plastics, Proc. Phys. Soc. 1953, B66, 728–737.
74. Lancaster, J.K. Relationships between the wear of polymers and their mechanical properties, Proc. Instn. Mech. Eng. 1968–1969, 183, 98–106.
75. Ratner, S.B.; Farberova, I.I.; Radyukevich, O.V.; Lure, E.G. Correlation between the wear resistance of plastics and other mechanical properties, Sov. Plast. 1964, 7, (in Russian)37–40.
76. Makinson, K.R.; Tabor, D. The friction and transfer of polytetrafluoroethylene, Proc. R. Soc. 1964, A281, 49–61.
77. Sviridenok, A.I.; Bely, V.A.; Smurugov, V.A.; Savkin, V.G. A study of transfer in frictional interaction of polymers, Wear 1973, 25, 301–308.
78. Tanaka, K.; Uchiyama, Y.; Toyooka, S. The mechanism of wear of PTFE, Wear 1973, 23, 153–172.
79. Jain, V.K.; Bahadur, S. Surface topography changes in polymer–metal sliding, . Proceedings of International Conference on Wear of Materials, Dearborn, 581–588.
80. Pratt, G.C. Bearing materials. Encyclopedia of Materials Science and Engineering; Bever, M.B., Ed.; Pergamon Press: Oxford, 1986, 281–288.
81. Tsukruk, V.V.; Nguen, T.; Lemieux, M.; Hazel, J.; Weber, W.N.; Shevchenko, V.V.; Klimenko, N.; Shedulko, E. Tribological properties of modified MEMS surfaces, Bhushan, B., Ed.;Tribology Issues and Opportunities in MEMS; Kluwer Academic Publishers: Dodrecht, 1998, 607–614.
82. Bhushan, B. Tribology and Mechanics of Magnetic Storage Devices, 2nd ed. IEEE Press, Springer: New York, 1996; 1125 pp.

4

Visualization and Characterization of Bulk and Surface Morphology by Microthermal Analysis and Atomic Force Microscopy

Scott Edwards, Nguyen Duc Tran, Maria Provatas, Namita Roy Choudhury, and Naba Dutta
Ian Wark Research Institute, University of South Australia, Mawson Lakes, South Australia, Australia

I. INTRODUCTION

Modern materials are usually blends or composites of complex morphologies that determine their intrinsic properties. The performance of any complex material is mainly controlled by its component's suprastructure design and properties or morphology. Polymer morphology deals with the arrangement of macromolecules into amorphous or crystalline regions, the form and structure of these regions, and the manner in which they are organized into more complex units. Morphology of a multicomponent system or blend defines the spatial arrangement of the component phases. Microscale morphology is thus a major determinant of bulk and surface properties.

As materials become increasingly complex and the knowledge of the exact state and interactions between regions becomes more demanding, the challenge to characterize these complex surfaces and systems in terms of their microscopic structure, not only in two but also in three dimensions, becomes apparent. To meet this, the development of instrumentation to provide microstructural information has occurred. Such instruments include the powerful imaging spectroscopes based on secondary ion mass spectroscopy (SIMS), X-ray photoelectron spectroscopy (XPS), and Auger spectroscopy. Despite the many advances that have been made in recent years to analyze such complex materials, there are still many unresolved problems. These include the effect and the interaction of ion, electron, or x-rays from the instrument source with the analyte, the cracking and degradation of the material and surface roughness, etc.

Thermal methods have been widely used to identify the existence of phases in multicomponent systems (morphology) and the properties of interfaces between phases. Once phases have been shown to be present, the questions regarding their spatial distribution and even their individual properties become of paramount importance. When the consideration of composites is included with that of pure macromolecules, it is easily seen that results from thermal analysis of materials are averages of many phases, regions, or areas. Conventional thermal methods give no information about the distribution of phases within a multicomponent system or how they are aligned in space. Additionally, none of the aforementioned techniques has achieved the status of a general, reliable, and simple-to-use tool. This is due to a variety of factors that include resolution, specificity, difficulties in sample preparation, and the amount of time required acquiring high-resolution images. Furthermore, distinct problems can occur when using the current thermal methods. Firstly, experiments are time-consuming, most noticeably in thermoanalytical measurements. The second difficulty is related to sampling. Analytes are often too small, too thin, or embedded within a larger component from which it is hard to extract. The most fundamental problem is that any information obtained is not spatially resolved. In addition, the bulk thermal response is often dominated by the higher concentration of the matrix or substrate, and thus it is difficult to obtain detailed characterization of dilute components, contaminants, or minor components without physically altering the samples. Current material science is limited when attempting to characterize or analyze high-performance systems [1].

The key questions that can arise are:

1. What are the sizes of the domains?
2. Do smaller domains have different properties to large ones?
3. What is the physical state of the material trapped in the domain, etc.?
4. Does the surface vary in properties from the bulk?
5. Do interactions between the surface and coating change the properties of the coating?

II. ATOMIC FORCE MICROSCOPY AND MICROTHERMAL ANALYSIS

On a molecular level, a macroscopically homogeneous system exhibits heterogeneity. In order to determine how a material's phases are three-dimensionally distributed, microscopy must be used. Electron and optical microscopy as well as existing forms of analytical spectroscopy (IR, Raman, and secondary ion mass spectroscopy) can be successfully used in mapping the spatial distribution of phases, although they fail to show which features correspond to which phase. Among all microscopic techniques, atomic force microscopy (AFM) allows one to perform complex analyses of material surfaces and near-surface regions with a nanometer-scale resolution; this includes surface topography imaging and measurements of stiffness, elasticity, hardness, and nanotribological characteristics. With multimode AFMs, the force at the tip could be adjusted to obtain more specific information of the sample.

Two AFM techniques, contact mode and TappingMode®, are commonly used to image samples depending on the sample properties and the experimental objectives. In contact mode, the sharp tip (silicon nitride) is rastered across the sample surface. However, at higher tip-sample force, the material may deform resulting in a height image that represents a map of the surface stiffness rather than that of the original surface topography. In contact mode experiments, force-induced elastic surface corrugations were found [2] at different dimensional scales. In such multilayered materials, this phenomenon can be attributed to different local stiffness of atoms, chain blocks (oriented crystalline materials), or layers of different density. The dragging motion of the lateral force, combined with adhesive force between the tip and the surface, may cause substantial damage to the sample surface and the tip, particularly when the mechanical stiffness is comparable to or less than the stiffness of the AFM probe.

Under ambient conditions, a several nanometers thick layer of condensed water vapor and other contaminants can be found covering most surfaces. As the AFM tip comes in contact with this layer, capillary action causes a meniscus to form, producing a degree of surface tension between the cantilever tip and the sample, which further contributes to the adhesive force (Fig. 1). This downward pull increases the overall force on the sample which, when combined with the lateral shear forces caused by the scanning motion, can distort measurements and lead to the movement or tearing of surface features [3]. TappingMode® has recently been developed to minimize such interactions between the tip and the sample.

In TappingMode®, the fast-oscillating probe tip (forced to oscillate at a frequency close to its resonance), with adjustable amplitude, lightly taps over the surface allowing only intermittent contact with the scanned surface. This approach prevents damage by virtually eliminating the lateral force inherent to contact mode AFM. At low force or in light tapping ($A_{sp}/A_0 \sim 0.8$, where A_0 is the free oscillation amplitude and A_{sp} is the set point [4,5]), the tip-sample contact area is minimal, providing the best resolution when imaging nonperiodic topographical features. Changes in the oscillation amplitude to the cantilever tip serve as a feedback signal for measuring the topographic variations across the surface [6]. In addition, height images recorded via light tapping can accurately reproduce a soft sample's true topography without damaging any sample feature. By mapping the phase of oscillation with respect to the piezoelectric driver cantilever, a scan of the phase image can be obtained (Fig. 2). The phase image detects variations in composition, adhesion, friction, and viscoelasticity and is very important in differentiating between the component phases of composite materials [7–10].

The only problem of TappingMode® AFM is identifying the difference between the real surface topography and the apparent one due to lateral variations of the indentation depth of the tip [11]. Nevertheless, this unique measurement mode has detected nano-particles in subsurface region [12] and characterized heterogeneous catalysts of radiation-induced metal clusters associated with polymer matrices.

Such advances in scanning probe microscopic techniques, and the integration of these with thermal analysis capabilities, have produced a new vista of instruments that can provide both spatial and calorimetric data collection capabilities. The introduction of micro-thermal analysis (μTA™) by TA Instruments in 1998 is a major advancement in the area of thermal analysis instrumentation. μTA™ provides potential advantages in morphological characterization of complex materials, composites, and blends. It is a very rapid form of thermal analysis that can overcome many of the current sampling problems while also providing spatially resolved thermal analysis. μTA™ combines the capabilities of advanced thermal analysis with AFM [13–19]. It has shown considerable promise for nanostructure/property characterization of complex materials. This technique is particularly useful

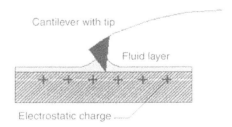

Figure 1 AFM tip–surface contaminants interaction.

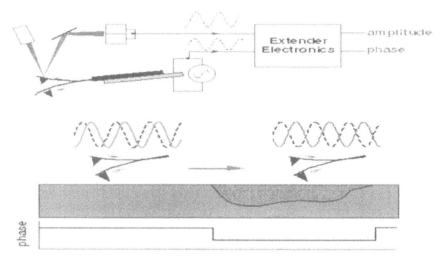

Figure 2   Phase imaging using tapping mode.

for characterizing those complex materials, which are difficult to analyze by the existing techniques.

μTA™ is able to image a sample and then obtain spatially resolved measurements on small areas, thus overcoming the limitations of conventional techniques. With a resolution at a submicron level [20], any point on the image can be selected for characterization of its calorimetric and mechanical properties. An important feature of this new technique is its ability to perform very fast experiments on this small scale using heating rates 2 orders of magnitude greater than that used in conventional calorimetry. This opens up the possibilities of rapid analysis of very large numbers of samples. A comparison of μTA™ to other techniques can be found in Table 1.

Topographical information is obtained in μTA™ by rastering the probe across the surface of a material, similar to AFM. The primary difference between μTA™ and AFM, however, is the probe [13–19]. In μTA™, the conventional AFM probe is replaced with a thermal probe (Fig. 3). The thermal probe consists of a 5-μm diameter Wollaston wire, containing a platinum filament enclosed in a silver sheath. The wire is bent to a certain degree, and the platinum is etched to reveal the silver sheath. The thermal probe forms an electrical bridge circuit, allowing it to have dual functionality as a heat sensor for measuring the resistance of the platinum and as a localized heater by applying a voltage to the Wollaston wire. The mirror on the cantilever acts as a laser spot, which forms part of the feedback loop of the microscope. Images, viewed as a function of thermal conductivity (d.c. thermal image), are acquired by measuring the power required to maintain the probe tip at a constant temperature as it rasters over the sample's surface. Additionally, by modulating the isothermal temperature and measuring the response to the temperature modulation, an image is obtained as a function of thermal diffusivity (a.c. thermal image). The thermal diffusivity images are similar to the thermal conductivity images since they are directly related to each other through density and heat capacity. Therefore an experiment usually consists of the d.c. and a.c. thermal imaging, as well as the topography of a sample's surface. Characterization of a surface occurs by selecting specific locations on the surface image, where upon the probe heats linearly with time to perform localized thermal analysis (LTA) at these marked locations. Data such as glass transition, crystallization, and melting tem-

Table 1 Comparison of μTA to Other Techniques

	AFM	μTA/LTA	Microscopy	Macro-TA
Sample preparation	Flat sample, no other sample preparation	No sample preparation	Time-consuming, cause artifacts (e.g., coating or staining)	Little sample preparation
Material property measured	Mechanical property mainly adhesion, hardness, friction. Tip could be chemically modified to study tip/surface interaction.	Thermal conductivity, diffusivity, morphology, crystallinity, degradation by μMDTA, μTMA	Morphology/distribution, no chemical/mechanical information	Thermal and mechanical properties, morphology
Atmosphere	Vibration-free bench, can be done in liquid and air	Vibration-free bench, a range of temperature (−70°C to 300°C) with sub-ambient accessories, ambient to 500°C (standard), 0–1500°C/min heating/cooling rate	Mainly vacuum chamber or ambient condition	Controlled flow with inert gases or air for DMA, or reactive gases for TMA and DSC; TMA: temperature range −150°C to 1000°C; heating rate: 0.01°C/min to 200°C/min; DSC: temperature range ambient to 725°C [−150°C, subambient accessories]; DMA: temperature range −145°C to 600°C; heating Rate: 0.1°C/min to 50°C/min; cool rate 0.1°C/min to 10°C/min
Interaction	Nondestructive; however, in contact mode, scratching may occur	μTA if conducted at high temperature is destructive. At room temperature, nondestructive.	Beam sample interaction may change or damage the sample	Destructive
Resolution/sensitivity	Lateral resolution a few nanometers, vertical scan range 7 μm, lateral scan range 160×160 nm	Lateral resolution of micrometer order, vertical scan range 10 μm, lateral scan range 2×2 μm	High resolution for TEM, SEM; medium resolution for optical	TMA sensitivity 100 nm; DSC sensitivity 0.2 μW (rms); DMA: tan delta resolution 0.00001, tan delta sensitivity 0.0001, strain resolution 1 nm

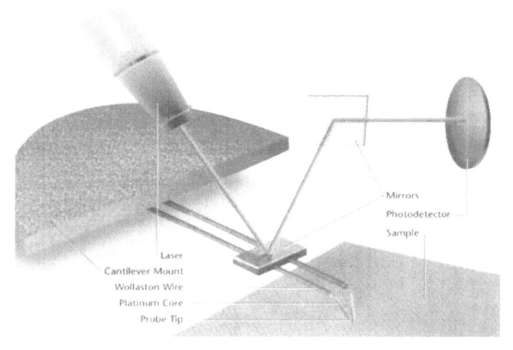

Figure 3   Schematic of a thermal probe. (Courtesy of TA Instruments.)

peratures can be collected by LTA, with a precision of approximately ± 3 K [19]. In addition, LTA can also be performed under a superimposed temperature modulation [13–20].

The current state of knowledge of this instrument is in its early stage, and thus few references are available outlining the principle and modes of its operation. To date, most studies conducted using the technology of the μTA™ have been applied in the realm of pharmaceutical chemistry [21] and biological sciences [12,14,22,23]. Since this is a new technique, case studies constitute an essential step in its understanding. As a number of important properties of polymers are controlled by their surface chemistry and morphology, the following cases are considered in this chapter:

- Bulk morphology: investigating surface defects or morphology/supramolecular structure of multicomponent systems (primarily rubber-toughened polymers)
- Surface morphology of coatings (plasma-polymerized coating)

III. BULK MORPHOLOGY

A. Rubber-Toughened Polymers and Their Morphology

Strong interest in polymer blends stems from the knowledge that unique properties can be achieved as a result of distinctive structural developments during processing. Understanding of phase interactions allows one not only to gain an insight into the nature of the morphology generated due to flow or deformation history, but also to control them at a very early stage of their processing.

It is well known that the toughness of glassy polymers can be dramatically improved by the addition of a dispersed rubber phase. Such morphology allows significant energy dissipation, primarily through the development of large, stable craze fields formed within the matrix or by cavitations of the rubber particles to facilitate shear yielding. Toughness can be controlled by a variety of factors such as:

1. The amount of added rubber
2. Rubber type
3. Degree of cross-linking
4. Levels of adhesion
5. Particle size and distribution
6. Graft levels
7. Molecular weight of the respective phases.

It is the interplay of all these variables that determines the material's ultimate morphology. Consequently, it is very difficult to decouple them to isolate and quantify their effects.

Among all, the rubber particle size and distribution are critical variables [24]. In addition, the effect of the rubber phase on a part's surface appearance is significant. The incorporation of smaller-sized rubber particles often results in a better surface finish. This is due to the enhancement of relaxation and recovery effects that take place in the melt during solidification. Smaller particles can pose a problem, however, as they are inefficient and ineffective in initiating and terminating crazes, respectively, thus increasing the rates of creep [25]. The incorporation of larger-sized particles will enhance a material's impact strength but produce the opposite effect on the quality of the surface finish, hence creating the need of a particle size compromise.

Thus the ability of a particular rubber to toughen a plastic drops significantly below a critical size. In addition, such a range of size enhances the degree of agglomeration, resulting in a poor surface finish. The relationship between the amount of rubber and the surface finish of injection-molded parts has been well documented by Chang [26,27]. Materials with decreasing rubber levels will increase the recoverable shear strain, yet decrease shear stress. These two factors have the combined effect of improving the overall surface finish. In any case, cross-linking can control the integrity of the rubber phase. Often, bimodal particle size distributions offer synergistic toughening as a result of combined shear yielding and crazing.

In typical multiphase thermoplastics, such as acrylonitrile–butadiene–styrene (ABS) or acrylonitrile–styrene–acrylate (ASA), a minor portion of rubber (5–20%) is incorporated as a dispersed phase into a rigid plastic matrix. For improved coupling to the styrene–acrylonitrile (SAN) continuous phase, the rubber phase is grafted with the same monomeric building blocks of which the resin phase consists [25]. Thus the final product will consist of both bound and unbound rubber phases. The size and the distribution of rubber particles contained within the matrix may determine such properties as impact strength, surface finish, and ease of processing.

The application of microthermal analysis to the morphological study of such a polycarbonate (PC)/ASA blend is illustrated in Fig. 4. The resulting relative conductivity image as well as the corresponding localized thermal profiles of a microtomed section of the compression molded sample clearly demonstrates how µTA™ visualizes and characterizes the morphology in a way that is different from any other technique available today. Analysis is consistent with the structural model of rubber-modified thermoplastics, consisting of varying sizes of discrete oval and circular-shaped rubber zones of lower conductivity embedded in a continuous matrix of higher conductivity. Clusters and grafted zones of the gel are clearly visible throughout the image. Closer inspection of the nine LTA curves reveals the presence of both SAN and PC, with glass transition temperatures determined to

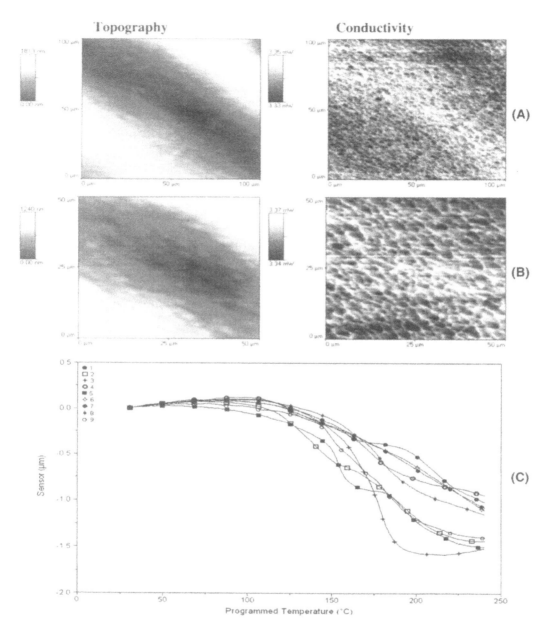

Figure 4 Topography and relative conductivity images and corresponding localized thermal profiles of a microtomed section of compression molded PC/ASA.

range between 100–124°C and 139–159°C, respectively. At the interface between the rubber zones and the matrix, the relative conductivity is higher compared to the particles themselves, indicating the coexistence of phases and partial mixing. This explains why ranges of glass transitions are calculated from LTA, as within these confined regions, mobility of the phase is greatly restricted. Low-temperature rubber transitions cannot be seen as without a cold stage, they are beyond the detection limit of the probe. In any case, µTA™ clearly identifies the immiscible, multiphase nature of the PC/ASA blend.

Microthermal analysis is capable of analyzing a variety of areas such as the surface, bulk, or interface. Figure 5 demonstrates the application of µTA™ to a tribology study of an injection-molded specimen consisting of the same PC/ASA blend studied previously in Fig. 4. Both topography and relative conductivity scans were performed on a part, molded at a high-material temperature (275°C), subjected to a wear analysis under a load of 50 N for 2 hr at 40°C using a temperature-controlled, reciprocating pin-on-disk wear machine. The effect of wear by the steel pin is clearly seen on the conductivity image. Comparison with the earlier conductivity image of the compression-molded specimen (Fig. 4) reveals a similar morphology, although with obvious differences. An interfacial wear mechanism has deformed the rubber regions, elongating them in the direction of pin motion. The boundaries between the phases have broken down either by the injection-molding process itself, the wear process, or by both. This has resulted in the agglomeration of phases into localized zones.

Topography of the sample must also be taken into account when investigating relative conductivity images. Areas of darkness, inconsistent with the conductivity throughout the rest of the image, may be partly due to topographical effects. This is due to variations in heat flow resulting from changes in topography. While the probe is in a valley, it is primarily surrounded by the sample, and when it is on a peak, it is surrounded by air. Thus the thermal homogeneity of the sample will vary the image contrast and the measured thermal conductivity increases and decreases, respectively [28]. While a comparison of topography and conductivity from Fig. 5 may show small similarities to one another, the thermal profile shows a significantly different morphology, and thus influences from the height of the sample are only small and are limited to regions of extreme low conductivity, seen as black

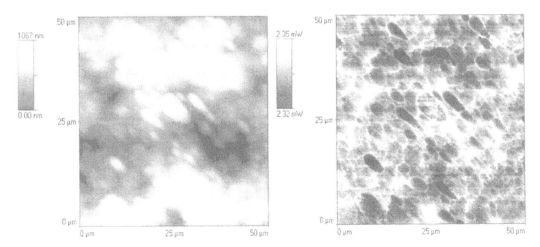

Figure 5 Topographical (left) and relative conductivity (right) images of a PC/ASA surface, subjected to wear by a metal pin under load.

spots. Another consideration is that since the sample has been sheared during the wear process, topography itself may provide hints on the morphology of the sample. Various phases within the plastic deform under shear in different ways and will relax to varying amounts, manifesting itself in the topography image. This is believed to have occurred in the sample shown in Fig. 5.

B. Understanding Structural Development and Surface Finish of a Blend During Processing

In a heterogeneous polymer blend, microscale morphology is a major determinant of many bulk and surface properties. While there has been a considerable amount of work investigating the various types of morphologies prevalent within thermoplastics and their blends [29–31], there is a distinct lack of documentation linking this characteristic to processing and surface finish. Traditionally, morphology has been studied through scanning electron microscopy (SEM) of fractured surfaces and transmission electron microscopy (TEM). These methods, while successful, are often hindered due to a lack of contrast between the phases [32]. Chemical etching by alkaline and acid solutions can provide a good method for complementary morphology observations of polymeric blends [33].

The formation of surface defects during injection molding originates from such sources as the machine, mold, or operator and can often be solved through in-house experience and expertise [34]. The defects produced directly from material influences are considerably more difficult to solve, as a fundamental understanding of the polymer's composition, rheology, and morphology is often required. Figure 6 illustrates the consid-

**INFLUENCES OF
SURFACE FINISH**

POLYMER   **METAL**

- Structure/Morphology
- Rubber particle size/agglomeration
- Phase separation
- Insulating (poor thermal conductivity)
- Part geometry
- Chemical structure
- Wetting/melt adhesion (under
  temperature, shear)
- Melt rheology/flow properties/creep
  flow
- Interface growth/skin formation (in-
  situ morphology development)

- Micro-structure/hardness
- Roughness
- Morphology
- Thermal conductivity
- Gate geometry/mould design
- Chemical structure
- Wettabilty/adhesion/shear
- Wear
- Coatings

Figure 6 The large variety of polymeric and metal properties that may influence the surface finish of an injection-molded part.

erable variety of material properties of both the polymer and the metal mold that may influence the development of surface defects. The shear number of material and processing variables listed truly unravel the scope of the dilemma. Direct visualization of the polymer melt inside a mold cavity is very effective in providing in-depth understanding of the process. However, it does not allow one to probe the in situ developed morphology. The application of μTA^{TM} is very appropriate in such cases.

Many of the common surface defects that appear regularly on molded parts (from toughened plastics) are understood and have been well characterized. Defects such as sink marks, weld lines, and heterogeneous fiber distributions are examples of such surface blemishes that have been extensively analyzed [35]. There are, however, a variety of other defects that are less understood and have not yet been clearly defined, particularly in reference to their morphologies. These defects include flow marks and flow lines, variations in gloss levels, varying heterogeneity of surface roughness, and stick–slip conditions. Many studies have sought to objectively quantify these defects and their mechanisms of formation and have produced a variety of viable theories.

Large majorities of surface defects come under the category of flow marks. Such defects are described as halos, tiger stripes, rings, webs, haze patterns, chatter marks, groves, spirals, flow lines, and sharkskins. Often, these visual flaws can be seen when injecting through both capillary dies and injection molds, although more complex flow geometries and steep thermal gradients complicate the latter [36]. In general [37], these various types of flow marks can be categorized into three groups according to its surface appearance (Fig. 7):

1. Alternate dull and glossy surface
2. Synchronous dull and glossy surface
3. Semicircular, microgrooved patterns

Early theories have contributed the formation of surface defects to critical levels of shear stress or the breakdown of adhesion at the polymer/metal interface, interpreted as a stick–slip condition [38–41]. Other contrary studies have reported that stick–slip promoters (i.e., fluoropolymers) actually eliminate defects, indicating that a poor finish was the result of stress levels within the melt and stretching at the exit of the die [42–44]. However, there is no existing technique currently available that can look into such developed morphologies.

Poor aesthetics due to gloss and color variations, primarily on parts consisting of multiphase polymer systems, can also be attributed to secondary flows in injection molds. This varying interpretation, suggested by Salamon and Donald [45], is significantly different than those discussed thus far, as they attribute variations in surface gloss to shear heating in the runner system. Secondary flows can be attributed to the varying degree to which the polymer melt experiences heating as it enters the cavity. As the molten material travels the runner system, the polymer adjacent to the wall experiences high shear rates and is subsequently hotter than the interior melt due to its poor conduction. Thus a region of hot polymer develops around a relatively cooler core. As this material is injected into the mold, the melt stream expands and the hotter component experiences a different flow characteristic compared to that of the core. The surface gloss of a blend is also related to the refractive index difference of the component polymers. However, for polymer blends, the composition of the surface is often different from that of the bulk, resulting in phase-segregated skin-core morphology. Once injected, such polymer expands into the mold cavity and produces two independent flows (Fig. 8). The hotter, lower viscosity component is concentrated on the edges of the part and experiences variations in flow front velocity and thermal boundaries to that of the main flow at the center of the cavity. Subsequently, as

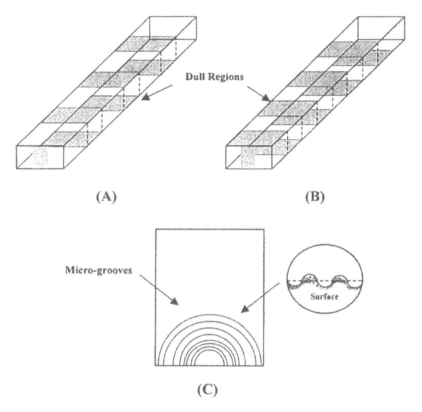

Figure 7 Categories of flow marks on injection-molded parts. (A) Synchronous dull and glossy surface, (B) alternate dull and glossy surface, and (C) semicircular, microgrooved patterns.

transcription to the mold cavity will vary with the two flows, varying regions of gloss are generated.

Chang [26,27] has extensively studied the surface appearance of modified ASA polymers. He observed flow marks as a series of low gloss bands called "chevrons," periodically distributed across a globally glossy surface. In general, regions of high gloss display a smooth surface, whereas dull bands show a distinct surface roughness. As the polymer melt encounters a step increase in cavity thickness, a slip condition may occur. This geometric discontinuity initiates a flow instability, which subsequently causes a series of low gloss bands, out of phase with each side of the molded part.

Studies by Hobbs [36] on polymeric blends of polycarbonate and ABS also reported on the formation of alternate dull and glossy bands as being generated by a wall slip mechanism. Ideas on the formation mechanics of these observed flow marks are similar to what was encountered during capillary extrusion, with high shear rates and high shear stresses generating a breakdown in fountain flow and oscillations in flow front velocities. Above a critical shear stress, "stick–slip" flow is initiated and the flow front begins to oscillate across the two surfaces of the cavity wall.

Atomic force microscopy has been performed on injection-molded PC/ASA plaques in order to understand the mechanics of gloss variations with reference to such processing variables as material temperature, injection time, and packing pressure [46] (Table 2). Flow

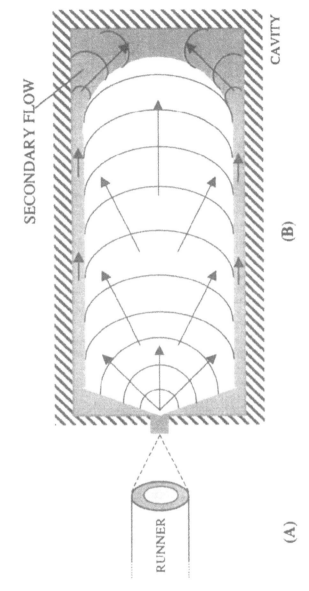

Figure 8   Schematic illustration of (A) shear heating in the barrel and runner system producing variation in melt temperature and (B) the resulting formation of secondary flows within the cavity.

Table 2 Injection-Molding Conditions for Engineered Surface Defected Samples, Under Ideal Injection Time

Injection time (sec)	Pack time (sec)	Material temp. (°C)	Pack pressure (bar)	Condition
2.61	10	260	55	Ideal
2.62	10	260	100	Ideal, pack pressure high
2.60	10	275	55	Ideal, material temp. high

patterns were found to vary between areas of differing surface gloss, with luster and dull regions showing a semicircular flow of molten material and a right to left flow direction, respectively (Fig. 9). The flow pattern, as seen in Fig. 9 for the dull area, indicates retraction of the melt front from the mold surface before complete solidification. Melt elasticity, high shear stress in the melt, and wall slippage at high shear stress can be possible reasons for this flow direction [46]. García-Ayuso et al. [47] reported on the variables such as rate, temperature, and pressure on the quality of barrier coatings on plastic film. In addition, surface roughness increased in the dull area, which was also observed by García-Ayuso et al. [47] and Salamon et al. [48] who further reported that surface roughness was aligned with the direction of flow.

Surface defects such as sink marks showed variation of flow around its surrounding area. The inner edge of the sink mark showed striations or "hills and valleys" (Fig. 10), similar to other studies [11,49]. These striations were a result of the melt front undergoing an unstable, distorted flow. Topography variance for the sink mark highlights that at an area of surface defect, there is an irregularity in phase dispersion.

Microthermal analysis on various key areas across a defected, injection-molded PC/ASA strip reveals the varying morphology that exists not only across the part, but also compared with the original unprocessed material seen in Fig. 4. Both relative conductivity images and LTA profiles were performed on microtomed sections from three separate

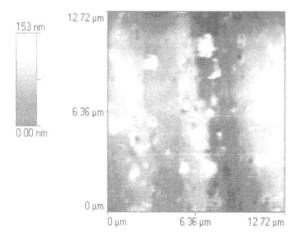

Figure 9 AFM image of the dull area under ideal injection-molding conditions.

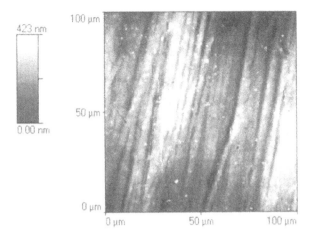

Figure 10 AFM image showing striations at the sink mark area.

locations on the same part and are displayed in Fig. 11. Measurements were performed on a glossy part near the gate, a nondefected glossy region at the center of the strip, and a region far from the gate. A clear difference can be noticed on three locations. The sections from both the gate and glossy regions reveal a more densely packed morphology, indicative of the effect of processing. Boundaries between phases have been reduced significantly and point to the formation of a highly networked skin-layer with little phase separation [50].

This is further confirmed through localized thermal analysis. Glass transition temperatures have increased significantly and identify a predominantly polycarbonate-rich surface. Preferential phase segregation during injection molding causes a dominant PC-skin layer on the surface of the parts. This phenomenon is related to each material's flow and viscosity characteristics, which promotes preferential phase segregation. In the blend, PC is the lower molecular weight component and thus flows well when compared with ASA. This produces a secondary flow phenomenon with PC concentrated at the edge of the part during processing, consequently forming the skin layer. Separate studies on flow marks and internal structure of PC/ABS blends [51] have drawn similar conclusions. Additionally, surface preference is believed to occur as a result of surface energy difference of polymer components. Therefore the polymer component that has a lower surface energy segregates on the surface in an effect to lowering the polymer–air surface tension [52].

It must also be mentioned that the phase morphology of a blend system changes as a function of its composition, intrinsic viscosity of the components, shear rate, processing conditions, etc. Examination by μTA™ of the core morphology of injection-molded PC/ASA under different processing conditions (Table 2) shows a multiphase system as indicated by the different values of conductivity [50]. Under ideal conditions, the morphology is densely packed and consists of varying oval- and circular-shaped lower conductivity areas (Fig. 12). Increase in pack pressure results in further densification of the morphology and a distinct alignment of particles can be seen. A further increase in pack pressure causes agglomeration and deformation of lower conductivity particles (Fig. 13). Detailed examination of processing conditions on the effect on surface roughness and morphology was performed on the prepared plaques. Gloss areas showed wave-like features that did not represent flow direction. Microflow marks, similar to the one that Yoshii et al. [53] observed,

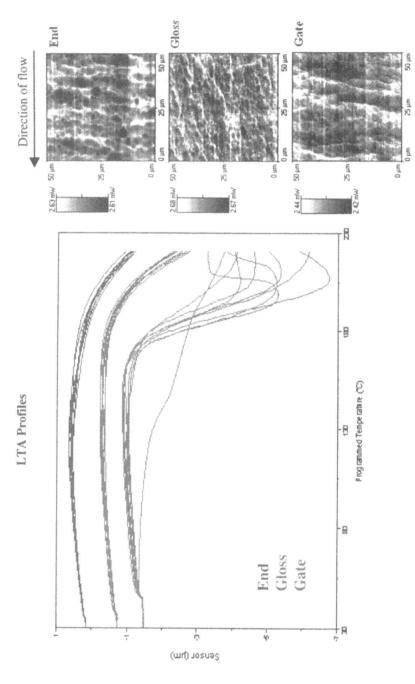

Figure 11 Microthermal analysis on three key locations on a PC/ASA part: localized thermal analysis (LTA) and relative conductivity scans of the gate, glossy, and end regions.

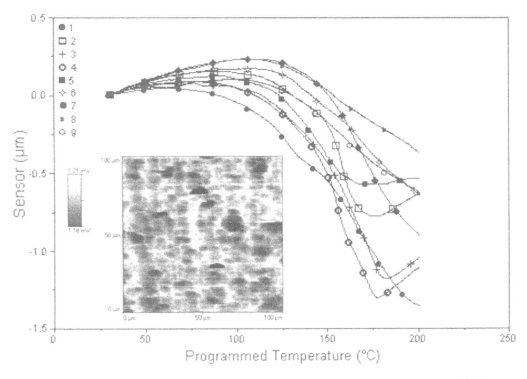

Figure 12 Thermal conductivity image under ideal injection-molding conditions, with its corresponding LTA.

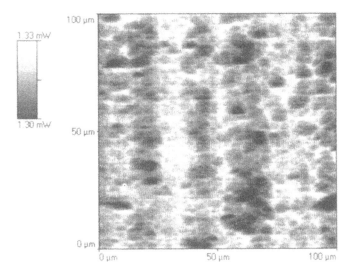

Figure 13 Thermal conductivity image of PC/ASA blend under high pack pressure.

were detected on the dull area (Fig. 14). However, microflow marks were not observed on the sample under high-material temperature since flow properties are improved at higher temperatures. LTA for the plaque samples showed a dominant PC phase at the surface.

Furthermore, morphology is dependent on the thermodynamic properties of the components during processing. The viscosity of each component [54] during processing generally determines and/or dictates which phase is continuous. However, it is quite evident that systems with greater differences in component melt viscosities show larger variations in skin layers during processing. As with earlier work [50,51,55], a disruption to the homogeneity of the formed PC skin layer is associated with the appearance of defects. This is illustrated in the LTA curves in Fig. 11. Flow marked part (near to the gate) shows lower glass transition temperatures, indicating quantities of SAN on the surface. In addition, systematic investigations and evaluations on polymer structure and morphology using both solution and solid state NMR, GPC, and microscopy show that the terpolymer composition has significant influence on the morphology and hence defect formation.

As LTA analysis is a highly localized technique, high values of glass transition temperatures are often seen and signify a restriction in mobility at the point of analysis. This is illustrated in Fig. 15, where the presence of rubber particles in the blend, both bound and unbound, will affect the mobility of the matrix chain segments that lay within an "entanglement" or "entrapment" zone. Those polymeric chains, in close proximity to a particle, will exhibit a higher T_g value than those further away (Fig. 16). Depending on the injection-molding conditions, various phases may coexist on the surface or subsurface of the

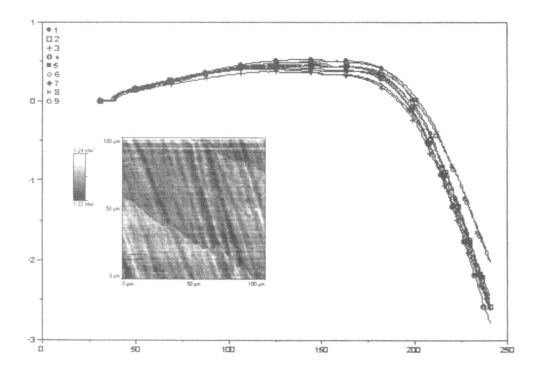

Figure 14 Thermal conductivity image of microflow marks for dull area PC/ASA blend under high pack pressure, with its corresponding LTA.

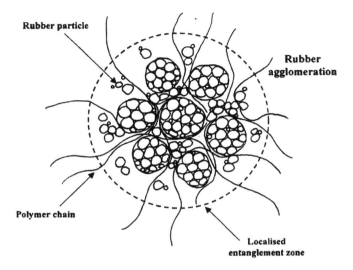

Figure 15 Illustrative diagram representing the restriction in mobility that chains experience near regions of particle agglomeration due to processing.

molded part. Therefore T_gs will be governed by the weight fraction of each phase/ component (ω) present in that particular location and can be expressed by the following equation:

$$T_{g_B} = \omega_1 T_{g_1} + \omega_2 T_{g_2} \tag{1}$$

where $\omega_1 + \omega_2 = 1$, T_{g_B}, T_{g_1}, and $T_{g_2} = T_g$s of the blend, component, and components 1 and 2, and w_1, w_2 = weight fractions of components 1 and 2.

Gloss-related studies by Salamon et al. [48] have attributed defect formation to postcavity surface contact phenomena. Analysis of halo surface defects, formed during the injection molding of polymeric blends, using both center- and end-gated molds, has suggested that temperature gradients along the feed system are the primary cause. The

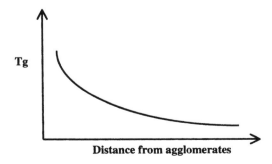

Figure 16 Trend of glass transition temperatures vs. distance from particle agglomerates.

authors reported that the level of gloss on a part could be related to the extensional strain rate at the melt flow front. This strain rate is reduced when there is a recovery of stress caused by a transition from a colder to hotter melt. This process results in an increase in the size of the skin layer, perpendicular to the direction of flow. To remain within the boundaries of the cavity, the skin layer must fold or wrinkle, causing variations in the surface roughness levels.

The physiochemical approach to morphology determination involves selectively removing one phase by etching the polymer surface and subsequently examining its topography. A solution of chromic acid will oxidize the rubber particles within a thermoplastic such as ABS, while leaving the SAN matrix untouched [56]. The resulting SEM images reveal phase orientation due to the molding cycle, establishing a processing fingerprint. Although etching techniques for morphological investigations have been available for some time and are relatively simple to do, there is little attention devoted to its application to rubber-modified blends of PC and ABS or ASA in literature. Both acid and alkaline etching techniques, developed by Bucknall et al. [33] and Eastmond and Smith [32], respectively, have been utilized in characterizing the morphology of self-prepared thermoplastic blends [56]. Few studies have investigated the effect of injection molding on the quality of commercially available blends.

Hamada and Tsunasawa [51] have successfully used the oxidation of polybutadiene to correlate the formation of flow marks to the internal structure of a commercial PC/ABS blend. Chromic acid was applied to both luster and glossy regions of an injection-molded part, revealing a varying internal structure. Dull regions revealed a rough morphology, indicative of a heavily etched surface, whereas gloss regions showed an isotropic landscape. These results suggest that the glossy, defect-free region is rich in polycarbonate, while the flow mark surface contains both polycarbonate and ABS phases.

IV. SURFACE MORPHOLOGY

A. Tailoring Surfaces by Plasma

Many crucial material properties such as adhesion and biocompatibility are governed by the molecular state at the topmost surface. Tailoring polymer surfaces by plasma engineering is an emerging technology that has revolutionized the conception and implementation of thin film coatings. The use of plasma is fundamental to all areas of thin film processing, ranging from computer chips to hard coatings, food packaging, biomaterial, polymers, and architectural glass coatings [57,58]. This technology is versatile and can be used for synthesis of conducting and semiconducting organic films, controlling the hydrophobic/hydrophilic (oleophobic/oleophilic) or biocompatible nature of a surface, and modifying the transport properties of membranes, improving adhesion, wettability, printability, dye uptake, and other technologically important surface characteristics [58–60].

Plasma is a partially or fully ionized gas resulting from an electrical discharge (d.c., a.c., radio, or microwave voltage) that can consist of free radicals, photons, and neutral and ionized species. The most energetic species contained within the complex entity of the glow discharge are electrons, which can initiate reactions on the substrate in the plasma state. The chemistry and physics of plasma and its interaction with an exposed surface are very complex and have been extensively reviewed [61].

Plasma modification can be classified into four broad categories: (1) surface preparation by breakdown of the surface and loose contaminants; (2) etching and creating new topographies; (3) surface activation by creation/grafting of new functional species; (4) and deposition of monolithic, adherent surface coatings by polymerization of monomeric species onto the surface. In plasma surface engineering, these reactions often occur simultaneously. The unique advantages of plasma modifications are the following: (1) treatment is limited to a very thin layer of the surface and, consequently, any surface modification can be achieved with minimum impact to the bulk character of the substrate polymer; (2) nearly all polymers, regardless of their chemical nature and reactivity (in a conventional sense), can be rendered to surface modification; (3) offers a wide spectrum of choice such as etching, implantation of atoms, radical formation, and thin polymer coating formation; (4) modification is very fast, in many instances taking only seconds; (5) plasma processes are commercially attractive, "dry," and the deposition of thin film coatings can be achieved with a minimum impact on the environment; and (6) ability to control film thickness and composition regardless of substrate geometry, while keeping energy consumption and waste to a minimum.

B. Polymer Surface Coatings by Plasma Deposition

Plasma polymerization is thin film-forming process, where thin films deposit directly on the surface and produce pore-free, uniform films of superior physical, chemical, electrical, and mechanical properties. The surface functionality of polymeric materials could be engineered by varying the process parameters such as the use of discharge gas type (inert, reactive, or their mixture), type of monomer, duration of treatment, power density, gas pressure, gas flow rate, etc. [62,63]. Furthermore, the effect of operational parameters on plasma polymerization can vary with types of reactor. Monomer molecules are activated by plasma to form mono- and diradicals with fragmentation and rearrangement; recombination of the radicals forms larger molecules. Glow characteristics are a crucial important factor in plasma polymerization and are dependent on both the monomer flow rate and the discharge power. The W/FM parameter, where W, F, and M are the discharge power in J/s, the monomer flow rate in mol/s, and the molecular weight of the monomer in kg/mol [61,64], respectively, is the apparent input energy per unit of monomer molecule in J/kg. Therefore the magnitude of the W/FM parameter is considered to be proportional to the concentration of activated species in plasma. The rate of plasma polymerization is increased by raising the W/FM parameter (while keeping the concentration of activated species lower than that of the monomer molecules present in the plasma). At faster flow rates, higher wattages are necessary for the system to remain at "full glow." In low W/FM, incomplete glow, plasma-induced polymerization of the monomer takes place with the formation of the plasma polymer, which is hybrid in nature. The wide diversity of choice in plasma-processing conditions may also be construed as an important drawback, and optimization of plasma polymerization process and quality of the film formed is of fundamental importance.

C. Characterization of Plasma Coating

In general, the performance of any plasma coating depends on the coating material, deposition rate, coating thickness, morphology, substrate type, and the level of adhesion at the substrate–coating interface. In spite of the proliferation of applications, there are many unresolved questions regarding the most efficient use of plasma processing and the subsequent characterization of the formed coating. This is largely due to the inherent complexity of the plasma state and the difficulties encountered in analyzing the formed film. Among the extensive list of surface-specific techniques available for plasma polymer surface analysis (Table 3), FTIR, XPS, and SIMS are the most popular techniques; those have been

Table 3 Principal Techniques Used for Surface Analysis of Plasma-Modified Polymers

Technique	Sample probe (energy)	Detected species	Sampling depth	Information derived
FTIR/PAFTIR[a]	IR radiation (5000–400 cm^{-1})	Bond vibration	1 nm to bulk	Chemical groups, molecular orientation
SEM/EDS[b]	Electron	Electron, x-ray	1–2 μm	Morphology, rapid qualitative, or with adequate standards, quantitative analysis of elemental composition, maps or line profiles, elemental composition of material
LD-FTMS[c]	Laser	Ablated fragments	~20 nm (depends on experimental material and laser used	Fragmentation pattern, relative molecular weight, and cross-linking density
XPS[d]	X-rays (1.2–1.5 keV)	Photoelectrons	3–5 nm	Elemental composition, chemical environment, molecular orientation
RBS[e]	He ions (2 MeV)	Backscattered He ions	20 to 10^4 nm	Elemental composition, depth profiles
SSIMS/TOFSIMS[f]	Ions (3–5 keV)	Sputtered ions	<0.5 nm	Fragmentation pattern, relative molecular weight, and cross-linking density
AFM/STM[g]	A physical probe is raster-scanned across the sample to produce a 3-D image	Surface force	Top surface (depends on experimental material and condition)	Topographical information down to the Angstrom level, a line profile with height measurements; additional properties such as thermal conductivity, magnetic and electrical field strength, and sample compliance on the nanometer scale can be simultaneously obtained using specialty probes
Micro-TA[h]	Heat	Thermal conductivity, diffusivity	Top surface bulk (depends on experimental material and condition	Topography, local thermal analysis, dynamic mechanical analysis

[a] Fourier-transform infrared spectroscopy/photoacoustic Fourier-transform infrared spectroscopy.
[b] Scanning electron microscopy/energy dispersive x-ray spectroscopy.
[c] Fourier-transformed mass spectrometry.
[d] X-ray photoelectron spectroscopy.
[e] Rutherford backscattering spectroscopy.
[f] Static secondary ion mass spectroscopy, time of flight SIMS.
[g] Atomic force microscopy/scanning tunneling microscopy.
[h] Microthermal analysis.

extensively employed in identifying the chemistry of the plasma-engineered surface [63]. For example, the shift in the core-binding energies of the C1s levels of fluoropolymer systems is broad and well documented, therefore providing suitable means to establish the structural features of such coatings on substrate [65–70]. Time of flight secondary ion mass spectroscopy (TOF-SIMS) provides a unique approach to the analysis of plasma films—to analyze both the chemistry at the surface as well as the spatial distribution of the deposited layer [71]. Traditional techniques such as optical microscopy and SEM have been extensively used in characterizing film topography; however, in general, they show limited success in determining the nanoscale morphology and roughness.

Combining the powers of AFM and μTA™ to the characterization of plasma-engineered surface coatings provides the opportunity to both visualize and intimately study the complexity of the coated film beyond the capabilities of the aforementioned techniques. In addition, the complementary features of SEM, AFM, and μTA™ can provide direct quantification of a coating's surface morphology, characteristics, and performance over an unprecedented scale of resolution. Plasma depositions of various fluoro and silicone polymers onto glass and flexible elastomer surfaces by these three techniques are discussed in this section to illustrate the unique information that could be derived using a combination of these complementary methods.

D. Investigation of Plasma Polymer Film by Scanning Electron Microscopy

The ability to engineer Teflon-like films of low surface energy is very important for many everyday applications such as protective barriers, biocompatible layers, stain-resistant textiles, low dielectric coatings for electronics, low friction automotive films, and other industrial applications. Such highly hydrophobic and frictionless films can be deposited from mixtures of CF_4/acetylene and other fluorine-containing monomers by plasma polymerization. Plasma polymerization of tetrafluoroethylene and hexafluoromethane has been successfully examined by Buzzard et al. [72] and Masuoka and Yasuda [73], respectively. Other fluorinated monomers, such as perfluoropyridine [74], isomeric difluoroethylenes, perfluorinated aromatics [75,76], alicyclics, heterocyclics [77], fluoronaphthalenes [78], and mixtures of fluoronaphthalene and hydrogen [79,80], have also been studied. Sandrin et al. [81] recently described the deposition of plasma polymers from octofluorocyclobutane, hexafluoropropylene, and trifluoroethylene and reported their potential use as interlayer dielectrics for next-generation multilayer integrated circuits (ICs) due to their very low permittivity and their compatibility with copper and copper processing. In addition, long-terminal perfluoroalkyl chains, renowned for their propensity to confer liquid repellency (oleophobicity as well as hydrophobicity), have recently been examined by Coulson et al. [82]. Their work demonstrated that the fluorinated alkenes of the type $C_nF_{2n+1}CH{=}CH_2$ (e.g., $C_{10}F_{21}CH{=}CH_2$), subjected to pulsed plasma deposition, underwent preferential activation, reacting at the polymerizable carbon/carbon double bond, while leaving the low surface energy perfluoroalkyl chains structurally intact.

The growth and the change in the surface morphology and the structure of a perfluoro(methylcyclohexane) plasma-polymerized coating, formed on a vulcanized ethylene propylene diene monomer (EPDM) substrate, are presented here using SEM. Flat sheets of EPDM were prepared by hot-pressing the gum elastomer with a vulcanizing agent under heat and pressure in a Teflon-coated mold. As an extremely clean surface is crucial for effective plasma film deposition, the EPDM surface has been precleaned in argon plasma (Ar pressure: 0.1 mbar, flow rate: 10 ml/min, 13.5 MHz RF on parallel electrode) using a capacitively coupled radio-frequency plasma reactor.

Pretreatment with argon plasma (Fig. 17) clearly demonstrates that the scrubbing process not only removed and volatilized loosely attached surface residues (poorly adhered fragments and contaminants), scouring down to the base polymer, but also modified the surface topography. The image reveals a microroughened stable surface, suitable for strong interaction. It must also be pointed out that the noble feed gases tend to initiate cleavage without grafting to the surface. Treatment times are typically short to facilitate an effective, useful surface and hence reduce any unwanted species or features that may be present. The etching parameters are controlled by plasma pressure, plasma power, and the electrical bias used.

The sequence of SEM micrographs, displayed in Fig. 18, clearly indicates the overall surface morphology of the plasma film. The deposition of a perfluoro(methylcyclohexane) plasma polymer coating is kinetically fast, producing a thin film that covers the elastomer surface with an exposure time of only 20 sec (Fig. 18A). The appearance of a rough topography indicates that uneven deposition is occurring at the early stages. Active surface regions (e.g., cracks, defects, or pores) present on the surface may provide possible sites for the formation of polymer islands on the surface of the film. Close examination of the plasma polymer film under higher magnification (inset, Fig. 18A) reveals the dendritic nature of the film. Complete coverage of the base polymer and the formation of a thin, uniform coating occur after an exposure time of ~ 1 min. At closer observation, ripple-like patterns are seen, displaying a periodic nature on a nanometer scale. After 2 min of treatment (Fig. 18B), relatively smooth coating with characteristic features on the surface is observed, demonstrating that the cyclic structure of perfluorocarbons is very effective in enhancing plasma polymerization. It has been reported [64] that plasma polymerization of perfluorocarbons is generally slower than that of hydrocarbons and the presence of multiple bond(s) or a cyclic structure within a monomer is necessary in order to achieve high-enough polymerization. Continuation of plasma exposure to the substrate creates additional layers of polymerized coating above the previously formed film. With increase in deposition time, patch-wise

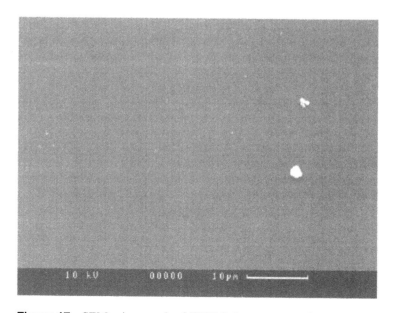

Figure 17   SEM micrograph of EPDM elastomer sample treated with argon plasma.

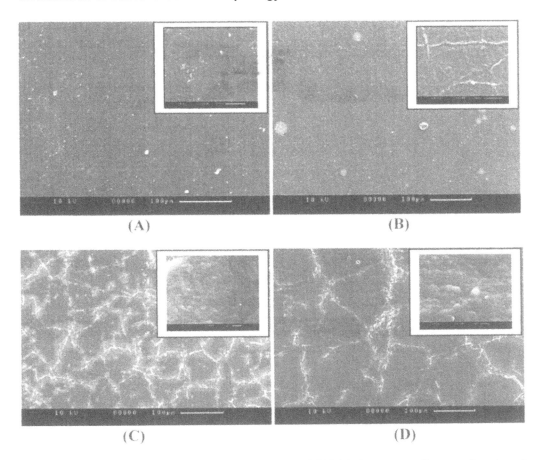

Figure 18 SEM micrograph of plasma polymer-coated EPDM elastomer surface as a function of perfluoro(methylcyclohexane) plasma reaction time at 60-W power; (A) 20 sec, (B) 2 min, (C) 8 min, and (D) 16 min. Detailed morphology in higher resolution is shown as inset.

deposition continues with finally coalescence of the film front to make a continuous film of overlapped patch-wise deposition of homogeneous distribution. At 2 min of treatment time, a layered pattern, as well as the presence of a film front, is clearly visible on the substrate. At 8 min (Fig. 18C), the development of the layered morphology is well advanced, having progressed to a densely packed polymer structure with granular or nodular-like features. Further exposure up to 16 min (Fig. 18D), the plasma polymer surface morphology transforms from an intricate sausage-like morphology to a rougher granular/nodular morphology.

In summary, the series of SEM micrographs demonstrate that the deposition of thin film coatings on an EPDM substrate is a kinetically fast process. Separate evaluation on the plasma polymer film by wetting studies confirmed large increase in the contact angle at treatment times as low as 20 sec. Longer times see the growth of this coating and the formation of a multilayered composite film. However, this does not significantly alter the contact angle, as complete coverage has already occurred at very short exposure times. This has been confirmed by Denes et al. [83] who stated that a fluoropolymer coating is stable and does not change its hydrophobic nature, even after prolonged exposure.

E. Investigation of Plasma Polymer Film by Microthermal Analysis

While SEM allows visualization of overall film surface morphology, thickness, and growth behavior, very limited information is obtained about the nature and the intrinsic properties of the film. μTA^{TM} overcomes the limitation of conventional microscopic methods. It allows not only imaging the surface or bulk morphology in terms of thermal conductivity/ diffusivity contrast, but also any point on the imaged surface can be characterized for its calorimetric or mechanical properties. Such unique ability of μTA^{TM} is applied in this study of plasma polymer films. To our knowledge, this is the first μTA^{TM} study on the polymer film prepared by plasma polymerization.

Investigation of both the topography and the relative conductivity images (100×100 μm scan) of the untreated vulcanized EPDM substrate by μTA^{TM} is displayed in Fig. 19. Analysis of the topography image clearly illustrates the smooth, featureless nature of the precleaned surface (Fig. 19A) with few hills and valleys (mold marks). The thermal conductivity image (Fig. 19B) reconfirms the overall homogeneity of the substrate. The observed differences on few places may be accounted for topographic reflections and not the chemistry at the surface. Microthermal analysis on the perfluoro(methylcyclohexane) plasma polymer film (same film as discussed above with 6-min plasma exposure time) is displayed in Fig. 20. As observed in SEM micrographs (Fig. 18), large differences in topography between the untreated and plasma-treated surfaces can be seen. While an analysis of topography confirmed the presence of deposited film (Fig. 20A) of granular-like structure on the surface, examination of the conductivity image exposed the uniform profile with some topographic contribution (with topographic reflection) of a homogeneous film (Fig. 20B).

The evolution of polymer film during plasma polymerization, accurately described by the series of SEM micrographs (Fig. 18), has also been studied by μTA^{TM}. Experiments show that images of relative conductivity reveal a uniformity of compositional distribution, irrespective of plasma polymerization time; only the roughness of the surface increases with exposure time. Relative surface atomic concentrations resulting from various plasma

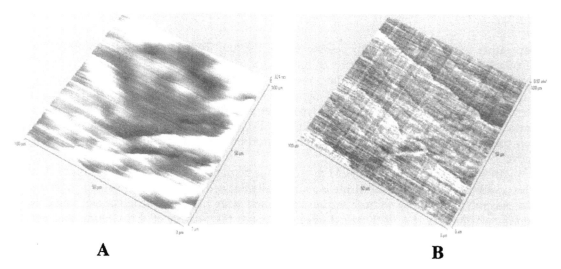

A   **B**

Figure 19 3-D representation of the μTA^{TM} topography of EPDM substrate (A) and its conductivity image (B).

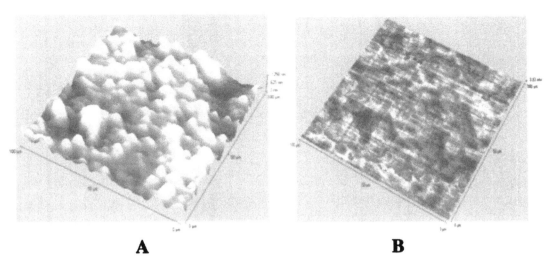

A   **B**

Figure 20 3-D representation of the μTA™ topography of the perfluoro(methylcyclohexane) plasma polymer onto the EPDM surface at power 60 W for 6 min (A) and its conductivity image (B).

exposure times, calculated using, FTIR, XPS, and TOF-SIMS, also confirm that treatment times as low as 30 sec and higher result in almost identical surface functionality and atomic levels. The high fluorine content of all plasma-treated specimens detected by these techniques identifies the presence of an effective coating, confirming that a consistent reaction mechanism controls the deposition process.

While μTA scans of relative thermal conductivity may not indicate the evolution and the degree of coating on a substrate's surface, an analysis of topography can. Figure 21 shows the topographic image of the plasma polymer film of perfluoro(methylcyclohexane) and tetramethyldisiloxane on a smooth glass surface. While the height image of the plasma polymer film on the rubber surface reveals relatively smooth edge granular morphology (Fig. 20), the plasma film on glass surface shows a sharp edge spiky morphology of the plasma coating (Fig. 21A). Tetramethyldisiloxane plasma polymer film appears to be rougher compared to the perfluoro(methylcyclohexane) film (Fig. 21B). For uniform coverage, the substrate/monomer interaction has to be higher than the monomer–monomer interaction, and, often, due to the complexity of their mutual affinity, the growth of the plasma polymer film occurs in a nodular fashion. From the figures above, it is clearly seen that both the nature of the substrate and the monomer used for plasma polymerization drastically modify the surface organization and deposition and play significant role on the nature and the quality of the plasma polymer film deposition.

F. Microthermal Probing of Local Thermal Properties of Thin Film

Local probing of surface thermal properties with a submicron resolution is possible using local microthermomechanical analysis (μTMA). The μTMA results of ultrathin plasma polymeric film deposited on a glass substrate are shown in Fig. 22. Experimental results are shown in six different locations to indicate the current capabilities and sensitivity range of μTA™ when probing a nanofilm.

All measurements were conducted from room temperature to 450°C. A heating rate of 10°C/sec was used for the visualization of the transition points on μTMA. A macro-DSC

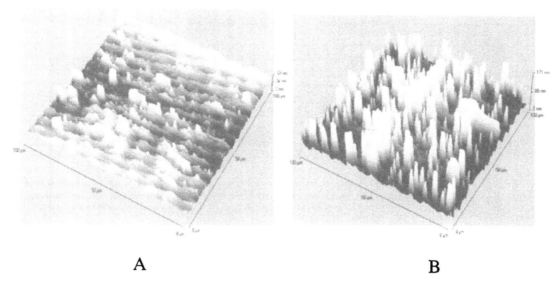

A   B

Figure 21 3-D representation of the µTA™ topography of the perfluoro(methylcyclohexane) plasma polymer onto the glass surface (A) and the topography of the tetramethyldisiloxane onto the glass surface (B).

and TMA measure the average response for the entire sample, and µTMA is able to measure a localized response from a few cubic microns of material. The small area measured, coupled with very fast heating rates, permits rapid investigation of changes in material properties across the sample surface. The glass transition (T_g) and melting points may be determined from the experimental data as an onset of a rapid probe deflection. The T_g of the per fluorinated plasma film appears to be ~140°C and its melting point occurs at ~407°C. As

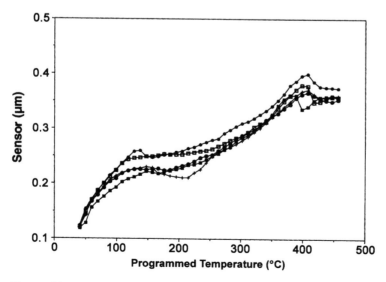

Figure 22 Microthermomechanical results for the plasma polymer film.

far as our knowledge is concerned, this is the first time μTMA has been successfully carried out on such thin plasma polymer film.

The microthermal responses in all the locations show the substrate to expand upon heating; however, the change in slope at glass transition and melting is still clearly envisaged. It is observed from the micro-TMA curves that they overlay very well; the small variation observed is ascribed to subtle inhomogeneity in the sample. It has been demonstrated that, under optimal conditions, transition temperatures measured with μTMA are close to that detected using DSC for the same material. TMA is not quantitative currently; however, using the relative value of x-axis provides quantitative information about the transition behaviors.

From the temperature of the transition, we can conclude that the film is made of homogeneous high polymer with Teflon-like characteristics. The high thermal stability (>340°C) of plasma polymer film of dodecafluorocyclohexane has been demonstrated by Denes et al. [83] using bulk thermal analysis (DTA-TG). Recently, Gorbunov et al. [84] reported the ability of micro-TA to precisely measure the glass transition temperature of ultrathin films (<25 nm) and has demonstrated that T_g drops significantly when the film thickness decreases below 200 nm with ultrathin polymer film deposited on solid substrate with weak film–substrate interactions. For very thin films (~25 nm), T_g of PS decreased by 20°C from the bulk value.

G. Investigation of Morphology and Surface Roughness by Atomic Force Microscopy

While SEM provides the overall view of the morphology and μTA™ provides information at the submicron level, AFM is used to acquire detailed morphology, structured in nanometer level. AFM also provides information on surface roughness values at a level presently unmatched by the other techniques. In general, the roughness is directly correlated with wetting properties. This relationship is generally described by Wenzel's equation: $\cos \theta' = r \cos \theta$ [85], where θ is the contact angle on a smooth surface, θ' is the contact angle on the rough surface, and r is a roughness factor. The roughness factor is described as r = actual surface/geometric surface. The actual surface refers to a rough solid and the geometric surface refers to the same solid assuming it was smooth. Thus for a nonwetting surface, a rough solid should be more strongly water-repellent than a smooth solid. Only AFM can provide the quantitative data in such detail for the determination of surface roughness, height, diameter, and volume distribution of grains and features on the surface (depth of holes and cracks, etc.). AFM also allows the observation of the adhesive surface and the characterization of the adhesive behavior at a nanoscale level [86].

Figures 23A and 24A display the representative topographic image of 1-μm² area of a plasma polymer film of perfluoro(methylcyclohexane) and tetramethyldisiloxane on EPDM substrate. Tapping mode TM AFM™ (in air) image of plasma polymer film deposited on EPDM substrate reveals smooth corrugation pattern of nanometer period. The detail of the roughness analysis is shown in Figs. 23B and 24B. The heights of the corrugation are in the range of 3.5 and 4.8 nm, respectively. The width of the corrugation varies in the ~150–250 nm range. AFM analysis gave rms roughness value of 3.5 for perfluoro(methylcyclohexane) and 4.7 for the tetramethylsiloxane onto the EPDM substrate. Coulson et al. [82] observed an rms roughness of 11 nm for 1H,1H,-2H-perfluoro-1-dodecene pulsed plasma coating on glass substrate. Hong et al. [87] used AFM for the surface morphology of water-repellent coated layer on safety glass for automobile. Morphological surface characterization of plasma polymer films on transparent gas barrier coating on plastic [47] has been discussed by García-Ayuso et al. using AFM. They reported the relationship between the quality of

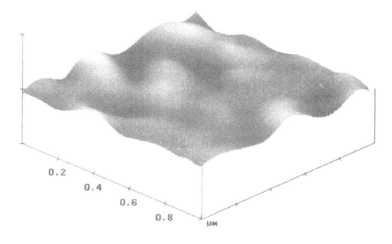

A

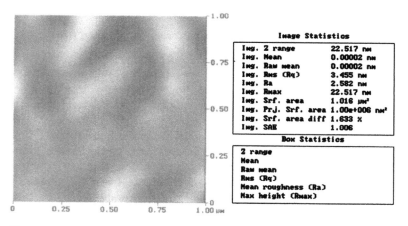

B

Figure 23 AFM topographic image acquired in tapping mode of the fluoro plasma polymer deposited onto the EPDM surface at power 60 W for 6 min (A) and surface roughness analysis of the coating (B).

deposition and surface morphology and the water vapor diffusivity of barrier coating on polyethylene terphthalate (PET). Films with roughness smaller than that of PET substrate show a low diffusivity; however, film rougher than PET substrate presents high permeability. Denes et al. [83] used AFM images to demonstrate the quality of dodecafluorocyclohexane-RF-plasma coated on Whatman-1 paper. Hartley et al. [88] recently described a novel surface-masking technique for the determination of plasma polymer coating film thickness of nanometer resolution using AFM. This method has also allowed accurate measurement of the kinetics of the deposition of plasma polymer films over a range of exposure times.

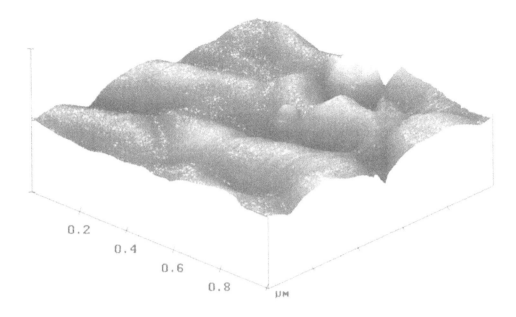

A

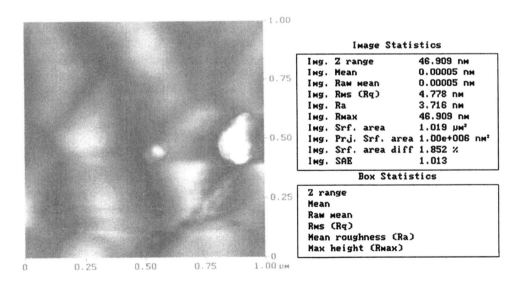

B

Figure 24 AFM topographic image acquired in tapping mode of the disiloxane plasma polymer coated on the EPDM surface at power 40 W for 4 min (A) and surface roughness analysis (B).

Hozumi and Takai [89] prepared ultrarepellent film using tetramethylsilane (TMS) and fluoroalkyl silane by microwave plasma-enhanced CVD. They produced film of root mean square (rms) roughness 11.3 to 60.8 nm, which had maximum contact angle of 160°, and they observed that water repellence increases with surface roughness. Youngblood and McCarthy [90] also observed the dependence of wettability of ultrahydrophobic PTFE plasma-polymerized film on the size scale, topology, and roughness characterized by AFM and SEM. García-Ayuso et al. [47] concluded that roughness measured using larger scanned area, such as optical interferometry, may lead to misleading data. Thus it clearly appears that a combination of these various microscopic analyses has the potential to fully visualize and characterize the film, which was never possible before without chemical etching.

V. FURTHER APPLICATIONS

A. Ionomers

Recent studies have commenced using micro-TA to examine the morphology of an ionomer blend. Blending of ionomers improves dispersion, and Fig. 25 shows the thermal conductivity image of an ionomer blended with clay that had been modified by cetyl trimethyl ammonium bromide $[CH_3(CH_2)_{15}N(CH_3)_3Br]$. Phases of differing thermal conductivities are observed consisting of particles with discrete circular shape and others where coalescence occurred resulting in elongated oval formations. Melting point measured for the modified ionomer was within the range of 85–92°C.

B. Biological Material

AFM has been used to analyze structures and surfaces of biological material such as cellulose fibril [91], crystalline cellulose [92], and starch granule [93]. Of late, Kutti et al. [94] reported the characterization of aged thermoplastic starches by applying both AFM and PFM. Friction images showed that the surface of the thermoplastic starches changed from a homogeneous surface to a heterogeneous surface with aging. Additionally, surface energy was found to increase with aging, leading to phase separation of aged starch.

C. Pulse Force Microscopy

Pulse force microscopy (PFM) is a new technique still in its infancy and more experimental research is required to gain a fuller understanding of this new technique. PFM was used to investigate the morphology of PS-PMMA blends. Increasing temperature phase separation of the polymer blend was seen. High-resolution images were obtained. The equivalent topography image showed no change indicating that the phase separation observed was not due to a topographical artifact. Therefore PS and PMMA phases were identified. As the temperature increases, the pull off force of PS shows changes, indicating that the retained phase is PS [95].

D. Packaging

Food and surgical industries packaging, commonly polyethylene terphthalate (PET), must protect the packed material from moisture, oxygen, and light [47] which has been achieved by the development of a barrier coating composed of aluminum oxide and/or silicon oxide.

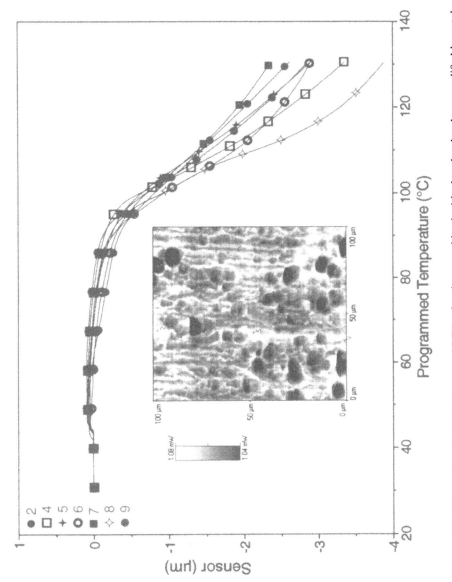

Figure 25 Thermal conductivity image and LTA of an ionomer blend with clay that has been modified by cetyl trimethyl ammonium bromide.

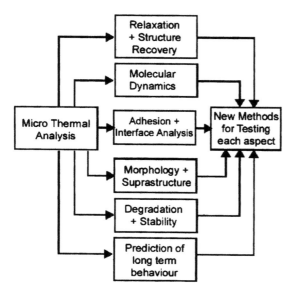

Figure 26 Micro-TA capability.

The relationship between barrier properties and coating morphology has previously been studied by optical interferometry [96]; however, AFM with its improved resolution was employed. Flat and homogeneous coatings had good barrier properties. These flat and homogeneous surfaces at high resolution were commonly grained. Coating of rough morphology comprised of poor barrier properties.

VI. CONCLUDING REMARKS

The ability of one system to be able to visualize surface morphology and then to characterize is phenomenal. μTA allows for a new world of thermal analysis and is currently being modified to obtain as much information of a material by combining micro-TA to PA-FTIR, mass spectrometry, TGA for evolved gas analysis, and DMA. In this chapter, we demonstrated the micro-TA and AFM capabilities (Fig. 26), studying polymeric materials over a wide range of applications. The high-resolution probing capability of AFM when combined with local property measurement and local compositional mapping of micro-TA takes materials science far beyond the current for recognizing the structure/morphology/property relationship of advanced materials in advanced application.

ACKNOWLEDGMENTS

The research work on this manuscript was supported by the Australian Research Council under collaborative research grant mechanisms. Financial supports of Bridgestone TG, Australia and Schefenacker Australia for research on plasma polymerization and surface analysis by micro-TA are highly appreciated. Thanks are also due to Dr. Kristen Bremmell for her help in carrying out a few AFM images presented here. Two of the authors (N.K. Dutta and N. Roy Choudhury) are thankful to Ankit K. Dutta for his patience, understanding, and cooperation during the preparation of the manuscript.

REFERENCES

1. Paul, D.R.; Bucknall, C.B. *Polymer Blends Volume 2: Performance*; John Wiley & Sons: New York, 2000.
2. Whangbo, M.H.; Magonov, S.N.; Bengel, H. Tip-sample force interactions and surface stiffness in scanning probe microscopy. Probe Microsc. 1997, *1* (1), 23–42.
3. Prater, C.B.; Maivald, P.G. 'Digital Instruments, Application note, Tapping Mode Imaging: Applications and Technology', USA, 2002.
4. Bar, G.; Thomann, Y.; Brandsch, R.; Cantow, H.J. Factors affecting the height and phase images in tapping mode atomic force microscopy. Study of phase-separated polymer blends of poly-(ethene-co-styrene) and poly(2,6-dimethyl-1,4-phenylene oxide). Langmuir 1997, *13*, 3807–3812.
5. Brandch, R.; Bar, G. On the factors affecting the contrast of height and phase images in tapping mode atomic force microscopy. Langmuir 1997, *13*, 6349–6353.
6. Zhong, Q.; Inniss, D.; Kjoller, K.; Elings, V.B. Fractured polymer/silica fiber surface studied by tapping mode atomic force microscopy. Surf. Sci. 1993, *290*, L688–L692.
7. Babcock, K.L.; Prater, C.B. Digital Instruments, Application note, Tapping Mode Imaging: Applications and Technology, USA, 2002.
8. Magonov, S.; Heaton, M.G. Atomic force microscopy, part 6: Recent developments in AFM of polymers. Am. Lab. (Shelton, Conn.) 1998, *30* (10), 9–16.
9. Magonov, S.; Godovsky, Y. Atomic force microscopy, part 8: visualization of granular nanostructure in crystalline polymers. Am. Lab. (Shelton, Conn.) 1999, *31* (8), 52–58.
10. Brice, J.C. Crystals for quartz resonators. Rev. Mod. Phys. 1985, *57*, 105–146.
11. Knoll, A.; Magerle, R.; Krausch, G. Tapping mode atomic force microscopy on polymers: Where is the true sample surface? Macromol. 2001, *34*, 4159–4165.
12. Feng, J.; Weng, L.T.; Chang, C.M.; Xie, J.; Li, L. Imaging of sub-surface nano-particles by tapping mode atomic force microscopy. Polymer 2001, *42*, 2259–2262.
13. Song, M.; Hourston, D.J.; Grandy, D.B.; Reading, M. An application of micro-thermal analysis to polymer blends. J. Appl. Polym. Sci. 2001, *81*, 2136–2141.
14. Price, D.M.; Reading, M.; Lever, T.J. Applications of micro-thermal analysis. J. Therm. Anal. Calorim. 1999, *56*, 673–679.
15. Häßler, R.; zur Mühlen, E. An introduction to µTA™ and its application to the study of interfaces. Thermochim. Acta 2000, *361* (1–2), 113–120.
16. Hammiche, A.; Pollock, H.M.; Reading, M.; Claybourn, M.; Turner, P.H.; Jewkes, K. Photo-thermal FT-IR spectroscopy: A step towards FT-IR microscopy at a resolution better than the diffraction limit. Appl. Spectrosc. 1999, *53* (7), 810–815.
17. Royall, P.G.; Kett, V.L.; Andrews, C.S.; Craig, D.Q.M. Identification of crystalline and amorphous regions in low molecular weight materials using microthermal analysis. J. Phys. Chem. B 2001, *105* (29), 7021–7026.
18. Reading, M.; Price, D.M.; Grandy, D.B.; Smith, R.M.; Bozec, L.; Conroy, M.; Hammiche, A.; Pollock, H.M. Micro-thermal analysis of polymers: Current capabilities and future prospects. Macromol. Symp. 2001, *167*, 45–62.
19. Moon, I.; Androsch, R.; Chen, W.; Wunderlich, B. The principles of micro-thermal analysis and its application to the study of macromol. J. Therm. Anal. Calorim. 2000, *59*, 187–203.
20. Hammiche, A.; Reading, M.; Pollock, H.M.; Song, M.; Hourston, D.J. Localized thermal analysis using a miniaturized resistive probe. Rev. Sci. Instrum. 1996, *67* (12), 4268–4274.
21. Royall, P.G.; Craig, D.Q.M.; Price, D.M.; Reading, M.; Lever, T.J. An investigation into the use of micro-thermal analysis for the solid state characterisation of an HPMC tablet formulation. Int. J. Pharm. 1999, *192*, 97–103.

22. Sanders, G.H.W.; Roberts, C.J.; Danesh, A.; Murray, A.J.; Price, D.M.; Davies, M.C.; Tendler, S.J.B.; Wilkins, M.J. Discrimination of polymorphic forms of a drug product by localized thermal analysis. J. Microsc. 2000, *198* (2), 77–81.
23. Xie, W.; Liu, J.; Lee, C.W.M.; Pan, W.P. The application of micro-thermal analysis technique in the characterization of polymer blend. Thermochim. Acta 2001, *367–368*, 135–142.
24. Paul, D.R.; Bucknall, C.B. *Polymer Blends Volume 1: Formulation*; John Wiley & Sons: New York, 2000.
25. Bucknall, C.B. *Toughened Plastics*; Applied Science Publishers: London, 1977.
26. Chang, M.C.O. Proceedings of 52nd Annu. Tech. Conf.-Soc. Plast. Eng., USA, 1994; 312–315 pp.
27. Chang, M.C.O. Surface defect formation in the injection molding of acrylonitrile–styrene–acrylate. Int. Polym. Process. 1996, *XI*, 76–81.
28. Royall, P.G.; Kett, V.L.; Andrews, C.S.; Craig, D.Q.M. Identification of crystalline and amorphous regions in low molecular weight materials using microthermal analysis. J. Phys. Chem. B 2001, *105* (29), 7021–7026.
29. Benson, C.M.; Burford, R.P. Morphology and properties of acrylate styrene acrylonitrile/polybutylene terphthalate blends. J. Mater. Sci. 1995, *30* (3), 573–582.
30. Benson, C.M.; Burford, R.P. Morphology and properties of ASA/PET blends. J. Mater. Sci. 1996, *31* (6), 1425–1436.
31. Wildes, G.; Keskkula, H.; Paul, D.R. Fracture characterisation of PC/ABS blends: Effect of reactive compatibilization, ABS type and rubber concentration. Polymer 1999, *40*, 7089–7107.
32. Eastmond, G.C.; Smith, E.G. An etch technique for morphological studies of multiphase polymers containing polycarbonates. Polymer 1973, *14*, 509–514.
33. Bucknall, C.B.; Drinkwater, I.C.; Keast, W.E. An etch method for microscopy of rubber-toughened plastics. Polymer 1973, *13*, 115–118.
34. Bryce, D.M. Plastic Injection Molding. Society of Manufacturing Engineers, USA, 1997; 331–342 pp.
35. Lacrampe, M.F. Defects in surface appearance of injection molded thermoplastic parts—A review of some problems of gloss distribution. J. Inj. Molding Technol. 2000, *4* (4), 167–176.
36. Hobbs, S.Y. The development of flow instabilities during the injection molding of multicomponent resins. Polym. Eng. Sci. 1996, *32* (11), 1489–1494.
37. Yokoi, H.; Nagami, S.; Kawasaki, A.; Murata, Y. Visual analyses of flow marks generation process using glass-inserted mold—Part 1. Micro-grooved flow marks. Proceedings of 52nd Annu. Tech. Conf. - Soc. Plast. Eng., USA, 1994; 368–372 pp.
38. Kurtz, S.J. *Advances in Rheology*; UNAM Press: Mexico, 1984; Vol. 3, 399.
39. Ramamurthy, A.V. Wall slip in viscous fluids and influence of materials of constructions. J. Rheol. 1986, *30* (2), 337–357.
40. Ramamurthy, A.V. Adv. Polym. Technol. 1986, *6*, 489.
41. Kalika, D.S.; Denn, M.M. Wall slip and extrudate distortion in linear low-density polyethylene. J. Rheol. 1987, *31* (8), 815–834.
42. Rudin, A.; Blacklock, J.E. Fluorocarbon elastomer aids polyolefin extrusion. Plast. Eng. 1986, *42* (3), 63–66.
43. Athey, R.J.; Thamm, R.C.; Souffie, R.D.; Chapman, G.R. Proceedings of 44th Annu. Tech. Conf.—Soc. Plast. Eng. 1986, *32*, 1149.
44. Beaufils, P.; Vergnes, B.; Agassant, J.F. Characterization of the sharkskin condition defect and its development with the flow conditions. Int. Polym. Process. 1989, *4* (2), 78–84.
45. Salamon, B.A.; Donald, R.J. Characterising and controlling secondary flows in injection molds. J. Inj. Molding Technol. 1997, *1*(1), 36–43.
46. Edwards, S.A.; Provatas, M.; Choudhury, N.R.; Matisons, J.G. Surface finishes in injection molding of polymeric materials and composites. Int. J. Mater. Prod. Technol. in press.
47. García-Ayuso, G.; Vázquez, L.; Martínez-Duart, J.M. Atomic force microscopy (AFM) morphological surface characterization of transparent gas barrier coatings on plastic films. Surf. Coat. Technol. 1996, *80*, 203–206.

48. Salamon, B.A.; Koppi, K.A.; Little, J. Halo surface defects on injection molded parts. Proceedings of 56th Annu. Tech. Conf.—Soc. Plast. Eng., USA, 1998; 515–519 pp.

49. Carprick, R.W.; Sasaki, D.Y.; Burns, A.R. Nanometer-scale structural, tribological, and optical properties of ultrathin poly(diacetylene) films. Polymer Preprints 2000, *41* (2), 1458–1459.

50. Provatas, M.; Edwards, S.A.; Choudhury, N. Roy. Surface finish of injection molded materials by micro-thermal analysis. Thermochim. Acta. in press.

51. Hamada, H.; Tsunasawa, H. Correlation between flow mark and internal structure of thin PC/ABS blend injection moldings. J. Appl. Polym. Sci. 1996, *60*, 353–362.

52. McEvoy, R.L.; Krause, S.; Wu, P. Surface characterization of ethylene-vinyl acetate (EVA) copolymers using XPS and AFM. Polymer 1998, *39* (21), 5223–5239.

53. Yoshii, M.; Kuramoto, H.; Kawana, T. The observation and origin of micro flow marks in the precision injection molding of polycarbonate. Polym. Eng. Sci. 1996, *36* (6), 819–826.

54. Kresge, E.N. Elastomeric blends. J. Appl. Polym. Sci. 1984, *39* (Chem Technol Rubber): 37–57.

55. Edwards, S.A.; Choudhury, N.; Provatas, M.; Matisons, J.G. Visualisation of surface and subsurface morphology: Effect of processing on a rubber-modified thermoplastic. J. Appl. Polym. Sci. submitted for publication.

56. Dong, L.; Greco, R.; Orsello, G. Polycarbonate/acrylonitrile-butadiene-styrene: 1. Complementary etching techniques for morphology observations. Polymer 1993, *34* (7), 1375–1382.

57. Rossnagel, S.M., Cuomo, J.J., Westwood, W.D., Eds.; *Handbook of Plasma Processing Technology, Fundamentals, Etching, Deposition, and Surface Interaction*; Noyes Publications: Westwood, 1990.

58. Feast, W.J., Munro, H.S., Eds.; *Polymer Surface and Interface*; Wiley & Sons: New York, 1989.

59. Garbassi, F.; Morra, M.; Occhiello, E. *Polymer Surfaces from Physics to Technology*; Wiley & Sons: New York, 1996.

60. Shul, R.J., Pearton, S.J., Eds.; *Handbook of Advanced Plasma Processing Techniques*; Springer: Berlin, 2000.

61. Inagaki, N. *Plasma Surface Modification and Plasma Polymerisation*; Technomic. Pub. Co. Inc.: Lancaster, 1996.

62. Pereira, M.R.; Fonseca, J.L.C. The kinetic of growth of plasma polymer films. Eur. Polym. J. 1999, *35*, 41–45.

63. Egitto, F.D.; Matienzo, L.J. Plasma modification of polymer surfaces for adhesion improvement. IBM J. Res. Develop. 1994, *38*, 423–439.

64. Yasuda, H.; Hsu, T.S. Some aspects of plasma polymerization of fluorine-containing organic compounds. J. Polym. Sci. Polym. Chem. Ed. 1978, *15*, 2411–2425.

65. Yasuda, H.; Hsu, T.S.; Brandt, E.S.; Reilley, C.N. Some aspects of plasma polymerization of fluorine-containing organic compounds. II. Comparison of ethylene and tetrafluoroethylene. J. Polym. Sci. Polym. Chem. Ed. 1978, *16*, 415–425.

66. Eufinger, S.; van Ooij, W.J.; Conners, K.D. DC-Plasma polymerization of hexamethyldisiloxane part II. Surface and interface characterization of film deposited on stainless-steel substrate. Surf. Interface Anal. 1996, *24*, 841–855.

67. Anderson, H.R. Jr.; Fowkers, F.M.; Hielscher, F.H. Electron donor–acceptor properties of thin polymer films on silicon. II. Tetrafluoroethylene polymerized by RF glow discharge techniques. J. Polym. Sci. Polym. Phys. Ed. 1976, *14*, 879–895.

68. Morosoff, N.; Yasuda, H. Plasma polymerization of tetrafluoroethylene. II. Capacitive radio frequency discharge. J. Appl. Polym. Sci. 1979, *23*, 3449.

69. Nakajima, K.; Bell, A.T.; Shen, M.; Millard, M.M. Plasma polymerization of tetrafluoroethylene. J. Appl. Polym. Sci. 1979, *23*, 2627–2637.

70. Dilks, A.; Kay, E. Plasma polymerisation of ethylene and the series of fluoroethylenes: plasma effluent mass spectrometry and ESCA studies. Macromolecules 1981, *14*, 855–862.

71. Tsai, Y.M.; Boerio, F.J.; van Ooij, W.J.; Kim, D.K.; Rao, T. Surface characterization of novel plasma-polymerized primes for rubber-to-metal bonding. Surf. Interface Anal. 1995, *23*, 261–275.

72. Buzzard, P.D.; Soong, D.S.; Bell, A.T. Plasma polymerization of tetrafluoroethylene in a field-free zone. J. Appl. Polym. Sci. 1982, *27*, 3965–3985.

73. Masuoka, T.; Yasuda, H. Plasma polymerization of hexafluoroethane. J. Polym. Sci. Polym. Chem. Ed. 1982, *20*, 2633–2642.

74. Clark, D.T.; Abu-Shbak, M.M. Plasma polymerization. IX. A systematic investigation of materials synthesized in inductively coupled plasma excited in perfluoropyridine. J. Polym. Sci. Polym. Chem. Ed. 1983, *21*, 2907–2919.

75. Clark, D.T.; Abrahman, M.Z. Plasma polymerization. V. A systematic investigation of the inductively coupled RF plasma polymerization of the isomeric tetrafluorobenzenes. J. Polym. Sci., Polym. Chem. Ed. 1981, *19*, 2689–2703.

76. Clark, D.T.; Abrahman, M.Z. Plasma polymerization. IV. A systematic investigation of the inductively coupled RF plasma polymerisation of pentafluorobenzene. J. Polym. Sci. Polym. Chem. Ed. 1982, *19*, 2129–2149. Plasma polymerization. VII. An ESCA investigation of the RF plasma polymerization of perfluorobenzene and perfluorobenzene/hydrogen mixtures' 1982; *20*, 1717–1728.

77. Clark, D.T.; Shuttleworth, D. Plasma polymerization. II. An ESCA investigation of polymers synthesized by excitation of inductively coupled RF plasma in perfluorobenzene and perfluorocyclohexane. J. Polym. Sci. Polym. Chem. Ed. 1980, *18*, 27–46. Plasma polymerization. III. An ESCA investigation of polymers synthesized by excitation of inductively coupled RF plasmas in perfluorocyclohexa-1,3-and 1,4-dienes, and in perfluorocyclohexene. J. Polym. Sci. Polym. Chem. Ed. 1980, *18*, 407–425.

78. Munro, H.S.; Till, C. An ESCA and plasma emission study of the plasma polymerization of 1-fluoronaphthalene and octafluoronaphthalene. J. Polym. Sci. Polym. Chem. Ed. 1986, *24*, 279–286.

79. Munro, H.S.; Till, C. An ESCA investigation of the inductively coupled RF plasma polymerization of fluoronaphthalene and fluoronaphthalene/hydrogen mixtures. J. Polym. Sci. Polym. Chem. Ed. 1985, *23*, 1621–1629.

80. Munro, H.S.; Till, C. ESCA and optical emission study of the inductively coupled RF plasma copolymerization of naphthalene and octafluoronaphthalene mixtures. J. Polym. Sci. A Polym. Chem. 1987, *25*, 1065–1071.

81. Sandrin, L.; Silverstein, M.S.; Sacher, E. Fluorine incorporation in plasma-polymerized octofluorocyclobutane, hexafluoropropylene and trifluoroethyiene. Polymer 2001, *42* (8), 3761–3769.

82. Coulson, S.R.; Woodward, I.S.; Badyal, J.P.S.; Brewer, S.A.; Willis, C. Plasma chemical functionalization of solid surfaces with low surface energy perfluorocarbon chains. Langmuir 2000, *16*, 6287–6293.

83. Denes, F.; Hua, Z.Q.; Simonsick, W.J.; Aaserud, D.J. Synthesis and characterization of Teflon-like macromolecular structures from dodecafluorocyclohexane and octadecafluorodecalin under RF-cold-plasma conditions. J. Appl. Polym. Sci. 1999, *71*, 1627–1639.

84. Gorbunov, V.; Fuchigami, N.; Luzinov, I.; Tsukruk, V.V. Microthermal analysis of ultrathin polymeric films with scanning thermal microscopy. Polymer Preprints 2000, *41* (2), 1493–1494.

85. Wenzel, R.N. Resistance of solid surfaces to wetting by water. Ind. Eng. Chem. 1936, *28*, 988–994.

86. Paiva, A.; Sheller, N.; Foster, M.D.; Crosby, A.J.; Shull, K.R. Study of the surface adhesion of pressure-sensitive adhesives by atomic force microscopy and spherical indenter test. Macromolecules 2000, *33*, 1878–1881.

87. Hong, B.S.; Han, J.H.; Kim, S.T.; Cho, Y.J.; Park, M.S.; Dolikhanyan, T.; Sung, C. Durable water-repellent glass for automobiles. Thin Solid Films 1999, *351*, 274–278.

88. Hartley, P.G.; Thissen, H.; Vaithianathan, T.; Griesser, H. Surface masking techniques for the determination of plasma polymer film thickness by AFM. Plasmas Polym. 2000, *5* (1), 47–60.

89. Hozumi, A.; Takai, O. Preparation of ultra water-repellent films by microwave plasma-enhanced CVD. Thin Solid Films 1997, *303*, 222–225.

90. Youngblood, J.P.; McCarthy, T.J. Ultrahydrophobic polymer surfaces prepared by simulta-

neous ablation of polypropylene and sputtering of poly(tetrafluoroethylene) using radio frequency plasma. Macromolecules 1999, *32*, 6800–6806.

91. Hanley, S.J.; Giasson, J.; Revol, J.F.; Gray, D.G. Atomic force microscopy of cellulose microfibrils: comparison of transmission electron microscopy. Polymer 1992, *33*, 4639–4642.
92. Kutti, L.; Peltonen, J.; Pere, J.; Teleman, O. Identification and surface structure of crystalline cellulose studied by atomic force microscopy. J. Microsc. 1995, *178*, 1–6.
93. Thomson, N.H.; Miles, M.J.; Ring, S.G.; Shewry, P.R.; Tatham, A.S. Real-time imaging of enzymatic degradation of starch granules by atomic force microscopy. J. Vac. Sci. Technol. 1994, *B12*, 1565–1568.
94. Kutti, L.; Peltonen, J.; Myllärinen, P.; Teleman, O.; Forssell, P. AFM in studies of thermoplastic starches during ageing. Carbohydr. Polym. 1998, *37*, 7–12.
95. Grandy, D.B.; Hourston, D.J.; Price, D.M.; Reading, M.; Silva, G.G.; Song, M.; Sykes, P.A. Microthermal characterization of segmented polyurethane elastomers and a polystyrene–poly(methyl methacrylate) polymer blend using variable-temperature pulsed force mode atomic force microscopy. Macromolecules 2000, *33*, 9348–9359.
96. Gavitt, I.F. Vacuum coating application for snack food packaging. Society of Vacuum Coaters 36th Ann. Tech. Conf. Proc., USA. 1993; 254–259 pp.

5

Macromechanics and Micromechanics of Ceramics

Besim Ben-Nissan
University of Technology, Sydney, Sydney, Australia

Giuseppe Pezzotti
Kyoto Institute of Technology, Kyoto, Japan

Wolfgang H. Müller
Technische Universität, Berlin, Berlin, Germany

I. SYNOPSIS

The interest in using ceramic materials in advanced engineering applications has grown considerably during the last two decades. This is not surprising because ceramics offer many of the physical properties required for the realization of high-tech applications. Examples of such properties include high-temperature endurance, extreme wear resistance, nontoxicity, and biocompatibility. However, the well-known brittleness and low fracture resistance of ceramic materials can become a serious shortcoming, unfortunately, not only from an objective point of view but also for subjective reasons that prevent many potential users from choosing ceramics before metals for a specific application. The latter is very often due to insufficient knowledge of the appropriate use and design of ceramics. It is the authors' objectives to ease these unnecessary fears and to help move toward unbiased and reliable engineering with brittle substances.

It is true that unlike metals, ceramics do *not* yield plastically under sudden load and impact. They are usually highly susceptible to scratches and flaws arising during production or use. Consequently, special attention must be paid by the development engineer to avoid high-peak tensile stresses in the structure. If this proves impossible, use must be made of a hybrid design instead. Obviously, only visually "flawless" specimens should be used. However, the term "flawless" needs to be quantified because a macroscopically integer specimen will still contain microscopical flaws, rendering it insufficient for a certain application. Moreover, due to microscopical *flaw size variation*, the strength within a batch of ceramic specimens can vary considerably. This must be taken into account especially during large-scale production. Another problem is the "fatigue" of ceramics. It turns out that the performance and the reliability behavior of ceramics are time-dependent in the following sense. A ceramic part can fail over time due to stress–corrosion cracking (i.e., subcritical growth of initially microscopical cracks inside the stressed ceramic material resulting from water vapor or other environmental influences). These flaws will grow over time and eventually turn "critical" even if the tensile stresses are always kept below an initially critical level.

Our aim is to provide the design engineer with a basis on which to guarantee the reliability of a fully ceramic structure, or that of a part made of a brittle material within a larger structure. To this end, suitable strength and toughness data are required. In what follows, we shall explain the procedures on how these can be determined and how they are used in practice. Moreover, it is imperative to realize that these quantities vary statistically within a particular batch and how this variation can be quantified experimentally. In addition, the state of stress in the component during use needs be analyzed thoroughly, either analytically or numerically, by performing a suitable stress analysis. Summarizing, we repeat the three key ingredients for a successful design with brittle matter as follows:

Suitable material data acquisition
Stress analysis
Statistical approach to strength variation and lifetime prediction.

These are combined in a so-called strength–probability–time (SPT) diagram, which can then be used to guarantee longevity and reliability in ceramic materials for a wide range of applications.

II. INTRODUCTION: BASIC PROPERTIES OF METALS COMPARED TO CERAMICS

Metals are intrinsically "soft." When a metal yields in a tensile test, dislocations move through its structure. Each different test, in its own way, measures the difficulty of moving dislocations in the material. When atoms are brought together to form a metal, each atom loses one or more electrons to the gas of free electrons, which move freely around the ion cores. The bridging energy comes from the general electrostatic interaction between the positive ions and the negative electron gas, and the bonds are not localized. If a dislocation passes through the structure, it displaces the atoms above its slip plane and those that lie below. However, this has only a small effect on the electron–ion bonding. Because of this, there is a slight drag on the moving dislocations.

In contrast, most ceramics are intrinsically "hard." Ionic or covalent bonds present an enormous lattice resistance to the motion of dislocations. The covalent bond is localized and the electrons are concentrated in the region between the bonded ions. When a dislocation moves through the structure, it must break and reform these bonds as it moves. Most ionic ceramics are hard and the ionic bond is electrostatic. If the crystal is sheared on the 45° plane, then alike ions remain separated. This sort of shear is relatively easily accomplished (the lattice resistance opposing is small). However, the horizontal shear does not carry ions over ions and the electrostatic repulsion between like ions opposes this strong attachment.

Lattice resistance is high in a polycrystalline material and, consequently, the hardness of a polycrystalline ionic ceramic is usually high. Generally, ceramics at room temperature have a very large lattice resistance. The stress required for dislocation movement is a large fraction of Young's modulus. To illustrate this point, consider Fig. 1, wherein the force between two atoms is depicted schematically. Obviously, at very short distances, the atoms will repel each other (negative force), whereas at very large distances, they will no longer

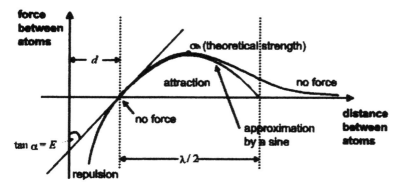

Figure 1 Theoretical fracture strength.

"feel" each other. In between, there is a "maximum of attraction." We approximate the force (stress)/distance relationship by a sine:

$$\sigma = \sigma_{th} \sin\left(\frac{2\pi x}{\lambda}\right) \approx \sigma_{th}\left(\frac{2\pi x}{\lambda}\right) \tag{1}$$

On the other hand, Hooke's law requires that:

$$\sigma = E\frac{x}{d} \tag{2}$$

Combining both equations yields for the theoretical strength:

$$\sigma_{th} = \frac{E\lambda}{2\pi d} \tag{3}$$

In order to evaluate this equation, we need to eliminate λ and link it to some physically observable parameter. To this end, we assume that the work necessary to separate the atoms is completely converted into surface energy. This work is represented by the area under the sine and therefore by integration:

$$\int_{x=0}^{\lambda/2} \sigma dx = -\sigma_{th}\frac{\lambda}{2\pi}\left(\cos\left[\frac{2\pi}{\lambda}\frac{\lambda}{2}\right] - \cos 0\right) = \sigma_{th}\frac{\lambda}{\pi} = 2\gamma_0 \tag{4}$$

where γ_0 denotes the specific surface energy (in J/m^3) and the factor 2 stems from the fact that *two* surfaces are created when the atoms are separated. The elimination of λ from the last two equations yields:

$$\sigma_{th} = \frac{E}{2\pi d}\frac{2\gamma_0\pi}{\sigma_{th}} \quad \Rightarrow \quad \sigma_{th} = \sqrt{\frac{E\gamma_0}{d}} \tag{5}$$

For a brittle material such as glass, we find that:

$$\left.\begin{array}{l} E = 70 \times 10^9 \text{ Pa} \\[2mm] d = 2 \times 10^{-10} \text{ m} \\[2mm] \gamma_0 = 1 \text{ J/m}^2 \end{array}\right\} \rightarrow \sigma_{th} = \sqrt{\frac{7}{2}\frac{10^9 \times 1}{10^{-10}}} \text{ Pa} \approx 6000 \text{ MPa} \tag{6}$$

Consequently, the "lattice resistance" is between $E/3$ and $E/10$ compared to $E/10^3$, or less for soft metals such as copper. This gives rise to yield strengths of ceramics on the order of 5 GPa. This enormous hardness is exploited in high-performance engineering usage, for abrasives, and in wear resistance-required applications.

III. FRACTURE IN CERAMICS VS. METALS

In ceramics, crack growth involves truly brittle fracture (sequential bond rupture at the tip of an atomically sharp crack). In metals, ductility accompanies fracture prior to total failure. It is, therefore, perfectly adequate to make use of "linear elastic fracture mechanics" (LEFM) in ceramics. The penalty that must be paid for choosing a material with a large lattice resistance is brittleness. Then the so-called "fracture toughness K_{Ic}" is low. At the tip of the crack, where stress is intensified, the lattice resistance makes slip very difficult. It is the crack tip plasticity which gives metals their high toughness (energy is absorbed in the plastic zone, making the propagation of the crack much more difficult). Although some plasticity can occur at the tip of a crack in a ceramic, too, it is very limited, the energy absorbed is small, and the fracture toughness is low. The result is that ceramics have values of K_{Ic} that are roughly one-fiftieth of those of good ductile materials. In addition, they almost always contain cracks and flaws, at least at a microscopical level.

Although brittleness is not an inherent barrier to the mechanical reliability of ceramic materials, it typically complicates the problem because of the resultant sensitivity to stress concentrators. This leads to statistical variations in strength due to the statistical distributions of critical stress concentrators that cause failure at stress levels that are often well below the desired design stresses. The stress concentrators typically are between 1 and 100 μm in comparison to metals that might have flaw sizes of several millimeters or larger.

The cracks originate in several ways, but most commonly by the following:

(i)   Manufacturing and production methods involving consolidation and high-temperature sintering leave small holes in sintered products. Generally, these will contain angular pores on the scale of the powder (or grain) size (Fig. 2). These are due to viscoelastic stresses during the consolidation of powders and differential shrinkage rates around agglomerates during sintering.
(ii) Thermal stresses caused by cooling or thermal cycling can generate small cracks.
(iii) Machining, finishing, or the environment used can create cracks on the surface.
(iv) Cracks could appear during the loading of a brittle solid, nucleated by the elastic anisotropy of the grains, or by slip on a single slip system.

At ambient temperatures, *failure almost invariably occurs from pre-existing flaws*. The crack propagation characteristics of the material and the statistics and mechanics of such flaws constitute the essential physical properties that dictate mechanical strength and reliability.

The sensitivity of strength to defects can be expressed by the strength relation for brittle cracks [1]:

$$\sigma = A K_{Ic} c^{-1/2} \tag{7}$$

where K_{Ic} is the fracture toughness, c is the dimension of the largest crack, and A is a dimensionless constant dependent on crack geometry. Once determined, the toughness

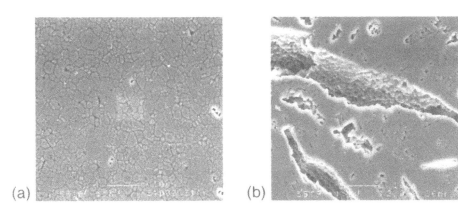

Figure 2 High-magnification scanning electron microscopy (SEM) micrographs showing (a) dense and (b) porous (12.4% porosity) grain structures. Specimens thermally etched at 1350°C for 1 hr. (From Ref. 88.)

provides a measure of the material's potential resistance to machining damage [2], impact damage [3], and thermal shock damage [4].

Ceramics with a relatively high toughness are intrinsically more reliable by virtue of their relative immunity to strength-degrading damage. The main factors that influence the mechanical properties of ceramics and their usage in engineering applications are:

> Their molecular bonding and structure: covalent, ionic, or mixed ionic–covalent bonding
> Defects or stress concentrators
> Mechanical and thermal stresses during fabrication
> Statistical variations of strength
> Working environment.

There are various design issues that should be considered for the application of engineering ceramics:

Engineering design aspects:

> (i) Adequate initial mechanical capability.
> (ii) Suitable service life.

Long-term service life and reliability:

> (i) The material might "change" during its service life.
> (ii) The flaws that are originally present may grow as a result of interaction between service stress and environment, resulting in failure before the component reaches its desired service life.

Common sources of stress concentrators in ceramics as stated earlier depend on many factors: ceramic consolidation and production methods (e.g., cold pressing, hot pressing, hot isostatic pressing, pressureless sintering, reaction sintering, slip casting, extrusion, tape casting etc.), relative heating and cooling rates, machining and surface finishing methods, mechanical loading, and working environment.

The improvement of the reliability of ceramics is one of the most important issues of current advanced ceramics and can be achieved by three approaches:

(i)   Identify the sources of the strength-degrading flaws and then develop new processing methods to eliminate the defects and thereby produce ceramics with very uniform and very high strengths, so that the design can be kept below the failure stress [5].

(ii)   Design ceramic microstructures with improved resistance to fracture, and hence some tolerance to defects (e.g., transformation-toughened zirconia and reinforced composites) [2,6].

(iii)   Learn to live with the brittleness (low KIc) of ceramics and develop a fundamental understanding of the micromechanics of failure, so that reliability can be assured by developing statistical and/or experimental analyses, or methods for nondestructive evaluation of specific accurate design parameters such as strength or lifetime [7].

Currently produced advanced wear-resistant ceramics have become a reality in a wide range of applications and are expected to grow rapidly through the end of the decade. The overall U.S. market for ceramic wear-resistant products totaled US$144 million in 1991. The U.S. abrasive products demand was US$3750 million by 1994 and US$4810 million by 1999, and is expected to grow to US$5810 million by 2004. The use of engineering ceramics in specific wear-related areas is increasing by an annual rate of approximately 4–5%. As a consequence of the potential economics associated with their use, a great deal of research has been funded in an effort to develop sound, widely applicable models on which a comprehensive design methodology can be based.

According to past production success and failure history, the design approach in ceramics can be divided into several categories [8].

(i) Empirical (trial and error): The empirical design concept is a trial-and-error approach, which is based on interactive fabrication and testing methods. It is used where the mechanical loads are minimal and/or when the available data for the material are too limited for the more analytical approaches.

(ii) Deterministic (safety factor): Deterministic design is a standard "safety factor" approach. The maximum stress in a compound is first mathematically calculated and a material is then selected, which has a strength with a reasonable margin of safety. The latter is usually determined from prior experience. Deterministic design does not account for the effects of flaw size and distribution and variation in strength with volume and area stressed. When a deterministic design approach is used with highly stressed ceramic components, large safety factors are required to assure a reasonable low risk of failure.

Proof testing, which is routinely used for both metal and ceramic component design and production, involves loading a component to a stress higher than the known service stress such that all components having flaws more severe than some critical value will fail during the test. Proof testing can eliminate much of the uncertainty associated with a reliance on the statistics of fracture.

(iii) Probabilistic approach: The first probabilistic approach used to account for the scatter in fracture strength of brittle materials is due to Weibull [9] and based on the weakest link theory. It assumed a unique strength cumulative distribution for the uniaxial fracture data obtained from simple specimen tests. The weakest link theory is analogous to pulling a chain where catastrophical failure occurs when the weakest link in the chain is broken. The Weibull approach is very popular because of its ease of application. The approach was introduced by Weibull in 1939 and based on an earlier weakest link model attributed to Midgley and Pierce in 1926, who originally proposed it while testing yarn.

The aforementioned empirical and deterministic design approaches may be adequate for most ceramic applications, but are limited in cases where high stresses and complex stress distributions are present. Thus, other more precise methods should be used and will be explained in more detail in the following sections:

(iv) Fracture mechanics concepts
 (v) Analytical and finite element stress/strain analysis
(vi) Combinations of the above.

In addition, an experimental method for microscopical stress determination will be described, which is based on either Raman or fluorescence microprobe spectroscopy (piezospectroscopy method) This stress analysis method is relatively new and yet in embryo, but it may, in the near future, become a very powerful tool for the experimental stress analysis of ceramic materials.

IV. ESSENTIALS OF LINEAR ELASTIC FRACTURE MECHANICS

It is perfectly legitimate to apply LEFM in order to characterize the fracture mechanics behavior of monolithic and (some) composite ceramics provided temperatures stay below 650°C. At higher temperatures, nonlinear effects will become more and more relevant. These can be due, for example, to the presence of glassy phases from sintering additives, which, if liquefied, exhibit viscoelastic material behavior. In what follows, we will refrain from such complications and present a summary of the basic concepts of LEFM, such as Griffith's fracture criterion for brittle materials and Irwin's concept of stress intensity factors (SIFs). To assist during daily practice, the standards for determining LEFM data for glasses and ceramics will also be summarized in this section.

A. Griffith's Fracture Criterion

Griffith in 1920 was the first to provide a convincing mathematical model that allowed quantifying the onset of instability of a sharp crack in a brittle material [10]. To model a crack, Griffith considered a sequence of elliptic holes within a plane made of a brittle material. In the limit of a vanishing minor axis $b\rightarrow 0$, the ellipse will turn into a sharp slit, the so-called Griffith crack. In the simplest case, the plane is subjected to a tensile stress σ at infinity, normal to the crack flanks (Fig. 3). In fracture mechanics terms, this type of loading is also known as a mode I loading condition (see below).

Griffith was the first to balance energy relevant to the presence of a straight crack in a brittle material. However, Griffith first erred by a factor of 2 as was already mentioned, however, without too much of an explanation. Several decades later, Sih and Liebowitz [11] provided a detailed proof of Griffith's original results according to which the difference in energy contained in the undamaged and damaged planes is given by:

$$U + S = -\frac{\pi\sigma^2 a^2}{E'} + 4\,a\gamma \tag{8}$$

The presence of an elliptic hole will be associated with a change of the state of stress in the body. The corresponding difference in elastic energy is accounted for by the first term U in

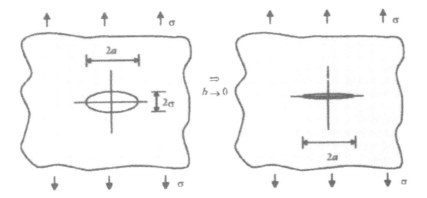

Figure 3   The transition from an elliptic hole to a Griffith crack.

this equation. Moreover, free surfaces, such as the two flanks of length $2a$, are associated with surface energy. This contribution results in second term S because γ denotes the specific surface energy and two crack surfaces must be taken into account, which explains the factor "4." Finally, E' is related to the elastic constants of the material. In the case of plane stress, it is equal to Young's modulus E, whereas for plane strain, it is given by $E/(1-v^2)$, where v denotes Poisson's ratio.

The interaction between both energies U and S is shown in Fig. 4 as a function of crack length parameter a. The change in surface energy S is positive and *increases linearly* with crack length parameter a. On the other hand, the stressed body relaxes and elastic energy is released due to the presence of the crack. Consequently, U is released, and what is more, the energy release is *quadratic* in a.

Consequently, we may argue that a small crack "consumes" more surface energy in order to grow by the amount of Δa, which can be gained from elastic relaxation. In other words, such a crack is stable and will close instead of growing any further. On the other hand, the elastic energy release during the extension of a large crack is much greater than that required for the creation of additional surface. Such cracks are unstable and will

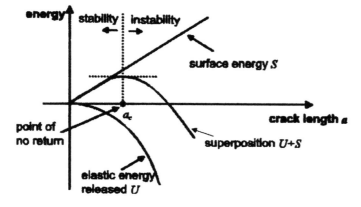

Figure 4   Graphical representation of Griffith's fracture criterion.

inevitably grow catastrophically. Obviously, there must be a critical crack length $2a_c$, beyond which catastrophical fracture must occur. In terms of Fig. 4, this crack length is characterized by the maximum of the net energy $U + S$. It can be computed by putting the first derivative of Eq. (8) with respect to the crack length parameter a equal to zero:

$$\frac{d(U + S)}{da} = 0 \Rightarrow -\frac{2\pi\sigma^2 a}{E'} + 4\gamma = 0 \tag{9}$$

leading to:

$$a_c = \frac{2\gamma E'}{\pi\sigma^2} \tag{10}$$

However, we may also look at the situation in an alternative way as shown in Fig. 4. Instead of keeping the external load fixed and changing the crack length, we may just as well consider a crack of finite size subjected to an increasing load, which eventually will become so large that the crack extends catastrophically. The corresponding critical load σ_c can also directly be obtained from Eq. (9) as follows:

$$\sigma_c = \sqrt{\frac{2E'\gamma}{\pi a}} \tag{11}$$

Summarizing, we may say that a certain combination of crack length and applied load will characterize the onset of instability and catastrophical fracture. This combination is already hidden in Eq. (9). As we shall see Sec. IV.B, it was the merit of the fracture mechanics scientist Irwin to bring this fact to the world's attention.

B. Irwin's Concept of Stress Intensity Factors

As it was shown in Sec. IV.A, Eq. (9) can be interpreted in terms of a catastrophical fracture criterion. As it was demonstrated above, it can be used to calculate the maximum crack length, or the maximum stress allowed without causing instability of the system as a whole. Alternatively, we may say that instability occurs if the product $\sigma^2 a$ reaches a critical value. This was pointed out by Irwin [12] and led him to the concept of a critical stress intensity factor of fracture toughness K_{Ic}. The series of arguments considered are as follows. First, the stresses at a point (r, ϕ) around the tip of a Griffith crack (cf., Fig. 5) can be written as:

$$\sigma_{xx} = \frac{K_I}{\sqrt{2\pi r}} \cos\frac{\phi}{2}\left(1 - \sin\frac{\phi}{2}\sin\frac{3\phi}{2}\right)$$

$$\sigma_{yy} = \frac{K_I}{\sqrt{2\pi r}} \cos\frac{\phi}{2}\left(1 + \sin\frac{\phi}{2}\sin\frac{3\phi}{2}\right) \tag{12}$$

$$\sigma_{xy} = \frac{K_I}{\sqrt{2\pi r}} \sin\frac{\phi}{2}\cos\frac{\phi}{2}\sin\frac{3\phi}{2}$$

Here again, reference is made to mode I loading conditions (i.e., the principal load is applied normal to the plane of the crack). For obvious reasons, the quantity K_I is known as the (mode I) SIF.

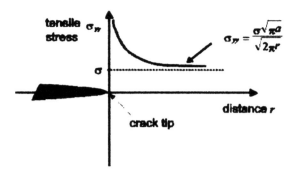

Figure 5 Crack tip geometry.

It was shown by Hahn [13] in 1977 that for the ideal geometry (infinite plane) and normal tensile loading conditions depicted in Fig. 3, the SIF is given by the root of the aforementioned product characterizing instability:

$$K_I = \sigma \sqrt{\pi a} \tag{13}$$

If a finite size specimen is concerned a correction function, Y needs to be introduced [14], which depends on its characteristic geometrical parameters (indicated by "..."):

$$K_I = \sigma \sqrt{\pi a} \, Y(\ldots) \tag{14}$$

Lists of such correction functions are complied in specialist handbooks [15,16]. The correction functions for three-point and four-point bend specimens will be provided in Sec. IV.C.

We may now rewrite Eq. (9) in the following way:

$$K_I = \sqrt{2\gamma E'} \tag{15}$$

The left-hand side of this equation is specimen-specific and loading-specific. Its explicit form follows by solving the linear elasticity problem (e.g., the one originally studied by Griffith, or more complicated ones as detailed in handbooks). The right-hand side of this equation, however, is material-specific and obviously a measure of the resistance of a body to brittle fracture (i.e., its resistance to the creation of new free surfaces). It is commonly known as fracture toughness K_{Ic} and allows the fracture criterion of Eq. (9) to be rewritten in the following concise form:

$$K_I = K_{Ic} \tag{16}$$

Expressed verbally, this means that the material will break in a brittle manner as soon as the stress intensity factor reaches its critical value.

C. Experimental Determination of Fracture Toughness

The fracture criterion (Eq. (16)) can easily be exploited to predict critical stresses, critical crack length(s), critical specimen dimensions, or a combination of the above once two items are known. First, it is required to provide a mathematical relation for the stress intensity factor K_I, pertaining to the specimen geometry and loading situation in question. This can either be achieved by solving the corresponding problem of linear elasticity (which is not necessarily a trivial task) or, more simply, by using solutions provided in appropriate handbooks. Second, the material-specific fracture toughness K_{Ic} must be known. In principle, a numerical value for K_{Ic} could be obtained from Eq. (15) provided that Young's modulus and the specific surface energy of the material are known. The latter quantity, however, is by no means easy to determine experimentally, and it is therefore not surprising that such a "direct" approach to measure K_{Ic} is not used in daily practice.

Usually, fracture toughness is determined by making use of the fracture criterion (Eq. (16)) in context with a standard fracture specimen geometry made of the material of interest. Typically, for ceramics, three-point or four-point bending test specimens are used. These are notched bars of rectangular cross-sections as indicated in Fig. 6. Now for a well-defined geometry (typically span $l = 40$ mm, width $d = 4.5$ mm, and height $B = 3.5$ mm), the fracture toughness K_{Ic} is obtained by measuring the load to fracture P and by evaluating the following equation (Eq. (15)):

$$K_{\mathrm{Ic}} = \sigma \sqrt{a}\, Y(X) \tag{17}$$

where $Y(X)$ denotes the following correction function:

$$Y(X) = \frac{1.99 - X(1 - X)(2.15 - 3.99X + 2.7X^2)}{(1 - 2X)(1 - X)^{3/2}}; \quad X = \frac{a}{B} \tag{18}$$

and σ is the maximum tensile stress in three-point and four-point bending bars, which, according to the elementary beam theory given as:

$$\sigma_{3\mathrm{PB}} = \frac{3Pl}{2dB^2}, \quad \sigma_{4\mathrm{PB}} = \frac{3Pe}{dB^2} \tag{19}$$

Fig. 7 shows a typical four-point bending arrangement.

In order to avoid subcritical crack growth (see below), relatively high loading rates of 12 N/sec (R. Ramme, personal communication, 1991) are being used during the test. The initial crack length a is typically around 1 mm and the crack should be as acute as possible. In this context, the introduction in the specimen of a pop-in precrack arrested into a stabilized jig has been suggested to guarantee a very sharp and pointed crack tip [17]. Although this is a standard procedure during K_{Ic} testing of metals, it is not a common practice for ceramics because the precrack itself may be affected by *microstructural factors* such as crack wake toughening mechanisms and crack face frictions, leading to a nonconservative assessment of the fracture toughness value. Typically, the width of the notch should not exceed 100 µm in order to obtain a "reliable" K_{Ic} value, which is a true material characteristic and specimen, as well as test-independent [18,19]. This can be achieved, within reasonable effort, using commercially available diamond saws. However, it has been reported that the notch root has to be sharpened to a radius of less then 10 µm to achieve a reliable toughness measurement [20]. Ideally, between 10 and 20 bars should be used to account for data scattering within a batch of specimens.

Although conceptually simple, K_{Ic} bending bar tests can still result in a fair amount of experimental work. Therefore, as a less labor-intensive alternative, the so-called indentation

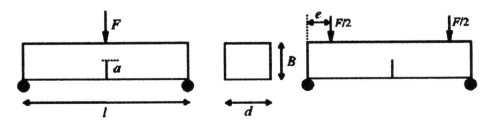

Figure 6 Three-point and four-point notched bending specimens.

Figure 7   Four-point bending rig.

methods were established. They allow for a very simple yet relatively accurate determination of the fracture toughness of brittle materials. As indicated in Fig. 8, a Vickers indenter is used to create an indent with cracks originating at its corners. The length of these cracks $(c-a)$ is then related to K_{Ic} by semi-empirical relation, such as the one developed by Evans and Charles [21]:

$$K_{Ic} = 0.16 \, H_V a^{0.5} \left(\frac{c}{a}\right)^{-3/2} \tag{20}$$

or the one quoted by Lawn et al. [22]:

$$K_{Ic} = 0.028 \left(\frac{E}{H_V}\right)^{0.5} H_V \, a^{0.5} \left(\frac{c}{a}\right)^{-3/2} \tag{21}$$

where $2a$ stands for the indenter width and H_V stands for the Vickers hardness. Clearly, indentation tests are prone to large scatter and errors due to local inhomogeneities in the

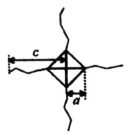

Figure 8   Determination of K_{Ic} from a Vickers indent.

material. Consequently, it should be repeated suitably often and used as an initial screening test rather than as a method for determining the "continuum" fracture toughness of a brittle material.

Due to their efficiency, indentation techniques for toughness measurements of brittle substances are a popular ongoing topic in the literature, from the late 1970s until today (in addition to the aforementioned references, see Anstis et al. [23], Chantikul et al. [24], Marshall et al. [25], Evans and Fuller [26], Laugier [27], and Shaobo et al. [28]). In particular, more recently, the aim seems to provide detailed theoretical foundations to an originally empirical approach [28,29].

Table 1 shows typical mechanical properties including toughness data of commonly used brittle engineering materials [30].

D. Fracture Criteria for Brittle Materials

Very often, Eq. (16) is rewritten as follows:

$$G = G_c \tag{22}$$

where G denotes the so-called energy release mode (the letter was chosen to acknowledge Griffith's pioneering contribution to fracture mechanics) and G_c is the critical energy release

Table 1 Mechanical Properties and Fracture Toughness of Various Materials

	Young's modulus (GPa)	Compressive strength (MPa)	Tensile strength (MPa)	Poisson's ratio	Density (g/cm³)	Fracture toughness (MPa m$^{1/2}$)	Hardness (Knoop)
Metals							
Titanium alloy (Ti–6Al–4V)	114	450–1850	900–1172	0.34	4.43	44–66	3200
Co–Cr–Mo	210	480–600	400–1030	0.29	8.3	120–160	3000
Stainless steel (316L)	193		515–620	0.30	8.0	20–95	
Ceramics							
Alumina	420	4400	282–551	0.27	3.98	3–5.4	2300
Zirconia (TZP)	210	1990	800–1500	0.31	5.74–6.08	6.4–10.9	1400
Silicon nitride (HPSN)	304	3700	700–1000	0.30	3.3	3.7–5.5	1600
Hydroxyapatite (3% porosity)	7–13	350–450	38–48		3.05–3.15		
Human tissues							
Cortical bone	3.8–11.7	88–164	82–114	0.28	1.7–2.0	2–12	130–170
Cancellous bone	0.2–0.5	23	10–20	0.32			
Cartilage	0.002–0.01		5–25				
Other							
PMMA	2.24–3.25	80	48–72		1.19	0.7–1.6	
UHMWHD polyethylene	0.69	20	38–48	0.20	0.94		

rate (i.e., the one beyond which catastrophical fracture will inevitably occur). For a crack subjected to tensile loads normal to its plane, we may simply write:

$$G = \frac{K_I^2}{E'} \tag{23}$$

The advantage of this relation does not become apparent until we note that it can easily be generalized to cover situations where a crack is not only loaded perpendicularly to its flanks but also in-plane and out-of-plane shear loads are applied, as indicated in Fig. 9. It is customary to distinguish these three basic types of loading by subscripts I, II, and III, and they are also referred to as the three basic modes of failure. It can be shown that the total energy release rate for a crack growing in its own plane due to the combined action of all three loading modes is additive and is given by [14]:

$$G = \frac{K_I^2}{E'} + \frac{K_{II}^2}{E'} + \frac{K_{III}^2}{2\mu} \tag{24}$$

where μ refers to the shear modulus of the material containing the crack.

Similarly, as in the mode I case, the specific mathematical form of the stress intensity factors K_{II} and K_{III} depends on the specimen geometry and the way the shear loading is prescribed. Explicit solutions can be found in the aforementioned handbooks.

On the other hand, for the case of mode I fracture, we may write the material-dependent right-hand side of Eq. (22):

$$G_c = \frac{K_{Ic}^2}{E'} \tag{25}$$

where K_{Ic} is determined experimentally in the method outlined above. However, knowledge of K_{Ic} is insufficient.

Clearly, a complete description of the fracture mechanical behavior of brittle substances should describe the material's resistance to failure under *pure* mode II or mode III loading, or even to a combination of two or even all three failure modes. For pure mode I or mode II, we may write:

$$G_c = \frac{K_{IIc}^2}{E'}; \qquad G_c = \frac{K_{IIIc}^2}{\mu} \tag{26}$$

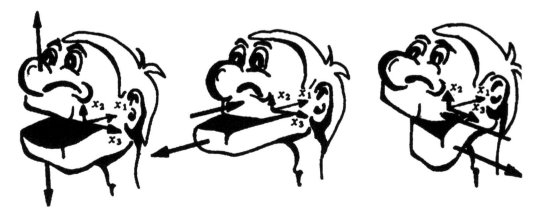

Figure 9   Three modes of fracture. (From Ref. 89.)

First attempts have been made to obtain reliable data for the fracture toughness of ceramics with respect to pure mode II and mode III loading conditions [31]. According to this work, which was based on alumina, K_{IIc} is comparable to the mode I fracture toughness K_{Ic}, whereas for K_{IIIc}, much higher values were reported.

In the second step, it is necessary to establish a suitable relationship between K_{Ic}, K_{IIc}, and K_{IIIc} for true mixed-mode loading conditions. Such "multiaxial fracture envelopes" have been established for metals and a summary of these can be found in the work by Richard [32]. However, these criteria need to be carefully checked before their application to brittle materials. Due to the limited amount of reliable data even for standard ceramics, no attempt will be made in this chapter to recommend a particular multiaxial fracture criterion. From a practical point of view, it seems appropriate to apply suitable safety factors in order to account for multiaxial loading and its effect on fracture in brittle structures.

V. SUBCRITICAL CRACK GROWTH

So far, we have been dealing with the description of *catastrophical fracture*. This is helpful if the intention is to find out whether a certain loading condition is impossible to bear by the brittle material the structure is made of. Clearly, we have to know the maximum crack size in that specimen to employ the fracture criterion and we also need to know the stress distribution within the ceramic. Both is not easy to assess, however, even if this assessment is carefully done and the outcome is that the structure will not fail during typical in-service conditions; a catastrophical fracture analysis alone is still insufficient. This is essentially due to the phenomenon of *subcritical crack growth*.

To appreciate and to understand this fact, we note that every structure is exposed to a more or less corrosive environment [33,34]. Simplistically speaking, water vapor or other chemical agents will penetrate into the material and diffuse toward the tips of the microflaws (pores, small cracks, inclusions, "loose" grain boundaries, etc.) that are always present in ceramics or glasses as a consequence of sintering, surface finishing, or, simply, daily use. By adsorption of the chemical, the surface energy of the material will be reduced. Thus, the crack will extend slowly over time until, finally, critical conditions are reached and the specimen breaks catastrophically. This explains the technical term for the phenomenon: *stress corrosion cracking*. The diffusion as well as the adsorption and cracking process will even be enhanced and accelerated by the enormous stresses present in the neighborhood of the crack (cf., Eq. (12)).

A detailed atomistic interpretation of stress corrosion cracking in brittle materials can be found in the pioneering work by Charles [33,35,36] and Wiederhorn [37,38]. In here, we shall present more phenomenally oriented arguments, which allow us to assess the consequences of stress corrosion cracking from the pragmatical perspective of a design engineer. To this end, we will present an equation capable of predicting subcritical crack growth in brittle materials in terms of two engineering constants, the so-called subcritical crack growth parameters. Experimental methods for determining these constants will then be presented where, as always, the focus is on industrial feasibility toward a swift and yet reliable lifetime analysis of brittle structures.

A. Subcritical Crack Growth Phenomenon

In order to study fatigue crack growth in metals, special phenomenal relationships, such as the Paris law, are frequently used, where the change in stress intensity ΔK_{I} (for mode I

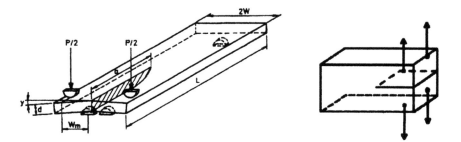

Figure 10 Double torsion (left) and double cantilever beam specimen (right).

loading conditions) during cycling around a certain load level dictates the amount of crack growth da per cycle dN, as shown:

$$v \equiv \frac{\mathrm{d}a}{\mathrm{d}N} = C\Delta K_I^m \tag{27}$$

In here, C and m are what might be called phenomenological constants, which need to be determined experimentally.

Similarly, the subcritical crack growth velocity v is governed by the currently active stress intensity K_I at the crack tip. The relationship (v,K_I) is normally determined using either double torsion [39] or double cantilever [37] specimens, as shown in Fig. 10. In principle, the measurement requires only recording of the applied force and the resulting displacement of the leading points as a function of time. This sounds as a very simple approach. However, in practice, this direct approach is quite laborious and extremely time-consuming, especially if the data for long-term subcritical crack growth are of interest. For experimental details, the reader is referred to the articles by Evans [39], Evans and Wiederhorn [40], and Li et al. [41].

A schematic and a typical real (v,K_I) curve (for soda lime glasses at different levels of humidity) are presented in Fig. 11. They are depicted in double logarithmical representation

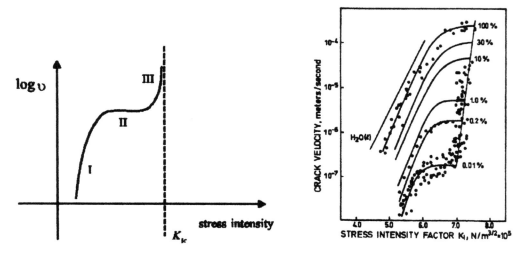

Figure 11 Subcritical crack growth in ceramic materials, schematically (left) and for soda lime glass (from Ref. 38) for different levels of humidity (right).

to account for a wide range of crack velocities. Three characteristic regions can easily be distinguished.

From a technical perspective, region I is the most important branch. It covers the field of small and medium crack velocities, which dictate the long-term lifetime of an in-use ceramic structure. Here the crack velocity depends exponentially on the applied crack driving force K_I. Physically speaking, the rate at which chemicals are transported to the crack tip is now greater than the rate at which they can be consumed. In region II, the crack speed reaches a certain saturation level and depends no longer on the applied force. The chemicals are consumed during the reaction more rapidly than they can be supplied. Finally, crack propagation in region III is related to the onset of catastrophical fracture and is independent of the environment the specimen is exposed to. More information regarding the physical interpretation of trimodal (v, K_I) curves as well as their kinetic modeling can be found in Wiederhorn [37,38], Charles and Hillig [42], Hillig and Charles [43], and Brown [44].

As indicated above, only region I is of immediate interest to an engineer. The lifetimes associated with the other two regions are generally much too short to be of technical importance. Thus, their influences will be neglected completely in all that follows. In mathematical terms, the behavior within region I can be described in a manner analogous to Eq. (27) (see Evans and Fuller [45], Evans et al. [46], Jakus et al. [47], Jakus and Ritter [48], Pabst [49], Richter [50], and Tradinik et al. [51]):

$$v \equiv \frac{da}{dt} = AK_I^n \tag{28}$$

where t denotes time. For obvious reasons, the two coefficients A and n are known as *subcritical crack growth parameters*. These need to be determined experimentally. As it was mentioned before, a possible method is to record the complete (v, K_I) curve of a ceramic material and to curve-fit region I according to the relationship shown in Eq. (28). This, however, is not very cost-effective in terms of both time and experimental skills involved. Consequently, a more "industrially feasible" approach to their determination would be welcome.

The solution to this predicament is as follows. We shall see shortly that the coefficient n by itself determines the lifetime of a macroscopically intact (i.e., *unnotched*) but intrinsically microflawed specimen made of a brittle material subjected to static or dynamic loading. Consequently, from the standpoint of a lifetime-oriented reliability analysis, it suffices to determine only that coefficient. Moreover, as will be demonstrated, unnotched bar specimens tested at different loading rates can be used for that purpose provided the appropriate statistics are used. This decreases the effort of specimen preparation considerably and, depending on the loading rates chosen, also minimizes the time required for the actual testing.

Subcritical crack growth due to cyclical loading will not be treated here in detail. Further information with regards to this topic can be found in Refs. 45, 52, and 53. According to experiments reported in these papers, no evidence that cycling enhances the slow crack growth observed in ceramic materials exists.

B. Subcritical Crack Growth and Lifetime of Notched Specimens Subjected to Constant Load

The arguments presented in this section and follow closely the work by Evans and Wiederhorn [40], Jakus et al. [47], Pabst [49], and Richter [50]. We will start examining the lifetime or time to failure t_f of a *notched* bend test specimen (e.g., Fig. 6) under subcritical crack growth. The initial crack length is denoted by a_i and the applied load can be characterized by a (constant) stress σ. The lifetime can then be obtained by integration of the equation defining crack growth velocity. The limits of integration are between the initial crack and the final crack length a_c (which, for obvious reasons, must be the critical crack length leading to catastrophical fracture):

$$\frac{da}{dt} = v \Rightarrow dt = \frac{da}{v} \Rightarrow t_f = \int_{a_i}^{a_c} \frac{da}{v} \tag{29}$$

It is useful to introduce into this equation the crack driving force (i.e., the stress intensity currently active at the crack tip). To this end, we may use the general form shown in Eq. (17), which, by virtue of the chain rule, yields:

$$K_I = \sigma\sqrt{a}\,Y\!\left(\frac{a}{B}\right) \Rightarrow dK_I = \left(\sigma\frac{1}{2\sqrt{a}}\,Y\!\left(\frac{a}{B}\right) + \sigma\sqrt{a}\,\frac{1}{B}\,Y'\right)da \tag{30}$$

For most specimen types, the correction function depends only weakly on its argument and departs considerably from one only if the ligament B is almost completely broken (i.e., beyond values $a/B \approx 0.6$; Fig. 12). Consequently, we neglect the second term in Eq. (30) and write:

$$dK_I = \frac{\sigma^2 Y^2}{2\sigma\sqrt{a}\,Y}\,da \Rightarrow da = \frac{2K_I}{Y^2\sigma^2}\,dK_I \tag{31}$$

This can now be inserted into Eq. (29):

$$t_f = \int_{K_{Ii}}^{K_{Ic}} \frac{2K_I}{vY^2\sigma^2}\,dK_I \tag{32}$$

where the limits of integration now correspond to the initial crack intensity K_{Ii} and to the fracture toughness K_{Ic} characteristic of catastrophical failure. Note that for other specimen configurations, it might not be allowed to neglect the derivative of the correction function, and numerical integration could become necessary to take this term into account.

We make now use of the Evans–Wiederhorn [40] relationship for subcritical crack growth shown in Eq. (28) and perform the integration:

$$t_f = \frac{2}{AY^2\sigma^2} \int_{K_{Ii}}^{K_{Ic}} \frac{dK_I}{K_I^{n-1}} = \frac{2}{\sigma^2 Y^2 A(n-2)} \left(K_{Ii}^{2-n} - K_{Ic}^{2-n}\right) \tag{33}$$

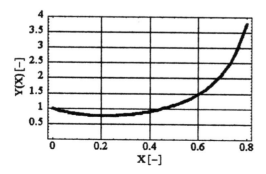

Figure 12   Behavior of the SIF correction function $Y(a/B)$ of a bending bar specimen.

This relation can still be written in a more concise form. As we shall see shortly, numerical values for the subcritical crack growth parameter n are typically around 20 and more. Moreover, we have:

$$K_{Ic} > K_{Ii} \tag{34}$$

and, therefore, the second term in the last equation can be neglected. Taking this into account and applying the general relation for an SIF shown in Eq. (17) to the initial crack configuration yield:

$$t_f = C \frac{a_i^{(2-n)/2}}{\sigma^n}; \qquad C = \frac{2}{Y^n A(n-2)} \tag{35}$$

This equation is now applied to two identical ceramic specimens, I and II (i.e., these are made of the same material and have the same initial crack length a_i). However, they are subjected to two different loads σ^I and σ^{II}. By elimination of the constant C, we obtain:

$$\frac{t_f^I}{t_f^{II}} = \left(\frac{\sigma^{II}}{\sigma^I}\right)^n \Rightarrow t_f \sigma^n = \text{const.} \tag{36}$$

We conclude that a certain product between lifetime and applied load σ is a constant. As one would expect, the higher the load is, the lower is the lifetime, and its duration decreases with increasing crack growth parameter n. The latter is also quite sensible because, according to the Evans–Wiederhorn law, the crack growth velocity increases dramatically with increasing n-value.

C. Subcritical Crack Growth and Lifetime of Notched Specimens Subjected to a Dynamic Load

The objective of this section is to explore how the lifetime relation shown in Eq. (36) is affected if, instead of a static load $\sigma = $ constant, dynamic loading conditions $\sigma = \sigma(t)$ are used instead. In this context, we are going to assume that the load increases linearly in time, which will simplify the integration and which is also relatively easy to achieve experimentally (e.g., by using a loading device the cross-head weight of which moves at a constant velocity):

$$\sigma(t) = \dot{\sigma}t; \qquad \dot{\sigma} = \text{const.} \tag{37}$$

A comparison between static and dynamic loading conditions is shown in Fig. 13.

Under dynamic loading conditions, failure due to subcritical crack growth will occur at time t_f^{dyn} and at a corresponding stress level σ_f, which is supposed to be identical to the applied static load that will lead to failure at time t_f^{st}. In order to determine how static and dynamic lifetimes are related, we combine Eqs. (17), (28), and (37) and separate variables a and t:

$$a^{-n/2}da = A(\dot{\sigma}Y)^n t^n dt \tag{38}$$

Both sides can now be integrated between initial and final conditions. If the general relation for the SIF shown in Eq. (17) is taken into account, we obtain:

$$t_f^{dyn} = (n+1) \frac{2}{\sigma_f^2 Y^2 A(n-2)} \left([\sigma_f \sqrt{a_i} Y]^{2-n} - [\sigma_f \sqrt{a_c} Y]^{2-n} \right) \tag{39}$$

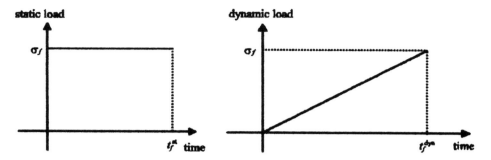

Figure 13 Comparison between static and dynamic loadings.

This can now directly be compared with the result for static lifetime shown in Eq. (33). The following simple relation results, which is worth some comments:

$$t_f^{\text{dyn}} = (n+1)t_f^{\text{st}} \tag{40}$$

Intuitively, one may have suspected that the dynamic lifetime is shorter than the static one. However, this is not so. Rather, the dynamic lifetime is longer by a factor of $n+1$. This is simply because during the static lifetime experiment, the load σ_f has been applied from the very beginning, whereas during dynamic loading conditions, this load is reached only gradually. Consequently, the crack does not find optimal growth conditions at all times, resulting in a longer lifetime.

D. Indirect Experimental Determination of the *n*-Value

The last two subsections have shown that static as well as dynamic lifetimes are critically dependent on only one of the two crack growth parameters appearing in the Evans–Wiederhorn [40] relationship, namely, the *n*-value. Hence, the question arises as to how to determine this quantity experimentally in a swift, yet accurate, manner. As it was pointed out before recording the full (v,K_I) curve in situ (i.e., measuring the crack length and the stress intensity as a function of time) using double torsion or other notched bend specimens is possible but impractical. Such tests require extensive specimen preparation, manual skill, and, in particular, time, all of which make them rather expensive and nonsuitable in an industrial environment.

Alternatively, a *statistical* approach has been suggested as follows [47,49]. The basic idea is to perform three-point or four-point bend tests using batches of *unnotched* bars that are broken at different loading rates. Two different strategies can be pursued: increasing rank analysis, or median fracture strength method, which will be explained once we have established a suitable mathematical relation, which forms the basis of both analyses.

1. The Fundamental Relation Between Lifetime and Stress

We evaluate Eq. (37) at the time of failure and combine it with Eq. (39) to obtain:

$$\sigma_f = \dot{\sigma}(n+1)t_f^{\text{st}} \tag{41}$$

Now if Eq. (35) is taken into account, the constancy of the following ratio can easily be established:

$$\frac{\sigma_f^{n+1}}{\dot\sigma} = C'; \qquad C' = \frac{2(n+1)a_i^{(2-n)/2}}{Y^n A(n-2)} = \text{const.} \tag{42}$$

Next consider two specimens (I and II), with the same initial flaw size, that are subjected to two different loading rates, so that:

$$\frac{\sigma_f^I}{\sigma_f^{II}} = \left(\frac{\dot\sigma^I}{\dot\sigma^{II}}\right)^{\frac{1}{n+1}} \tag{43}$$

Note that so far, in parallel with subcritical crack growth, we have always used prenotched bar specimens. This, however, is not really necessary. Rather, it suffices to require that, on a statistical average, the flaw distribution and their orientation within specimens I and II are the same. Then Eq. (43) should still be applicable. Now if the logarithm is taken on both sides, this equation will turn into the following *linear* relationship:

$$\ln \sigma_f^I = \ln \sigma_f^{II} + \frac{1}{n+1}\ln\left(\frac{\dot\sigma^I}{\dot\sigma^{II}}\right) \tag{44}$$

This will now be used to determine the *n*-value by means of two different experimental techniques.

2. Increasing Rank Analysis

Consider two batches, 1 and 2, of the same total number N of specimens. Typically, N should be on the order of 20–30. The two batches are then broken at two different loading rates $\dot\sigma^I$ and $\dot\sigma^{II}$, and the corresponding strength data σ_j^I and σ_j^{II} $(j=1,2,\ldots,N)$ are recorded. Clearly, the latter will typically scatter and not be arranged in any order. However, for each batch, the data points are now ranked in *increasing order* such that:

$$\left(\sigma_1^I, \sigma_1^I, \ldots, \sigma_i^I, \sigma_{i+1}^I, \ldots, \sigma_N^I\right), \left(\sigma_1^{II}, \sigma_1^{II}, \ldots, \sigma_i^{II}, \sigma_{i+1}^{II}, \ldots, \sigma_N^{II}\right), \sigma_i^{I/II} > \sigma_{i+1}^{I/II} \tag{45}$$

These sorted data are now plotted in a $(\ln \sigma^I, \ln \sigma^{II})$ diagram. According to the fundamental relations shown in Eq. (43), the data should group a linear line with unit slope, the intercept of which, with the vertical coordinate axis, is given by $1/(n+1)$. Due to logarithmical representation, evaluation is practically performed such that, first, the "center of gravity" of the data point is calculated and made the basis of a new coordinate system. The line of unit slope will also intersect with its vertical axis. The value of this point of intersection can easily be determined and directly be related to $1/(n+1)$. Figure 14 shows an example of this procedure.

3. Median Fracture Strength Plot

Alternatively, we now consider several batches i, each of which consists of N_i specimens. This number may vary from specimen to specimen, which should be at least five and ideally

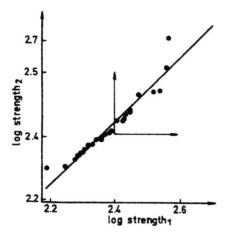

Figure 14   Increasing rank plot.

vary between 20 and 30 from batch to batch. Each batch is tested at a different loading rate (i.e., the specimens are broken at small, medium, and large cross-head speeds, which, preferably, are increased in units of 10). Then average strength can be calculated from the strength data of each batch as follows:

$$\bar{\sigma}_i = \frac{1}{N_j} \sum_{j=l}^{N} \sigma_j^i \tag{46}$$

Plotting the logarithm median values as a function of the logarithms of loading rates (or cross-head speeds), a $(\log \sigma, \sigma)$ diagram is obtained. According to the fundamental relation (43), the data points should group around a straight line, the slope of which is proportional to $1/(n+1)$. An example of this approach is provided in Fig. 15.

  Clearly, if properly applied, in both techniques, increasing rank analysis as well as the method of median fracture strengths should lead to the same results.

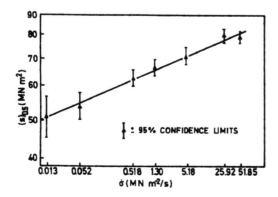

Figure 15   Median fracture strength plot of soda lime glass. (From Ref. 47.)

VI. STATISTICAL VARIABILITY OF THE STRENGTH OF BRITTLE SUBSTANCES

The two methods used for the experimental determination of the subcritical crack growth parameter n clearly are based on the fact that the strength σ within a batch of specimens made of a brittle material *varies* statistically. This is not surprising because as a consequence of the sintering and surface finishing process, specimens within a test batch will contain flaws of different size and severity. Consequently, their strength will vary within a certain range. Intuitively, we would state a ceramic to be of "good quality" if this variation is kept as small as possible. Obviously, we now face the problem of quantifying this criterion for quality. The key to solving this problem is the so-called *Weibull statistics*.

Moreover, as it has been pointed out before, our final goal is to arrive at a lifetime prediction for components made of brittle materials. Intuitively speaking, lifetime will, among other things, certainly be governed by the strength of the material. Because the strength data of brittle materials show strong variability, an appropriate statistical approach is necessary to combine lifetime with strength. A decisive step in this direction is the so-called Weibull analysis, which will be explained in subsequent sections.

A. The Weibull Fit for Strength Data of Brittle Materials

A typical example of the variation of strength data resulting from four-point bending tests of ceramic specimens is shown in Fig. 16. The data are presented in the form of a histogram (i.e., grouped in strength categories, e.g., 350–400, 400–450 MPa, etc.). Obviously, the strength centers around a median value: some specimens of the batch failed at low loads, others are capable of withstanding very high loads, and most of them fall into an intermediate strength category. As it was mentioned above, the strength of high-quality ceramics should not show too much variability and their strength distribution should be narrow, showing as sharp a transition as possible. In fact, during industrial quality assurance, the (sudden) occurrence of a broad distribution may indicate not only a less advanced material but also an inherent processing fault. Due to the arbitrariness of strength categories chosen, such information can hardly be obtained from a histogram alone. In this

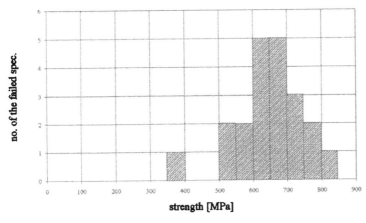

Figure 16   Fracture strength histogram of ceramic four-point bend specimens.

context, cumulative distributions of strength diagrams have proven to be particularly useful.

A cumulative strength diagram of data shown in Fig. 16 is presented in Fig. 17. Note that the ordinate has been normalized by the total number of specimens in the batch N, thus providing a measure for the probability of failure P_i. Most simply, we could write:

$$P_i = \frac{i}{N} \tag{47}$$

where i denotes the total number of specimens that failed below or at a certain strength level. However, for statistical reasons, other discrete measures of probability are frequently used (e.g., ASTM C 1293-00 [54], DIN 51110-Teil3 [56], Nadler [56], and Nemeth et al. [57]):

$$P_i = \frac{i - 0.5}{N}; \qquad P_i = \frac{i - 0.3}{N + 0.4} \tag{48}$$

The cumulative strength diagram has a characteristic S-shape. Now in order to characterize the quality of the ceramic, we may say that the sharper the S is, the "better" the ceramic should be. Clearly, the location of the "center" of the S is also of importance because we also aim for a "high-strength" material.

Both of these features, sharpness and median strength, can easily be characterized if we curve-fit the cumulative strength data. This can be done reasonably well by using a two-parameter Weibull function:

$$P(\sigma) = 1 - \exp\left(-\left\{\frac{\sigma}{\sigma_0}\right\}^m\right) \tag{49}$$

Suggestively speaking, we may say that the sharpness of the distribution is characterized by the parameter m, which, for obvious reasons, is also known as the *shape parameter* or *Weibull modulus*. The greater the m is, the smaller the variability in strength will be. High-quality ceramics show m-values of 10 and more. The *scale parameter* or *median*

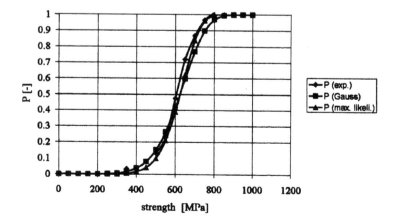

Figure 17 Cumulative distribution of strength with two-parameter Weibull fits.

strength value σ_0 corresponds roughly to the strength at the 63rd percentile of probability of failure:

$$P(\sigma_0) = 1 - \frac{1}{e} = 0.632 \ldots \tag{50}$$

Typical Weibull fits of the cumulative strength data are also shown in Fig. 17. In practice, m and σ_0 are determined from four-point bending experiments with unnotched specimens (see Sec. IV.C for typical specimen geometries and loading rates). Various techniques can be used to process the resulting strength data. Most commonly, the method of least squares (Gaussian method) or the maximum likelihood procedure is used [54,55,58]. The least squares method is based on the following equations:

$$y_i = \ln \ln \frac{1}{1 - P_i}; \qquad x_i = \ln \sigma_i; \qquad A = -m \ln \sigma_0 \tag{51}$$

$$m = \frac{N \sum_{i=1}^{N}(x_i y_i) - \sum_{i=1}^{N} x_i \sum_{i=1}^{N} y_i}{\sum_{i=1}^{N} x_i^2 - \sum_{i=1}^{N} x_i \sum_{i=1}^{N} x_i}; \qquad A = \frac{\sum_{i=1}^{N} y_i \sum_{i=1}^{N} x_i^2 - \sum_{i=1}^{N} x_i \sum_{i=1}^{N}(x_i y_i)}{N \sum_{i=1}^{N} x_i^2 - \sum_{i=1}^{N} x_i \sum_{i=1}^{N} x_i}$$

The maximum likelihood method, however, leads to an implicit equation which, in general, must be solved numerically, for example, using a Newton–Raphson solver:

$$\frac{N}{m} + \sum_{i=1}^{N} \ln(\sigma_i) - N \frac{\sum_{i=1}^{N} [\sigma_i^m \ln(\sigma_i)]}{\sum_{i=1}^{N} \sigma_i^m} = 0; \qquad \sigma_0 = \left(\frac{1}{N} \sum_{i=1}^{N} \sigma_i^m \right)^{\frac{1}{m}} \tag{52}$$

From a practical standpoint, it is important to know that the maximum likelihood method tends to reduce the influence of outliers among data on the final results, whereas the method of least squares takes all data into account equally. For a more complete discussion regarding the advantages and disadvantages of both methods, in particular from a mathematical point of view, see Nemeth et al. [57] and Jayatilaka [60].

It should also be noted that the two-parameter Weibull distribution is the simplest fit that can be used. Sometimes it is required to acknowledge other features of the strength distribution (e.g., the so-called threshold strength σ_{th}, below which no failure can occur). This can be achieved by using the three-parameter Weibull distribution, as follows:

$$P(\sigma) = \begin{cases} 1 - \exp\left(-\left\{\dfrac{\sigma - \sigma_{th}}{\sigma_0}\right\}^m\right) & \text{if } \sigma > \sigma_{th} \\ 0 & \text{if } \sigma \leq \sigma_{th} \end{cases} \tag{53}$$

The benefits arising from the use of this third parameter are questionable. Most brittle materials show a threshold strength that is relatively small and, consequently, difficult to determine reliably without using an unreasonably large amount of specimens. From a more conservative point of view, it seems reasonable to assume σ_{th} to be zero and, in what follows, the two-parameter Weibull distribution will be applied throughout. However, for a more complete treatment of this subject, the reader is referred to Weibull [9], Marshall et al. [59], and Jayatilaka [60].

B. The Weibull Distribution and Its Dependence on Specimen Geometry and Loading Conditions

1. Problem Definition

It must be pointed out that the strength data that lead to the histogram and cumulative strength plots shown in Fig. 17 depend considerably on the specimen type (i.e., its volume and surface size), as well as on the loading conditions employed. The mean strength of specimens made of brittle materials depends on the *specimen size*, the *specimen surface*, and the *loading conditions*. For example, a more voluminous specimen is more prone to failure because, compared to a smaller one, it is more likely to contain a flaw of critical size. A similar remark holds for its surface, which, during the surface finish process, is more likely to be affected negatively. Moreover, if, due to the loading process, the region of tensile stresses is large, then it is more likely that failure will occur in high-tensile region. For example, a four-point bending specimen will break more easily than a three-point bend specimen of the same volume. This can clearly been seen from the moment diagrams shown in Fig. 18, which indicate that four-point bending will lead to a much wider region of large tensile stresses. From elementary beam theory, it is known that the maximum tensile stress arising in these specimens is given by:

$$\sigma_{\max} = \frac{M_{\max}}{w}; \qquad w = \frac{dB^2}{6} \tag{54}$$

where w denotes the section modulus, which was explicitly specified for a rectangular cross-section of width d and height B.

Clearly, the Weibull distribution introduced in Sec. VI.A should reflect *geometry* and *loading* dependencies. In fact, it was this very question that originally prompted Weibull to conceive the distribution named after him. In what follows, we shall summarize Weibull's line of reasoning.

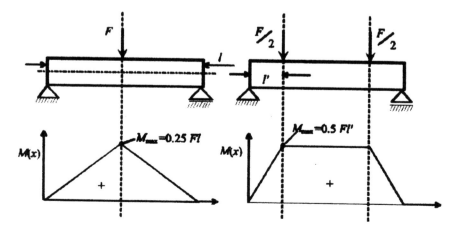

Figure 18   Moment diagrams and stress distribution in three-point and four-point bend specimens.

2. Basic Assumptions Leading to the Weibull Distribution

In order to answer the problem of size and loading dependence of the Weibull distribution, we follow the original papers of Weibull [9] and Tradinik et al. [51] and accept the following three key properties regarding the strength of a brittle material:

A brittle material is statistically homogeneous and isotropic (i.e., on average, the specimen strength does not depend upon position or direction).

Subvolumina and subsurfaces of a specimen are statistically independent (i.e., in order to obtain the probability of failure of the whole specimen, the probabilities of failure of its subsets can simply be multiplied).

Subvolumina and subsurfaces of a specimen follow the weakest link concept (i.e., the weakest part of a specimen determines the strength of the whole).

It can be shown that on the basis of these assumptions, the probability of failure for a body subjected to a spatially varying, but uniaxial, state of stress $\sigma(x,y,z)$ can be written as:

$$P = 1 - \exp\left(-\frac{1}{V_0} \int \int_V \int \left[\frac{\sigma(\underline{x})}{\sigma_0}\right]^m dV\right); \qquad \underline{x} = (x, y, z) \tag{55}$$

when volumetrical flaws dominate, and as:

$$P = 1 - \exp\left(-\frac{1}{A_0} \int \int_A \left[\frac{\sigma(\underline{x})}{\sigma_0}\right]^m dA\right) \tag{56}$$

if surface flaws govern the fracture strength of the specimen.

In these equations, V_0 and A_0 denote a suitable reference volume or reference surface. Note that symbols V and A refer only to the part of the specimen volume or surface that is subjected to *tensile* stress. This is because brittle materials are most susceptible to failure due to tensile (positive) stresses. Certainly, they will also fail under compression (negative stresses), but this occurs at a stress level several times higher. Negative stresses are thus completely ignored in lifetime predictions of brittle materials. Details of the proof can be found in Refs. 9 and 51.

Furthermore, note that, for ceramics, it is commonly accepted that the influence of surface flaws can be neglected for specimens in as-received or surface-finished condition [56]. However, the opposite is true in glass science.

3. Analytical Evaluation of the Weibull Integral

It is fair to say that a closed form evaluation of the volume or surface integrals shown in Eqs. (55) and (56) is mathematically "nontrivial," even if only very simple geometries and loading conditions are considered. As an example, we will look in this section at three-point and four-point bend specimens with circular or rectangular cross-sections. It is interesting to note that the one-dimensional stress distributions of Eqs. (55) and (56) can be mapped onto a single function, which depends only on a single integration variable $f(s)$, even though they depend on the position vector. This is the so-called stress–density function, which was introduced by Nadler [56,61,62] and helps to simplify the integration process considerably. As outlined in Nadler's work, we may write:

$$P = 1 - \exp\left(-\frac{V}{V_0}\left[\frac{\sigma}{\sigma_0}\right]^m \int_{s_{min}}^1 s^m f(s) ds\right) \tag{57}$$

for volumetrical flaws, and:

$$P = 1 - \exp\left(-\frac{A}{A_0}\left[\frac{\sigma}{\sigma_0}\right]^m \int_{s_{\min}}^{1} s^m g(s) \mathrm{d}s\right) \tag{58}$$

for surface flaw-governed specimens. It must be noted that σ denotes the maximum tensile stress in the uniaxially stressed specimen (i.e., the stress used during the collection of strength data; cf., Eq. (54) for bending bar specimens). Moreover, by definition:

$$s_{\min} = \frac{\Sigma}{\sigma} \tag{59}$$

where Σ is the smallest tensile stress in the system (which may or may not be equal to zero). Finally, the stress–density functions $f(s)$ and $g(s)$ can be obtained from the following relations:

$$\int_s^1 f(\bar{s})\mathrm{d}\bar{s} = \frac{V(s)}{V} \Rightarrow f(s) = -\frac{\mathrm{d}}{\mathrm{d}s}\frac{V(s)}{V}$$

$$\int_s^1 g(\bar{s})\mathrm{d}\bar{s} = \frac{A(s)}{A} \Rightarrow g(s) = -\frac{\mathrm{d}}{\mathrm{d}s}\frac{A(s)}{A} \tag{60}$$

Because Σ denotes any tension smaller than the maximum tensile stress occurring in the system, and $V(s)$ and $A(s)$ denote the volume and the surface of the specimen subjected to stress Σ, respectively.

Tables 2 and 3 present a list of volumetrical and surface flaw-governed stress–density functions for bending bars of circular or rectangular cross-section subjected to various types of loading. The advantage of stress–density functions is obvious. Instead of having to solve complicated two-dimensional or three-dimensional integrals, it now becomes necessary only to evaluate one single integral, which can be achieved by elementary integration schemes.

Table 2 Volumetrical Stress–Density Functions $f(s)$ of Various Specimen Types

Specimen type	Stress–density function $f(s)$
(1) Pure moment, rectangular c.s.	$1/2$
(2) Three-point bending, rectangular c.s.	$-1/2 \ln s$
(3) Pure moment, circular c.s.	$\dfrac{2}{\pi}\sqrt{1 - s^2}$
(4) Three-point bending, circular c.s.	$\frac{2}{\pi}(\cos\mathrm{h}^{-1}(\frac{1}{s}) - \sqrt{1 - s^2})$
(5) Four-point bending, rectangular c.s.	$\dfrac{l_1}{l}f_1(s) + \dfrac{l_2}{l}f_2(s)$
(6) Four-point bending, circular c.s.	$\dfrac{l_1}{l}f_3(s) + \dfrac{l_2}{l}f_4(s)$
(7) Shrinkage fit of ceramic annulus	$\dfrac{\varepsilon(1+\varepsilon)}{1-\varepsilon}\dfrac{1}{[(1+\varepsilon)s - \varepsilon]^2}$

c.s. = Cross-section; l = length of lower support length; l_1 = length of upper support; $l_2 = l - l_1$ (for other symbols, see text).

Table 3 Surface Flaw Stress–Density Functions $g(s)$ for Common Engineering Samples

Specimen type	Stress–density function $g(s)$
Pure bending, rectangular c.s.	$0.25[1 + \delta(s)]$
Three-point bending, rectangular c.s.	$0.25[1 - \ln(s)]$

c.s. = Cross-section; $\delta(s)$ = Dirac's delta function.

Also note that the superposition principle can be used in order to obtain the density functions of more complicated loading cases from more simple ones. As an example, consider a bar subjected to four-point bending. Its stress–density function can be obtained by superposition from the pure moment and from the three-point bending case as follows (see the moment diagrams shown in Fig. 18 and Tables 2 and 3 for the indices):

$$f_{5/6}(s) = \frac{l - 2l'}{l} f_{1/3}(s) + \frac{2l'}{l} f_{2/4}(s) \tag{61}$$

We shall now illustrate the use of stress–density functions by two examples. First, consider a three-point bending specimen of rectangular cross-section as shown in Fig. 6 (without a crack). The only nonvanishing stress component is given by Timoshenko and Goodier [63] as:

$$\sigma_{xx} = -\frac{6P}{dB^3} y \left(\frac{1}{2} - |x| \right) \tag{62}$$

where x and y refer to the horizontal and vertical coordinate axes, respectively, counting from the center of the specimen. Using this relation for the tensile stresses, we now determine the parameter s according to Eq. (60) and find that:

$$s = -\frac{2y}{B} \left[1 - \frac{2|x|}{l} \right] \tag{63}$$

The volume $V(s)$ required for evaluating Eq. (60) is found by integration as follows:

$$V(s) = 2d \int_0^{|x| = (1-s)l} \left[\frac{B}{2} + y(x) \right] dx = (1 - s + s \ln s) \frac{Bld}{2} \tag{64}$$

The total specimen volume V is given by Bld, and Eq. (60) can now be used to obtain the stress–density function as follows:

$$f(s) = -\frac{d}{ds} \frac{1}{2} (1 - s + s \ln s) = -\frac{1}{2} \ln s \tag{65}$$

As a second example for the use of stress–density functions, we consider another geometry frequently encountered in engineering applications, namely, a ceramic annulus as illustrated in Fig. 19. For instance, the annulus could be used in context with a shrinkage fit onto an axle of circular cross-section. Due to the misfit ζ, stresses will result, which, for convenience, are expressed in polar coordinates (r, θ) as follows [64]:

$$\sigma_{rr} = -\gamma \left(\frac{r_i^2}{r^2} - \varepsilon \right); \qquad \sigma_{\theta\theta} = \gamma \left(\frac{r_i^2}{r^2} + \varepsilon \right) \tag{66}$$

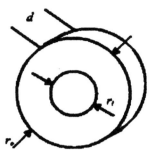

Figure 19 Geometry of a ceramic annulus.

where:

$$\varepsilon = \frac{r_i^2}{r_o^2}; \quad \gamma = \frac{E_2\zeta}{\left(1 - \dfrac{E_2(1 - v_1)}{E_1(1 - v_2)}\right)(1 - v_2) + (1 - v_2) + \dfrac{E_2(1 - v_1)}{E_1}} \tag{67}$$

The symbols E_1, E_2, v_1, and v_2 denote Young's modulus and Poisson's ratio of the annulus (index 1) and the axle (index 2), respectively. Finally, r_i and r_o denote the inner and outer radii of the annulus, respectively.

It is intuitively clear and can immediately be confirmed by Eq. (66) that the radial stresses σ_{rr} are *compressive* in nature, whereas the tangential stresses $\sigma_{\theta\theta}$, act in a *tensile* mode. The maximum tensile stress can be found at the inner surface of the annulus. Thus, the parameters s and s_{min} become:

$$s = \frac{\varepsilon}{1 + \varepsilon}\left(1 + \frac{r_0^2}{r^2}\right); \quad s_{min} = \frac{2\varepsilon}{1 + \varepsilon} \tag{68}$$

As in the case of the bar specimen, we now determine the volumes $V(s)$ and V:

$$V(s) = \pi\left(r - r_i^2\right)d; \quad V = \pi r_0^2(1 - \varepsilon)d \tag{69}$$

where d refers to the thickness of the annulus. By applying Eq. (60), it is now straightforward to determine the stress–density function of the case of a shrinkage fit annulus:

$$f(s) = \frac{\varepsilon(1 + \varepsilon)}{1 - \varepsilon} \frac{1}{[(1 + \varepsilon)s - \varepsilon]^2} \tag{70}$$

It is worth mentioning that in the case of the annulus, the stress distribution is, strictly speaking, no longer uniaxial. Rather, there are three principal stresses: the first one σ_{rr}, which is compressive; the second one $\sigma_{\theta\theta}$, which is tensile; and the third one σ_{zz}, which is equal to zero (for plane stress conditions). Indeed, as we shall see shortly, the approach presented in this section is consistent with a Weibull representation of failure under multiaxial stress conditions.

4. Conversion of Weibull Distributions for Different Specimen Types and Loading Conditions

In this section, we are going to provide an answer to the following problem. Suppose the Weibull parameters m and σ_0^A are known from a curve fit of experimental strength data using an A-type specimen and loading configuration. Is it then possible to predict the variability in strength (i.e., the Weibull distribution for a B-type specimen) and loading configuration made of the same material provided its state of stress is known?

We are going to answer this question first for the case of one-dimensional stress distributions, and following the equation for the Weibull fit (Eq. (49)), we write for A-type specimens:

$$P_A = 1 - \exp\left[-\left(\frac{\sigma_{max,A}}{\sigma_{0,A}}\right)^m\right] \tag{71}$$

where $\sigma_{0,A}$ is the scale parameter and $\sigma_{max,A}$ is the maximum tensile stress encountered in the A-type specimen. This will now be compared with the Weibull distribution for A-type specimens, which, according to Eq. (55), reads:

$$P_A = 1 - \exp\left[-\frac{1}{V_0}\int_{V_A}\left(\frac{\sigma_A(x)}{\sigma_0}\right)^m dV\right] \tag{72}$$

We conclude that:

$$V_0\sigma_0^m = (\sigma_{0,A})^m \int_{V_A}\left(\frac{\sigma_A(x,y,z)}{\sigma_{max,A}}\right)^m dV \tag{73}$$

This can be substituted into the analogue of Eq. (72) for B-type specimens:

$$
\begin{aligned}
P_B &= 1 - \exp\left[-\frac{1}{V_0}\int_{V_B}\left(\frac{\sigma_B(x)}{\sigma_0}\right)^m dV\right] \\
&= 1 - \exp\left(\frac{\int_{V_B}\left(\frac{\sigma_B(x)}{\sigma_{max,B}}\right)^m dV}{\int_{V_A}\left(\frac{\sigma_A(x)}{\sigma_{max,A}}\right)^m dV}\right)\left(\frac{\sigma_{max,B}}{\sigma_{0,A}}\right)^m
\end{aligned}
\tag{74}
$$

Therefore, provided that the one-dimensional stress distributions $\sigma^B(\underline{x})$ and $\sigma^A(\underline{x})$ are known, and m and $\sigma_{0,A}$ have been determined from A-type experiments, the one-dimensional B-type distribution can now be predicted by computation. Note that all Vs in that equation must be replaced by As if surface flaws are dominant.

Following the arguments of Sec. VI.B.3, we may replace the three-dimensional integrals in the last equation by one-dimensional integrals involving stress–density functions, which reduces the computational effort considerably:

$$P_B = 1 - \exp\left(-\frac{V_B}{V_A}\frac{\int_0^1 s^m f_B(s)\,ds}{\int_0^1 s^m f_A(s)\,ds}\left(\frac{\sigma_{max,B}}{\sigma_{0,A}}\right)^m\right) \tag{75}$$

Needless to say that by replacing the volumetrical stress–density functions $f_{A/B}(s)$ with $g_{A/B}(s)$, the corresponding relation for specimens governed by surface flaws can be obtained.

In a similar manner, the average strength of B-type configurations can be predicted from A-type measurements. To establish the conversion rule, we start from the definition of the mean value of strength $\bar{\sigma}$:

$$\bar{\sigma} = \int_0^\infty \sigma\frac{\partial P}{\partial \sigma}\,d\sigma \tag{76}$$

and conclude that:

$$\frac{\overline{\sigma}_A}{\overline{\sigma}_B} = \sqrt{\frac{V_B}{V_A} \frac{\displaystyle\int_{s_{min}}^1 s^m f_B(s)\,ds}{\displaystyle\int_{s_{min}}^1 s^m f_A(s)\,ds}} \tag{77}$$

We now turn to the case of a ceramic part subjected to *multiaxial* tensile stresses. In order to link the Weibull distribution to specimen volume and to the multiaxial state of stresses, Eq. (55) must be suitably changed. It was Weibull himself who first made a suggestion for the multiaxial case. According to him, tensile, and only tensile, principal stresses σ_1, σ_2, and σ_3 contribute to the probability of failure P in an *additive* manner as follows:

$$P = 1 - \exp\left(-\frac{1}{V_0} \int\int\int_V \left[\frac{\sigma_1^m + \sigma_2^m + \sigma_3^m}{\sigma_0^m}\right] dx\,dy\,dz\right) \tag{78}$$

However, Weibull's suggestion clearly neglects interactions between the principal stress components. Thus, it might lead to nonconservative predictions.

Note that, due to the complexity of multiaxial problems, the resulting stresses will usually no longer be available in analytical form. Rather, they need to be determined using suitable numerical methods, such as finite element analysis [65–67].

Various other attempts have been made to improve the assessment of multiaxially stressed ceramic components. Of special merit besides Weibull's original work are the flaw density and orientation concept developed by Batdorf and Crose [68] and Batdorf and Heinisch [69], the elemental strength approach of Evans [3], and the hypothesis of positive principal strains developed by Beierlein [70]. A critical comparison of the predictions of various multiaxial statistical fracture theories for brittle materials has been published by Chao and Shetty [71]. Also compare the articles (and references cited therein) by Lamon [72] and Nemeth et al. [57,65]. No further attempt is made here to present a unified statistical theory of multiaxial failure. This field of reliability analysis of brittle materials is still the subject of ongoing research. Experimental verification of the predictions of the various theories is sparse, and, consequently, the reader is referred to the aforementioned and other current literature.

However, if use is made of Weibull's hypothesis, then a conversion of the Weibull distribution between two types of multiaxially stressed type of specimens can be performed as follows (cf., Eqs. (78) and (74) for an explanation of the symbols used):

$$P_B(\sigma) = 1 - \exp\left(-\frac{\displaystyle\int\int\int_{V_B} \left[\frac{\sigma_{1,B}^m(\underline{x}) + \sigma_{2,B}^m(\underline{x}) + \sigma_{3,B}^m(\underline{x})}{\sigma_{max,B}^m}\right] dV}{\displaystyle\int\int\int_{V_A} \left[\frac{\sigma_{1,A}^m(\underline{x}) + \sigma_{2,A}^m(\underline{x}) + \sigma_{3,A}^m(\underline{x})}{\sigma_{max,A}^m}\right] dV} \left[\frac{\sigma_{max,B}}{\sigma_{0,A}}\right]^m\right) \tag{79}$$

5. Examples of Conversions Between Weibull Distributions

Figure 20 presents a comparison of conversions between Weibull plots for two relatively simple loading cases. Three-point and four-point bending experiments were performed using standard zirconia bars of rectangular cross-section, and the resulting strength data

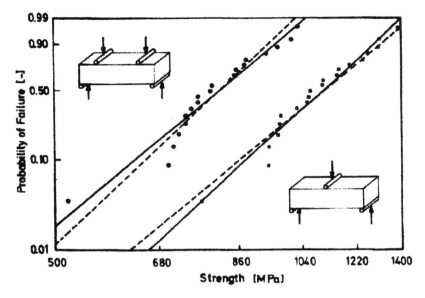

Figure 20 Conversion of Weibull distributions between three-point and four-point bending test specimens of the same volume.

were curve-fitted (solid lines) using a two-parameter Weibull distribution as shown in Eq. (49).* Moreover, the resulting shape parameters were used to theoretically predict Weibull distributions using Eq. (75), together with the appropriate stress–density functions of Table 2. Note that the dimensions of the specimens used for both types of tests were kept the same. Consequently, the figure allows assessing as to whether a conversion between two different one-dimensional loading cases can be represented reasonably accurate by the theory presented before. The agreement between both sets of results is fairly good.

Figure 21 presents the conversion between four-point bending and ring-on-disk tests performed with silicon nitride by Katayama and Hattori [73]. The predictions were based on Eqs. (75) and (79). Clearly, this is an example of a conversion between uniaxially and biaxially stressed bodies. In the same context, the work by Giovan and Sines [74] should be mentioned. The agreement between prediction and experiment is still fair, indicating that Weibull's hypothesis certainly captures the important features of strength distribution even in the case of multiaxially stressed ceramics.

Finally, Fig. 22 presents a conversion between silicon dies used for telephone card applications (i.e., a cubic anisotropic material) of different thickness (0.130 and 0.185 mm) tested in a double-ring test rig [75]. Obviously, the influence on specimen thickness is subtle and the theory (after taking the anisotropy into account by using suitable averages of Young's modulus) is not convincing although pointing in the right direction.

Clearly, additional experiments using brittle specimens of complex shape and loading conditions showing isotropic as well as anisotropic material behavior must be performed in

---

*All curves are represented in so-called Weibull format, which follows naturally by applying the logarithm twice to Eq. (75). Consequently, the ordinate is logarithmical and the abscissa is represented as $\ln \ln (1/1-P)$.

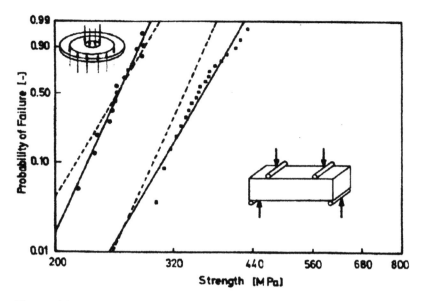

Figure 21 Conversion of Weibull distributions between four-point bending and expanded ring tests. (From Ref. 73.)

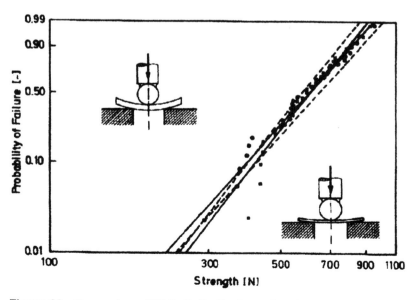

Figure 22 Conversion of Weibull distributions of surface-treated Si dies examined in a modified double-ring test. (From Ref. 75.)

order to gain complete confidence in the theoretical approach and to open Weibull theory to new "applications" (e.g., for the materials used in the biomedical devices and microelectronic sectors).

VII. LIFETIME PREDICTIONS

As indicated in Fig. 23, we will now combine the techniques and results of previous sections to construct a lifetime diagram for components made of brittle materials. In other words, we will construct a diagram that combines maximum applied static load S, probability of failure P, and time to failure T. For obvious reasons, this diagram is also known as SPT representation.

A. Derivation of the SPT Relation

Similar to Sec. V.C, we consider a static load, which, in magnitude, is equal to the dynamically applied stress to failure σ of the Weibull distribution shown, for example, in Eq. (49) or to $\sigma_{\max,\mathrm{B}}$ of Eqs. (74), (75), and (79). For conciseness, we shall use the symbol σ from now on. In order to get a foothold onto the SPT relation, we start from Eq. (36). We conclude that the following relationship must hold between the time to failure t of a

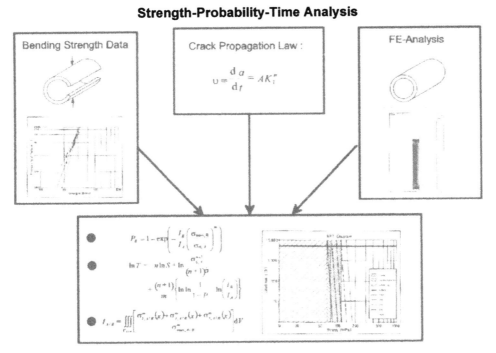

Figure 23 "Flow chart" illustrating the approach to lifetime prediction of structures made of brittle materials.

specimen subjected to a maximal load σ and the arbitrary time to failure T corresponding to an arbitrary static load S:

$$\frac{t}{T} = \left(\frac{S}{\sigma}\right)^n \tag{80}$$

According to Eq. (41), we may write:

$$\sigma = \dot{\sigma}(n+1)t \tag{81}$$

where $\dot{\sigma}$ is the loading rate of the bending tests used to determine the two Weibull parameters m and σ_0^A. Taking the logarithms of Eq. (80) and substituting the last equation, it follows that:

$$(n+1)\ln \sigma = n \ln S + \ln T + \ln(n+1)\dot{\sigma} \tag{82}$$

We now rearrange the Weibull distribution shown in Eqs. (73), (74), and (78):

$$\ln \ln \frac{1}{1-P} = \ln \frac{I_B}{I_A} + m \ln \sigma \dot{-} m \ln \sigma_{0,A} \tag{83}$$

where I_A and I_B denote the Weibull integrals necessary for conversion between test and actual specimen types A and B, respectively (i.e., for one-dimensional stress configurations):

$$I_{A/B} = \int_{V_A} \left(\frac{\sigma_{A/B}(\underline{x})}{\sigma_{\max,A/B}}\right)^m dV \tag{84}$$

or:

$$I_{A/B} = \int_0^1 s^m f_{A/B}(s) ds \tag{85}$$

and for multiaxial states of stress according to Weibull's hypothesis:

$$I_{A/B} = \int\int\int_{V_{A/B}} \left[\frac{\sigma_{1,A/B}^m(\underline{x}) + \sigma_{2,A/B}^m(\underline{x}) + \sigma_{3,A/B}^m(\underline{x})}{\sigma_{\max,A/B}^m}\right] dV \tag{86}$$

We now eliminate the stress σ from Eqs. (82) and (83) to yield:

$$\ln T = -n \ln S + \frac{(n+1)}{m}\left\{\ln \ln \frac{1}{1-P} - \ln\left(\frac{I_B}{I_A}\right)\right\} + \ln \frac{\sigma_{0,A}^{n+1}}{(n+1)\dot{\sigma}} \tag{87}$$

This is the SPT relationship, which allows predicting the lifetime of a B-type specimen using the material parameters m and $\sigma_{0,A}$, which are known from experiments using a simple, inexpensive, A-type configuration.

B. An Example of an SPT Analysis

As an example of an SPT analysis, consider the case of a ceramic pipe made of alumina used as part of a transmitter tube, as indicated in Fig. 24. The pipe is soldered onto a copper base as indicated in the FE mesh shown in Fig. 25. During operation, the temperature increases considerably, and due to the thermal mismatch inherent to the system, tensile stresses arise in the ceramic pipe that lead to its cracking over time.

Figure 24   Transmitter tubes involving use of a ceramic pipe.

In the first step, an FE analysis was performed to assess the state of stress and, in combination with a postprocessor, to evaluate the Weibull integrals shown in Eq. (86).

Moreover, C-rings were prepared from remnants of broken pipes and used to determine all material parameters essential to an SPT lifetime analysis (Fig. 26). The Weibull parameters m and $\sigma_{0,A}$, as well as the subcritical crack growth parameter n were determined in C-ring strength tests [76]. As an example, the Weibull fit is shown in Fig. 27 together with 90% confidence intervals [77]. Clearly, in order to evaluate Eqs. (86) and (87) properly, the stresses within the C-rings (which, in the present case, is the A-type specimen) had to be taken into account.

Figure 28 finally presents the SPT lifetime diagram (i.e., a graphical representation of Eq. (87)). It allows, for a to-be-guaranteed average lifetime, the prediction of the maximal static stress (i.e., in the present case, change in temperature), which can be applied without risking a certain percentage of failed specimens.

C. Automated SPT Analysis Through Use of Software

As it was demonstrated in Sec. VII.B and summarized in Fig. 23, various steps are involved in the generation of an SPT diagram, ranging from strength data analysis and processing to a (sometimes) sophisticated stress analysis of the brittle component. Consequently, the numerical effort involved varies between the elementary and the sophisticated. However, in any case, it is tedious and time-consuming if performed manually. In order to guarantee an accurate and swift analysis, such work is preferably performed using suitable computer software. Not surprisingly, various professional software tools are available. However, before these are discussed in more detail, it should be mentioned that the processing of strength data, Weibull curve fitting, and subcritical crack growth parameter analysis is easy to implement and visualize, for example, by using universal software, such as MS-EXCEL.

An example of professional reliability software to be used for reliability analysis of brittle components is the CARES package, which is an acronym for Ceramics Analysis and Reliability Evaluation of Structures. The CARES software was developed at the NASA Lewis Research Center and it can be used by professionals to optimize the design and manufacture of brittle material components.

CARES is an integrated package that predicts the probability of a monolithic ceramic component's failure as a function of its time in service. It can be coupled with commercial finite element programs, which determine stress distribution within the structure, with reliability evaluation and fracture mechanics routines for modeling strength-limiting

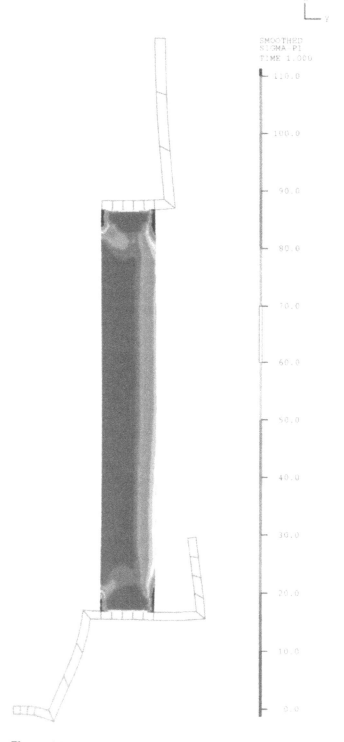

Figure 25   FE mesh of the ceramic pipe soldered onto a copper base (shown in part).

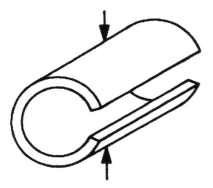

Figure 26 A C-ring specimen.

defects, the way it was outlined in Sec. VII.B. Further information can be obtained from the CARES web page http://www.grc.nasa.gov/WWW/RT1996/5000/5250n.htm), or by consulting the manuals and papers of the developers [57,65,66].

Another example of reliability software of brittle materials is STAU (an acronym for STatistische AUswertung or statistical evaluation) developed at the Institut für Materialforschung of the Forschungszentrum in Karlsruhe, Germany. Using multiaxial Weibull theory, the program determines the probability of failure based on a previous finite element stress analysis. STAU uses FE interpolation formulae and does not require any additional mesh refinement to perform statistical analysis. Static as well as dynamic loading conditions of the brittle structure can both be taken into account. Typical applications involve severely stressed components in turbine machinery, chemically inert parts used in chemical engineer-

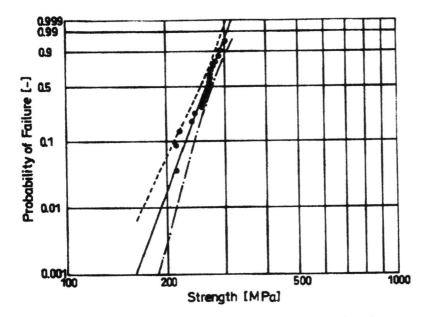

Figure 27 Weibull fit of the strength data obtained for alumina pipes.

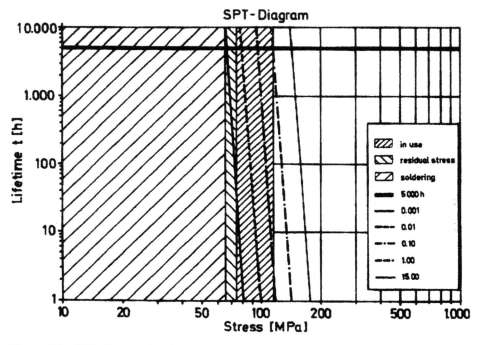

Figure 28   SPT diagram for alumina pipes.

ing showing low mechanical strengths, as well as biomedical applications such as dental ceramics [78].

Another alternative is the IBM-PC-driven software CERAMTEST, which was developed by Müller. CERAMTEST is basic, menu-driven, graphics-oriented software for reliability analysis of brittle materials, which allows Weibull fits and (simple) lifetime prediction calculations. The analysis is based on the methods outlined in previous sections. Stress–density functions of relevant uniaxially stressed engineering specimens have been implemented (cf., Tables 2 and 3), which, by using material data obtained for an A-type specimen configuration, allow to predict the lifetime of B-type configurations.

VIII. EXPERIMENTAL STRESS ANALYSES: MICROSTRESSES

A. New Piezospectroscopical Techniques Using Laser Microprobe Excitons

Raman and fluorescence spectroscopies using laser microprobes are becoming widely used to study stress distributions in solids because of their potential for high spatial resolution (e.g., in microelectronic device structures [79], biomedical-related ceramic coatings [80], and hybrid organic/inorganic composites [81]). The high spatial resolution that becomes nominally available with conventional lasers has led to the techniques being applied for characterizing microstress distributions that vary significantly within small (micron-scale) volumes. Typically, the diameter of the focussed laser beam is 1 μm and the penetration depth can vary from tens of nanometers to several millimeters, according to the transparency characteristics of the material. Therefore, a potentially wide volume range of the stressed structure can be probed. Each point within the probed volume scatters light with a wave number characteristic of the local stress at that point. However, the interpretation of the spectrum shift upon stress is material-dependent and complications arise from two factors:

(i)   Both depth and distribution of light intensity in the probed volume affect the peak shift.

(ii)   All components of the stress tensor may, in principle, contribute to the observed peak shift (according to unknown fractions), thus making the deconvolution of a scalar peak shift value in terms of individual tensorial stress components difficult.

As for solving the above point:

(i)   Spatially resolved experimental stress analysis methods, aided by theoretical calculations of Raman spectra, have been recently proposed and applied to the field of microelectronics [82].

(ii)   Attempts to deconvolute deviatoric stress components from hydrostatic ones in terms of observed shift and width of fluorescence (ruby) lines have also been presented for rationalizing residual stress fields in sintered alumina [83]. However, the details of peak shift deconvolution depend on the type of material and are not straightforward for complex stress states.

Experimental difficulties have so far hindered a fully quantitative characterization of the stress tensor in ceramics using Raman or fluorescence microprobe spectroscopy. Several key issues remain unsolved in this kind of microscopical fracture analysis; challenges including a full three-dimensional (quantitative) map of the microstress field. However, these new experimental assessment methods present the following useful features:

(i)   Two-dimensional maps of Raman or fluorescence peak shift can be automatically collected and transformed in maps of the trace stress tensor over relatively large areas in the neighborhood of stressed regions or cracks.

(ii)   Experimental procedures can be attempted to deconvolute contributions to fluorescence peak shift arising from residual (i.e., preexisting) and externally applied stresses.

(iii)   The dependence of the stress field on the laser spot size (and depth) can be systematically analyzed, thus varying also the penetration depth of the laser in the specimen subsurface.

The above piezospectroscopical characterizations of microscopical stress fields should be compared to more traditional fracture mechanics analyses, such as the assessment of fracture toughness and rising R-curve behavior, and crack opening displacement (COD) in toughened ceramics. It is also of importance to note that Raman microprobe assessments may enable one to characterize, concurrently to microstresses, local phase transformation processes, which may originate from the microstresses themselves.

B. Theoretical Background for Piezospectroscopical Stress Analysis

Monitoring the fluorescence spectrum of ceramic materials can provide quantitative information on the local stress state because, under strain, the energy of electronic excitation levels is perturbed. Therefore, the energy of the emitted photons, when excited electrons return to the ground state, is strain-dependent, which manifests with a fluorescence line shift. Stress is proportional to strain, if experiments are conducted in the elastic regime, as it applies to the majority of ceramics. Fluorescence of chromium (i.e., ruby Cr^{3+}) in the alumina (Al_2O_3) lattice, as well as in other oxygen crystal environments (e.g., MgO and $Mg\,Al_2O_4$), manifests with sharp, intense lines whose hydrostatic stress dependence has been long known and characterized [84,85]. Another spectroscopical technique that can also provide stress information in ceramic materials is based on the Raman effect. In this case, stress (rather than strain) information is directly achieved because the stress changes the frequency of the vibrational modes of the ceramic lattice. In other words, stress alters the Raman wave numbers. Raman bands of many ceramics are known (e.g., zirconia, ZrO_2, silicon carbide, SiC, aluminum nitride, AlN, silicon nitride, Si_3N_4, etc.). However, unlike the fluorescence lines of Cr^{3+} in some oxides, the Raman lines of most ceramics of technological interest have a rather low intensity and a strong chemical dependence. Thus, provided that the stress dependence of such weak Raman bands is detectable, any piezospectroscopical characterization that aims to use a Raman effect in ceramics must start with precise, case-by-case calibration procedures.

In a first-order approximation, the stress dependence of a fluorescence band position is governed by a linear tensorial equation:

$$\Delta v = \Pi_{ij}\sigma_{ij}$$
(88)

where Δv is the fluorescence peak frequency shift, σ_{ij} is the stress tensor, and Π_{ij} is a matrix of piezospectroscopical coefficients.

In the case of polycrystalline samples, the tensorial expression loses its meaning because micron-sized crystallites, which possess individual orientations, are present in the probed volume. However, an approximate expression may still hold, which relates the measured frequency shift Δv to a mean stress $\langle\sigma\rangle_h$, which is assumed to be of a hydrostatic nature (index h) within the probed volume:

$$\langle\Delta v\rangle = \mathrm{tr}\underline{\underline{\Pi}}\langle\sigma\rangle_h$$
(89)

where $\mathrm{tr}\underline{\underline{\Pi}}$ is the trace of the piezospectroscopical tensor. If a simple uniaxial stress $\langle\sigma\rangle_{uni}$ is applied to the polycrystalline body, Eq. (89) can be expressed using an average of the diagonal piezospectroscopical tensor $\Pi_{uni} = \mathrm{tr}\underline{\underline{\Pi}}/3$:

$$\langle\Delta v\rangle = \Pi_{uni}\langle\sigma\rangle_{uni}$$
(90)

Clearly, error is involved in using Eq. (89), which arises from neglecting the dependence of the piezospectroscopical tensor on the crystallographical orientation of individual crystallites. In the case of a general three-dimensional stress state, additional error may arise from considering the stress state as simply equitriaxial. Obviously, the impact of these errors on the final outcome of the piezospectroscopical analysis depends on the ratio between the probing size and the characteristic scale of the analyzed phenomenon. However, the actual magnitude of this error must be estimated case by case by preliminary calibrations to properly assess the reliability of the microscopical stress determination. Π_{uni} and $\mathrm{tr}\Pi$ coefficients of Al_2O_3, for example, can be calibrated from the shift of the R_2 fluorescence line, which is less sensitive to the (unknown) crystallographical orientation of individual grains. For this line, the following is assigned $\Pi_c = -2.80$ cm^{-1}/GPa and $\Pi_a = -2.16$ cm^{-1}/ GPa for the a-axis and c-axis of the sapphire single-crystal, respectively [86]. In nontextured polycrystals, the value $\Pi_{uni} = \mathrm{tr}\Pi/3 = (\Pi_c + \Pi_a)/3$ is usually accepted. The error involved in neglecting the crystallographical orientation of the Al_2O_3 grains has been assessed to be less than 10% [83].

In a two-phase composite, the spatial average of stress $\langle\sigma\rangle_h$ within the probed volume arises from superposition of two residual stress contributions; first, the isotropic stress contribution $\langle\sigma\rangle_i$, due to thermal expansion mismatch between the alumina and the metallic phase; second, the deviatoric stress contribution $\langle\sigma\rangle_a$, due to thermal expansion anisotropy between the a-axis and c-axis of Al_2O_3:

$$\langle\sigma\rangle_h = \langle\sigma\rangle_i + \langle\sigma\rangle_a \tag{91}$$

The stress components can be represented through their corresponding fluorescence peak shifts [86]:

$$\langle\Delta v\rangle_i = (2\Pi_a + \Pi_a)\langle\sigma\rangle_i \tag{92}$$

$$\langle\Delta v\rangle_a = -(\Pi_a - \Pi_c)\langle\sigma\rangle_a \tag{93}$$

The stress component $\langle\sigma\rangle_a$ is grain size-dependent and can be experimentally assessed by determining the shift $\langle\Delta v\rangle_a$ from the wave number difference in peak position between the polycrystalline body (e.g., a single-crystal sapphire). The shift $\langle\Delta v\rangle_a$ can be experimentally determined using a monolithic alumina body and compared to results obtained with those from a sapphire single crystal. In the absence of externally applied stresses, this comparison allows one to separate $\langle\Delta v\rangle_a$ from $\langle\Delta v\rangle_i$.

An externally applied stress will produce an additional shift of the fluorescence peak, thus complicating the microprobe stress analysis. In many cases of interest, the overall stress state is given by the superposition of an externally applied (e.g., uniaxial) stress and a three-dimensional residual stress field. The shift due to the external field is thus given by:

$$\langle\Delta v\rangle_{\text{ex}} = \frac{2\Pi_a + \Pi_a}{3}\langle\sigma\rangle_{\text{ex}} \tag{94}$$

If one assumes an equitriaxial state for the residual stress component, the piezospectroscopical relation for the total peak shift $\langle\Delta v\rangle_{\text{tot}}$ can be expressed as:

$$\begin{aligned}
\langle\Delta v\rangle_{\text{tot}} &= \langle\Delta v\rangle_{\text{ex}} + \langle\Delta v\rangle_i + \langle\Delta v\rangle_a \\
&= (\Pi_a + \Pi_a)\left(\frac{\langle\sigma\rangle_{\text{ex}}}{3} + \langle\sigma\rangle_i\right) - (\Pi_a + \Pi_a)\langle\sigma\rangle_a
\end{aligned} \tag{95}$$

Because the experimentally measurable shift is only $\langle\Delta v\rangle_{\text{tot}}$, Eq. (95) shows that it is not possible by a single peak shift measurement to deconvolute individual contributions by both residual and external stress components. Therefore, two independent peak shift measurements, in the absence and presence of external (or residual) stress, are required.

C. Application of the Piezospectroscopical Microprobe Technique

As an example of the assessment of microscopical stress field by microprobe piezospectroscopy, the microscopical stress field developed in the neighborhood of a propagating crack is shown in this section. The crack is monitored in a model composite consisting of an alumina matrix dispersed with metal molybdenum particles. A stably propagated crack was arrested after an extension of several hundred microns and after observation by an optical microscope, with the entire bending jig placed into the fluorescence microprobe spectroscopy apparatus. Determination of local stresses by ruby fluorescence peak shift was performed both at critical external load and upon unloading, at zero external load. The crack could be concurrently observed by monitoring the optical microscope image. Figure 29(a) represents a low magnification optical image of the equilibrium crack. Figure 29(b) and c presents microscopical stress maps obtained from the observed area (cf., the total area in inset in (a)), with a laser spot of 1 µm, in the unloaded and critically loaded configurations, respectively. These maps were obtained from experimentally determined peak shifts, via Eq. (95), with

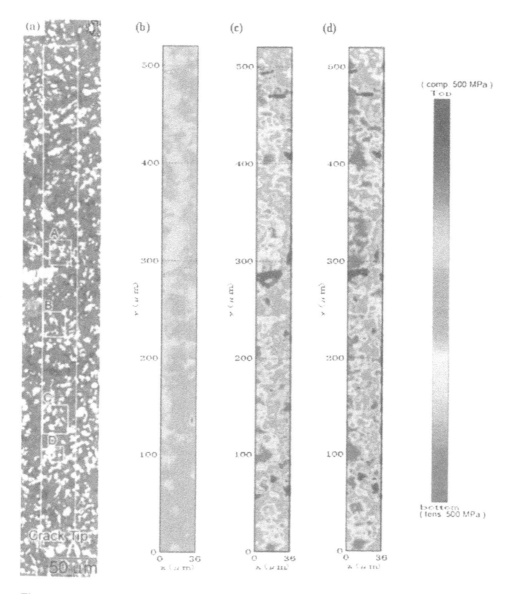

Figure 29 Optical micrograph (a) of a crack propagating in the Al$_2$O$_3$/Mo composite with related microstress maps (collected with a laser spot of 1 μm) at zero external load (b) and at critical external load for crack propagation (c). (d) represents the stress field obtained from the algebraical overlapping of the stress fields in (b) and (c). Locations A, B, C, and D are typical regions in which crack face bridging by metallic particles was observed.

residual and external stresses treated as equitriaxial and uniaxial stresses, respectively. Figure 29(d) represents the algebraical sum of the maps in (b) and (c), thus being the net stress field arising from the superposition of an external uniaxial load. A reference zero-stress spectrum for building stress maps in (b) and (c) could be obtained either from fine alumina powder, or from a finely crushed single-crystal sapphire. A difference of less than 3% was found in the stress assessment between the above two choices.

The residual stress map (b) reveals main regions of negligible stress with minor compressive areas of weak magnitude. At critical load, a tensile stress field—up to several hundreds of megapascals—was observed along the crack path. This tensile stress field arises from the presence of unbroken metal particles subjected to tractions in the crack wake. Therefore, microscopy evidence provided demonstrates clearly a crack bridging effect by metallic particles. This mechanism could be responsible for the rising R-curve effect observed in this material [87]. The maximum bridging traction monitored over the crack length was measured as about 400 MPa. The magnitude of the maximum (tensile) bridging traction was not dependent on the observed metal particle size and a total number of 50 bridging molybdenum particles were observed.

The stress maps in Fig. 29 give a vivid semiquantitative representation of the stress state in the neighborhood of an advancing crack, showing the functionality and usefulness of the fluorescence microprobe technique in revealing stress intensification around the stress concentrators such as cracks and notches.

IX. CONCLUSIONS

In order to assess the quality and lifetime of ceramic materials and products, three major steps need to be performed:

> Analysis of strength data using Weibull statistics and subcritical crack growth behavior
>
> Analytical/experimental or numerical stress analysis of the brittle material/component/structure
>
> Combination of the above in a strength–probability–time diagram.

The state of the art allows for an accurate lifetime and reliability analysis of "simple" monolithic specimens made of glass or ceramic (i.e., one-dimensional or two-dimensional loading conditions for elementary geometries, such as bars, rods, and circular structures). Although the theory is general enough to handle more complex forms and loads, all predictions are to be viewed with care and still need to be validated or at least complimented by experimental work. The same is true for ceramic composite materials as well as for brittle anisotropic substances used in the biomedical and semiconductor industries.

X. SELECTED NOTATION

a	Major axis of Griffith ellipse; crack length parameter; indentation radius
A	First subcritical crack growth parameter; specimen surface; index of A-type specimens
A_0	Reference surface
a_c	Critical crack length
b	Minor axis of Griffith ellipse
B	Height of bar specimen; index of B-type specimens

c	Crack length plus radius of Vickers indent
Δa	Incremental increase of crack length
γ	Specific surface energy
d	Width of bar specimen; thickness of ceramic annulus
e	Lever arm of four-point bending specimen
E	Young's modulus
$f(s)$	Stress–density function for volumetrical flaw distributions
ϕ	Angle in front of crack tip
$g(s)$	Stress–density function for surface flaw distributions
G	Energy release rate
G_{Ic}	Critical energy release rate
H_V	Vickers hardness
K_I	Stress intensity factors for mode I
K_{II}	Stress intensity factors for mode II
K_{III}	stress intensity factors for mode III
K_{Ic}	Fracture toughness
l	Lower support of bending specimen
m	Weibull modulus
μ	Shear modulus
n	Second subcritical crack growth parameter (n-value)
N	Number of specimens within a batch
v	Poisson's ratio
P	Load applied to bar specimens; probability of failure
r	Radial distance in front of crack tip
r_i	Inner radius of ceramic annulus
r_o	Outer radius of ceramic annulus
s	Stress–density integration parameter
S	Surface energy per unit length; arbitrary static stress
σ	Stress applied normal to crack plane
σ_c	Critical stress value
σ_{min}	Threshold strength
σ_0	Weibull scale parameter
$\bar{\sigma}$	Mean value of strength
$\dot{\sigma}$	Loading rate
Σ	Arbitrary (maximum) tensile stress
t	time
t_f	Time to failure
t_f^{st}	Static lifetime
t_f^{dyn}	Dynamic lifetime
T	Lifetime associated with arbitrary static stress
U	Change in elastic energy due to presence of crack (per unit length)
V	Specimen volume
V_0	Reference volume
Y	Correction function of stress intensity factor

REFERENCES

1. Lawn, B.R.; Wilshaw, T.R. *Fracture of Brittle Solids*; Cambridge University Press: Cambridge, 1975.
2. Marshall, D.B. Strength characteristics of transformation toughened zirconia. J. Am. Ceram. Soc. 1986, *66* (4), 277–283.

3. Evans, A.G. A general approach for the statistical analysis of multiaxial fracture. J Am. Ceram. Soc. 1978, *61* (7–8), 302–308.
4. Hasselman, D.P.H. Unified theory of thermal shock fracture initiation and crack propagation in brittle ceramics. J. Am. Ceram. Soc. 1969, *52* (11), 600–607.
5. Lange, F.F. Sinterability of agglomerated powders. J. Am. Ceram. Soc. 1984, *67* (2), 83–89.
6. Evans, A.G.; Faber, K.T. Crack-growth resistance of microcracking brittle materials. J. Am. Ceram. Soc. 1984, *67* (4), 255–260.
7. Hartsock, D.L.; McLean, F. What the designer with ceramics needs. Am. Ceram. Soc. Bull. 1984, *63* (2), 266–270.
8. Richerson, D.W. Applications of modern ceramic engineering. Mech. Eng. 1982, *104* (12), 24–33.
9. Weibull, W. A statistical theory of the strength of materials. Ingenieurvetenskapsakademien Handlingar, Stockholm 1939, *151*, 5–45.
10. Griffith, A.A. The phenomena of rupture and flow in solids. Philos. Trans. Ser. A 1920, *221*, 163–198.
11. Sih, G.C.; Liebowitz, H. On the Griffith energy criterion for brittle fracture. Int. J Sol. Struct. 1967, *3*, 1–22.
12. Irwin, G.R. Analysis of stresses and strains near the end of a crack traversing a plate. J Appl. Mech. 1957, *24*, 361–364.
13. Hahn, H.G. Bruchmechanik, Einführung in die theoretischen Grundlagen. Mechanik Band, 30, Teubner Verlag: Stuttgart, 1977.
14. Anderson, T.L. *Fracture Mechanics—Fundamentals and Applications*; CRC Press: Boca Raton, 1991.
15. Tada, H.; Paris, P.C.; Irwin, G.R. *The Stress Analysis of Cracks Handbook*; Paris Productions, Inc.: St. Louis, MO, 1985.
16. Murakami, Y., Ed.; *Stress Intensity Factors Handbook*; Pergamon Press: Oxford, 1988.
17. Fujii, T.; Nose, T. Evaluation of fracture toughness for ceramic materials. ISIJ Int. 1989, *29*, 717–725.
18. Bertolotti, R.L. Dependence of fracture strength on strain rate of polycrystalline alumina. Ceramurgia Int. 1975, *1* (1), 33–34.
19. Kriz, K. Einfluß der Mikrostruktur auf die langsame Rißausbreitung und mechanische Eigenschaften von heißepreßtem Siliziumnitrid zwischen Raumtemperatur und 1500°C. Ph.D. Thesis, Universitt Stuttgart, 1983.
20. Nishida, T.; Hanaki, Y.; Pezzotti, G. Effect of notch-root radius on the fracture toughness of a fine-grained alumina. J. Am. Ceram. Soc. 1994, *77* (29), 606–608.
21. Evans, A.G.; Charles, E.A. Fracture toughness determinations by indentation. J. Am. Ceram. Soc. 1976, *59* (7–8), 371–372.
22. Lawn, B.R.; Evans, A.G.; Marshall, D.B. The median/radial crack system. J. Am. Ceram. Soc. 1980, *63* (9–10), 574–581.
23. Anstis, G.R.; Chantikul, P.; Lawn, B.R.; Marshall, D.B. A critical evaluation of indentation techniques for measuring fracture toughness: I. Direct crack measurements. J. Am. Ceram. Soc. 1981, *64* (9), 533–538.
24. Chantikul, P.; Anstis, G.R.; Lawn, B.R.; Marshall, D.B. A critical evaluation of indentation techniques for measuring fracture toughness: II. Strength method. J. Am. Ceram. Soc. 1981, *64* (9), 539–543.
25. Marshall, D.B.; Tatsuo, N.; Evans, A.G. Simple method for determining elastic modulus to hardness ratios using Knoop indentation measurements. J. Amer. Ceram. Soc. 1982, *65* [N-10], C175–C176.
26. Evans, A.G.; Fuller, E.R. Crack propagation in ceramic materials under cyclic loading conditions. Metall. Trans. 1974, *5*, 27–33.
27. Laugier, M.T. The elastic/plastic indentation of ceramics. J. Mater. Sci. Lett. 1985, *4*, 1539–1541.
28. Shaobo, X.; Guangxia, L.; Changchun, L. Application of indentation technique in determining fracture toughness of ceramics. Eng. Fract. Mech. 1988, *31* (2), 309–313.

29. Zeng, K.; Border, K.; Rowcliffe, D.J. The Hertzian stress field and formation of cone cracks: I. Theoretical approach. Acta Metall. Mater. 1992, *40* (10), 2595–2600.

30. Lutton, P.; Ben-Nissan, B. The status of biomaterials for orthopaedic and dental applications: Part I. Materials. Mater. Technol. 1997, *12* (2), 59–63.

31. Suresh, S.; Shih, C.F.; Morrone, A.; O'Dowd, N.P. Mixed-mode fracture toughness of ceramic materials. J. Am. Ceram. Soc. 1990, *73* (5), 1257–1267.

32. Richard, H.A. Bruchvorhersagen bei überlagerter Normal- und Schubbeabspruchung sowie reiner Schubbelastung von Rissen. Habilitation Thesis, Fachbereich Maschinenwesen, Universität Kaiserslautern, 1984.

33. Charles, R.J. Static fatigue of glass, II. J. Appl. Phys. 1958, *29* (11), 1554–1560.

34. Hillig, W.B. Sources of weakness and the ultimate strength of brittle amorphous solids. In *Modern Aspects of the Vitreous State*; MacKenzie, J.D., Ed.; Butterworth: Washington, 1962, 152–194 pp.

35. Charles, R.J. Static fatigue of glass. I. J. Appl. Phys. 1958, *29* (11), 1549–1553.

36. Charles, R.J. A review of glass strength. In Progress in Ceramic Science; Burke, J.E.; Pergamon Press: Oxford, 1961, 1–38 pp.

37. Wiederhorn, S.M. Moisture assisted crack growth in ceramics. Int. J. Fract. Mech. 1968, *4* (2), 171–177.

38. Wiederhorn, S.M. Mechanisms of subcritical crack growth in glass. In Fracture Mechanics of Ceramics; Bradt, R.C. Hasselman, D.P.H., Lange, F.F., Eds; Plenum: New York, 1978; Vol. 2, 549–580.

39. Evans, A.G. A method for evaluating the time-dependent failure characteristics of brittle materials and its application to polycrystalline alumina. J. Mater. Sci. 1972, 7, 1137–1146.

40. Evans, A.G.; Wiederhorn, S.M. Crack propagation and failure prediction in silicon nitride at elevated temperatures. J. Mater. Sci. 1974, *9* (2), 270–278.

41. Li, L.; Weick, J.M.; Pabst, R.F. Bemerkungen und Ergebnisse zur unterkritischen Rißausbreitung an Doppel-Torsionsproben. Dtsch. Keram. Ges. 1980, *57*, 5–9.

42. Charles, R.J.; Hillig, W.B. The kinetics of glass failure by stress corrosion. *Compte Rendu of Symposium sur la Résistance Mécanique du Verre et les Moyens de l'Améliorer*; 1961; 511–527.

43. Hillig, W.B.; Charles, R.J. Surfaces, stress-dependent surface reactions, and strength. In High-Strength Materials—Present Status and Anticipated Developments, Proceedings of the Second Berkeley International Materials Conference, San Francisco; Zackay, V.F., Eds.; 1964; 682–705.

44. Brown, S.D. Multibarrier kinetics of brittle fracture: I. Stress dependence of the subcritical crack velocity. J. Am. Ceram. Soc. 1979, *62* (9–10), 516–524.

45. Evans, A.G.; Fuller, E.R. Crack propagation in ceramic materials under cyclic loading conditions. Metall. Trans. 1974, *5*, 27–33.

46. Evans, A.G.; Jones, R.L. Evaluation of a fundamental approach for the statistical analysis of fracture. J. Am. Ceram. Soc. 1978, *61* (3–4), 156–160.

47. Jakus, K.; Coyne, D.C.; Ritter, J.E. Jr. Analysis of fatigue data for lifetime predictions for ceramic materials. J. Mater. Sci. 1978, *13*, 2071–2080.

48. Jakus, K.; Ritter, J.E. Jr. Lifetime prediction for ceramics under random loads. Res. Mech. 1981, *2* (1), 39–52.

49. Pabst, R.F. Möglichkeiten zur Charakterisierung der langsamen Riausbreitung bei keramischen Werkstoffen. Ber. Dtsch. Keram. Ges. 1980, *57*, 1–12.

50. Richter, H. Langsame Rißausbreitung und Lebensdauerbestimmung. Vergleich zwischen Rechnung und Experiment. Ber. Dtsch. Keram. Ges. 1980, *57* (1), 10–12.

51. Tradinik, W.; Kromp, K.; Pabst, R.F. A combination of statistics and subcritical crack extension. Mater. Sci. Eng. 1982, *56*, 39–46.

52. Evans, A.G. Fatigue in ceramics. Int. J. Fract. 1980, *16* (6), 485–498.

53. Lauf, S. Über die Anwendbarkeit und Problematik von dynamischen und zyklodynamischen Versuchen zur Lebensdauerberechnung bei Hochleistungskeramik. Ph.D. Thesis, MPI, Stuttgart, 1986.

54. American Society for Testing Materials C 1293-00. Standard practice for reporting uniaxial strength data and estimating Weibull distribution parameters for advanced ceramics. Book of Standards: Philadelphia, PA, 2001; Vol. 15.01.

55. *DIN 51110: Teil 3. Prüfung von keramischen Hochleistungswerkstoffen, 4-Punkt-Biegeversuch, Statistische Auswertung, Ermittlung der Weibull-Parameter*; Deutsche Norm: Berlin, 1993; 1–16 pp.

56. Nadler, P. Beitrag zur Charakterisierung und Berücksichtigung des spezifisch keramischen Festigkeitsverhaltens. Ph.D. Thesis, Berkakademie Freiberg, 1989.

57. Nemeth, N.N.; Manderscheid, J.M.; Gye-kenyesi, J.P. Ceramics analysis and reliability evaluation of structures (CARES). NASA Technical Paper 1990; 2916 pp.

58. Hosokawa, T. Statistical distribution of modulus of rupture for ceramic materials. Ph.D. Thesis, University of Washington, 1978.

59. Marshall, D.B.; Lawn, B.R.; Evans, A.G. Elastic/plastic indentation damage in ceramics: the lateral crack system. J. Am. Ceram. Soc. 1982, *65*, (11), 561–566.

60. Jayatilaka, A.D. *Fracture of Engineering Brittle Materials*; Applied Science Publishers: London, 1979.

61. Nadler, P. Die Anwendung der statistischen Festigkeitstheorie in der keramischen Werkstoffprüfung, Teil 1. Hermsd. Tech. Mitt. 1972, *33*, 1031–1038.

62. Nadler, P. Die Anwendung der statistischen Festigkeitstheorie in der keramischen Werkstoffprüfung, Teil 2. Hermsd. Tech. Mitt. 1974, *40*, 1262–1271.

63. Timoshenko, S.; Goodier, J.N. *Theory of Elasticity*; McGraw-Hill Book Company, Inc.: New York, 1951.

64. Müller, W.H.; Schmauder, S. Interface stresses in fiber-reinforced materials with regular fiber arrangements. Compos. Struct. 1992, *24* (1), 1–21.

65. Nemeth, N.N.; Manderscheid, J.M.; Gyekenyesi, J.P. Designing ceramic components with the CARES computer program. Am. Ceram. Bull. 1989, *68* (12), 2064–2072.

66. Szatmary, S.A.; Gyekenyesi, J.P.; Nemeth, N.N. Calculation of Weibull strength parameters: Batdorf flaw density constants and related statistical quantities using PC-CARES. NASA Technical Memorandum 103247, 1990.

67. Ben-Nissan, B. Review: reliability and finite element analysis in ceramic engineering design. Mater. Forum 1993, *17*, 105–125.

68. Batdorf, S.B.; Crose, J.G. A statistical theory for the fracture of brittle structures subjected to ⋅ nonuniform polyaxial stresses. J. Appl. Mech. 1974, *41*, 459–465.

69. Batdorf, S.B.; Heinisch, H.L. Jr. Weakest link theory reformulated for arbitrary fracture criterion. J. Am. Ceram. Soc. 1978, *61* (7–8), 355–358.

70. Beierlein, G. Festigkeitsverhalten keramischer Werkstoffe unter mehrachsiger mechanischer Beanspruchung. Ph.D. Dissertation, Ingenieurhochschule Zwickau, 1988.

71. Chao, L.Y.; Shetty, D.K. Equivalence of physically based statistical fracture theories for reliability analysis of ceramics in multiaxial loading. J. Am. Ceram. Soc. 1990, *73* (7), 1917–1921.

72. Lamon, J. Statistical approaches to failure for ceramic reliability assessment. J. Am. Ceram. Soc. 1988, *71* (2), 106–112.

73. Katayama, Y.; Hattori, Y. Effects of specimen size on strength of sintered silicon nitride. J. Am. Ceram. Soc. 1982, *65* (10), C164–C165.

74. Giovan, M.N.; Sines, G. Biaxial and uniaxial data for statistical comparisons of a ceramic's strength. J. Am. Ceram. Soc. 1979, *62* (9–10), 510–515.

75. Müller, W.H.; Hansen, O.; Kümmel, M.; Eckert, A.; Fischbach, R. Statistische Bruchmechanik anisotroper Sprödwerkstoffe. Fortschr.ber. Dtsch. Keram. Ges. 1995, *10* (3), 148–162.

76. Popp, G.; Heider, W. Der C-Ringtest, ein Prüfverfahren für keramische Rohre aus Si–SiC. Z. Ind. Qualitätssicherung 1983, *28* (10), 289–292.

77. Abernethy, R.B.; Breneman, J.E.; Medlin, C.H.; Reinman, G.L. *Weibull Analysis Handbook*; Air Force Wright Aeronautical Laboratories (AFSC): Wright Patterson Air Force Base, OH, 1983.

78. Rufin, A.C. A study of statistical failure prediction models for brittle materials subjected to multiaxial states of stress. Ph.D. Thesis, University of Washington, 1981.

79. De Wolf, I. Stress measurements in Si microelectronics devices using Raman spectroscopy. J. Raman Spectrosc. 1999, *30*, 877–883.

80. Sergo, V.; Sbaizero, O.; Clarke, D.R. Mechanical and chemical consequences of the residual stresses in hydroxyapatite coatings. Biomaterials 1997, *19*, 477–482.

81. Cooper, C.A.; Young, R.J. Investigation of structure/property relationships in particulate composites through the use of Raman spectroscopy. J. Raman Spectrosc. 1999, *30*, 929–938.

82. Atkinson, A.; Jain, S.C. Spatially resolved stress analysis using Raman spectroscopy. J. Raman Spectrosc. 1999, *30*, 885–891.

83. Ma, Q.; Clarke, D.R. Piezospectroscopic determination of residual stresses in polycrystalline alumina. J. Am. Ceram. Soc. 1994, *77* (2), 298–302.

84. Forman, R.A.; Piermarini, G.J.; Barnett, J.D.; Block, S. Pressure measurement made by the utilization of ruby sharp line luminescence. Science 1972, *176*, 284–285.

85. Grabner, L. Spectroscopic technique for the measurement of residual stress in sintered Al_2O_3. J. Appl. Phys. 1978, *49* (2), 580–583.

86. He, J.; Clarke, D.R. Determination of the piezospectroscopic coefficients for chromium-doped sapphire. J. Am. Ceram. Soc. 1995, *78* (5), 1347–1353.

87. Sbaizero, R.; Pezzotti, G. Tailoring the microstructure of a metal-reinforced ceramic matrix composite. J. Eng. Mater. Technol. 2000, *122* (7), 363–367.

88. Latella, B.A.; Liu, T.-S. Performance characteristics of porous alumina ceramic structures. J. Australas. Ceram. Soc. 2000, *36* (2), 13–21.

89. Lemaitre, J.; Chaboche, J.-L. *Mechanics of Solid Materials*; Cambridge University Press: Cambridge, 1994.

6
Scuffing and Seizure: Characterization and Investigation

Tadeusz Burakowski
Radom Technical University, Radom, Poland

Marian Szczerek and Waldemar Tuszynski
Institute for Terotechnology, Radom, Poland

I. INTRODUCTION

In the construction and operation of machines, the prevention of scuffing of their sliding pairs has to be especially considered. The problem becomes more and more important due to increasing load and relative speed of sliding elements, as a result of technical development. These factors make the risk of scuffing higher, which is one of the main reasons for serious damage of, for example, toothed gears [1].

Examples of surface damage to a lubricated test specimen (ball), after an experiment carried out using a four-ball tribotester under extreme pressure (EP) conditions, are given in Fig. 1.

Scuffing produces irregular deep grooves and decohesion areas in the surface layer of tribosystem elements (Fig. 1(a)). In the macroscale, their surface is very rough that even traces of metal plastic flow may take place. Thin layers of metals are distributed on the surface of sliding pair elements (Fig. 1(b)). Due to significant damages to their surfaces, the clearance between them decreases. This in turn leads to seizure (Fig. 1(c)), which very often results in a stopping of the relative movement between elements of the sliding pair.

II. SCUFFING NATURE

A. Terminology and Definitions of Scuffing

Despite years of research on scuffing, its terminology has not been systemized so far; the term *scuffing* has many synonyms. There are, for example, *seizure*, *scoring*, *galling*, and *seizing*.

(a)

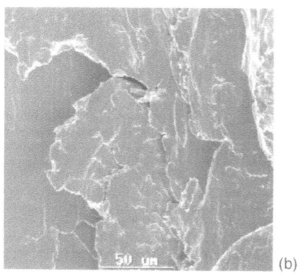

(b)

Figure 1 The surface damage to a lubricated test specimen (ball) after an experiment carried out using a four-ball tribotester under extreme pressure conditions: (a) damage produced by scuffing; and (b) magnified area of material transfer onto the surface; and (c) damage produced by seizure.

The variety of terms presented here makes interpretation of literature data difficult. Moreover, many terms are very often used as equivalents (e.g., *scuffing* and *seizure*).

The lack of consensus on the origin of scuffing as well as its symptoms is reflected by numerous definitions of the phenomenon. Some of them associate scuffing with wear; others associate it with friction. The former are definitions given or quoted by, for example, Enthoven and Spikes, Dyson, Ludema, and Sadowski. Among the latter, one can classify the definition given, for example, by Nosal.

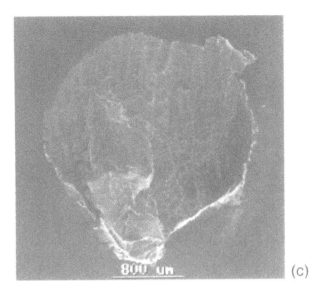

(c)

Figure 1 (*Cont.*)

Enthoven and Spikes [2] have quoted the definition suggested by the Organization for Economic Cooperation and Development, characterizing scuffing as "localized damage caused by the occurrence of solid-phase welding between sliding surfaces, without local surface melting."

Dyson [3] has adopted the definition presented by the Institution of Mechanical Engineers stating that scuffing is "gross damage characterized by the formation of local welds between the sliding surfaces."

Ludema [4] has considered scuffing to be "a roughening of surfaces by plastic flow, whether or not there is material loss or transfer."

Sadowski [5] has interpreted scuffing as "a process of the interference with stabilized wear and a boundary example of such wear."

But according to Nosal [6], scuffing is "a collection of phenomena taking place in the sliding pair, localized mainly deep inside the surface layer, producing an increased and unstabilized friction which is likely to result in seizure."

B. Scuffing Initiation

Practical lubrication of most sliding pairs (e.g., high-speed toothed gears) is neither a purely hydrodynamic (HD) process nor an elastohydrodynamic (EHD) process. Despite either HD or EHD film supporting most loadings applied to the tribosystem elements, collisions between the highest surface asperities cannot be excluded. Thus, besides fully lubricated friction, dry and boundary friction may also appear. Their common action is called *mixed friction* or *mixed lubrication* [7].

The model of mixed lubrication given by Stachowiak and Batchelor [8] is shown in Fig. 2.

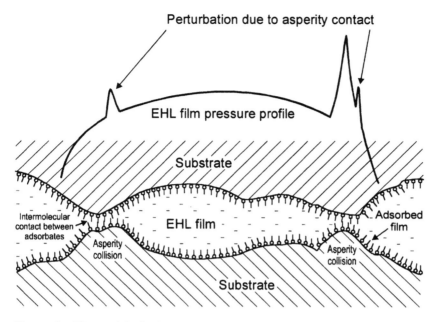

Figure 2   The model of mixed lubrication. (From Ref. 8.)

The lubricating film being produced at mixed friction, compared to such created at fully lubricated friction, is significantly thinner. Furthermore, collisions of surface asperities give rise to a local load increase, lubricating film collapse, and scuffing. This is a serious practical problem because scuffing can appear in tribosystems creating apparently good conditions for lubrication.

A kinetic model, proposed by Nosal [6], presents scuffing as a sequence of the following phenomena:

Collapse of the lubricating film
⇓
Removal of oxides layers in microareas of contact, leading to direct metal–metal contacts
⇓
Appearance of adhesive bonds due to temperature increase or plastic deformation
⇓
Development of adhesive bonds deeper and deeper inside the surface layer
⇓
Shearing of adhesive bonds, tearing out, and transfer of metal particles from one element to the other due to their relative movement
⇓
Rapid development of the above phenomena
⇓
Macroscopical range of destruction (scuffing)

The kinetic model of scuffing assumes that the pathway from stabilized friction to scuffing requires some time, rejecting the belief that it takes place very rapidly (i.e., during $t \to 0$).

The kinetic model of scuffing is presented graphically in Fig. 3.

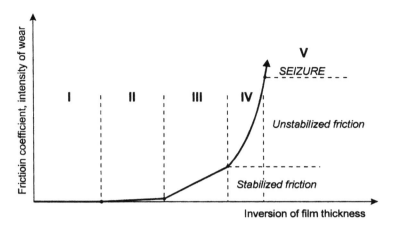

Figure 3　The transfer stages, from lubricated friction to seizure for a sliding pair. (From Ref. 6.)

From Fig. 3, it can be seen that due to the exposure of a sliding tribosystem to, for example, increasing load, some transfer stages from fully lubricated friction to seizure can be distinguished, namely:

Fully lubricated friction (I)

⇓

Boundary lubrication (II)

⇓

Mixed lubrication (III)

⇓

Scuffing (IV)

⇓

Seizure (V)

An increase in friction and intensity of wear, and a decrease in lubricating film thickness accompany the above stages.

Stage I (fully lubricated friction) does not result in any wear. During stage II (boundary lubrication), tribochemical wear appears. In stage III (mixed lubrication), adhesive wear dominates. Stage IV represents adhesive scuffing. Deep destruction of the surface layer takes place, leading to stage V (seizure).

Changes of wear character, observed for a lubricated specimen (ball) in a four-ball tribotester at continuously increasing loads, are illustrated in Fig. 4.

At the first stage of the experiment (at a load lower than 3000 N), mixed lubrication took place (Fig. 4(a)). Traces of local adhesive bonds shearing can be observed. At a load of about 3000 N, areas of surface topography typical of scuffing appeared (Fig. 4(b)); damage developed from the microscale to the macroscale. The further increase in loading led to deep destruction of the whole area of the wear scar—areas of plastic flow are also visible (Fig. 4(c)). If an increase in loading had been continued, it would have inevitably resulted in seizure of the tribosystem studied.

C.　Hypotheses on Metal–Metal Adhesive Bonding

It has already been stated that scuffing is a result of adhesive bonds between surface asperities. Adhesive bonding of metals is a complex phenomenon, thus a complete under-

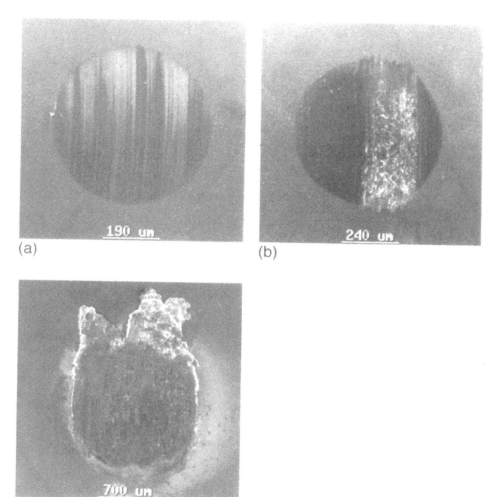

(a)
(b)
(c)

Figure 4 The destruction of a lubricated specimen (ball) surface in a four-ball tribotester at continuously increasing load: (a) 1500 N; (b) 3000 N; and (c) 7000 N.

standing of it is yet to be achieved. Attempts to explain it are known in the literature as *hypotheses on metal–metal adhesive bonding*.

Nosal [6], on the basis of literature review, has quoted the following hypotheses:

Mutual solubility hypothesis
Hypothesis considering a crystal structure type
Metal bonding hypothesis
Thin layers hypothesis
Recrystallization hypothesis
Diffusive hypothesis
Energetic hypothesis.

According to the mutual solubility hypothesis, the inclination of metals to adhesive bonding depends on their ability to create liquid solutions. Metals of mutual solubility are able to form strong bonds, whereas metals that are unable to produce solutions are less prone to adhesion.

The hypothesis considering a crystal structure type relates adhesion ability to crystal structure. This becomes apparent when comparing behaviors of various allotropic forms of the same metal; after allotropic transformation at an elevated temperature, adhesive bonding ability may increase.

According to the metal bonding hypothesis, adhesive bonds are caused by the creation of metal bonding between atoms of metals in contact. This is the reason for experiments having been performed to associate adhesive interactions and bonding ability with electron structure of metals. Despite the efforts, studies on the ability of metals for adhesive bonding, depending on their electron structure, have not given an explicit solution so far.

Following the thin layers hypothesis, the creation of adhesive bonds is possible if the metal surfaces in contact are free from oxides and any artificial layers. Now, it is known that it is only an indispensable, but not a sufficient, condition for adhesive bonding to take place.

The recrystallization hypothesis attributes the adhesive bonding of metals to recrystallization in the primary interphase. Recrystallization takes place due to the plastic deformation energy of microareas in contact and an increase in their temperature caused by friction. A large strain lowers the recrystallization temperature, making possible adhesive bonding of metals even at room temperature.

According to the diffusive hypothesis, adhesive bonding takes place due to diffusion causing an intergrowing of grains through the primary interphase. Diffusion can be initiated, apart the temperature, by elastic and plastic deformation; it multiply increases the coefficient of diffusion.

According to the energetic hypothesis, adhesive bonding of metals is based on their durable connection due to metallic bonding. The contact of metals (i.e., reduction of surface–surface distance to a value comparable with parameters of the crystal structure) enabling the creation of the metallic bonds is not enough for adhesive bonding to occur. For adhesive bonding, it is necessary that the energy of the atoms present in a crystal structure should increase up to a level called the *energy barrier of bonding*. Factors activating this process are plastic deformation, increase in temperature, compressive stress, and radiation.

D. Nanoscale Phenomena Under Extreme Pressure Conditions

The kinetic model of scuffing [6] analyzes the phenomenon in the microscale. However, interesting observations can be also obtained from an analysis in the nanoscale under extreme pressure conditions. Three phenomena having an influence on scuffing will be discussed: creation of plasma in the atomic scale, surface melting, and emission of low-energy exoelectrons (EEEs).

Plaza [9] has quoted literature data stating temperatures of 1000°C and pressures of >5 GPa being reached in a short time in the contact zone of surface asperities at boundary lubrication. In the atomic scale, under such conditions, there are changes affecting the state of electrons, chemical bonds, and atom packing; the matter becomes plasma.

Thiessen et al. [10] have explained elementary processes accompanying friction by means of the magma–plasma model. It assumes that plastic deformation, made by mechanical interaction, results in disturbance of the surface layer of tribosystem elements; the energy of the crystal structure reaches a high level. Due to this, the material becomes magma, which, owing to a high concentration of energy, for a short time converts to plasma (Fig. 5).

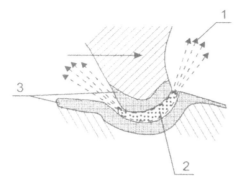

Figure 5 The magmà–plasma model of elementary processes accompanying friction: (1) atoms, electrons, and photons; (2) plasma; and (3) hard coating after solidification. (From Ref. 10.)

Emission of electrons is additionally accompanied by various chemical reactions and other processes influencing friction.

Interesting results can be also obtained from an analysis of a phenomenon called *surface melting* [11]. It is defined as a self-covering of some crystalline surfaces by a thin layer (measured in the atomic scale) of liquid, when their temperature reaches the point of bulk melting. The situation is illustrated in Fig. 6.

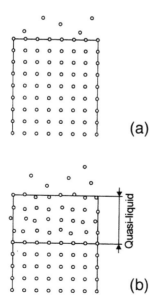

Figure 6 The crystalline structure of a solid heated up to the melting temperature: (a) surface still remaining crystalline; and (b) surface self-covered by a thin layer of quasi-liquid. (From Ref. 11.)

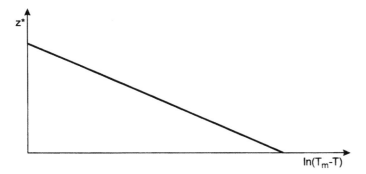

Figure 7 An example of dependence between the thickness of the quasi-liquid layer (Z^*) and value of $\ln(T_m-T)$. (From Ref. 11.)

The disturbed layer (Fig. 6(b)), strictly speaking, is not a liquid. Being a layer of finite thickness, it "feels" the presence of a crystalline order below, which makes it oriented to some extent. It is the reason why the melted layer is called the *quasi-liquid*. The thickness of the quasi-liquid layer increases when temperature T applied to the solid reaches its melting temperature T_m (Fig. 7).

It can be assumed that the produced quasi-liquid layer exerts a significant influence on processes taking place during friction. Physico-chemical characteristics of materials, such as diffusion, adhesion, and chemical reactivity, are changed. In addition, mechanical properties change; a decrease in shear strength makes the friction coefficient lower.

Another interesting phenomenon being considered in the nanoscale is the so-called *emission of low-energy exoelectrons* [12,13]. It takes place during the disturbance of the surface layer, resulting in plastic deformation, fatigue cracking, or phase transition. Due to exoelectron emission, places with excess positive charges arise on the surface (Fig. 8) [14].

(a)

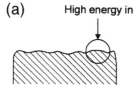

(b)

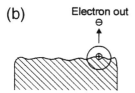

Figure 8 The mechanism of exoelectron emission: (a) absorption of friction energy; and (b) emission of exoelectron. (Reprinted from Ref. 14.)

The phenomenon of low-energy exoelectron emission was applied by Kajdas [15] to formulate the model of a *negative ion radical action mechanism* (NIRAM). According to the model, exoelectron emission sets the beginning of the following chain of events:

Absorption of exoelectrons by additives in the lubricant

⇓

Creation of negative ions and radicals

⇓

Reactions between ions, radicals, and the metal surface

⇓

Creation of an antiwear (AW) layer of metalo-organic compounds

⇓

Formation of a layer of inorganic compounds, protecting from scuffing

The role played by inorganic compounds in scuffing protection will be explained later in this chapter.

E. Scuffing: Process Leading to Seizure

In context of the scuffing definition given by Nosal [6], seizure is "a stopping of the mechanism as a result of high friction localized deep in the surface layer, which causes its permanent damage." The author describes scuffing as a process leading to seizure. A similar approach has been proposed by Park and Ludema [16]: Seizure (seizing) is "severe damage of sliding surfaces where the driving system cannot provide sufficient force to overcome friction—the sliding pair ceases to slide." In addition, Stachowiak and Batchelor [8] have suggested such an approach, giving its graphical interpretation (Fig. 9).

Work identifying seizure as a stopping of the relative sliding movement between elements of a tribosystem can be also found, for example, in Ref. 17, which describes an analysis of scuffing propagation in a sliding pair of nonconformal contact. Scuffing propagation was estimated at a continuously increasing normal load (P_n), from the shape of the curves of friction force (P_t) and sliding speed (v_s) (Fig. 10). It can be observed that due to scuffing, the friction force increases rapidly and the sliding speed loses its stability. At the moment of seizure, the sliding speed reaches zero, which means that the relative movement between the elements of the sliding pair stops.

It has to be mentioned that in some works, seizure (similarly to scuffing) is associated not with friction but with wear. For instance, according to Lawrowski [18], seizure is "de-

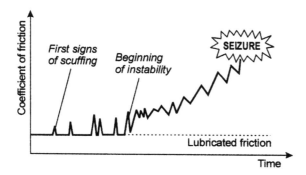

Figure 9 Scuffing as a process leading to seizure. (From Ref. 8.)

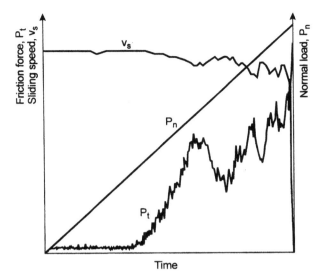

Figure 10 Examples of the curves of traction force (P_t) and sliding velocity (v_s), obtained for a block-on-roller tribosystem at continuously increasing normal load (P_n). (Reprinted from Ref. 17; Elsevier Science.)

struction of sliding surfaces, resulting from catastrophic processes of wear." The destruction makes elements of a sliding pair useless for further work.

F. Criteria of Scuffing

There are a significant number of factors responsible for the initiation of scuffing. In practice, they are presented as the so-called *criteria of scuffing*. Similar to the terminology and definitions of scuffing, there are many such criteria.

The most popular criteria of scuffing can be systemized in the following groups:

Minimum thickness of lubricating film
Temperature
Speed
Tension
Friction power intensity
Plastic strain
Wear debris present in a friction zone.

A brief characteristic of the above criteria is given below.

The checking of scuffing initiation possibility, taking place according to the mechanism of minimum lubricating film thickness, is based on the calculation of thickness (h_0) from the EHL theory, followed by the calculation of dimensionless film thickness (λ) (Eq. (1)) [19]:

$$\lambda = \frac{h_0}{\sigma_z} \tag{1}$$

where σ_z is the composite root mean square (RMS) surface roughness of two solids in contact.

The value of the dimensionless lubricating film thickness (λ) allows the establishment of a kind of lubrication occurring under given conditions and, in turn, checking if there is any possibility for scuffing. It is assumed that if $\lambda \leq 1$, then scuffing is likely to occur [16].

Undoubtedly, the most popular is the criterion given by Blok [20], based on the domination of thermal processes in lubricating film collapse. Owing to the simplicity of calculations, the procedure is applied in modern methods for the verification of resistance of toothed gears to scuffing. It can be written that:

$$T = T_m + T_f \tag{2}$$

where T is the surface contact temperature, T_m is the bulk temperature of the contacting solids, and T_f is the flash temperature in the contact zone.

The verification of scuffing possibility is based on a comparison between the surface contact temperature and the critical temperature T_{cr} determined experimentally. If $T > T_{cr}$, scuffing will initiate.

The criterion of speed has been proposed by Bushe and Fyodorov [21]. According to them, the critical sliding speed v_c, which initiates scuffing, can be determined from the following formula:

$$v_c = D_f \frac{Ch}{\tau} \tag{3}$$

where D_f is the diffusion coefficient at friction, C is a constant, h is the average height of surface asperities, and τ is the friction force per length unit.

The higher the diffusion coefficient is, the less is the possibility of scuffing at a given speed.

The tension criterion was formulated by Niemann [22]. Niemann suggested the calculation of the relation between the contact stress at the scuffing load, determined using an FZG gear machine (k_f), and the real contact stress (k_w), given by Eq. (4):

$$S_F = \frac{k_f}{k_w} \tag{4}$$

Scuffing will appear if the critical value of ratio S_{Fcr} is not exceeded:

$$S_F < S_{Fcr}$$

Another criterion is associated with friction power intensity, which is calculated by relating the power of friction (dw_{ww}) to the area of contact (ds) [23]. According to this criterion, the exceeding of the critical value of friction power intensity is the condition for scuffing initiation:

$$\frac{dw_{ww}}{ds} > \left(\frac{dw_{ww}}{ds}\right)_{cr}$$

It is accompanied by a rapid increase in wear (Z), which means that $dZ/dt \gg 0$. The situation is illustrated in Fig. 11.

The criterion of plastic strain has been developed by Park and Ludema [16] on the basis of the plasticity index criterion. The authors have rejected the opinion that a small 2% plastic strain (implied in the plasticity index) of contacting surface asperities is enough for adhesion, which in turn is necessary for scuffing initiation. In their opinion, the prerequisite condition for scuffing is a large-strain, even plastic flow of material.

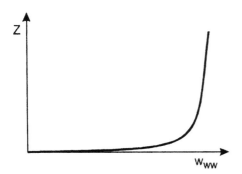

Figure 11 Graphical interpretation of the friction power intensity criterion. (From Ref. 23.)

A little known contribution to the development of scuffing criteria is the criterion taking into consideration the appearance of wear debris, as proposed by Ludema [4]. It associates scuffing initiation with the presence of wear debris in a friction zone. Despite verification of the criterion by Enthoven and Spikes [2], there is still some uncertainty on the interpretation of a mechanism of scuffing initiation. Ludema has claimed that wear debris, carrying a part of load, makes the oil pressure decrease below a level that is necessary for creating a stable lubricating film. But according to Enthoven and Spikes, the accumulation of wear debris in the contact inlet makes sufficient lubrication of its rear part impossible, which unavoidably leads to scuffing. Yet another mechanism has been proposed by Rapoport [24] for relatively soft steels. Rapoport has suggested that the main reason for scuffing of such materials is the blocking of relative motion between elements of a sliding pair, caused by wear debris. This leads to an increase in the coefficient of friction and then temperature. On exceeding its critical value, the initiation of scuffing by adhesive bonding is likely to occur.

Different from the above discussed approach is work presented by Sadowski [5], who has claimed that exceeding the critical value by only one parameter of friction (e.g.,

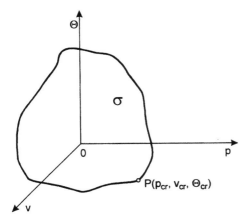

Figure 12 Graphical interpretation of the scuffing criterion given by Sadowski. (From Ref. 5.)

temperature) is not enough for scuffing initiation. Scuffing will start if the pressure (p), speed (v), and contact temperature (θ) reach simultaneously their critical values:

$$p = p_{cr}, \qquad v = v_{cr}, \qquad \Theta = \Theta_{cr}$$

A graphical interpretation of the discussed criterion is given in Fig. 12.

Scuffing initiation takes place if the point (P) of the coordinates representing particular parameters lies on an envelope σ. Inside the envelope, there is a space containing parameters typical of stabilized wear.

In conclusion, scuffing criteria, creating possibilities to predict conditions of its initiation, are important for the verification of engineering assumptions from the point of view of scuffing prevention.

Other ways to prevent scuffing (the technological one and one taking place during a run of a tribosystem) will be discussed in the last part of this chapter.

III. INVESTIGATION OF SCUFFING IN LUBRICATED TRIBOSYSTEMS

A. Selection of Tribotesters for Scuffing Investigation

The basic criterion concerning the selection of tribotesters for scuffing investigation is the possibility of reaching of the so-called *extreme pressure conditions*. Boruta and Wachal [25] have proposed using the *friction power intensity* (JMT) for the numerical estimation of conditions in a tribosystem. JMT is calculated from Eq. (5):

$$\text{JMT} = \mu\, pv \tag{5}$$

where JMT is the friction power intensity [MW/m^2], μ is the friction coefficient, p is the pressure [MN/m^2], and v is the sliding speed [m/sec].

The results of JMT calculations for various tribotesters are presented in Table 1. To simplify the analysis, Boruta and Wachal assumed that $\mu = 1$ in all cases.

From a comparison of values obtained, it can be concluded that only tribotesters of a nonconformal contact (point, line) are able to create EP conditions necessary for scuffing initiation.

In the group of tribotesters for scuffing investigation (of a point contact), the following instruments can be distinguished:

SRV machine (ball-on-disk geometry, operating at high-frequency oscillating movement)
Four-ball tribotester
Machine of a three-roller-cone geometry.

Table 1   The Values of the Friction Power Intensity (JMT) Obtained for Tribotesters of Different Contact Geometries

	JMT (MW/m^2)
Tribotesters—conformal contact	Up to 500
Tribotesters—nonconformal (line) contact	Up to 4776
Tribotesters—nonconformal (point) contact	Up to 6862

Source: Ref. 25.

In the group of linear contact tribotesters used for scuffing investigation, one can find the following:

Falex machine having a pin-and-vee block tribosystem
Ring-on-ring stands, simulating pressure, and speed conditions existing in toothed
   gears along the drive length
Timken apparatus (block-on-ring)
FZG gear machine (toothed gear working in a circulating power system).

The schemes of the abovementioned tribosystems used for scuffing investigation are shown in Fig. 13.

Kulczycki and Stryjewski [26] have analyzed experimental data for automotive and industrial gear oils, obtained using a four-ball tribotester, Timken apparatus, and FZG

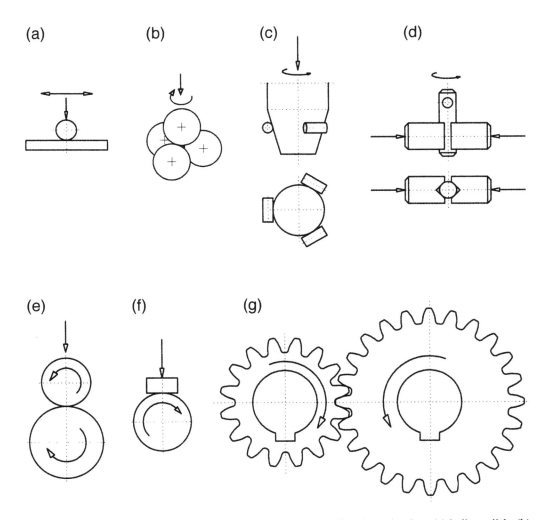

Figure 13 The tribosystems most frequently applied for scuffing investigation: (a) ball-on-disk; (b) four-ball; (c) three-roller-cone; (d) pin-and-vee block; (e) ring-on-ring; (f) block-on-ring, and (g) toothed gear.

machine. They have pointed on the necessity for simultaneous application of various tribotesters to obtain a reliable assessment of lubricant quality. On the other hand, Dyson [3] has postulated the investigation of lubricants under conditions similar to the working ones, possible only during component testing. Thus, for gear oils, Dyson has recommended a gear machine. Dyson has stated that it is not sufficient to rely only on results obtained using bench machines (e.g., a four-ball tribotester) because experimental conditions offered by such tribotesters are far from real conditions of.lubrication in practice.

Unfortunately, an important barrier to a wider application of component tests is its high cost in comparison with bench testing. Some sources have estimated the costs of the former to be even 100 times higher than the latter [27]. For this reason, some researchers recommend the substitution of incomparably less expensive bench testing for very costly component tests. For example, van de Velde et al. [28,29] have suggested the use of four-ball testing instead of FZG gear tests.

B. Test Methods for Scuffing Identification

The lack of widely recognized scuffing definitions and, consequently, scuffing criteria enabling its clear identification leads to a variety of experimental methods in use. Below, one can find a review of the most popular tribological methods. Most of them are standardized.

1. Methods Based on the Friction Criterion

In these methods, the identification of scuffing is based on the determination of load, causing either a sudden increase* in the friction force (friction torque or coefficient of friction), or exceeding of a certain value of these measures. The pathway of scuffing identification, according to the friction criterion, is shown schematically in Fig. 14. The load at which scuffing initiation occurs is called the *critical load* or *scuffing load* and is denoted as P_{cr}. During an experimental run, the load increases either continuously (Fig. 14(a)) or gradually (Fig. 14(b)).

Three examples of standardized test methods for scuffing identification, based on the friction criterion, are presented below.

PN-76/C-04147[†] standard [30], concerning tests using a four-ball tribotester, points to a significant increase in the friction torque as a symptom of lubricating film collapse. The load responsible for this, applied to the tribosystem, is called the *scuffing load*. In this method, the load increases continuously (Fig. 14(a)).

From an analysis of ASTM D 2981-71 standard [31], concerning tests performed under oscillating block-on-ring contact conditions, it follows that exceeding of a certain value of the coefficient of friction should be assumed as scuffing initiation. A similar procedure can be found in ASTM D 5706-97 standard [32], concerning testings under oscillating ball-on-disk contact conditions, using an SRV machine. In both these methods, the load increases gradually (Fig. 14(b)).

2. Methods Based on the Wear Criterion

The pathway of scuffing identification, according to the wear criterion, is shown in Fig. 15. The loading of the tribosystem is realized gradually. Wear, obtained at a given value of load, is measured after a test run. It is assumed that scuffing takes place at a load (the so-called

---

* *Sudden increase* is not a strict term; in the standards, it has not been defined precisely.
[†] The symbol *PN* means Polish standard.

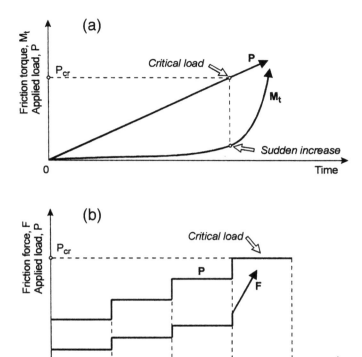

Figure 14 The determination of load causing scuffing initiation according to the friction criterion: (a) continuously increasing load; and (b) gradually increasing load.

critical load or *scuffing load*, in the figure denoted as P_{cr}) that causes either a sudden increase in wear, or exceeding of a certain value of its measures.

Examples of standardized test methods for scuffing identification based on the wear criterion are presented below.

In ASTM D 2783-82 standard [33], concerning four-ball tests, the wear criterion is associated with the wear scar diameter measured on stationary balls. The highest load at which wear does not exceed a certain limiting amount is called the *last nonseizure load*.

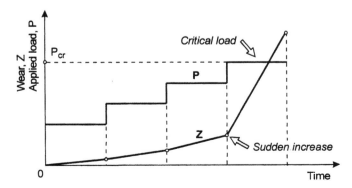

Figure 15 The determination of load causing scuffing initiation according to the wear criterion.

From an analysis of PN-83/H-04302 standard [34], concerning testing in a three-roller-cone tribosystem, it follows that scuffing identification requires measurements of the roller wear. Scuffing takes place at the so-called *critical load*, causing a rapidly increasing wear depth.

The British standard IP 334/86 [35], concerning tests using FZG gear machines, recommends the application of visual scuffing identification. Scuffing takes place when the summed total width of the damaged areas on all the pinion gear teeth faces is estimated to exceed a certain value; the load at which it takes place is called the *failure load stage*. The German standard DIN 51 354, Part 2 [36] proposes the application of another method for scuffing identification. It is a method based on the determination of the weight loss of the pinion between the subsequent stages of loading. The failure load stage is one at which a rapid increase in wear is noticed.

It has to be underlined, as far as the wear criterion is concerned, that wear is measured on a so-called *specimen*, whereas the *counterspecimen* is rejected. In the standards, it is clearly defined which one from the elements of a tribosystem is to be subjected to assessment. Despite this, some researchers (e.g., Willermet and Kandah [37]) have proven, in the case of a four-ball tribosystem, that not only assessment of the specimen wear but also the wear of the counterspecimen is highly advisable.

3. Complex Methods

Many researchers (e.g., Boruta and Wachal [38]) have shown that a simultaneous application of a few test methods instead of one, for scuffing investigation, is more appropriate (e.g., measurements of the friction torque together with wear). These researchers have proposed their own method realized in a four-ball tester at constant load. The basic criteria of lubricant assessment in the method are: the time measured from the beginning of a run until scuffing initiation (manifested by a sudden increase in the friction torque), wear, scuffing duration, and coefficient of friction. Figure 16 presents an example of the friction torque curve obtained using this method, together with the characteristic time ranges. The line-marked area reflects scuffing.

In addition, Dyson [3] has postulated a simultaneous consideration of the friction torque and the wear for scuffing identification. Dyson has assumed a simultaneous appearance of a sudden increase in the friction torque and significant wear (the worn area should have a surface topography typical of scuffing) as symptoms of scuffing in toothed gears.

A similar approach can be found in ASTM D 2782-77 standard [39], concerning testing based on the Timken method (block-on-ring tribosystem). It recommends a simul-

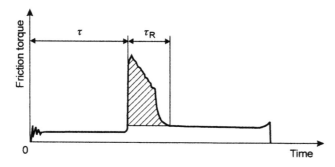

Figure 16   The friction torque versus run duration; τ—time to scuffing initiation; τ_R—duration of scuffing. (From Ref. 38.)

Table 2 The Coefficient of Friction and Wear Rate During Transition Between Different Lubrication Modes

	Friction coefficient	Wear rate $(\times 10^{-6} \text{ mm}^3/\text{N m})$
(Partial) EHL—boundary lubrication	0.1–0.35	1–5
Boundary lubrication—scuffing	>0.35	5–500

Source: Ref. 40.

taneous assessment of the friction torque and wear at gradually increasing loads. Assessment of the friction torque is realized indirectly; its increase is registered as an instability of the spindle rotational speed and increasing noise level. If it is accompanied by wear of the specimen (block) characteristic of scuffing surface topography, then scuffing is dealt with. The minimum load at which it takes place is called the *score value*.

Test methods based on simultaneous observations of both friction and wear were applied (e.g., by Begelinger and de Gee) to study the influence of load and sliding speed on the change in lubrication mode in a concentrated contact [40,41]. They have determined the values of the friction coefficient and wear rate (wear related to load and sliding distance) typical of a transition from one lubrication mode to another (Table 2). Special attention should be paid to the very high wear rate during scuffing.

By determining the load and speed at which the coefficient of friction and wear rate reached their limiting values given in Table 2, the abovementioned researchers have obtained the so-called *transition diagram* (Fig. 17).

It can be noticed from Fig. 17 that scuffing takes place at high values of both load and sliding speed.

4.   Other Methods

Among others different from the abovedescribed test methods for scuffing identification, one can distinguish the method based on measurements of temperature in the friction zone, using infrared and optical methods. The latter is realized by means of an optical microscope coupled with a TV camera and video recorder. These methods were used in the works of Enthoven and Spikes [2]. Figure 18 shows a scheme of the sliding pair (ball-on-flat) applied by them, together with the scientific equipment for measurement and recording.

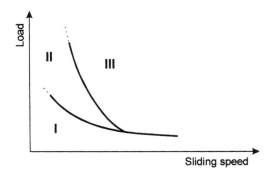

Figure 17 The transition diagram: I—partial EHL; II—boundary lubrication; III—scuffing. (Reprinted from Ref. 41; Elsevier Science.)

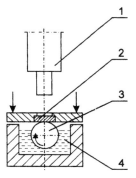

Figure 18 The sliding pair and scientific equipment applied: (1) optical/infrared microscope; (2) sapphire window; (3) steel ball; and (4) lubricant. (From Ref. 2.)

An analysis of contact temperature value, measured by means of infrared microscopy, was applied for scuffing identification; a rapid increase in temperature could indicate collapse of the lubricating film.

A greater advantage was taken of the video recording of the friction zone during different test periods; it was possible to observe both scuffing initiation and its propagation. Figure 19 shows changes in the friction zone appearance, as a result of the test duration and gradually increasing load.

The contact zone at the beginning of the test (the lowest load) can be seen in Fig. 19(a)). With increasing load, some places of wear particles accumulation appear (top left corner in Fig. 19(b)). On a further increase in load (it reached a maximum value), the friction zone during different test periods looks like the ones presented in Fig. 19(c)–(f). The very large quantity of wear debris accumulated in the inlet area (c) leads to the onset of scuffing (d); a

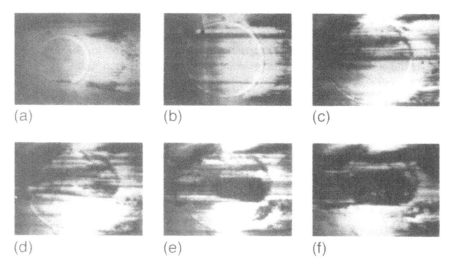

Figure 19 The contact zone during different test periods. ((a)–(f) are described in the text.) (From Ref. 2.)

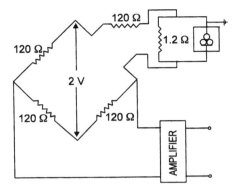

Figure 20 The schematic diagram of the instrument for measuring contact resistance in a four-ball tribosystem. (Reprinted from Ref. 42; Elsevier Science.)

small black spot can be seen on the center line of the contact area, almost at the edge of the contact exit. Scuffing propagates rapidly (e, f), resulting in seizure of the sliding pair.

Test methods based on measurements of electrical resistance of a sliding contact, followed by monitoring of changes in the coefficient of friction and intensity of wear, are also widely used. In these methods, lubricating film collapse, due to collisions between surface asperities of contacting solids, results in a rapid decrease in the contact electrical resistance, as well as an increase in the coefficient of friction and wear intensity. Such methods were applied by, for example, Kuo et al. [42], who investigated the influence of load and sliding speed on the change in the lubrication mode in a four-ball tribosystem. A schematic diagram of the contact resistance instrument is shown in Fig. 20. In practice, electrical resistance is determined as a function of the measured voltage after its amplification.

The results obtained by the abovementioned researchers are demonstrated in Fig. 21. They are similar to the results reported by Begelinger and de Gee [41]. The difference is only in the identification of "pure" EHL region by Kuo et al. It can be noticed that this takes place only at low loads and high sliding speeds.

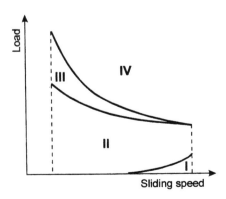

Figure 21 The lubrication mode diagram: I—EHL; II—partial EHL; III—boundary lubrication; IV—scuffing (this region is called *seizure* in the source material). (Reprinted from Ref. 42; Elsevier Science.)

Other groups of test methods used for scuffing identification are based on measurements of electrostatic charges generated during friction. They make use of the electrostatic charging of a friction zone due to collisions between contacting surface asperities, taking place after partial or total collapse of the lubricating film and removal of oxides from the surface. The high sensitivity of these methods makes them very useful for on-line machinery condition monitoring.

Tasbaz at al. [43] applied (at the laboratory scale) the method for electrostatic charging measurements aiming at very early identification of scuffing initiation in a lubricated ball-on-disk sliding pair. A scheme of the sliding pair used is shown in Fig. 22.

Feeding with oil was realized drop by drop before a run, up to the moment of achieving an adequate film thickness on the disk. During the run, the oil feed was switched off to eliminate effects from oil droplets charging. Information on electrostatic charge was obtained from two sensors: the primary and the secondary sensors. The first one measured charging in the contact zone, whereas the second one measured charging far away from the zone.

Figure 23 shows data on experimental values of electrostatic charging (measured using the primary sensor) as well as the coefficient of friction obtained by the abovementioned researchers.

Figure 23(a) and (b) concerns the first period of the experimental run. The practical lack of signals coming from the electrostatic sensor can be noticed (the signal level at the first period of the run can be adopted as a background). The relatively small values of the coefficient of friction can be observed; however, they are a little bit higher than the value typical of boundary lubrication (i.e., 0.1). At very early scuffing initiation, the situation changes dramatically (Fig. 23(c) and (d)). Electrostatic charge appears in the friction zone and the coefficient of friction significantly increases. Based on additional investigations using profilometry and energy-dispersive spectrometry (EDS), Tasbaz et al. have proven that the increase in charging at very early scuffing initiation is likely to be a result of increasing disk wear intensity, as well as phase transitions taking place in the materials of the specimens (e.g., diffusion of carbon from the oil into the surface layer).

It should be also added that signals coming from the secondary sensor were much weaker in comparison with those from the primary one and are believed to be evidence that charge generated at the contact area exists long enough to be detected at a remote location.

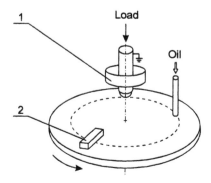

Figure 22 The schematic diagram of the pin-on-disk sliding pair with location of electrostatic sensors: (1) primary sensor; and (2) secondary sensor. (Reprinted from Ref. 43; Elsevier Science.)

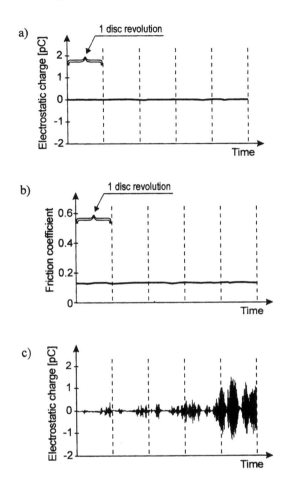

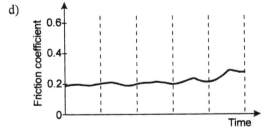

Figure 23 Data from the primary electrostatic sensor and friction coefficient versus time: (a) changes in electrostatic charge typical of the first stages of the run; (b) friction coefficient at the first stages of the run; (c) changes in electrostatic charge at very early scuffing initiation; and (d) friction coefficient at very early scuffing initiation. (Reprinted from Ref. 43; Elsevier Science.)

Another test method that allows observations of both scuffing initiation and its propagation is a method for analyzing acoustic emission [44]. Acoustic emission comes from the generation and propagation of elastic waves in solids and liquids. These waves are a result of microcracking, generation and annihilation of dislocations, and relative movement of contacting materials.

Marczak et al. [44] made a research on an oil containing lubricating additives. They used a four-ball apparatus. Load was increased gradually. After each load stage, the wear scar diameter on the stationary balls was measured. Acoustic emission was analyzed indirectly by measuring tribosystem vibrations during a run. An accelerometer was positioned on the stationary balls holder.

The results obtained are presented in Fig. 24. Two curves are plotted. One portrays a change in the wear scar diameter at increasing load, and the other one presents a change in the acoustic signal.

From Fig. 24, it is apparent that at a certain value of load, lubricating film collapse occurs, which is reflected by the catastrophically increasing wear; scuffing initiates. It is accompanied by a sharp increase in acoustic signal amplitude.

The acoustic emission method can be employed for on-line monitoring of the technical state of machines.

Another technique of scuffing identification is based on analyzing wear intensity, characterized by the amount of wear particles contained in the lubricating oil. Such methods are applied mainly for monitoring of friction couples of machines by assessment of oil condition. This is based usually on magnetical or optico-magnetical principles [45,46]. A block diagram of an optico-magnetical system for on-line monitoring of oil conditions is shown in Fig. 25.

The presented system makes possible an analysis of conditions of an oil lubricating tribosystem (test unit) 6. The optico-magnetical sensor 1 is applied for this purpose. Computer 5, controlling the system by means of units 2–4, allows the automatization of analyses.

C. Experimental Identification of Seizure

Contrary to numerous, more or less precise criteria suggested for scuffing identification, there is still lack of a clearly specified criterion for seizure. The definition presenting seizure as a stopping of the relative movement of sliding pair elements is not very precise because

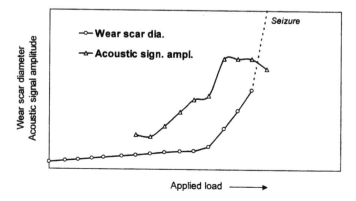

Figure 24   Changes in the wear scar diameter and acoustic signal versus increasing load. (From Ref. 44.)

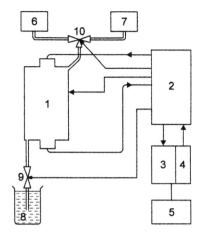

Figure 25 The block diagram of an optico-magnetical system for on-line monitoring of lubricating oil condition: (1) optico-magnetical sensor; (2) electronic block of the sensor; (3) analogue-to-digital converter; (4) I/O control unit; (5) computer; (6) test unit; (7) tank with clean oil; (8) drain tank; and (9, 10) electromagnetical valves. (From Ref. 45.)

stopping, or not, of the relative movement in the tribosystem depends on the so-called *excess of power* at the disposal of a drive source.

The problem of the excess of power was analyzed by Nosal [6]. Nosal has distinguished two possibilities: when a drive source has, or has not, an excess of power at its disposal. Nosal has not given any details, however, on how big it should be, which still makes comparisons of experimental data on scuffing, especially scuffing propagation, obtained at different values of the excess of power impossible.

In the light of the above, the precise definition of a required power of a drive source is reasonable. It has been done, for example, in PN-76/C-04147 standard [30], where the power of the four-ball tester is specified. It is very important if one applies, contained in the standards, procedures for the determination of the so-called *weld point*, being one of the basic parameters used for an assessment of the antiseizure properties of lubricants. Seizure is identified here with a stopping of the tribosystem, resulting from welding of their test elements (balls). In view of this, if the motor does not have enough power (less than recommended), a stopping of the spindle rotation could take place at a lower load, without possibility for welding of the elements. In such a case, determination of the weld point would be impossible.

ASTM D 2670-81 standard [47], concerning the investigation of lubricants in a pin-and-vee block tribosystem using a Falex machine, suggests a slightly different way to assure scuffing tests comparability. Breakage (shearing) of a special brass pin used for locking of the pin specimen, which takes place at a certain value of the friction torque, is applied as a criterion of seizure. Due to precise requirements on the kind of locking pin material and its heat treatment, seizure always takes place under conditions of the same friction torque; hence, experimental results will be comparable.

D. Critical Assessment of Test Methods for Scuffing Investigation

Due to the lack of a commonly accepted definition of scuffing, numerous test methods for its identification have appeared. These methods use an indirect identification of the lubricating layer collapse, by measuring the friction force or friction torque, intensity of wear, contact temperature, or resistance, and by applying certain critical values. Some methods are subjective, additionally bearing inaccuracy coming from an excess of the driving source power (e.g., methods based on analyzing noise levels or changes in the rotational speed of a spindle).

The methods mentioned above are mainly based on the determination of conditions of scuffing initiation and, in most cases, make an assessment of its propagation impossible. However, scuffing should be investigated not only by the determination of the moment of its initiation but also by an analysis of its propagation. The reason is that scuffing may propagate differently in the case of different lubricants, despite similar conditions of its initiation. This was confirmed during tribological investigations performed for two commercial hydraulic gear oils, denoted I and II.

The research was performed using a four-ball apparatus, according to PN-76/C-04147 standard [30]. The so-called *scuffing load* (P_t) was determined. This is the load that causes the collapse of the lubricating film, resulting in a sudden increase in friction torque.

During a run, the load was increased continuously, and the friction torque as well as the load were measured.

For both oils, the friction torque curves, obtained at continuously increasing load (P), are presented in Fig. 26.

If only the standardized scuffing load (P_t), causing scuffing initiation, was assumed as a measure of lubricating action, the difference between these two oils would be negligible; both of them will give the same P_t (2100 N) (Fig. 26). In practice, however, oil I caused seizure and destruction of a high-power toothed gear, whereas oil II enabled its faultless operation. Thus, a proposition of additional criteria for differentiation of such oils was highly required.

The difference between these oils can only be revealed through an analysis of the scuffing propagation reflected by friction torque curves (Fig. 26). Oil I gives the sudden increase in friction, leading to immediate seizure (stopping of the relative movement of balls). A completely different behavior is exhibited by oil II; the tribosystem does not stop until the end of the run, and the change in the friction torque is much gentler.

It was also observed that oil I gave a much larger wear scar diameter than oil II.

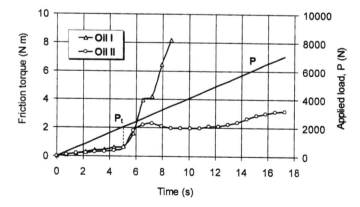

Figure 26 The friction torque curves versus time, obtained for oils I and II at continuously increasing load (P). P_t—scuffing load. (Reprinted from Ref. 48; Elsevier Science.)

Experience gained from this experiment enables the authors of this chapter to state that the scuffing propagation under conditions of continuously increasing load as well as wear can also be a lubricating oil assessment criterion.

Unfortunately, hardly can one find methods for analyzing scuffing propagation. One of them, a commonly known and standardized method for studying lubricants under extreme pressure conditions, is a four-ball test based on an analysis of changes in wear intensity at gradually increasing loads applied to the tribosystem, up to the moment of seizure (welding of the balls). The method proposes the so-called *load–wear index*, which allows assessments of lubricants studied. The described method is presented in ASTM D 2783-82 standard [33].

However, the realization of gradually increasing load in the abovementioned method is not satisfactory. According to the authors, a continuous increase in load would be a better solution for the reasons given below.

First of all, this kind of loading, starting always from zero, assures identical conditions for the creation of the surface layer during friction to be maintained. Thus, if tests of various lubricants are carried out at the same sliding speed, load-increasing speed, and initial bulk temperature of oils, and by applying the same tribosystem, one can compare the obtained tribological characteristics as well as lubricants with one another. On the contrary, at gradually increasing load, only characteristics obtained at the same load can be compared, which makes the evaluation of the lubricants more complicated.

Secondly, if at gradually increasing load its initial value is too high, the critical loading responsible for the collapse of the lubricating film and scuffing initiation takes place right after the start of a test run. Under such conditions, seizure appears immediately after scuffing initiation. The duration of the scuffing process is very short, so it is difficult to analyze.

Finally, the repeatability of the tests performed at gradually increasing load is reduced by a number of test specimens (at each load, a new set of specimens is required) and their resolution depends on the load increment.

Taking the above into consideration, elaboration of a new tribological test method for lubricants evaluation under scuffing conditions was necessary. The test should be performed at continuously increasing load, have good resolution, and give possibilities to obtain repeatable results. Apart from these, its rapidity and low cost would be desirable. Such a method should be based on recognition of scuffing as a process leading to seizure. Thus, the definition of seizure, taking into account its identification, is a prerequisite. Additionally, to make possible analyses of wear scar surfaces, test specimens at the moment of seizure must not be permanently joined, for example, by welding. It means that adequate test conditions should be selected.

These requirements are met in the new test method below described.

E. New Test Method for Lubricant Assessment Under Scuffing Conditions

The new test method has been discussed in detail, for example, in Refs. 48–51. This method together with, indispensable for its realization, a modified four-ball tribotester, was awarded the Gold Medal during the 25th International Fair of Inventions, New Techniques, and Products held in Geneva in 1997. The method has been patented [52].

1. Test Principles

According to the new test method, the three stationary steel balls are pressed against the upper one in the presence of a lubricant to be tested, at continuously increasing load (from zero). The upper ball rotates at the constant speed. During a test run, friction and load are measured. One should observe changes in the friction torque until seizure occurs.

The following seizure criterion should be taken: Seizure occurs at the moment of exceeding 10 N m in friction torque. The 10-N m value has been adopted, taking the life of the ball chuck into account.

The test conditions are as follows:

Rotational speed: 500 ± 20 rpm
Speed of continuous load increase: 409 N/sec
Initial applied load: 0 N
Maximum load: 7200 ± 100 N.

It should be added that experimental conditions were adjusted in a way that prevents the balls from welding at seizure. Owing to this, it is possible to analyze the wear scar surface afterward.

The maximum load, about 7200 N, was adopted by taking into consideration the dynamics of the load system.

Test balls are chrome alloy-bearing steel, with diameter of 12.7 mm (0.5 in.). Surface roughness is $R_a = 0.032$ μm and hardness is 60–65 HRC.

A friction torque curve (M_t) obtained at continuously increasing load (P) is shown in Fig. 27.

Scuffing initiation occurs at the time of a sudden increase in the friction torque (point 1). The load at this moment is called the *scuffing load* and is denoted as P_t. The scuffing load is a standard measure of the lubricating properties of the tested oil according to PN-76/C-04147 standard [30].

According to the new test method, the load still increases (over a value of P_t) until seizure occurs (i.e., friction torque exceeds 10 N m; point 2). The load at this moment will be called the *seizure load* and denoted P_{oz}. If 10 N m is not reached, the maximum load (ca. 7200 N) is considered to be the seizure load (although in such a case there is no seizure). P_{oz} should be determined with an accuracy of 100 N or better.

Examples of friction torque curves obtained for automotive gear oils of API GL-3 and GL-5 levels and a mineral base oil are presented in Fig. 28.

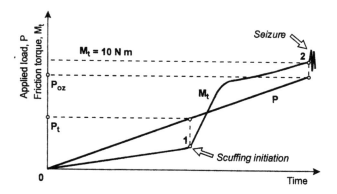

Figure 27   The simplified friction torque curve (M_t) obtained at continuously increasing load (P): (1) scuffing initiation; (1–2) scuffing propagation; and (2) seizure (exceeding 10 N m friction torque). (Reprinted from Ref. 48; Elsevier Science.)

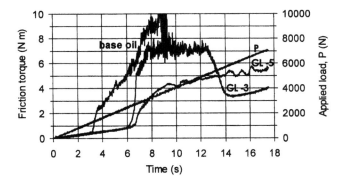

Figure 28 Examples of the friction torque curves obtained at continuously increasing load (*P*) for automotive gear oils of API GL-3 and GL-5 levels and a mineral base oil.

From Fig. 28, it is apparent that the seizure load for the base oil is about 3500 N. For the both gear oils, P_{oz} is about 7200 N, although in this case seizure did not appear.

The condition necessary to carry out the new test procedure is that the motor of a four-ball tester has enough power to avoid a stopping of the movement until a 10-N m friction torque (seizure) is attained. When 10 N m has been reached, the upper ball must be stopped either automatically (via a special cutoff device or adequately programmed motor controller), or manually (with a switch), on the basis of observed (on-line) changes in the friction torque.

On the basis of many experiments performed under scuffing conditions, the authors suggest that for every tested lubricant, the so-called *limiting pressure of seizure* (denoted p_{oz}) should be calculated. Its value reflects the lubricant behavior under scuffing conditions and equals the nominal pressure exerted on the wear scar surface at seizure, or at the end of the run (when seizure has not appeared). The limiting pressure of seizure is calculated from Eq. (6), derived, for example, in the work [48]:

$$p_{oz} = 0.52 \frac{P_{oz}}{d^2} \tag{6}$$

where p_{oz} is the limiting pressure of seizure [N/mm^2], P_{oz} is the seizure load [N], and d is the average wear scar diameter measured on the stationary balls [mm].

The average wear scar diameter (d) is calculated from the parallel (to the striations) and normal scar diameters measured on the stationary balls, with an accuracy of at least 0.1 mm.

The 0.52 coefficient results from the force distribution in the four-ball tribosystem.

The bigger the p_{oz} value is, the better is the action of the tested lubricant under scuffing conditions [48–51].

2. Modified Four-Ball Tester

The realization of the new test method is possible using a modified four-ball tester shown in Fig. 29.

A very important feature of the presented tester is the possibility of continuous increasing of load during a run, owing to modification of the load system; the additional independently controlled motor moves the weight along the lever. This unique feature is necessary for the realization of the new test procedure and has been patented [53].

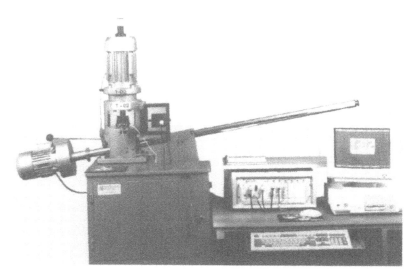

Figure 29 The modified four-ball tester.

The modified four-ball tester is controlled via a special microprocessor-aided controller, motor speed controller, and PC.

The microprocessor-aided controller allows taking control of the tester. The motor speed controller, by changing the frequency of electrical current, enables the reduction of the rotational speed from 1450 to 500 rpm, which is required in the new test method. This controller allows also an automatic stopping of the machine when the seizure criterion (friction torque 10 N m) is met.

Friction and load are measured using force transducers. The signals from the transducers are acquired by a digital amplifier. After amplification and conversion, the signals are sent to the computer disk, where they are saved for further processing.

3. Resolution of the Method

In Refs. 50 and 51, the authors have compared the resolution and the time consumption of the elaborated test method and other standardized four-ball tests. For this purpose, automotive gear oils of various API performance levels were investigated.

It has been proven that the limiting pressure of seizure p_{oz}, contrary to most standardized measures, is very effective in tribological differentiation between tested oils. One of the standardized indicators, the so-called *weld point*, also allows their differentiation. Its determination, however, takes almost 20 times longer time than that required for p_{oz} and, at seizure, test balls are welded, which makes analyses of the wear scar surface impossible. Under test conditions adopted in the new method, after tribological experiments, surface analyses of the wear scar are always possible—test balls never weld.

4. Precision

The precision of the new test method, expressed by repeatability and reproducibility, has been assessed on the basis of interlaboratory tests (round-robin comparison) performed at nine various laboratories, both at universities and petroleum refineries. More data are presented in Refs. 50 and 51.

It has been stated that repeatability (r) and reproducibility (R), for the average value of p_{oz}, can be calculated from Eqs. (7) and (8), respectively:

$$r = 0.17p_{oz} + 42 \tag{7}$$

$$R = 0.28p_{oz} + 74 \tag{8}$$

In the above equations, all values are expressed in Newtons per square millimeter.

The repeatability (r) and reproducibility (R), related to their average values and expressed in percent, were reported as 23% and 39%, respectively, according to the round-robin interlaboratory tests. The values of r and R, given in ASTM D 2783-82 [33], are 17% and 44%, respectively, valid for lubricants tested under scuffing conditions using the widely used standard procedure for the determination of the so-called *load–wear index*. This means that the precision of the new test method is similar to that achievable with standardized tribological investigations, but the new method is much more effective from a point of view of the time consumption, resolution, and cost.

5. Application of the New Test Method for Classification of Automotive Gear Oils

For the qualification of a given automotive gear oil to one of the API GL quality levels, the performance of full-scale component tests is necessary. For example, to meet the perform-ance specifications of API GL-5, apart from the assessment of some oil physico-chemical properties, one has to carry out also component testing advised by the Coordinating Research Council (CRC) (i.e., axle and gear tests) [54]. Such experiments are very expensive and the necessary equipment is not widely available. An economic matter is particularly acute here, especially when comparing the costs of component tests with the costs of bench ones. As it has already been mentioned, in some reports, the former are assessed to be up to 100 times more expensive than the latter [27]. The reason is that the component tests require real machine elements (hypoid gears, differentials, and spur gears), whereas bench tests are run on cheap, simple specimens, such as balls, rings, rollers, etc.

Taking the above remarks into consideration, some researchers (e.g., Bisht and Singhal [55]) have already successfully substituted the Amsler disk testing for very expensive component tests included in the API GL-4 performance procedure. Bisht and Singhal have proven that it is possible to choose such test conditions as the obtained results will be within intervals characteristic for API GL-4 and GL-5 levels. On that basis, one can classify a tested gear oil.

It has already been shown that the new test method (presented in section III.E.1) has good resolution and is very fast, and its precision is comparable to standardized tribological tests. Thus, it is well founded to employ the developed method to classify automotive gear oils with respect to API GL quality levels.

To verify such possibility, about 20 commercial automotive gear oils were tested. They were oils of API GL-3, GL-4, and GL-5 levels, used for the lubrication of automotive gearboxes (GL-3 and GL-4) and final drives (GL-5).

Experiments were carried out using the modified four-ball tester (presented in section III.E.2). For each of the tested oils, the limiting pressure of seizure p_{oz} was determined according to the new test method. The results were then grouped with respect to API GL levels, given by the producers.

The ranges between a minimum and maximum value of p_{oz} for particular API GL levels are shown in Fig. 30.

From Fig. 30, it is apparent that the new method has satisfactory resolution; the three ranges of p_{oz} do not overlap. Thus, the method allows differentiation between tribological

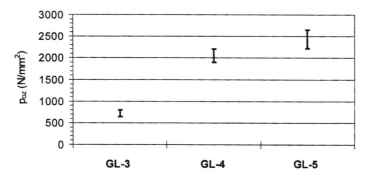

Figure 30 Ranges of p_{oz} obtained for particular API GL levels of the tested automotive gear oils.

properties of automotive gear oils, and, more importantly, the classification of such oils with respect to API GL quality levels.

This is of great economic importance because four-ball tests, substituting (at least at the preliminary stage) for very expensive component tests, allow the finding of an optimum oil formulation at a relatively low cost.

IV. PREVENTION OF SCUFFING AND SEIZURE

During scuffing (unstabilized friction), the intensity of wear of elements constituting a tribosystem is incomparably higher in comparison with that taking place at stabilized friction. This means that to prevent scuffing (and seizure), an adequately modified surface layer of tribosystem elements is of key importance.

There are generally two ways to constitute the surface layer of tribosystem elements: the technological one and the one taking place during a run of a tribosystem. The first one is associated with surface treatment technology. The second one is based on modification of the surface layer produced due to surface–lubricant interactions.

This problem has been well described by the model of transformation of the surface layer proposed by Burakowski and Marczak [56]:

$$\text{TWP}_{1,2} \xrightarrow[\text{MS (additives)}]{p, v, T} \text{EWP}_{1,2}$$

The model assumes that the so-called *technological surface layer* (TWP), produced due to technological processes, undergoes transformation into the "operation" *surface layer* (EWP) under working conditions in a tribosystem, such as pressure (p), speed (v), and temperature (T). The transformation is also an effect of interactions between the surface and the active additives in a lubricant (MS).

It should be added at this point that prevention of scuffing can be realized not only in the technological or "operational" way, but also by applying appropriate engineering assumptions for a sliding pair. The so-called *scuffing criteria*, discussed in the first part of this chapter, can be helpful here.

A. Technological Prevention

By analyzing the kinetic model of scuffing presented by Nosal [6], one can come to a conclusion that the effective prevention of scuffing has to be based on the breaking of the listed chain of events, in which a particular role is played by a properly modified surface layer. This problem has been exhaustively discussed in Ref. 57.

Thus, prevention of lubricating film collapse can be realized by adaptation of the surface layer to effective oil lubrication, by means of the shaping of the high bearing ratio of the surface, accompanied by microholes serving as reservoirs of a lubricant. It is also possible to shape the topographical pattern of the machined surface to increase the efficiency of the lubricating film creation. For example, Piekoszewski [58] proved that machining direction, perpendicular to the direction of the movement, significantly increases oil film thickness, eliminating the necessity for a costly reduction of surface roughness.

Contacts between physically clean asperities can be eliminated owing to the production of a near-surface zone, shielding surface forces of metal and characterizing itself by low shearing strength and high cohesion to the substrate. This role can be played by inorganic compounds (e.g., FeS) produced as a result of interactions between active lubricant additives and the surface of contacting elements. These problems will be discussed in the next part of this chapter.

The number of arising adhesion bonds can be reduced by effective heat transfer from the friction zone and by elimination of plastic deformations through hardening of the surface layer. It has to be added that thermo-chemical surface treatment (e.g., sulfonitriding and nitriding) has been proven to improve significantly the durability of the lubricating film in comparison with heat treatment (hardening) [59].

The growth of adhesive bonds, dependent on the ability of contacting metals to produce, can be prevented by an adequate selection of sliding materials and their surface treatment.

There are many technological methods for the creation of the technological surface layer of elements of tribosystems. They can be divided into the two main groups:

1. Techniques for surface modification
2. Techniques for deposition of coatings.

These techniques are summarized below.

1. Surface Modification Techniques

These techniques allow a modification of the near-surface zone of a material with possible addition of ingredients not originating from the modified material. The modified zone can be even up to some millimeters deep. The surface modification techniques are classified in Fig. 31.

As can be seen from Fig. 31, the surface modification techniques can be realized as follows:

Without modification of chemical composition of the material, by adequate forming only of the surface topography. These methods usually improve the durability of the surface owing to:

Plastic deformation of the surface: static (pressure burnishing) or dynamic (impact burnishing); peculiar fast growth is observed for the dynamic methods, particularly for the ball impact hardening, and also recently for the laser impact hardening; deformation of the base structure is usually accompanied by an increase in the surface roughness

Structural changes of the surface due to phase transitions, without its melting, made to obtain harder structures, for example, diffusion-free transition of austenite into martensite during fusion-free surface treatment: induction, flame, laser, electron, or plasma hardening

Structural changes of the surface caused by its melting and cooling, made, for example, due to surface treatment: laser, electron, or plasma hardening; the structural change results in a change of hardness, not always increasing

With modification of chemical composition of the material, to increase hardness (and other properties), with possibly small changes to surface topography, realized:

Without melting, by saturation (diffusive) alloying of the material surface with one or more nonmetallic elements (e.g., nitrogen and carbon), metals (chromium, boron, aluminum, and titanium), and their combinations, or their implantation (in ionic form) into the surface (reaching a depth of about 1 μm)

With melting, by laser, electron, or plasma alloying of the material surface with metallic and/or nonmetallic elements.

2. Techniques for Deposition of Coatings

Coatings are layers deposited on a substrate material (its surface may be previously modified), possibly including some substrate elements.

Techniques for deposition of coatings are classified in Fig. 32.

As can be seen from Fig. 32, coatings can be deposited from:

A solid medium, by cladding (applied, e.g., in segments of slide bearings) or friction plating (kind of friction welding, e.g., copper plating)

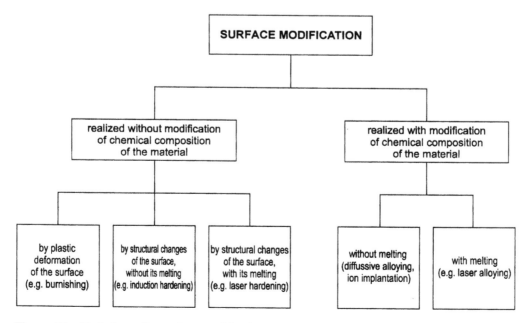

Figure 31 Techniques for surface modification.

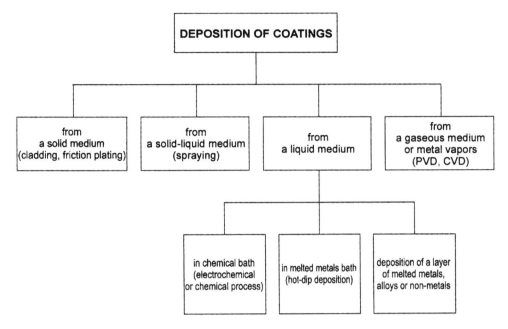

Figure 32 Techniques for deposition of coatings.

A solid–liquid medium, in the form of melted metals, melted metal alloys, or particles of metals, cermetals, and nonmetals heated up to the plastic state or melted; particles are hot-sprayed (arch, flame, plasma, or laser spraying) in a plastified form or even droplets onto metal or nonmetallic substrates, or electrodischarged onto metal substrates

A liquid medium:

  In the form of *chemical bath*; using this technique, common or alloy coatings (usually metallic) are deposited, with or without contribution of electrical current, by means of electrochemical (electrolytic and conversion coatings) or chemical reduction (chemical and conversion coatings), with thickness reaching some tens of micrometers

  In the form of *melted metals bath* (e.g., zinc, aluminum, or their alloys); elements to be deposited are dipped into the bath, producing hot dip coatings

  In the form of *melted metals or alloys* and possibly *nonmetals* (e.g., polymers)

A gaseous medium under low pressure (e.g., of nitrogen) or metal vapors (e.g., of titanium, chromium, and aluminum), usually taking advantage of physical support (e.g., glow discharge, electron beam, and extra accelerating voltage)—the so-called *physical vapor deposition* (PVD) process, and/or chemical reactions—the so-*called chemical vapor deposition* (CVD) process; this is the most popular way of producing hard antiwear coatings of some micrometers thick, made of nitrides, carbides, borides, and oxides (e.g., TiN, CrN, TiC, WN, TiB, and Al_2O_3).

According to their composition, coatings can be classified as (Fig. 33):

One-component coatings, composed of one material (e.g., Al, Cr, Mo, Cu, Ag, and Au), or *one-phase coatings*, composed of one phase (e.g., TiN and CrN)
Multicomponent coatings, composed of solid-state solutions of three or four metals and nonmetals, or *multiphase coatings*, composed of different phases of the same components (e.g., TiN/Ti_2N).

According to their structure and properties, the following coatings can be distinguished (Fig. 33):

Common coatings, for example, one-component, of the same properties along the whole cross-section of the coating (e.g., TiN)
Composite coatings, of the same or similar properties in various zones, however constituting a composite of phases or compounds (e.g., TiC/Al_2O_3)
Gradient coatings, of structure and properties changing gradually from the surface to the substrate [e.g., TiN/Ti(C, N)/TiC coatings].

According to the number of layers, the following coatings can be distinguished (Fig. 33):

Single-layer coatings (e.g., TiN)
Multilayer coatings, usually composed of some different layers, for example, three ($TiC/TiN/Al_2O_3$) or two or three very thin layers, deposited alternatively up to some tens of times (the so-called *compact coatings*).

Technological surface layers (i.e., modified surface and coatings) are usually of the multizone layered structure, making their properties (e.g., tribological) dependent on distance from the surface.

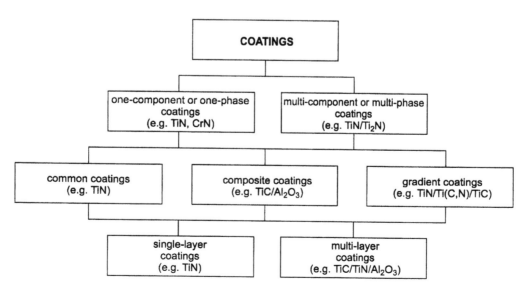

Figure 33 Kinds of coatings.

More often, for tribological purposes, technological surface layers are made of composite forms, usually composed of two or more different modified surface layers and coatings, seldom modified surface alone, never coatings as themselves. This improves hardness, adhesion to the substrate, and stress resistance, and makes it possible to obtain similar coefficients of heat expansion, to reduce inside stress, and to improve the antiwear properties of TWP.

Techniques used in surface engineering are reviewed exhaustively in Ref. 60.

B. Prevention During a Run of a Tribosystem

This kind of prevention of scuffing and seizure is associated with mechanisms of interactions between a lubricant and the surface of elements of a tribosystem.

There are many opinions on such mechanisms. Nowadays, two kinds of their description have emerged: mechanical and chemical [61]. According to the former, hydrodynamic lubrication, EHL, and micro-EHL account for most of the load-bearing mechanisms. In other words, loading is supported mainly by the oil film and partly by surface asperities of contacting solids, so a significant role is subscribed to the rheology of a lubricant (mainly its viscosity). The chemical school points on the creation of a surface layer, a result of chemical reactions, taking place in a friction zone, preventing surface asperities on direct contact. Such reactions are possible due to the so-called *lubricating additives* added to lubricating media [62–64]. These are antiwear additives (e.g., zinc dialkyl dithio phosphate, ZDDP*) or extreme pressure additives (e.g., organic S–P compounds).

1. Effect of Oil Viscosity

An important role of oil viscosity in the prevention of scuffing is confirmed by experimental results presented in Fig. 34. The figure illustrates the influence of the kinematic viscosity of base oils—mineral as well as synthetic ones (PAO 4 and PAO 6)—on the so-called *scuffing load* (P_t), at which collapse of the lubricating film and scuffing initiation occur. The experiments were carried out using the modified four-ball tribotester (described in section III.E.2) at continuously increasing load.

From Fig. 34, it can be noticed that an increase in oil viscosity significantly improves lubrication, postponing scuffing initiation.

The diagram shown in Fig. 35 [65] also illustrates the importance of oil viscosity for the load-bearing capacity of a tribosystem and for prevention of scuffing. Load-bearing capacity is expressed in the figure by means of two components: hydrodynamic and morphological (connected with surface phenomena).

From Fig. 35, it can be seen that in a lubricated contact, the hydrodynamic component, associated among others with oil viscosity, prevails. Thus, it is well founded to consider an increase in oil viscosity to be one of the methods for scuffing prevention (e.g., in toothed gears) [66].

It should be added, however, that oil viscosity influences only conditions of scuffing initiation. It does not have any effect on the propagation of the phenomenon. Scuffing propagation and seizure are influenced only by interactions between a lubricant and the surface of the elements of a tribosystem [48–51]. Hence, a great role is played here by a morphological component relating to surface phenomena (Fig. 35).

---

*In some works, ZDDP is considered to be partly an EP additive.

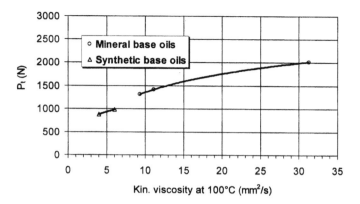

Figure 34 The effect of the kinematic viscosity of base oils on the scuffing load (P_t), causing collapse of the lubricating film and scuffing initiation.

2. Effect of Lubricant–Surface Interactions

   a.   Temperature in the Contact Zone Under Extreme Pressure Conditions.   The state of the surface layer of tribosystem elements during its operation, as has already been mentioned, is a result of interactions between a lubricant and the surface. The analysis on various factors affecting these interactions should start from the contact temperature because the heat generated in the contact zone, being the consequence of sliding speed and pressure, is a driving force for lubricant–surface interactions.

   Hebda and Janecki [67], as well as Barwell [68], have reported that during scuffing, the temperature of the material microvolume is likely to achieve the melting point; for steel, it is over 1400°C.

   The possibility of reaching very high temperatures in a friction zone, under severe conditions, has been confirmed in Ref. 69, where the temperature distribution on the surface of plates sliding one over the other at a speed of 300 m/sec, under a pressure of 1 MPa, has been determined. An infrared detector and a light pipe technology were employed to

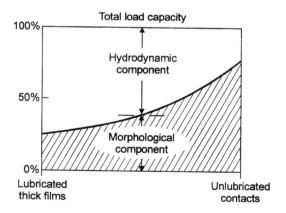

Figure 35 Weights of morphological and hydrodynamic components in different modes of lubrication. (From Ref. 65.)

measure the temperature. The highest measured value was 1100°C, whereas the lowest one was 877°C (Fig. 36).

Hsu et al. [61], basing on the mechanical model of friction modified by taking into consideration the plastic deformation of surface asperities, have proven that the asperity temperature reaches up to 410°C.

Castro and Seabra [70] have estimated the influence of increasing load on the total temperature of the surface of a tooth (FZG gear test machine). The test gear was lubricated with a base oil, pitch velocity was 8.6 m/sec, and initial oil temperature was 90°C. At the last 12th degree of loading, when scuffing was observed, the surface temperature was about 425°C.

The authors of this chapter have calculated the temperature at the contact of surface asperities, adopting the model of their thermoelastic deformations proposed by Wisniewski [71]. The calculations were performed for a lubricated (with an oil containing EP additives) four-ball tribosystem, assuming continuously increasing load.

The calculated values of the maximum temperature of contacting surface asperities are presented in Fig. 37. An increase in the temperature (ΔT_{max}) is given in relation to normal load (P_n) applied to one ball. At collapse of the lubricating film (P_n is about 1100 N), a step rise in the coefficient of friction has been assumed; hence, a step rise in the temperature appears.

From Fig. 37, it can be noticed that an increase in the temperature of contacting surface asperities at the moment of scuffing initiation reaches 500°C. This temperature increases up to 1000°C at the highest loading of the four-ball tribosystem. It stays in agreement with the previously quoted results, obtained by various researchers.

  b.   Influence of Test Conditions and Lubricating Additives on Mechanisms of Lubricant–Surface Interactions. Stachowiak and Batchelor [8] have analyzed the influence of temperature and load on mechanisms of lubricant–surface interactions. The results are gathered in Table 3.

Thus, under severe conditions (high temperature and high load), the prevention of scuffing action is realized due to the formation of a surface layer composed of inorganic compounds.

In the literature, one can find information on various mechanisms describing the formation of layers made of inorganic compounds. For example, the work of Forbes [72] can be mentioned. The researcher, basing on studies of a series of organic monosulfides and disulfides, has proposed a model describing their interactions with the iron surface at increasing contact temperature (T) (Fig. 38).

From Fig. 38, it can be seen that the disulfide is initially adsorbed on the iron surface and then cleavage of the sulfur–sulfur bond occurs so that iron mercaptide R–S–Fe is formed (R, hydrocarbon radical); this is the so-called *AW region*. The increasing

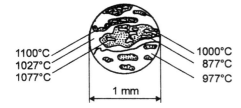

Figure 36   The temperature distribution for the surface studied. (From Ref. 69.)

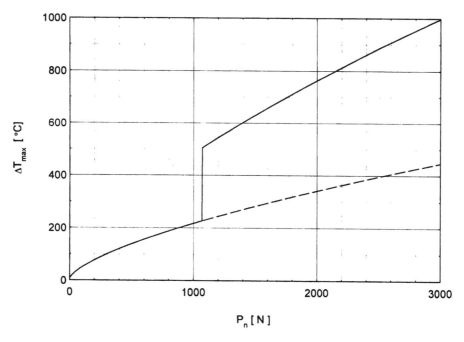

Figure 37 The increase in the maximum temperature (ΔT_{max}) at contacting surface asperities versus normal load (P_n) falling on one ball, with the contact lubricated with an oil containing EP additives.

Table 3 The Influence of Temperature and Load on Mechanisms of Lubricant–Surface Interactions Affecting Lubrication Mode

Temperature	Load	Lubrication mechanisms
Low	Low	Viscosity enhancement close to contacting surface, not specific to lubricant
	High	Friction minimization by coverage of contacting surfaces with adsorbed monomolecular layers of surfactants
High	Medium	Irreversible formation of soap layers and other viscous materials on worn surface by chemical reaction between lubricant additives and metal surface
		Surface-localized viscosity enhancement specific to lubricant additive and base stock
		Formation of amorphous layers of finely divided debris from reactions between additives and substrate metal surface
	High	Reaction between lubricant additives and metal surface
		Formation of sacrificial films of inorganic material on the worn surface, preventing metallic contact and severe wear

Source: Ref. 8.

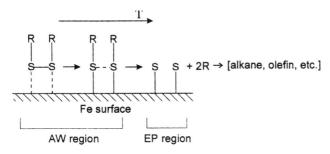

Figure 38 Graphical presentation of Forbes' hypothesis on the AW and EP actions of mono-sulfides and disulfides at increasing contact temperature. (Reprinted from Ref. 72; from Excerpta Medica, Inc.)

temperature (T) in the contact zone causes cleavage of the carbon–sulfur bond, giving an inorganic sulfur-containing layer (not necessarily pure iron sulfide) on the iron surface (so-called *EP region*).

It must be added that this hypothesis has been verified in later works [73]. Although it has resulted in a modification of the AW action, EP action has been confirmed.

Plaza [9] and Wachal [74] have proposed one similar to the above mentioned scheme, describing interactions between a lubricant and the surface, at increasing contact temperature:

Physical adsorption of organic compounds

⇓

Chemical adsorption/tribochemical reactions with the surface

⇓

Formation of inorganic products of reactions

Matveevsky and Buyanovsky [75] and Buyanovsky [76] also observed the antiscuffing action of inorganic compounds. They have proposed a scheme of relations between increasing temperature, state of the surface layer of contacting elements, and coefficient of friction (Fig. 39).

As can be seen in Fig. 39, at low temperatures, the friction coefficient rises. At the highest point of this section of the friction coefficient curve, the EP additives become

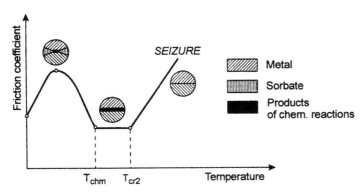

Figure 39 The relationship between increasing temperature, state of the surface layer, and friction coefficient; lubrication with an oil containing EP additives. (From Ref. 76.)

involved in chemical reactions with the surface, producing sulfur or phosphorus compounds. It results in the lowering of the friction coefficient until the so-called *chemical modification temperature* T_{chm} is reached. Owing to chemical modification of the surface, the friction coefficient remains constant and relatively low within the quite wide range of temperatures. However, a further increase in the temperature above the so-called *second critical temperature* (T_{cr2}) leads inevitably to seizure.

Matveevsky et al. [77] have verified the model, basing on experimental results for dibenzyl disulfide dissolved in a base oil, obtained using a low-speed four-ball tribotester. They analyzed changes in the composition of the surface layer, particularly the sulfur content, for the tribosystem elements exposed to increasing temperature. The researchers have found that an increase in the content of sulfur in the surface layer results in a friction decrease, which is necessary for scuffing and seizure prevention.

Johansson et al. [78], on the basis of tribological investigations and complex analyses of the worn surface, have formulated mechanisms of interaction between the surface of the tribosystem elements and two different additives, at increasing load. The employed additives were ZDDP (AW and partly EP additive) and TNPS* (EP additive).

In the case of ZDDP, the mechanism of its interaction with the surface is as follows:

Collapse of the lubricating film and contact between surface asperities

⇓

Formation of sulfides, phosphates, and oxides

⇓

High coefficient of friction and low wear intensity

⇓

Mechanical destruction of the surface

⇓

Formation of hard wear debris

⇓

Abrasive wear

⇓

Metal contact on a large area; slow chemical reactions

⇓

High adhesion

⇓

Rapid increase in the coefficient of friction

⇓

Seizure

In the case of TNPS, the mechanism of its interaction with the surface is as follows:

Collapse of the lubricating film and contact between surface asperities

⇓

Formation of sulfides, phosphates, and oxides

⇓

Low coefficient of friction and high wear intensity

---

* TNPS, ditert nonyl pentasulfide.

⇓

Mechanical destruction of the surface

⇓

Formation of soft wear debris

⇓

Metal contact on a large area; rapid chemical reactions

⇓

Large changes in adhesion

⇓

Slow increase in the coefficient of friction

⇓

Seizure

From the above models, it follows that the speed of chemical reactions between active additives and the surface of contacting elements is decisive for antiscuffing behavior, connected with a rise in the coefficient of friction. In the case of the AW and partly EP additives (ZDDP), reactions are not fast enough to prevent the rapid increase of friction coefficient, which inevitably leads to fast seizure of the tribosystem. A completely different behavior is observed for the EP additives (TNPS); due to fast reactions, the coefficient of friction increases slowly, which prevents quick seizure of the tribosystem.

Similar observations concerning the effectiveness of AW and EP additives, under extreme pressure conditions, are described in Refs. 48 and 49. These works determine to what extent the surface layer of the tribosystem elements can be changed due to interactions with a lubricant. Applying the modified four-ball tribotester, the authors performed tribological investigations of an oil containing a package of EP additives at a concentration of 10%. These additives were based on organic S–P compounds. During experiments, the load applied to the tribosystem increased continuously, making it possible to achieve extreme pressure conditions. After the run, the worn surface was analyzed using a scanning electron microscope (SEM) and an energy-dispersive spectrometer (EDS).

Figure 40 presents a scanning electron microscopy (SEM) image of the wear scar and maps of the surface distribution of sulfur and phosphorus.

Figure 40 illustrates that due to interactions between the additives studied and the surface of steel specimens (balls), a significant concentration of sulfur and phosphorous in the surface layer of the wear scar takes place. The presence of these elements confirms earlier quoted information that under extreme pressure conditions, chemical reactions between active compounds of additives (S–P compounds) and the steel surface result in the creation of a thin layer of probably inorganic compounds of S and P with iron.

Stachowiak and Batchelor [8] have quoted the experimental results of lubrication with S–P additives. They suggest that there is a relationship between the sliding speed and load, on one hand, and the sulfur (and phosphorous) content in the surface layer present there due to surface–lubricant interactions, on the other hand. At low sliding speeds and loads, in the surface layer of the wear scar, phosphorous predominates, whereas a significant increase in the severity of test conditions leads to an increase in the sulfur and a decrease in phosphorous content. It can be concluded that sulfides are responsible for the prevention of scuffing and seizure under EP conditions, whereas phosphates maintain low friction and wear intensity under normal operating conditions.

Interactions between active additives in lubricating oils and the surface of contacting elements have been pointed on so far. The interactions result in the formation of a thin layer of probably inorganic compounds (e.g., FeS), preventing the tribosystem from scuffing.

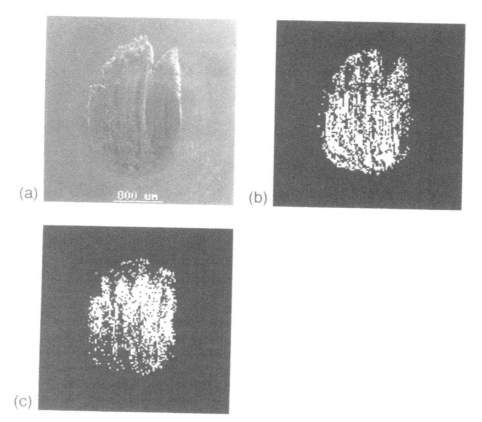

Figure 40 The SEM image of the wear scar on a stationary ball (a) and maps (EDS) of the surface distribution of sulfur (b) and phosphorus (c) in the wear scar; four-ball tribosystem lubricated with a mineral oil containing a package of EP additives, at a concentration of 10% (by weight).

However, the question concerning only their chemical origin arises. Such a question is justified by, as found earlier, high values of the temperature in the contact zone, often exceeding 1000°C, under extreme pressure conditions. It suggests the possibility of diffusion, which in research works is taken up rarely. This approach is justified when analyzing the following experimental data.

Tribological research was performed on a package of AW additives, dissolved in a mineral base oil at concentrations of 0.2% and 3%. This package was based on ZDDP. Experiments were carried out using the modified four-ball tribotester. During tests, loading applied to the tribosystem was continuously increasing, making it possible to achieve extreme pressure conditions. After the run, the surface of the wear scar was analyzed using an Auger electron spectroscope. Such instruments contain an ion gun, which allows etching of the surface with argon ion beam to study deeper and deeper layers of the material. It made possible the studying of the profile of element distribution in the wear scar surface layer.

Changes in the concentration of S, P, and Fe along the depth of the wear scar surface layer are presented in Fig. 41.

It can be observed that the structure of the surface layer depends on the concentration of the additive in the base oil. The increase in the additive concentration results in the rise in

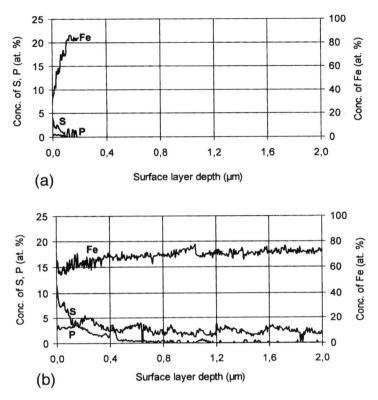

Figure 41 Changes in the atomic concentration of sulfur, phosphorus, and iron along the surface layer depth for two weight concentrations of the ZDDP-containing package in a mineral base oil: (a) 0.2%; and (b) 3%. (Reprinted from Ref. 48; Elsevier Science.)

the depth of diffusion of some elements (S and P) coming from ZDDP. For 0.2% concentration (Fig. 41(a)), the diffusion depth of S and P is only 0.2 μm; for 3%, it exceeds 2 μm (Fig. 41(b)). The authors have also noticed that the higher concentration of the additive in the base oil resulted in the larger content of sulfur and phosphorus in the surface layer.

The observations made point to the possibility of diffusion of some elements from lubricating oils under extreme pressure conditions. Such a possibility has been confirmed by Lauer et al. [79]. They applied Auger electron spectroscopy (AES) to study the influence of organic disulfides (EP additives) on the chemical composition of the surface layer formed on the steel disk in a ball-on-disk tribosystem. The researchers have noticed that the surface layer is composed of two zones: the outside zone (oxides) and the inside zone (FeS). In the inside zone, the sulfur content decreases exponentially, which confirms the occurrence of diffusion.

c. Models of the Surface Layer and Its Role in the Prevention of Scuffing and Seizure. It has been mentioned earlier that under extreme pressure conditions, due to interactions between the lubricant and the surface of tribosystem elements (chemisorption, chemical reactions, and diffusion), a surface layer is formed, preventing direct contacts between surface asperities. It is well illustrated by the surface layer models presented, for example, by Burakowski, Marczak, and Wachal.

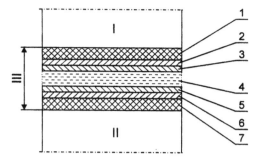

Figure 42 The model of the surface layer according to Burakowski and Marczak: I, II—engineering materials; III—"interbody"; (1, 7) zones of chemical compounds; (2, 6) monomolecular zones of chemisorbed compounds; (3, 5) multimolecular zones sorbed physically; and (4) oil. (From Ref. 56.)

Burakowski and Marczak [56] have elaborated on a model of the surface layer formed on elements of a sliding pair as a result of their interaction with a lubricant (Fig. 42).

It is worth to quoting here observations made by So and Lin [80], who have proposed a similar model of the surface layer. They report that the outside zone of the surface layer is constituted by a "gel-like" substance. The presence of such a layer was also found on the surface of a tribosystem elements by Guangteng and Spikes [81]. They have suggested that one or two molecular layers more viscous than a lubricant account for observed deviations in the lubricating film thickness measured using ultrathin film interferometry and predicted from classical EHL theory. It stays also in agreement with the results by Stachowiak and Batchelor presented earlier in Table 3.

Wachal [74] has confirmed that the structure of the surface layer in different micro-areas is inhomogeneous and dependent on local conditions (pressure and temperature). Figure 43 demonstrates a model of the surface layer according to Wachal. One can distinguish zones of chemisorbed compounds (of a complex structure) as well as zones of products of chemical reactions, they being simple compounds (e.g., FeS).

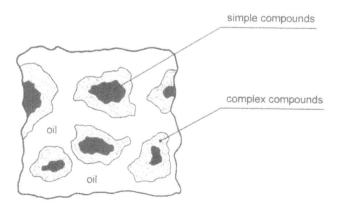

Figure 43 The model of the surface layer according to Wachal. (From Ref. 74.)

It has been mentioned previously that the formation of inorganic compounds (e.g., sulfides or phosphates) on the metal surface prevents scuffing. However, there is a lack of concise explanation on their antiscuffing action.

Pytko [63] has claimed that the factor decisive for the antiscuffing properties of sulfides and phosphates is their relatively high melting point. But according to data quoted by Godfrey [82], antiscuffing properties of inorganic iron sulfides may come from their five-times-lower shear strength and four-times-lower hardness compared to those of steel. This makes it easier to shear adhesive bonds and locate shear planes in the thin surface layer of FeS rather than deeper in the steel substrate, preventing destruction [83]. It makes possible the decrease not only of the wear intensity but the coefficient of friction as well, finally protecting tribosystems from scuffing and seizure under extreme pressure conditions.

REFERENCES

1. Pytko, S.; Sroda, P. Classification and evaluation of machines for investigation of materials for production of gear wheels. ZEM 1975, *1*, 39–58. *in Polish.*
2. Enthoven, J.; Spikes, H.A. Infrared and visual study of the mechanisms of scuffing. Tribol. Trans. 1996, *39*, 441–447.
3. Dyson, A. Scuffing—a review. Tribol. 1975 April, 77–87.
4. Ludema, K.L. A review of scuffing and running in of lubricated surfaces with asperities and oxides in perspective. Wear 1984, *100*, 315–331.
5. Sadowski, J. *Thermodynamic Aspects of Tribological Processes*; Radom Technical University: Radom, 1997. *in Polish.*
6. Nosal, S. *Tribological Aspects of Scuffing in Sliding Pairs*; Poznan Technical University: Poznan, 1998. *in Polish.*
7. Hebda, M.; Wachal, A. *Tribology*; WNT: Warsaw, 1980. *in Polish.*
8. Stachowiak, G.W.; Batchelor, A.W. *Engineering Tribology*, 2nd ed.; Butterworth-Heinemann, Boston, Oxford, Auckland, Johannesburg, Melbourne, New Delhi, 2001.
9. Plaza, S. *Physico-Chemistry of Tribological Processes*; Lodz Technical University: Lodz, 1997. *in Polish.*
10. Thiessen, P.A.; Meyer, K.; Heinicke, G. *The Basics of Tribochemistry*; Akademie-Verlag: Berlin, 1967. *in German.*
11. Wojtczak, L.; de Landa Castillo-Alvarado, F.; Rutkowski, J.H. Surface melting exemplified in case of aluminum samples. Tribologia 1996, *4*, 445–461. *in Polish.*
12. Nakayama, K.; Hashimoto, H. Triboemission of charged particles and photons from wearing ceramic surfaces in various hydrocarbon gases. Wear 1995, *185*, 183–188.
13. Nakayama, K. Triboemission of charged particles from various solids under boundary lubrication conditions. Wear 1994, *178*, 61–67.
14. Kajdas, C. Importance of anionic reactive intermediates for lubricant component reactions with friction surfaces. Lubr. Sci. 1994, *6–3*, 203–228.
15. Kajdas, C. A novel approach to lubrication mechanism under metalworking conditions. Proceedings of International Conference EUROMETALWORKING, Milan, 1994; 1–8.
16. Park, K.-B.; Ludema, K.C. Evaluation of the plasticity index as a scuffing criterion. Wear 1994, *175*, 123–131.
17. Kwasniak, E.; Pasierski, J.; Stupnicki, J. Some remarks concerning scuffing of ceramics slid against steel. Wear 1997, *209*, 219–228.
18. Lawrowski, Z. *Tribology. Friction, Wear and Lubrication*; PWN: Warsaw, 1993. *in Polish.*
19. Tallian, T.E. On competing failure modes in rolling contact. ASLE Trans. 1967, *10*, 418–439.
20. Blok, H. The flash temperature concept. Wear 1963, *6*, 483–494.
21. Bushe, N.A.; Fyodorov, S.V. The state and properties of friction contact of metals at friction in conditions of clutching. Frict. Wear 1991, *12–1*, 46–55. *in Russian.*

22. Niemann, G. *The Elements of Machines: Part 2. Gears*; Springer-Verlag: Berlin, 1960. *in German*.

23. Szczerek, M. *The Methodology Problems of Systematizing Experimental Tribotests*; ITeE: Radom, 1997. *in Polish*.

24. Rapoport, L. The influence of temperature on failure development in steels under transition to seizure. J. Tribol. 1996, *118*, 527–531.

25. Boruta, J.; Wachal, A. The analysis of working conditions of oils at sliding friction. Tech. Smarownicza Trybol. 1978, *3*, 77–82. *in Polish*.

26. Kulczycki, A.; Stryjewski, K. Assessment comparison of AW and EP properties in gear oils done with different model methods. Trybologia 1988, *6*, 9–11.

27. Alliston-Greiner, A.F. Test methods in tribology. Proceedings of the 1st World Tribology Congress, London, 1997; 85–93.

28. van de Velde, F.; Willen, P.; de Baets, P.; van Geetruyen, C. Substitution of inexpensive bench tests for the FZG scuffing test: Part I. Calculations. Tribol. Trans. 1999, *42*, 63–70.

29. van de Velde, F.; Willen, P.; de Baets, P.; van Geetruyen, C. Substitution of inexpensive bench tests for the FZG scuffing test: Part II. Oil tests. Tribol. Trans. 1999, *42*, 71–75.

30. PN-76/C-04147. Test for lubricating properties of oils and greases. *in Polish*.

31. ASTM D 2981-71 (82). Wear life of solid film lubricants in oscillating motion.

32. ASTM D 5706-97. Determining extreme pressure properties of lubricating greases using a high-frequency, linear-oscillation (SRV) test machine.

33. ASTM D 2783-82. Measurement of extreme-pressure properties of lubricating fluids (four-ball method).

34. PN-83/H-04302. The friction test in 3-rollers-cone system. *in Polish*.

35. IP 334/86. Load-carrying capacity tests for oils. FZG gear machine.

36. DIN 51 354 (Part 2). Testing of lubricants; mechanical testing of lubricants in the FZG gear rig test machine; gravimetric method for lubricating oils A/8.3/90. *in German*.

37. Willermet, P.A.; Kandah, S.K. Wear asymmetry—a comparison of the wear volumes of the rotating and stationary balls in the four-ball machine. ASLE Trans. 1983, *26*, 173–178.

38. Boruta, J.; Wachal, A. The experimental analysis of usefulness of four-ball tests for tribological research. Sci. Pap. Rzeszow Tech. Univ. Mech. 1987, *15*, 99–104.

39. ASTM D 2782-77. Measurement of extreme pressure properties of lubricating fluids. Timken method.

40. Begelinger, A.; de Gee, A.W.J. Thin film lubrication of sliding point contacts of AISI 52100 steel. Wear 1974, *28*, 103–114.

41. Begelinger, A.; de Gee, A.W.J. Failure of thin film lubrication—a detailed study on the lubricant film breakdown mechanism. Wear 1982, *77*, 57–63.

42. Kuo, W.-F.; Chiou, Y.-C.; Lee, R.-T. A study on lubrication mechanism and wear scar in sliding circular contacts. Wear 1996, *201*, 217–226.

43. Tasbaz, O.D.; Wood, R.J.K.; Browne, M.; Powrie, H.E.G.; Denuault, G. Electrostatic monitoring of oil lubricated sliding point contacts for early detection of scuffing. Wear 1999, *230*, 86–97.

44. Marczak, R.; Marczak, M.; Ranachowski, Z.; Senatorski, J.; Guzik, J.; Ludew, R. Investigation of acoustic emission of tribological process. Tribologia 2000, *3*, 475–485.

45. Markova, L.V.; Myshkin, N.K.; Semenyuk, M.S.; Kwon, O.K.; Kong, H. On-line optico-magnetic detector for monitoring friction units. Proceedings of the 5th International Symposium INSYCONT, Cracow, 1998; 117–123.

46. Myshkin, N.K.; Petrokovets, M.I.; Chizhik, S.A. Wear measurements and monitoring at macro- and microlevel. In *Fundamentals of Tribology and Bridging the Gap Between the Macro- and Micro/Nanoscales*; Bhusan, B., Ed.; NATO Science Series: Series II. Mathematics, Physics and Chemistry, Kluwer Academic Publishers: Dordrecht, 2001; Vol. 10.

47. ASTM D 2670-81. Measuring wear properties of fluid lubricants. Falex pin-and-vee block method.

48. Piekoszewski, W.; Szczerek, M.; Tuszynski, W. The action of lubricants under extreme pressure conditions in a modified four-ball tester. Wear 2001, *249*, 188–193.

49. Piekoszewski, W.; Szczerek, M.; Tuszynski, W. Method for scuffing propagation assessment. Tribotest 2001, *7* (3), 219–228.

50. Tuszynski, W. Investigation of antiwear action of lubricants. PhD Dissertation. WITPiS: Sulejowek, 1999. *in Polish.*

51. Szczerek, M.; Tuszynski, W. *Tribological Investigations. Scuffing*; ITeE: Radom, 2000. *in Polish.*

52. Polish Patent No. 179123 from July 31, 2000. A method for assessment of antiseizure properties of lubricants using a four-ball apparatus. Owner: ITeE. *in Polish.*

53. Polish Patent No. 177203 from October 29, 1999. A four-ball apparatus for assessment of tribological properties of lubricants. Owner: ITeE. *in Polish.*

54. Wlostowska, E. Automotive gear oils, Part 2. Paliwa Oleje Smary Eksploat 1995, *21*, 22–26. *in Polish.*

55. Bisht, R.P.S.; Singhal, S. A laboratory technique for the evaluation of automotive gear oils of API GL-4 level. Tribotest 1999, *6* (1), 69–77.

56. Burakowski, T.; Marczak, R. Operation surface layer and its investigation. ZEM 1995, *3*, 327–337.

57. Nosal, S. Technological methods for scuffing prevention. In *Some Aspects of Wear of Materials in Sliding Tribosystems*; Zwierzycki, W., Ed.; PWN: Warsaw, 1990; 222–267. *in Polish.*

58. Piekoszewski, W. Identification of lubrication conditions in a non-conformal contact of rough solids. PhD Dissertation. Poznan Technical University: Poznan, 1996. *in Polish.*

59. Kula, P.; Pietrasik, R. An effect of the surface layer on boundary lubrication. Sci. Pap. Lodz Tech. Univ. Mech. 1996, *85*, 85–92. *in Polish.*

60. Burakowski, T.; Wierzchon, T. *Surface Engineering of Metals*; CRC Press: Boca Raton, 1999.

61. Hsu, S.M.; Shen, M.C.; Klaus, E.E.; Cheng, H.S.; Lacey, P.I. Mechano-chemical model: reaction temperatures in a concentrated contact. Wear 1994, *175*, 209–218.

62. Kajdas, C. *The Basics of Fuelling and Lubrication in Cars*; WKL: Warsaw, 1983. *in Polish.*

63. Pytko, S. *The Basics of Tribology and Lubrication Engineering*; AGH: Cracow, 1989. *in Polish.*

64. Zwierzycki, W. *Lubricating Oils*; Glimar-ITeE: Gorlice, 1996. *in Polish.*

65. Heshmat, H.; Pinkus, O.; Godet, M. On a common tribological mechanism between interacting surfaces. Tribol. Trans. 1989, *32*, 32–41.

66. Watson, H.J. Gear failures. In *Tribology Handbook*; Neale, M.J. Ed.; Newness-Butterworth: London, 1973; E4.

67. Hebda, M.; Janecki, J. *Friction, Lubrication and Wear of Elements of Machines*; WNT: Warsaw, 1972. *in Polish.*

68. Barwell, F.T. *Bearing Systems*; WNT: Warsaw, 1984. *in Polish.*

69. Chichinadze, A.W., Ed. *The Basics of Tribology (Friction, Wear, Lubrication)*; Nauka i Tiechnika: Moscow, 1995, 352–357. *in Russian.*

70. Castro, J.; Seabra, J. Scuffing and lubricant film breakdown in FZG gears: Part I. Analytical and experimental approach. Wear 1998, *215*, 104–113.

71. Wisniewski, M. The tribological effect of the microstructure of the surface in concentrated contacts. *Fortschrittberichte der Verein Deutscher Ingeniure. Reihe 1. No. 204*; VDI-Verlag: Düsseldorf, 1991. *in German.*

72. Forbes, E.S. The load carrying action of organo-sulphur compounds—a review. Wear 1970, *15*, 87–96.

73. Forbes, E.S.; Reid, A.J.D. Liquid phase adsorption/reaction studies of organo-sulphur compounds and their load-carrying mechanism. ASLE Trans. 1973, *16*, 50–60.

74. Wachal, A. Analysis of boundary layer estimating criteria in lubricating oils investigations. ZEM 1983, *3*, 325–332.

75. Matveevsky, R.M.; Buyanovsky, I.A. *Antiseizure Properties of Lubricants Under Conditions of Boundary Lubrication*; Izd. Nauka: Moscow, 1978. *in Russian.*

76. Buyanovsky, I.A. Thermal–kinetic method of estimation of lubricants serviceability thermal limitation at heavy regimes of boundary lubrication. Frict. Wear 1993, *14* (1), 129–142. *in Russian.*

77. Matveevsky, R.M.; Buyanovsky, I.A.; Karaulow, A.K.; Mischuk, O.A.; Nosovsky, O.I. Tran-

sition temperatures and tribochemistry of the surfaces under boundary lubrication. Wear 1990, *136*, 135–139.

78. Johansson, E.; Hogmark, S.; Redelius, P. Surface analysis of lubricated sliding metal contacts: Part II. Formation and failure of tribochemical films during boundary lubrication. Tribologia 1997, *16*, 26–38. Finnish periodical.

79. Lauer, J.L.; Benoy, P.A.; Vlcek, B.L.; Calabrese, S.J. Formation of adsorbed and chemically reacted sulfur films on steel surfaces during sliding. ASLE Trans. 1990, *33*, 586–594.

80. So, H.; Lin, Y.C. The theory of antiwear for ZDDP at elevated temperatures in boundary lubrication condition. Wear 1994, *177*, 105–115.

81. Guangteng, G.; Spikes, H.A. Boundary film formation by lubricant base fluids. Tribol. Trans. 1996, *39*, 448–454.

82. Godfrey, D. Boundary lubrication. In *Interdisciplinary Approach to Friction and Wear*; Ku, P.M, Ed.; Southwest Research Institute: Washington, DC, 1968; 335–384.

83. Coy, R.C.; Quinn, T.F.J. The use of physical methods of analysis to identify surface layers formed by organosulphur compounds in wear tests. ASLE Trans. 1975, *18*, 163–174.

7
Wear Mapping and Wear Characterization Methodology

Christina Y. H. Lim and S. C. Lim
National University of Singapore, Singapore

I. INTRODUCTION

There are many ways of characterizing the friction and wear properties of tribological systems. The most common approach is to obtain appropriate values of frictional traction and rates of wear (in various formats, such as weight loss over sliding distance, volumetric loss over sliding distance, and so on) under specific conditions of tribological contact and environment. Although this method of characterization is precise for a pair of materials in tribological contact, it is limited in that it provides only information pertaining to the particular situation (such as the relative sliding speed and the contact pressure under which the tribological data are obtained). Information on changes to the tribological characteristics, such as the dominant wear mechanisms, when the contact or environmental conditions are varied, is not provided to the user. Such knowledge could be crucial to engineers' and designers alike. This lack could be compensated if the mapping approach is used.

In this chapter, we will describe how wear mapping could be an effective way for engineers and designers to gain a broader perspective of the overall tribological characteristics of different sliding systems. This should help them more effectively design against wear and improve the efficiency and reliability of mechanical systems. While much of this chapter will focus on the use of wear mapping as a means of wear characterization, a brief mention will also be made on the usefulness of friction maps. The application of the wear mapping technique to understand the wear of cutting tools and how this same approach could be extended to enable the effective use of coated inserts from the tool wear perspective will be described.

II. WEAR MAPPING AS A CHARACTERIZATION TOOL

Presenting wear data in a graphical form is not a new idea. One of the earliest presentations is the wear rate surface constructed by Okoshi and Sakai [1] for the sliding of steel components. The next significant development is the wear regime map for soft steels proposed by Childs [2], although there were a number of two- and three-dimensional representations of wear data [3,4]. The single factor that separates Childs' map from the rest is the use of a wide range of sliding conditions. This allowed the user a better global view of the wear characteristics of soft steels. Later on, Tabor [5] remarked that wear might be the result of interacting mechanisms with no single process dominating. He then suggested that wear mechanism maps could be developed to explore this much broader pattern of wear behavior. The proposed wear map is one that summarizes data and models for wear, showing how the mechanisms interface, and allowing the dominant mechanisms for any given set of conditions to be identified.

Following on this line of thought, the wear mechanism map for steels in unlubricated sliding was constructed based on experimental data (primarily from pin-on-disk tests) and theoretical models culled from a wide range of sources [6]. This map describes the unlubricated pin-on-disk wear behavior of steels over a wide range of sliding conditions. The map also predicts the field of dominance of one wear mechanism and when its contribution becomes less important. Within each field, contours of predicted normalized wear rates are superimposed. Expanding on the original map, a companion wear-mode map and a wear-transition map were developed to provide additional information that the first map could not conveniently present [7]. The latter two maps summarize the sliding conditions associated with mild and severe wear and where wear transitions occur for steels. The wear-transition map also shows how the various wear transitions reported in the literature could be related and harmonized with one another.

Earlier work on steels had shown that when sliding becomes more severe, the measured coefficient of friction depends on the sliding condition much more than on the surface properties (such as surface roughness), which were found to be more important during slower sliding. With such an observation in mind, it was considered that friction maps might be useful companions for the wear maps of similar sliding systems. An attempt was made, following the presentation of the wear maps for steels, to develop a friction map summarizing the effects of sliding conditions on the dry friction of steels [8]. During slow sliding (less than 1 m/sec), the coefficient of friction between dry steel surfaces is determined by surface roughness and by the plastic (and perhaps elastic) properties of the surfaces. At higher speeds (greater than 1 m/sec), the surface condition is modified by local heating (which can cause oxidation or even melting); then the coefficient of friction depends in a reproducible way on sliding speed and contact pressure. The friction regime map shows that this higher-speed regime can be further subdivided according to the extent of oxidation and degree of melting. More recently, a friction map showing changes in friction force with sliding velocity for thin liquid films in various tribological regimes was proposed by Luengo et al. [9]. This map shows that the film is Newtonian at both high and low velocities, but not in between. This map links changes in behavior with changing tribological contact conditions during boundary lubrication. In another study, Briscoe et al. [10] constructed a friction map relating the friction coefficient and surface deformation mechanisms to a wide range of contact strains for three commercially available polymers in scratching tests using various spherical and conical indenters. This map reveals that the friction coefficient is a reflection of the elastic–plastic response of a polymer to contact strain during scratching.

III. METHODOLOGY OF WEAR MAPPING

Researchers involved in the construction of wear mechanism maps would have their individually favored approach; the choice of these slightly different approaches is almost always a personal one. We have used the following steps to construct wear maps [11]:

1. For the pair of materials of interest, first decide on their mode of contact (e.g., unidirectional sliding), contact geometry (e.g., pin-on-disk), the environment

in which they are to interact (e.g., atmospheric condition), and lubrication condition.

2. Gather experimental data from the literature on wear rates and wear mechanisms pertaining to this sliding pair measured under conditions exactly like or very close to those specified in step 1. In-house tests would have to be carried out if data are lacking. Mathematical models describing wear behavior of this pair should be gathered as well.

3. Decide on the parameters to be used as axes for the map. One can construct a two- or three-dimensional wear diagram; so far, the majority of wear maps are two-dimensional. Also decide on the range of sliding conditions to be included in the map. It is desirable to select as wide as possible a range (further discussed below). For situations such as machining, the range should preferably be similar to that recommended for that particular group of tools whose wear behavior is to be mapped (further discussed below).

4. Construct the empirical wear maps. This is carried out by first grouping the wear data according to the mode and mechanism of wear. The wear rate and wear mechanism data, appropriately classified, are then plotted into the (usually) two-dimensional space defining the map. The field of dominance of each mechanism is then demarcated using field boundaries, and the approximate locations of the contours of constant wear rate are located. At this stage, the wear map is sufficiently informative and it should provide a summary of the global wear behavior of the sliding pair of interest.

5. Finally, introduce the appropriate mathematical models to describe the wear behavior of this sliding system. Where these are not available, new models would have to be developed. The calibrated model for each field is then used to calculate the projected wear rates for conditions in the field where no experimental data are available. These wear rate contours are then superimposed onto the map. Thus a complete wear mechanism map is generated.

It has to be cautioned that construction of wear maps is an extremely time-consuming undertaking. Collection of relevant experimental data alone invariably takes up a large portion of the time needed. To clearly understand these data, classification of the reported mechanisms of wear would be the next major task. Harmonizing all the data points to construct the empirical wear map, even a fairly straightforward two-dimensional one, could often be tricky. This is because one has to expect a certain amount of scatter in the experimental data collected from different sources. The principle to follow when handling such localized scatter is to look at the larger global wear rate trend. Very often, a close examination of the data would reveal a pattern in which these data could be related while accommodating a certain degree of scatter.

When dealing with the reported dominant wear mechanisms, it is probably best to go straight to the raw data, especially the reported photographic evidence. This is because different researchers could very well have used different descriptors to describe the same or similar wear characteristics. Our experiences in the construction of wear maps have shown that this is a fruitful approach to elucidate the various dominant wear mechanisms reported. A new descriptor, incorporating elements used by the various researchers, is usually used in the empirical map.

To develop appropriate mathematical models to predict the wear rates when a particular wear mechanism is dominant is the final, and almost always the most difficult, phase of the exercise. Several approaches could be taken. One is to test the various models available in the literature that describe this particular wear mechanism and pick the one that is able to most accurately predict the observed rates of wear. The other is to examine these models and combine key elements in them and develop a separate, and hopefully less complex, model. In the absence of any suitable model, new ones would have to be developed. Because tribological interactions invariably involve the generation of frictional heating, appropriate mathematical models to describe frictional heating should first be incorporated into the wear models used. Only in situations where frictional heating is expected to be insignificant, such as during very slow sliding, could the effect of frictional heating be ignored.

IV. OTHER WEAR MAPS

Wear maps for other materials in sliding or other tribological contact conditions have also been prepared. For example, a number of wear maps for aluminum alloys in dry sliding have been presented [12,13]. While Liu et al. [12] have focused on the dominant mechanisms, including the presentation of wear rate contours, Wilson and Alpas [13] have primarily concentrated on the transition between mild wear and severe wear. Besides metals, there are wear maps for various engineering ceramics [14–18], polymers [10,19], and metal–matrix composites [11,20–22]. Wear maps for several different tribological contact conditions have also been presented, including fretting maps [23–25], erosion and erosion–corrosion maps [26–28], and abrasion map [29]. For a more in-depth discussion on these wear maps, readers are kindly referred to an earlier publication [11].

V. USEFULNESS OF WEAR MAPS

One of the main features of wear maps is the display of the fields of dominance (defined by the operating conditions) of different wear mechanisms, giving an overview of the characteristics of tribological systems of which these maps represent individually. This will help designers and engineers to better design and maintain tribological systems to reduce and combat wear. Take, for example, if the oxidational wear mechanism is dominant under a particular sliding condition; designers should avoid selecting this sliding condition for steel components when the environment is one where oxygen is absent or is of limited supply, as the formation of oxide films on steel surfaces generally leads to mild wear [30,31]. At the same time, engineers maintaining the system should ensure that the sliding contact does not experience oxygen starvation. While it is generally agreed that effective lubrication should be able to separate the surfaces in relative motion, oxide films on mating steel surfaces should prevent catastrophic failure in the event of film breakdown or lubrication starvation. In such situations, the broad overview of tribological behavior provided by wear maps can be helpful.

VI. WEAR MAPS HAVE TO COVER A WIDE RANGE OF SLIDING CONDITIONS

It has been reported that with continued tribological interactions, the field boundaries demarcating the regions of dominance of different wear mechanisms could shift with time [32,33]. Therefore it is possible to transit from a mild wear condition to a severe wear condition with no change in the operating condition of a tribological system. A wear map is able to suggest the possible dominant wear mechanisms a tribological system might face should this occur. While designers are usually well acquainted with the dominant wear mechanism of the designed sliding conditions, they might not be familiar with the other mechanisms of wear the system might experience should the field boundaries begin to shift. Such information could be readily obtained from the wear map, and appropriate protective measures taken to prevent catastrophic and premature failures. The same could also be said of the unexpected overloading (beyond the designed operating conditions) that might occur during service. Again, information gleaned from a wear map would enable engineers and designers to take appropriate protective measures.

In view of this, wear maps should cover as large a range of sliding conditions as possible. The earlier map for steels [6] spans a very wide range of sliding speeds (from less than 1 mm/sec to nearly 1000 m/sec) while covering more than five decades of contact pressure. This allows users of this map to have a good understanding of the relationship between the various dominant wear mechanisms under different sliding conditions. In comparison, the availability of experimental data (at the point of construction) only allowed a considerably smaller range of sliding conditions to be used for the more-recent empirical wear mechanism map for Al–SiCp composites [11,21]. This map covers less than four decades of contact pressure and a smaller range of sliding speeds (from less than 0.1 m/sec to nearly 30 m/sec). Even with the reduced range, six dominant wear mechanisms could be represented in this map.

As an illustration, a hypothetical wear map, having only one decade of both contact pressure and sliding speed, is superimposed together with the map for Al–SiCp composites onto the skeletal wear map for steels [6] in Fig. 1. It is clear that the area covered by the hypothetical map (b) is only a small fraction of the total area covered by the wear map for steels. Although the area covered by the Al–SiCp map (a) is larger, it is still small in relation to the steel map. Nevertheless, it overlaps the range of operating conditions typically found in dry or marginally lubricated contacts in machines between steel surfaces [34], making it still reasonably useful. A map, such as the hypothetical one, covering a small range might be too "localized" to be useful for the proper design of components in tribological contact.

VII. WEAR MAP FOR COATED AND OTHER MODIFIED SURFACES

The usefulness of wear mechanism maps is not restricted to bulk materials. Borel et al. [35] suggested that it is also meaningful to employ these maps in understanding the wear of abradable coatings deposited on gas-turbine components. They proposed wear mechanism maps for two AlSi-plastic coatings tested under high temperatures, and indicated that these maps are useful for modeling wear mechanisms and designing coating systems for enhanced performance of gas turbines at elevated temperatures. In a later work, wear maps made it possible to quickly determine the effects of coating microstructure on abradability, culminating in the formulation of a general abradability model for aeroengine coatings [36]. More recently, Wang and Kato [37] explored the mechanisms by which wear particles are generated on a carbon nitride coating. The resulting mode transition maps were used to locate the range of sliding conditions under which no wear particles would be generated. The determination of this safe range of operation is crucial in microelectromechanical systems (MEMS), in which the loss of material would enlarge tight clearances, and where any loose wear debris, especially micron-sized ones, can lead to failure of micromechanical components. Other coating systems that have been characterized using the wear mapping approach include TiN [38,39] and Al_2O_3 [39]. In addition to coatings, wear maps may also be constructed for other types of modified surfaces. For example, Kato et al. [40] presented wear

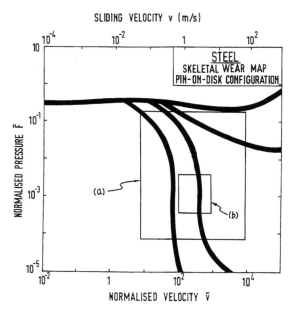

Figure 1 Superimposition of the range of sliding conditions covered by the empirical wear map for Ai–SiCp composites [12,21], indicated by the rectangular area (a), and a hypothetical wear map (having one decade of contact pressure and sliding speed), indicated by the rectangular area (b), onto the skeletal wear maps for steels [6]. It is clear that the area covered by the hypothetical map (b) is only a small fraction of the total area covered by the wear map for steels. Although the area (a) covered by the Al–SiCp map is larger, it is still small in relation to the steel map. Nevertheless, it overlaps the range of operating conditions typically found in dry or marginally lubricated contacts in machines between steel surfaces [34], making it still reasonably useful. A map, such as (b), covering a small range might be too "localized" to be useful for the proper design of components in tribological contact.

mechanism maps providing an additional dimension of information that was not possible to include in the wear mechanism map for steels [6].

VIII. DEVELOPMENT OF WEAR MAPS FOR MACHINING OPERATIONS

Wear mapping has also been extended beyond laboratory-type tests to study coating wear in practical machining operations, notably turning [41–46] and milling [47]. The rest of this chapter will describe the adaptation of the mapping methodology for the characterization of tool wear in turning, and illustrate how wear mechanism maps lead to a better understanding of the effective use of coatings on cutting tools.

Graphical representations of tool wear first appeared in the late 1950s when Trent [48] produced a series of machining charts. More than two decades later, this concept was revived by Yen and Wright [49], who proposed a qualitative wear map for cutting tools. Following on this, Kendall [50] attempted to relate tool wear to the wear mechanism map for steels [6] in a qualitative manner. The common feature shared by these diagrams is the presence of a range of machining conditions, called the safety zone in the two later maps, under which tool wear would not be excessive. Against this background, Lim et al. [51] explored the feasibility of constructing a series of empirical wear maps for different cutting tools. These quantitative maps should allow the user to select appropriate machining conditions for achieving the desired productivity (in terms of material removal rate) at an acceptable tool wear rate. A similar, albeit qualitative, approach was independently pursued by Quinto [52], who presented a tool failure mode diagram derived from Kendall's qualitative wear map [50], Trent's machining chart [48], and the wear mechanism map for steels [6]. The safe zone on this map is defined as a region of gradual wear associated with predictable and reliable tool performance. All these maps are intended to inform the user of the machining conditions that would give rise to the least amount of tool wear.

The effort to construct empirical wear maps for cutting tools has led to several maps for various uncoated and coated tools in the turning of steels [41–46,53]. The maps depict the global characteristics of tool wear over the recommended range of machining conditions for these tool–workpiece material combinations. The key difference between these and the qualitative maps [49,50,52] is that data (both wear rates and wear mechanisms) drawn from actual machining tests were used, thereby allowing the optimization of actual operations. The general methodology outlined earlier in this chapter was adapted for the development these maps; this process will be briefly described here.

IX. MAPPING METHODOLOGY FOR CUTTING TOOLS

The mapping of cutting tool wear began with identifying the parameter by which wear is measured in the chosen operation turning. For this and other types of machining, the useful life of a tool is determined by the extent of wear on the flank or rake face of the tool, known as flank wear and crater wear, respectively [54]. (Schematic illustrations of the turning operation, as well as flank and crater wear on a tool are given in Figs. 2 and 3, respectively.) With this in mind, relevant flank- and crater-wear rate and wear mechanism data were then gathered from the technical literature. Our earlier experiences with developing the wear maps have shown that the sheer number of variables that influence wear to varying degrees makes it extremely difficult, if not impossible, to produce a universal diagram that describes all materials or test configurations. For this reason, workpiece materials were restricted to plain and low alloy steels; machining operations other than turning, such as milling and drilling, as well as laboratory-type experiments, such as pin-on-disk, were also excluded. Data from tests

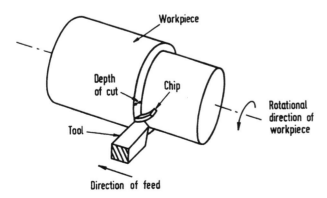

Figure 2 Schematic illustration of single-point turning.

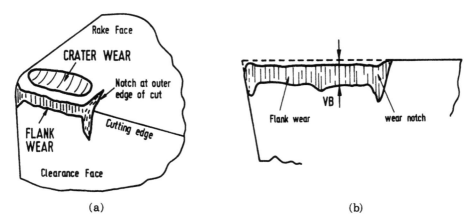

Figure 3 (a) Regions of tool wear on a cutting tool. (b) Measurement of flank wear.

in which a cutting fluid was used were omitted as well, because machining without cutting fluids represents the worst-case scenario, where tool wear is likely to be the most severe. Furthermore, the effects of different cutting fluids on tool wear are inconclusive [55,56].

The major variables that affect tool wear have been identified as cutting speed, feed rate, and the depth of cut [57]; of these, it has been suggested that the depth of cut is the least important [58]. In the interest of making the wear maps as user-friendly as possible, it was decided that two-dimensional maps should be constructed, with cutting speed (in m/min) and feed rate (in mm/rev) as the axes. Thus each point in the plane defined by these axes represents a unique machining condition, with which may be associated an experimentally measured tool wear rate or observed wear mechanism. This enabled the collected wear rate and wear mechanism data to be transcribed onto the maps. Separate maps were plotted for each class of tool material because the vastly different material properties and micro-structure would render hopeless any attempt at combining data for all types of tool materials in a single diagram. Flank and crater wear were also separately considered because these have no direct relationship with each other and, thus, it is not possible to merge the wear data in a meaningful manner.

The maps were then validated through carefully executed turning tests carried out in accordance with ISO 3685:1993 [54]. For coated tools, additional tests were carried out to provide tool wear data for conditions not reported in the literature, with the aim of covering as much as possible the range of recommended cutting conditions.

X. WEAR MAPS FOR CUTTING TOOLS

Wear maps describing the flank- and crater-wear characteristics of a number of types of tool materials have already been published [41–46,53]. Here the flank-wear maps of uncoated and TiC-coated cemented carbide tools will be used to illustrate their application in optimizing machining operations, as well as their usefulness in exploiting the full potential of coated tools.

The flank-wear map for uncoated carbide tools is shown in Fig. 4 [53]. Every point on the map represents a particular machining condition defined by cutting speed and feed

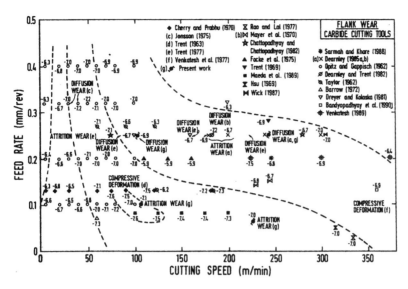

Figure 4 Map showing the rates and mechanisms of flank wear of uncoated carbide tools during dry turning of steels [53]. The numbers next to the points are $\log_{10}$(VB/cutting distance). The approximate machining conditions under which the three wear mechanisms are observed are also indicated.

rate; the value next to each shows the flank-wear rate measured under that condition. This dimensionless flank-wear rate is expressed as $\log_{10}$(VB per cutting distance), which has been found to be more meaningful than the conventional definition of flank wear, VB [54] (Fig. 3). Wear mechanisms observed under different machining conditions are also indicated on the map. The dashed boundaries demarcate different regimes, within which flank-wear rates of similar ranges in values are contained. The overall correlation of wear rate data does not appear to be affected, in any significant manner, by the depth of cut (which varied from 0.25 to 3 mm), thus lending support to the earlier decision to omit depth of cut in the maps.

The wear map may be more clearly visualized in Fig. 5, where the boundaries have been replotted with all data points removed and various regions appropriately shaded [53]. On this map, the area with the lowest wear rates (-7.5 to -7.9) is denoted the safety zone: This is the region in which the tools would experience the least amount of flank wear. The area with the next highest wear rates (-7.0 to -7.4, shaded with dots), known as the least-wear regime, is also of significance, because wear rates here are only slightly higher than those in the safety zone. Because this regime covers a much wider area than the safety zone, machining within this region would allow higher rates of material removal while maintaining reasonably low levels of flank wear.

When a thin coating of TiC is applied onto the carbide tool, the tool wear characteristics undergo considerable changes. Figures 6 and 7 show the flank-wear maps for TiC-coated carbide inserts during dry turning. It is readily apparent that the application of the TiC coating brings about an expansion in the range of machining conditions within which similar wear rates may be experienced. Wear mechanism maps describing the crater-wear characteristics of TiC-coated tools, as well as those constructed for uncoated [53] and TiN-coated, high-speed steel tools [41–44], also exhibit similar trends.

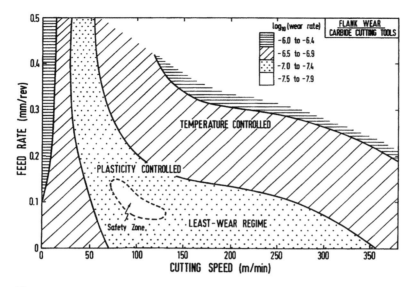

Figure 5 Map for the flank wear of uncoated carbide tools during dry turning of steels [53]. The regions where the different ranges of wear rates are observed are shaded accordingly. The safety zone is the region where flank wear rates are the lowest; the least-wear regime with the next highest wear rates is also indicated. The approximate locations of dominance of the two main wear mechanisms are also shown.

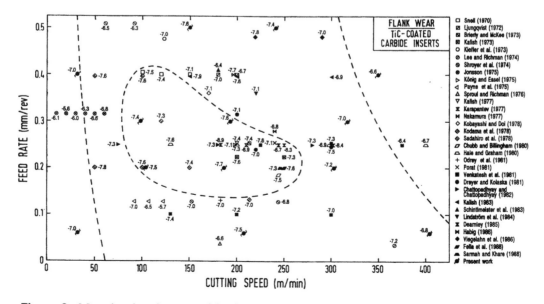

Figure 6 Map showing the rates of flank wear of TiC-coated carbide tools during dry turning of steels [46]. The numbers next to the points are $\log_{10}$(VB/cutting distance).

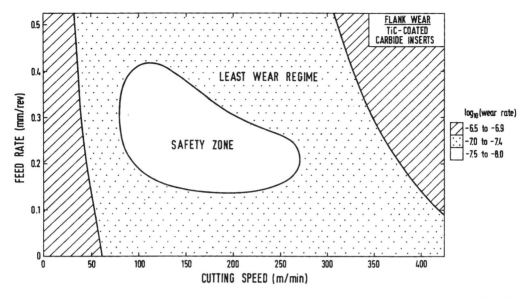

Figure 7 Map for the flank wear of TiC-coated carbide tools during dry turning of steels [46]. The regions where the different ranges of wear rates are observed are shaded accordingly. The safety zone is the region where flank wear rates are the lowest; the considerably larger least-wear regime is also indicated.

Our earlier experiences in constructing wear maps have taught us that local variations in wear rates are to be expected, and some conflicts may appear difficult to resolve. We acknowledge that such differences are inevitable, considering the wide-ranging sources from which these data were drawn; however, taken as a whole, the variations in wear rates with machining conditions can be adequately described, as suggested by the maps.

XI. EFFECTIVE USE OF COATED TOOLS

On its own, each wear map suggests the optimum machining condition for the specific tool–work combination based on the criterion tool wear. However, combining the maps in a suitable manner can provide useful information on how coated tools could be more effectively employed. The following section will demonstrate, using the flank-wear maps for carbide tools, that coated tools could perform as poorly as their uncoated counterparts if inappropriate machining conditions are selected [59].

Figure 8 shows the superimposition of the contours of the flank-wear maps for uncoated and TiC-coated carbide tools. The most striking feature is the much larger safety zone in the case of the coated inserts. The locations of the safety zones for the uncoated and coated tools do not coincide as very different processes are responsible for the wear of the uncoated carbide tool and the TiC layer [46,53]. It is also worth noting that a dramatic expansion of the least-wear zones of the uncoated tools is obtained with the application of the TiC layer, especially at higher speeds and feeds. In fact, TiC-coated carbide tools are able to machine under conditions that would otherwise cause failure in uncoated carbide tools, and yet manage to sustain only moderate wear.

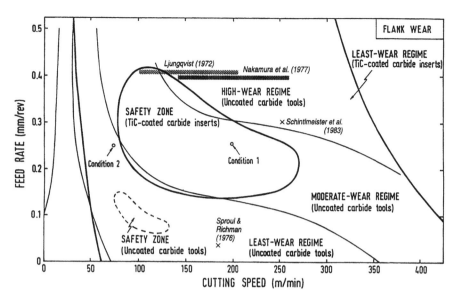

Figure 8 The effects of TiC coatings on the flank wear of carbide tools during dry turning, showing the expansion of the safety zone and the least-wear regime. Conditions 1 and 2, together with the conditions of other authors mentioned in the text, are indicated.

To better visualize the extent of the advantage offered by the TiC coating, an "enhancement factor" is introduced, which measures the increase in wear resistance (or conversely, the decrease in wear rate) brought about by the TiC coating. This enhancement factor is similar in concept to that used by Gu et al. [47] when comparing the improvement in tool life by coated tools in milling. From the overlap of the various zones on the two superimposed maps (Fig. 8), a new map showing the enhancement in flank-wear resistance as a result of the coating may be constructed. For example, the overlap of two zones with the same ranges of wear rates on both maps would combine to give a region with an enhancement factor of 1- to 2.5-fold. An overlap with a zone containing the next lowest wear rates would give a region with the next best enhancement factor of 1.3- to 8-fold. The highest enhancement factor of between 4- and 25-fold is obtained for the overlap of zones with ranges of wear rates a decade apart (e.g., −6.5 to −6.9 compared with −7.5 to −7.9). The enhancement factor is calculated by dividing the value of (VB/cutting distance) at the higher limit of the two overlapping zones by that at the lower limit. The ranges of the enhancement factors from such a comparison do not increase in a linear manner from one zone to another; this is due to the use of the logarithmic scale in representing the rates of flank wear on the wear maps.

The map showing the enhancement in flank-wear resistance as a result of the TiC coating is given in Fig. 9. The regions with the best enhancement factor of between 4- and 25-fold are identified as "peak enhancement zones": These are the ranges of machining conditions under which the TiC coating is most effective in improving flank-wear resistance. On the other hand, the coating has the least effect reducing flank wear in the regions that are hatched with diagonal lines.

It is evident from Fig. 9 that the TiC coating significantly improves flank-wear resistance over a wide range of machining conditions. The map shows that by selecting a

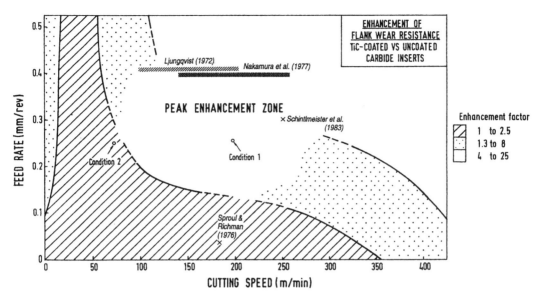

Figure 9 Map showing the enhancement in flank wear resistance as a result of the TiC coating on carbide tools during dry turning of steels.

TiC-coated tool over an uncoated one, and increasing cutting speeds and feed rates, productivity could even be raised in tandem with a reduction in tool wear. Hitherto, large improvements in wear rates as a result of a TiC coating have been reported by various researchers, testing within only a small range of conditions [60,61]. However, the present wear maps are able to clearly demonstrate that such benefits of TiC coatings can be realized over a much wider range of machining conditions.

It is interesting to see how, by studying tool wear in this manner, much information, apparently contradictory, falls into place in a consistent pattern. When the flank-wear rates of uncoated and TiC-coated carbide tools were compared under two different sets of machining conditions, it was found that while the coated inserts consistently wore less than the uncoated ones, the extent of reduction in wear rates varied from a high 4.4-fold to an almost negligible 1.3-fold. This alone suggests that it is possible to further maximize the benefit of the TiC coating. The two sets of conditions are highlighted in Figs. 8 and 9, and the wear rates measured under each condition are plotted as histograms in Fig. 10 to facilitate comparison. When the rates of crater wear were compared in the same fashion, improvements ranging from a modest 3.9-fold to an impressive 28.6-fold were registered [46].

Turning to Fig. 8, it may be seen that Condition 1 lies within the flank-wear safety zone of the coated inserts but inside the high-wear regime for uncoated tools. In Fig. 9, this condition is in the peak enhancement zone, so a sizable difference in wear rates of between 4- and 25-fold is expected. This is confirmed by Fig. 10, which shows that the uncoated tool wore 4.4 times more than the coated one. On the other hand, Condition 2 falls inside the least-wear regimes for both coated and uncoated tools (Fig. 8), placing it in the area with the poorest enhancement factor of up to only 2.5-fold (Fig. 9). Thus little difference in the wear rates is expected; Fig. 10 shows an improvement of only 1.3-fold over the uncoated insert.

Not only do Figs. 8 and 9 explain the varied amounts of reduction in wear rates in our own turning tests, they are also able to rationalize the different extents of improvement in

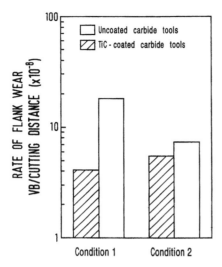

Figure 10 Comparison of the flank wear rates of uncoated and TiC-coated carbide tools under Conditions 1 and 2 (Figs. 6 and 7).

wear resistance reported by various researchers. For example, Ljungqvist [62] tested various grades of TiC-coated tools and found that within the speed range of about 100–200 m/min at a feed rate of 0.4 mm/rev (Figs. 8 and 9), the tool life for coated tools measured by flank wear criterion was between 2 and 10 times that of uncoated tools. The enhancement map (Fig. 9) suggests the same order of magnitude of improvement (1.3- to 25-fold increase in tool life).

In contrast, when uncoated and TiC-coated inserts were used in turning tests at 182.5 m/min and 0.035 mm/rev, Sproul and Richman [63] observed a difference in wear rates of only 1.7 times. Their condition falls within the least-wear regimes for both uncoated and coated tools (Fig. 8), a situation similar to Condition 2 discussed earlier. Likewise, this condition lies in the region with enhancement factor of up to a maximum of 2.5-fold (Fig. 9), and as such, the difference in wear rates is not expected to be large. This is in good agreement with the prediction of the wear maps, demonstrating again that the potential of TiC coatings to reduce tool wear could be considerably diminished if the machining condition is inappropriately chosen.

In another instance, turning tests carried out by Schintlmeister et al. [64] compared the performance of TiC-coated inserts to that of an uncoated carbide tool at 250 m/min and 0.3 mm/rev. Although no specific wear rates were given, they reported that, while the tool life of the uncoated carbide tool was reached after 10 min, there was no measurable flank wear on the coated tool after 20 min. As shown in Fig. 8, this condition lies inside the ultra-high-wear regime for the uncoated tools but only falls within the least-wear regime for coated tools; Fig. 9 predicts that coated tools are likely to wear up to 25 times less than the uncoated ones. It is not surprising then to find the coated tool hardly showing any sign of flank wear. The wear maps are again able to correctly account for this experimental observation.

In a further example, Nakamura et al. [65] noted that the flank wear of coated tools was one-fourth to one-fifth that of uncoated ones when turning between 141 and 254 m/min at a feed rate of 0.4 mm/rev (Figs. 8 and 9). Under these conditions, Fig. 9 shows that flank wear for the coated tools is expected to be 4–25 times less than that of uncoated tools, showing reasonable agreement with experimental data.

XII. REFINING CUTTING TOOL MAPS FOR SPECIFIC WORKPIECE MATERIALS

The wear maps for cutting tools presented earlier [41–46,53] were developed with the aim of depicting the overall tool wear characteristics during dry turning of a group of roughly similar steels—plain carbon and low-alloy steels. However, slight variations in the composition, hardness, and microstructure of steels are known to alter the severity of tool wear [66]. While hardness may be easily incorporated into these maps by the use of normalized variables, such as in the wear-mechanism maps for steels [6], the role of hardness in tool wear has been found to be secondary compared to those of composition and microstructure [58,67]. Unlike pin-on-disk sliding, for example, machining involves shearing and large deformations of the work material; the severity and onset of several wear mechanisms are influenced by such factors as the strain-hardening rate of the steel, heat treatment, as well as the presence and type of secondary phases and inclusions [58,67]. These parameters are difficult to include in the wear maps as their effects on wear cannot at present be predicted from mechanical properties. Therefore the wear maps that have been constructed are approximate guides to the general trends of tool wear behavior for a range of work material rather than precise statements for any particular tool–workpiece material combination. These maps would be of greatest interest to the engineer or designer who seek the global view. The machinist, on the other hand, might be primarily concerned with obtaining recommended operating conditions for specific tool–workpiece combinations. This type of information has traditionally been found in machining databases and handbooks [68], Trent's machining charts [48], or catalogs and manuals from tool manufacturers.

With this in mind, it is possible to produce wear maps that are work material specific simply by extracting the relevant information from the earlier, more general, maps, provided that sufficient data exist for each work material of interest. This has been carried out for uncoated and TiC-coated carbide tools turning two groups of steels, namely, a plain medium carbon steel (AISI 1045) and a low-alloy medium carbon steel (AISI 4340) [69,70]. Comparison of these material-specific maps with their more general counterparts reveals that interestingly, the location and contours of the safety zones did not significantly shift from map to map, despite moderate differences wear rates [69,70]. This suggests that the global approach of wear maps is useful in "ironing-out" local scatter and variations to reveal the underlying patterns of wear behavior common to most plain carbon and low-alloy steels. Thus work-material-specific wear maps may not always be necessary, unless extreme departures, such as the machining of highly alloyed stainless steels, cause a marked shift in the operating wear mechanisms. In such cases, separate wear maps should be constructed.

It is instructive at this juncture to revisit Trent's machining charts [48], which were constructed for specific combinations of tool and workpiece materials. Figure 11 shows the machining chart for the dry turning of EN8 (roughly the equivalent of AISI 1045) steel using carbide tools. The lines demarcate machining conditions under which a heavy built-up edge dominates and the onset of severe cratering. Trent suggested that the useful cutting range for this tool–work combination lies between build-up and rapid cratering; within this range of machining conditions, tool wear is expected to gradually progress in an even manner.

In practice, it is not possible to isolate a tool from wear at its flank face or rake face, as these processes simultaneously occur during machining. On Trent's machining chart (Fig. 11) [48], phenomena associated with wear at both flank (built-up edge) and rake (cratering) faces are reflected in the same diagram. However, with the mapping approach, separate maps were constructed to describe flank and crater wear. This is because not only do flank and crater wear occur at different locations on the tool, they are also measured by different

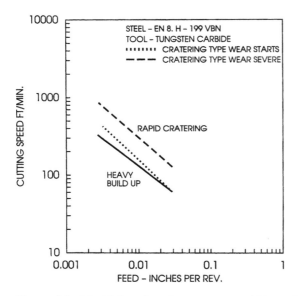

Figure 11 Machining chart for tungsten carbide tool turning EN8 steel. (From Ref. 48.)

parameters. Thus there is no direct relationship between the two and it would not be feasible to merge measured flank- and crater-wear rates in a meaningful manner on a single map. Nonetheless, it has been found that by defining an arbitrary unit that describes the "degree of wear damage" sustained by the tool, flank- and crater-wear-rate maps may be usefully combined [71]. An "overall wear damage map" is constructed by superimposing maps of flank and crater wear corresponding to the same material, and combining the two safety zones to give a region with the least overall wear damage; the overlap of one safety zone with the next least-wear region of the other map would give the next least-wear-damage region, and so on, with a certain amount of liberty taken to smooth out some of the boundaries. The overall wear damage map for uncoated carbide tools in the dry turning of AISI 1045 steels is shown in Fig. 12. The least-damage zone on this map is the range of machining conditions where the tool will sustain the least amount of both flank and crater wear.

Figure 13 shows the contours of this overall wear damage map superimposed with the boundaries from Trent's machining chart (Fig. 11 replotted in rectilinear coordinates) [48], as well as the nominal machining conditions recommended in one machining handbook for the same tool–work combination [68]. It is interesting to note that Trent's useful cutting range coincides with the least-damage and low-wear-damage regions on the overall wear damage map, while the recommended conditions from the machining data handbook show good agreement with the limits of low to moderate wear damage.

Although Trent's machining charts [58], machining databases [68], and wear maps all share the common aim of defining a safe operating region, the latter goes beyond the safety zone to encompass a much broader range of machining conditions. This, together with actual flank and crater wear rates, enables the user to make independent and informed decisions when weighing productivity against tool wear. Companion maps describing observed wear mechanism [41,44,46,53] provide engineers and designers with a better understanding of tool wear processes, their transitions and relationships with measured wear rates. Seen in this light, Trent's machining charts and data from machining handbooks may be considered subsets of the more universal wear map. Nonetheless, this does not in

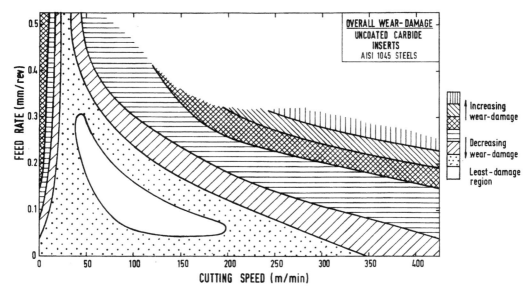

Figure 12 Overall wear damage map for uncoated carbide tools during dry turning of AISI 1045 equivalent steels.

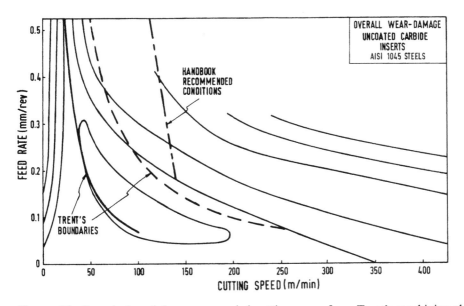

Figure 13 Boundaries of the recommended cutting range from Trent's machining chart (Fig. 9) [48] and machining handbook [68] superimposed on the contours of the overall wear damage map for uncoated carbide tools (Fig. 12).

any way diminish the importance of such charts and databases, the value of which lies in their wide coverage of a large number of specific tool and workpiece material combinations.

XIII. CONCLUSION

Wear maps are difficult to construct, invariably requiring a considerable amount of time and effort. However, once constructed, they are useful for wear (and sometimes, friction) characterization as well as serve as useful databases for designers and engineers to combat wear and to improve system reliability. We have shown that in the case of cutting tools, wear maps can be used not only to understand how machining conditions could affect tool wear, but also to help select the optimal machining conditions to strike a balance between productivity and tool wear. In the case of coated tools, application of the wear-mapping approach could potentially lead to more effective use of these tools.

ACKNOWLEDGMENTS

We would like to thank the National University of Singapore for the financial support and the provision of facilities to undertake the research reported here.

REFERENCES

1. Okoshi, M.; Sakai, H. Researches on the mechanism of abrasion. Report III, mechanism of abrasion of cast iron and steel. Trans. JSME 1941, 7, 29–47.
2. Childs, T.H.C. The sliding wear mechanisms of metals, mainly steels. Tribol. Int. 1980, 13, 285–293.
3. Welsh, N.C. The dry wear of steel. Part I. The general pattern of behaviour. Philos. Trans. R Soc. Ser. A 1965, 257, 31–50.
4. Egawa, K. Effects of the hardness of hardened steel on frictional wear (3rd report)-cubic model form of region of different forms of wear. J JSLE Int. Ed. 1982, 3, 27–30.
5. Tabor, D. Status and direction of tribology as a science in the 80's. Proceedings of International Conference on Tribology in the 80's, Cleveland, 1983; 1–17.
6. Lim, S.C.; Ashby, M.F. Wear-mechanism maps. Acta Metall. 1987, 35, 1–24.
7. Lim, S.C.; Ashby, M.F.; Brunton, J.H. Wear-rate transitions and their relationship to wear mechanisms. Acta Metall. 1987, 35, 1343–1348.
8. Lim, S.C.; Ashby, M.F.; Brunton, J.H. The effects of sliding conditions on the dry friction of metals. Acta Metall. 1989, 37, 767–772.
9. Luengo, G.; Israelachvili, J.; Granick, S. Generalized effects in confined fluids: new friction map for boundary lubrication. Wear 1996, 200, 328–335.
10. Briscoe, B.J.; Evans, P.D.; Pelillo, E.; Sinha, S.K. Scratching maps for polymers. Wear 1996, 200, 137–147.
11. Lim, S.C. Recent developments in wear-mechanism maps. Tribol. Int. 1998, 31, 87–97.
12. Liu, Y.B.; Asthana, R.; Rohatgi, P. A map for wear mechanisms in aluminum alloys. J Mater. Sci. 1991, 26, 99–102.
13. Wilson, S.; Alpas, A.T. Thermal effects on mild wear transitions in dry sliding of an aluminium alloy. Wear 1999, 225–229, 440–449.
14. Hsu, S.M.; Wang, Y.S.; Munro, R.G. Quantitative wear maps as a visualisation of wear mechanism transitions in ceramic materials. Wear 1989, 134, 1–11.

15. Hsu, S.M.; Lim, D.S.; Wang, Y.S.; Munro, R.G. Ceramics wear map: concept and method development. J STLE 1991, *47*, 49–54.
16. Kato, K. Tribology of ceramics. Wear 1990, *136*, 117–133.
17. Gautier, P.; Kato, K. Wear mechanisms of silicon nitride, partially stabilized zirconia and alumina in unlubricated sliding against steel. Wear 1993, *162–164*, 305–313.
18. Hsu, S.M.; Shen, M.C. Ceramic wear maps. Wear 1996, *200*, 154–175.
19. Briscoe, B.J.; Evans, P.D. The influence of asperity deformation conditions on the abrasive wear of γ-irradiated PTFE. Proceedings of the International Conference on Wear of Materials, Denver; American Society of Mechanical Engineers: New York, 1989; 449–457.
20. Wang, D.Z.; Peng, H.X.; Liu, J.; Yao, C.K. Wear behaviour and microstructural changes of SiCw–Al composite under unlubricated sliding friction. Wear 1995, *184*, 187–192.
21. Kwok, J.K.M. Tribological properties of metal–matrix composites. Ph.D. thesis, National University of Singapore, 1996.
22. Wilson, S.; Alpas, A.T. Wear mechanism maps for metal matrix composites. Wear 1997, *212*, 41–49.
23. Vingsbo, O.; Söderberg, S. On fretting maps. Wear 1988, *126*, 131–147.
24. Kassman, A.; Jacobson, S. Surface damage, adhesion and contact resistance of silver plated copper contacts subjected to fretting motion. Wear 1993, *165*, 227–230.
25. Wei, J.; Fouvry, S.; Kapsa, Ph.; Vincent, J. *Fretting behaviour of TiN coating*, Proceedings of the 10th International Conference on Surface Modification Technologies, SMT10, Singapore; The Institute of Materials: London, 1996; 24–36.
26. Hutchings, I.M. Transitions, threshold effects and erosion maps. In *Erosion of Ceramic*; Ritter, J.E., Ed.; Trans. Tech. Publications: Zurich, Chapter 4.
27. Stack, M.M.; Corlett, N.; Zhou, S. Construction of erosion–corrosion maps for erosion in aqueous slurries. Mater. Sci. Technol. 1996, *12*, 662–672.
28. Stack, M.M.; Corlett, N.; Zhou, S. Impact angle effects on the transition boundaries of the aqueous erosion–corrosion map. Wear 1999, *225–229*, 190–198.
29. Trezona, R.I.; Allsopp, D.N.; Hutchings, I.M. Transitions between two-body and three-body abrasive wear: influence of test conditions in the microscale abrasive wear test. Wear 1999, *225–229*, 205–214.
30. Quinn, T.F.J. The role of oxidation in the mild wear of steel. Br. J Appl. Phys. 1962, *13*, 33–37.
31. Quinn, T.F.J. Oxidational wear. Wear 1971, *18*, 413–419.
32. Yust, C.S.; DeVore, C.E. The friction and wear of lubricated $Si_3N_4/SiC_{(w)}$ composites. Tribol. Trans. 1991, *34*, 497–504.
33. Yust, C.S. Wear transition surfaces for long-term wear effects. In *Tribological Modeling for Mechanical Designers, ASTM STP 1105*; Ludema, K.C. Bayer, R.G., Eds.; American Society for Testing and Materials, West Conshohocken, PA. ASTM, 1991; 153–161.
34. Williams, J.A. Wear modelling: analytical, computational and mapping: a continuum mechanics approach. Wear 1999, *225–229*, 1–17.
35. Borel, M.O.; Nicoll, A.R.; Schläpfer, H.W.; Schmid, R.K. The wear mechanisms occurring in abradable seals of gas-turbines. Surf. Coat. Technol. 1989, *39–40*, 117–126.
36. Barbezat, G.; Clarke, R.; Nicoll, A.R.; Schmid, R. Proceedings of the International Symposium on Tribology, Beijing; International Academic Publishers: Beijing, 1993; 756–766.
37. Wang, D.F.; Kato, K. Effect of friction cycles on wear particle generation of carbon nitride coating against a spherical diamond. Tribol. Int. 2000, *33*, 115–122.
38. Wilson, S.; Alpas, A.T. Wear mechanism maps for TiN-coated high speed steel. Surf. Coat. Technol. 1999, *120–121*, 519–527.
39. Kato, K. Microwear mechanisms of coatings. Surf. Coat. Technol. 1995, *76–77*, 469–474.
40. Kato, H.; Eyre, T.S.; Ralph, B. Wear mechanism map of nitrided steel. Acta Metall. 1994, *43*, 1703–1713.
41. Lim, S.C.; Lim, C.Y.H.; Lee, K.S. The effects of machining conditions on the flank wear of TiN-coated high-speed-steel tool inserts. Wear 1995, *181–183*, 901–912.
42. Lim, S.C.; Lim, C.Y.H.; Lee, K.S. Crater wear of TiN coated high speed steel tool inserts. Surf. Eng. 1997, *13*, 223–226.

43. Lim, C.Y.H.; Lim, S.C.; Lee, K.S. The performance of TiN-coated high speed steel tool inserts in turning. Tribol. Int. 1999, *32*, 393–398.

44. Lim, C.Y.H.; Lim, S.C.; Lee, K.S. Crater wear mechanisms of TiN coated high speed steel tools. Surf. Eng. 2000, *16*, 253–256.

45. Lim, C.Y.H.; Lim, S.C.; Lee, K.S. Machining conditions and the wear of TiC-coated carbide tools. In *Wear Processes in Manufacturing, ASTM STP 1362*; Bahadur, S., Magee, J., Eds.; ASTM: West Conshohocken, PA, 1999; 57–70.

46. Lim, C.Y.H.; Lim, S.C.; Lee, K.S. Wear of TiC-coated carbide tools in dry turning. Wear 1999, *225–229*, 354–367.

47. Gu, J.; Barber, G.C.; Tung, S.C.; Gu, R.-J. Tool life and wear mechanism of uncoated and coated milling inserts. Wear 1999, *225–229*, 273–284.

48. Trent, E.M. Tool wear and machinability. J Inst. Prod. Eng. 1959, *38*, 105–130.

49. Yen, D.W.; Wright, P.K. Adaptive control in machining—a new approach based on the physical constraints of tool wear mechanisms. J Eng. Ind. 1983, *105*, 31–38.

50. Kendall, L.A. Tool wear and tool life. Metals Handbook, 9th Ed.; American Society for Metals: Metals Park, OH, 1989; Vol. 16, 37–48.

51. Lim, S.C.; Liu, Y.B.; Lee, S.H.; Seah, K.H.W. Mapping the wear of some cutting-tool materials. Wear 1993, *162–164*, 971–974.

52. Quinto, D.T. Technology perspective on CVD and PVD coated metal-cutting tools. Int. J Refract. Met. Hard Mater. 1996, *14*, 7–20.

53. Lim, S.C.; Lee, S.H.; Liu, Y.B.; Seah, K.H.W. Wear maps for some uncoated cutting tools. Tribotest 1996, *3*, 67–88.

54. Tool-Life Testing with Single-Point Turning Tools. ISO 3685:1993(E); International Organization for Standardization: Geneva, 1993.

55. Sproul, W.D.; Rothstein, R. High rate reactively sputtered TiN coatings on high speed steel drills. Thin Solid Films 1985, *126*, 257–263.

56. Gu, J.; Tung, S.C.; Barber, G.C. Wear mechanisms of milling inserts: dry and wet cutting. In *Wear Processes in Manufacturing, ASTM STP 1362*; Bahadur, S., Magee, J., Eds.; ASTM: West Conshohocken, PA, 1999; 31–47.

57. Shaw, M.C. *Metal Cutting Principles*; Clarendon Press: Oxford, 1984.

58. Tipnis, V.A. Cutting tool wear. In *Wear Control Handbook*; Peterson, M.B., Winer, W.O., Eds.; American Society of Mechanical Engineers: New York, 1980; 891–930.

59. Lim, S.C.; Lim, C.Y.H. Effective use of coated tools—the wear-map approach. Surf. Coat. Technol. 2001, *139*, 127–134.

60. Kodama, M.; Shabaik, A.H.; Bunshah, R.F. Machining evaluation of cemented carbide tools coated with HfN and TiC by the activated reactive evaporation process. Thin Solid Films 1978, *54*, 353–357.

61. Dearnley, P.A. Rake and flank wear mechanisms of coated and uncoated cemented carbides. J Eng. Mater. Technol. 1985, *107*, 63–82.

62. Ljungqvist, R. Development of titanium-carbide coated cemented carbide inserts. Proceedings of the 3rd International Conference on Chemical Vapour Deposition. American Nuclear Society; Hinsdale, IL, 1972; 383–396.

63. Sproul, W.D.; Richman, M.H. Some aspects of a sputtered interfacial eta-carbide layer and its effect on a TiC-coated WC-Co cemented carbide. Met. Technol. 1976, *3*, 489–493.

64. Schintlmeister, W.; Wallgram, W.; Kanz, J. Properties, applications and manufacture of wear-resistant hard material coatings for tools. Thin Solid Films 1983, *107*, 117–127.

65. Nakamura, K.; Inagawa, K.; Tsuruoka, K.; Komiya, S. Applications of wear-resistant thick films formed by physical vapor deposition processes. Thin Solid Films 1977, *40*, 155–167.

66. Finn, M.E. Metals Handbook, 9th Ed.; American Society for Metals: Metals Park, OH, 1989; Vol. 16, 666.

67. Trent, E.M.; Wright, P.K. *Metal Cutting*, 4th Ed.; Butterworth-Heinemann: Boston, 2000; 251–310.

68. *Machining Data Handbook,* 3rd Ed.; Machining Data Center, Metcut Associates Inc.: Cincinnati, 1980.

69. Lim, C.H.Y.; Lau, P.P.T.; Lim, S.C. The effects of work material on tool wear. Wear 2001, *250,* 345–349.

70. Lim, C.H.Y.; Lau, P.P.T.; Lim, S.C. Work material and the effectiveness of coated tools. Surf. Coat. Technol. 2001, *146/147,* 298–304.

71. Lim, S.C.; Lee, S.H.; Liu, Y.B.; Seah, K.H.W. Wear maps for uncoated high-speed steel cutting tools. Wear 1993, *170,* 137–144.

8
Measuring Technique and Characteristics of Thin Film Lubrication at Nanoscale

Jianbin Luo and Shizhu Wen
Tsinghua University, Beijing, China

I. INTRODUCTION

A. Advancements in Thin Film Lubrication

Thin film lubrication (TFL) [1,2], the lubrication regime between elastohydrodynamic lubrication (EHL) and boundary lubrication, has been well studied since 1990. The lubrication phenomena in this regime are different from those in EHL wherein the film thickness is strongly related to the speed, and from that in boundary lubrication wherein the film thickness is mainly determined by molecular dimension and characteristics of the lubricant. Since the early 1990s, more attention has been paid to TFL regime.

In lubrication history, researches have been mainly focused for a long period on two fields—fluid lubrication and boundary lubrication. In boundary lubrication, lubrication models proposed by Bowdeon and Tabor [3], Adamson [4], Kingsbury [5], Cameron [6], and Homola and Israelachvili [7] indicated the research progress in the principle of boundary lubrication and the comprehension about the failure of lubricant film. In fluid lubrication, the EHL proposed by Grubin in 1949 has been greatly developed by Dowson and Higginson [8], Hamrock and Dowson [9], Archard and Cowking [10], Cheng and Sternlicht [11], Yang and Wen [12], and so on. The width of the chasm between fluid lubrication and boundary lubrication has been reduced by these researches. The researches of micro-EHL and mixed lubrication have been trying to complete the whole lubrication theory system. Nevertheless, the transition from EHL to boundary lubrication is also an unsolved problem in the system of lubrication theory. Thin film lubrication bridges the EHL and boundary lubrication [13].

The TFL studied by Johnston et al. [14], Wen [15], Luo et al. [1,2,16–19], Tichy [20–22], Matsuoka and Kato [23], Hartal et al. [24], Gao and Spikes [25], and others has become a new research area of lubrication in the 1990s. Some significant progress has been made in this area. In the 1940s, it had been proved that by using X-ray diffraction pattern a fatty acid could form a polymolecular film on a mercury surface and the degree of molecular order increased from outside toward the metal surface [26]. Allen and Drauglis [27] proposed an "ordered liquid" model to explain the experimental results of Fuks on thin liquid film. However, they thought the thickness of ordered liquid is more than 1 μm which is too large compared to that shown in Refs. 2, 17, and 18. The surface force apparatus (SFA) developed by Israelachvili and Tabor [28] to measure the van der Waals force, and which later became more advanced [29], has been well used in the tribological test of thin liquid layer in molecular order. Using SFA, Alsten and Granick [30], Granick [31], and Luengo et al. [32] observed that the adsorptive force between two solid surfaces was strongly related to the distance between the two surfaces and the temperature of the lubricant, and the effective viscosity of liquid in the nano-gap is larger than the bulk. In 1989, Luo and Yian [33] proposed a fuzzy friction region model to describe the transition from EHL to boundary lubrication. In their model, the transition region was considered as a process in which the qualitative lubrication change and the quantitative parameters' change were interlaced. Johnston et al. [14] found that EHL phenomenon did not exist with films of less than 15 nm thick. Gupta and Sharma [34] discussed the relationship between the disjoining pressure isotherms and the brine film thickness, which was about 1 to 10 nm. The disjoining pressure consists of the contributions from (1) ionic–electrostatic action; (2) long-range dispersion and dipole interaction force between molecules; and (3) structural force that was generated due to entropic effects associated with molecular packing and orientation in very thin films. Tichy [20–22] proposed the models of thin lubricant film according to the improved EHL theory. Luo et al. [1,2,18,35] have proposed the physical model of TFL, the relationship between the transition thickness from EHL to TFL and the viscosity η, the time effect of lubricant film and the lubrication map of different lubrication regimes.

In the late 1990s, some other advancements have been made in experimental studies made by Gao and Spikes [36], Hartl et al. [37] and in theoretical studies by Thompson et al. [38], Hu and Granick [39]. However, there are many unsolved problems in the area of nanolubrication [40].

B. Definition of Thin Film Lubrication and Boundary Lubrication

The definition of different lubrication regimes is a historically unsolved problem [41]. In boundary lubrication, molecules will be absorbed on the solid surface of the tribopair and form a monomolecular absorbed layer as described by Hardy [42] as shown in Fig. 1(a). When film thickness is from a few nanometers to 100 nm or thicker, lubricant molecules act as shown in Fig. 1(b) as proposed by Luo et al. [1,2]. The layer close to the surfaces is the adsorbed film that is a monomolecular layer. The layer in the center of the gap or apart from the solid surface is the dynamic fluid film. It is formed by hydrodynamic effects and its thickness mainly depends on fluid factors, e.g., speed, lubricant viscosity, pressure. The layer between the adsorbed and dynamic fluid films is the ordered liquid film, which is formed during the tribological process. By the solid surface force, the shearing stress, and the

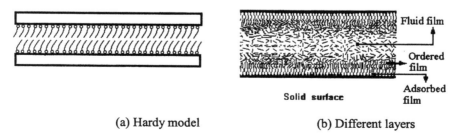

(a) Hardy model   (b) Different layers

Figure 1 Physical model of lubrication. (From Ref. 2.)

interaction of rod-shaped molecules with one another, the lubricant molecules form an ordered region between the static and dynamic films.

The molecules in the adsorbed layer are more ordered and solidlike than that in the fluid state because the adhesion force between the lubricant molecules and the solid surface is much stronger than that among lubricant molecules. Therefore the tribological property of ordered liquid is different from that of usual fluid. The molecular order degree of such layer becomes smaller along the direction apart from the solid surface. The tribological properties of such layer are different from that of boundary lubrication because they are also related to the lubricant viscosity and speed. Alternatively, such layers are also different from the hydrodynamic fluid layer because characteristics of these layers are related to the surface force, molecular polar character, molecular sizes, the distance from the surface, and so on. Therefore the lubrication is called thin film lubrication [1,2,20,24]. In the point contact region, the lubrication regime should be divided into four types, i.e., HDL, EHL, TFL, and BL. If the surface roughness in the contact region is very small compared to the film thickness, the transition between the lubrication regimes will take place according to the variation of the film thickness as shown in Fig. 1(b). When the film is very thick, the fluid film plays the leading role and the lubrication properties obey the EHL or DHL rules. As the film becomes thinner, the proportion of the ordered layer thickness to the total film thickness will become larger. When the proportion is large enough, the ordered liquid layer will play the chief role and the lubrication regime changes into TFL. After the layer is crushed, the monomolecular adsorbed layer will have the leading role and the lubrication belongs to boundary lubrication.

C. Lubrication Map

For lubrication engineers, it is very important to know which lubrication regime plays the leading role in the contact region. Two methods are usually used to distinguish different lubrication regimes. One is using Streibeck curve [4]. However, it is very difficult for engineers to confirm the lubrication regime in the contact region based on the friction coefficient of the lubricant without the entire Streibeck curve or its working condition because the friction coefficients of different kinds of lubricants are very different. Another method is using the ratio of film thickness h to the combined surface roughness R_a to judge the lubrication regime [43], i.e., the lubrication regime is EHL or HDL if the ratio h/R_a is larger than 3, mixed lubrication if the ratio ranges from 1 to 3, and boundary lubrication when the ratio is smaller than 1. This method is very useful for engineers, but it is not applicable for the ultraflat solid surfaces because even if the ratio h/R_a is much larger than 3, the lubrication regime is a typical boundary lubrication when R_a is 1 nm or smaller, e.g., the surface of monocrystal silicon and the thickness of only two molecular layers is 5 nm, e.g., $C_{15}H_{31}COOH$. Therefore the distinguishing method for different lubrication regimes should not only include the ratio h/R_a, but also the number of the molecular layers, or the ratio of film thickness to lubricant molecular size. Thus the method that uses the ratios h/R_a and h/R_g together to distinguish the lubrication regimes as shown in Fig. 2 has been proposed by Luo et al. [35] in which h is film thickness, R_a is the combined surfaces roughness, and R_g is the effective radius of lubricant molecules. The ratio h/R_g indicates the number of molecular layers in the gap of tribopair. When ratio h/R_a is larger than 3, three lubrication regimes can be formed separately according to the ratio h/R_g. The regime is boundary lubrication when ratio h/R_g is smaller than about 3. It is TFL when ratio h/R_g is larger than 3 and lower than 10 to 15, which is related to the surface energy of the solid surfaces and molecular polarity, and EHL or HDL when ratio h/R_g is higher than 15. The mixed lubrication will be found

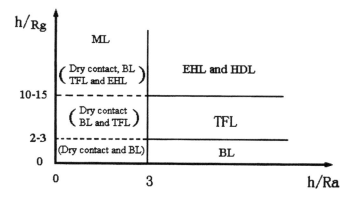

Figure 2 Lubrication map. Ra-combined surface roughness, h-film thickness, and Rg-effective molecular radius. (From Ref. 35.)

when ratio h/R_a is smaller than 3. Different types of mixed lubrication with different values of the ratio h/R_g are shown in Fig. 2.

The relationship of film thickness and its influence factors in different regions with different lubrication regimes is shown in Fig. 3 when film thickness is larger than three times the combined surface roughness in the contact region [35]. In the EHL region, film thickness changes with different parameters according to the EHL rules. As the speed decreases or pressure increases, films with different viscosity or molecular length will meet different critical points and then go into the TFL region. The pressure has a slight influence on the film thickness in such region. These critical points are related to the viscosity of lubricant, solid surface energy, chemical characteristics of lubricant molecules, etc. When the speed decreases and pressure increases further, the films will meet the failure points where the liquid film will collapse. If the polar additives have been added into the lubricants, the TFL regime will be maintained at lower speeds shown as the dotted lines in Fig. 3. So the speed influences film failure. If the speed is less than that of the failure point, the film thickness will suddenly decrease to a few molecular layers [35].

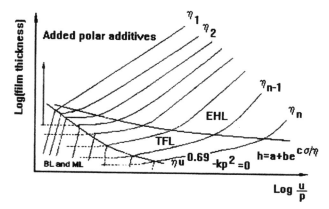

Figure 3 Film characteristics in different regimes. Viscosity: $\eta_1 > \eta_2 > ... > \eta_{n-1} > \eta_n, h/R_a > 3$. (From Ref. 35.)

II. MEASUREMENT OF FILM THICKNESS AT NANOSCALE

In the 1990s, researches on nanoproblems in tribology have become an important area which attracted much attention. The lubricant film thickness in tribopairs of ultraprecision instruments or machines usually ranges from a few to tens of nanometers under the condition of point or line contacts with heavy load, low speed, and low viscosity lubricant. The lubrication in such range belongs to TFL [1,2]. Several optical methods had been developed in the 1990s for the measurement of film thickness at nanoscale.

The optical method has been widely used to measure lubricant film thickness for many years. Gohar and Cammeron [44] successfully used the technique of optical interferometry to measure EHL film in the range from 100 nm to 1 μm in 1967. They measured the film thickness in terms of the colors of interference fringes and the resolution in the vertical direction is about 25 nm. Surface force apparatus (SFA) [27,45] allows normal and shear forces to be measured between atomically flat solid surfaces while their separation is determined to 0.1 nm by using interferometry. Because its tribomaterial, i.e., mica, is far from that in the realistic tribological process, it is widely used in the physical field to investigate the flowing characteristics of liquid in the nanoscale. Wedged spacer layer optical interferometry proposed by Johnston et al. in 1991 [14], using a combination of a solid spacer layer and spectrometer analysis of reflected light, can measure the film down to 1 nm thick. The resolution in the vertical direction is about 1 nm and that in the horizontal direction is related to the width of a narrow grate, which is about 100 μm. The method of relative optical interference intensity (ROII) proposed by Luo et al. in 1994 [1,2] can measure the liquid film between the surface of Cr and that of a steel ball with a resolution of 0.5 nm in the vertical direction and 1 μm in the horizontal direction. In 1999, Hartl et al. [24,37] developed the thin film colorimetric interferometry (TFCI) method which also can measure the thickness of TFL film well. These three methods are introduced in the following sections.

A. Wedged Spacer Layer Optical Interferometry

This method is developed by Johnston et al. in 1991 and well described in Ref. 14. The principle of optical interference is shown schematically in Fig. 4. A coating of transparent solid, typically silica, of known thickness, is deposited on top of the semireflecting layer. This solid permanently augments the thickness of any oil film present and is known as a "spacer layer." The destructive interference now obeys the equation:

$$n_{\text{oil}}h_{\text{oil}} + n_{\text{sp}}h_{\text{sp}} = \frac{(N + \frac{1}{2} - \phi)\lambda}{2\cos\theta} \quad N = 0, 1, 2 \tag{1}$$

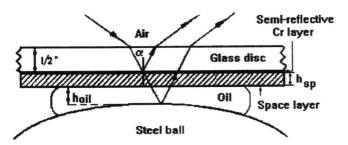

Figure 4 Spacer layer method. (From Ref. 14.)

and the first interference fringe occurs at a separation reduced by h_{sp}, where h_{sp} is the spatial thickness of the spacer layer. With a flat spacer layer, Westlake was able to measure an oil film of 10 nm thickness using optical interferometry [46]. Guangteng and Spikes [47] used an alumina spacer layer whose thickness varied in the shape of a wedge over the transparent flat surface in an optical rig. The method was able to detect oil film thickness down to less than 10 nm. A problem encountered using this technique was the difficulty of obtaining regular spacer layer wedges. Also, with low viscosity oils, requiring high speeds to generate films, high-speed recording equipment was needed to chart the continuously changing interference fringe colors.

The two limitations of optical interferometry, the one-quarter wavelength of light limit and the low resolution, have been addressed by using a combination of a fixed-thickness spacer layer and spectral analysis of the reflected beam. The first of these overcomes the minimum film thickness that can normally be measured and the second addresses the limited resolution of conventional chromatic interferometry.

A conventional optical test rig is shown in Fig. 5 [48]. A superfinished steel ball is loaded against the flat surface of a float-glass disk. Both surfaces can be independently driven. Nominally pure rolling is used as shown in Fig. 5, where the disk is driven by a shaft and the ball is driven by the disk [14].

The disk was coated with a 20-nm sputtered Cr semireflecting layer, a silica spacer layer was sputtered on top of the Cr. This spacer layer varied in thickness in the radial direction but was approximately constant circumferentially round the disk.

The reflected beam was taken through a narrow, rectangular aperture arranged parallel to the rolling direction, as shown in Fig. 6(a). It was then dispersed by a spectrometer grating and the resultant spectrum captured by a black-and-white video camera. This produced a band spectrum spread horizontally on a television screen, with the brightness of the spectrum at each wavelength indicating the extent of interference, as shown schematically in Fig. 6(b). The vertical axis of the spectral band mapped across the center of the contact as illustrated in Fig. 6(a). This image was screen-dumped in digital form into a microcomputer, which drew an intensity profile of the spectrum as a function of wavelength. This is shown in Fig. 6(c). In practice the spectrometer was arranged so that one digitized screen pixel corresponded to a wavelength change of 0.48 nm. A cold halogen white light source was employed and the angle of incidence was 0°.

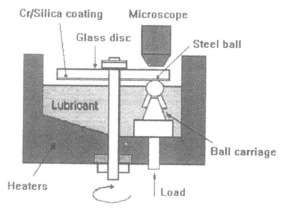

Figure 5 Schematic representation of optical EHD rig. (From Ref. 48.)

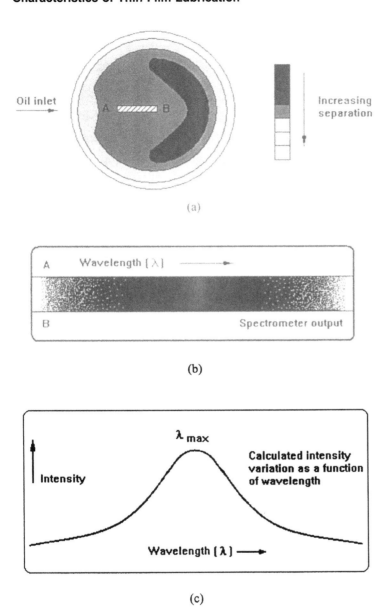

Oil inlet

Increasing
separation

(a)

A   Wavelength (λ)

B   Spectrometer output

(b)

λ max

Calculated intensity
variation as a function
of wavelength

Intensity

Wavelength (λ)

(c)

Figure 6 Schematic representation of screen display showing calculated intensity profile.

By this means, the wavelength of light which constructively interfered could be determined accurately for separating film thickness, thus permitting a highly resolved film thickness measurement.

The silica film thickness was determined, as a function of the disk radius, by an optical interference method, using the spectrometer. A steel ball was loaded against the silica surface to obtain an interference pattern of a central, circular Hertzian area with surrounding circular fringes due to the air gap between the deformed ball and flat. The thickness of the silica layer in the Hertzian contact was calculated from the measured wavelength at which maximum constructive interference occurred, $\lambda_{\max}$, by solving Eq. (2) for h_{sp}. N was known from the approximate thickness of the silica layer and Φ was taken to be 0.28 from the value measured previously in air. The refractive index of the separating medium is known to affect Φ by less than 2% [49]. The refractive index of the spacer layer, n_{sp}, was measured as 1.476 ± 0.001 by the method of Kauffman [50].

$$h_{sp} = \frac{(N - \Phi)\lambda_{\max}}{2n_{sp}} \tag{2}$$

Using the thin silica spacer layer, it was found that the two calibration methods did not agree, as shown in Fig. 7(a). This was tentatively ascribed to the effect of penetration of the reflecting beam into the substrate. With a very thin silica layer, the depth of penetration and thus the phase change would depend upon the thickness of the silica spacer layer and also upon that of any oil film present.

The solution to this problem was to use a space layer of thickness greater than the wavelength of visible light, which is above the limit of penetration of a reflected light beam. Because of the space layer or oil, any variation above this value will have no further effect on phase change. Figure 7(b) shows that there is a good agreement between talysurf and optical calibration methods with a thick space layer.

B. Relative Optical Interference Intensity Method

The ROII method was proposed by Luo et al. in 1994 [1,2]. The principle of optical interference is shown in Fig. 8. On the upper surface of the glass disk, there is an antireflective coating. There is an oil film between a super polished steel ball and the glass disk is covered by a semitransmitted Cr layer. When a beam of light reaches the upper surface of the Cr layer, it is divided into two beams—one reflected at the upper surface of the Cr layer and the other passing through the Cr layer and the lubricant film, and then reflected at the surface of the steel ball. As the two beams come from the same light source and have different optical paths, they will interfere with each other. When the incident angle is $0°$, the optical interference equation [51] is as follows:

$$I = I_1 + I_2 + 2\sqrt{I_1 + I_2}\cos\left(\frac{4\pi nh}{\lambda} + \phi\right) \tag{3}$$

where I is the intensity of the interference light at the point where the lubricant film thickness h is to be measured, I_1 is the intensity of beam 1 and I_2 is that of beam 2 in Fig. 1, λ is the wavelength of the monochromatic light, ϕ is the system's pure optical phase change caused by the Cr layer and the steel ball, and n is the oil refractive index.

I_1 and I_2 can be determined by the maximum interference intensity $I_{\max}$ and the minimum one $I_{\min}$ in the same interference order.

$$\begin{aligned} I_1 &= I_{\max} + I_{\min} + 2\sqrt{I_{\max}I_{\min}} \\ I_2 &= I_{\max} + I_{\min} - 2\sqrt{I_{\max}I_{\min}} \end{aligned} \tag{4}$$

If I_a and I_d are expressed as follows:

$$\begin{aligned} I_a &= \frac{I_{\max} + I_{\min}}{2} \\ I_d &= \frac{I_{\max} - I_{\min}}{2} \end{aligned} \tag{5}$$

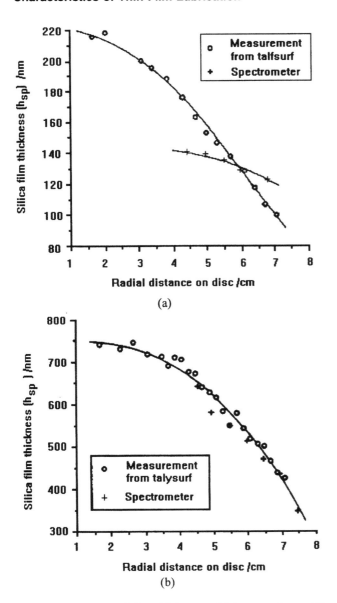

Figure 7 Measured film thickness—thick spacer layer. (From Ref. 14.)

We define the relative interference intensity as below:

$$\bar{I} = \frac{I - I_a}{I_d} = \frac{2I - (I_{max} + I_{min})}{I_{max} - I_{min}} \tag{6}$$

hence from Eqs. (3), (4), (5), and (6), the lubricant film thickness can be determined as below:

$$h = \frac{\lambda}{4n\pi} [\text{arc} \cos(\bar{I}) - \phi] \tag{7}$$

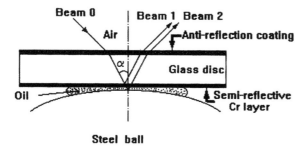

Figure 8 Interference lights. (From Refs. 1,2.)

If the ball contacts the surface of the glass disk without oil, $h=0$, and then the pure phase change ϕ of the system can be obtained as follows:

$$\phi = \text{arc cos}(\bar{I}_0)$$
$$\bar{I}_0 = \frac{I_0 - I_a}{I_d} \tag{8}$$

where I_0 is the optical interference intensity at the point where the film thickness is zero. It should be determined by experiments. Then Eq. (7) can be rewritten as:

$$h = \frac{\lambda}{4n\pi}\left[\text{arc cos}(\bar{I}) - \text{arc cos}(\bar{I}_0)\right] \tag{9}$$

A diagram of the measuring system is shown in Fig. 9 [52]. After the interference light beams reflected separately on the surfaces of the Cr layer and the steel ball pass through a microscope, they become interference fringes caught by a TV camera. The optical image is translated to a monitor and also sent to a computer to be digitized.

The experimental rig is shown in Fig. 10 [18]. The steel ball is driven by a motor through a belt, a shaft, a soft coupling, and a quill. The ball-mount is floating during the running process in order to keep the normal force constant. The micrometer is used to make the floating mount move along the radial direction of the disk and maintains a fixed position. The microscope can move in three dimensions.

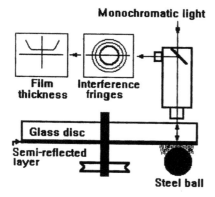

Figure 9 Diagram of the measuring system. (From Ref. 50.)

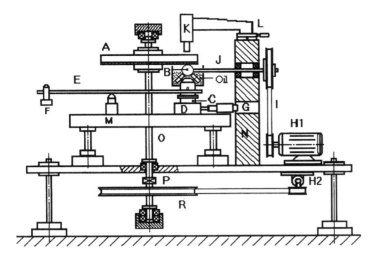

Figure 10 Experiment rig. A—glass disc with a self-reflecting coating; B—steel (52100) ball; C—ball mount; D—sliding mount; E—lever; F—weight; G—micrometer; H1, H2—motors; I—belt; J—shaft; K—microscope; L—three dimensional disc; M—support disc; N—wall; O—main shaft; P—shaft coupling; R—wheel. (From Refs. 2,18.)

The resolution of the instrument in the vertical direction depends upon the wavelength of visible light (450 to 850 nm), the oil refractive index, and the difference between the maximum and the minimum interference intensity as follows [2,18]:

$$\Delta h = \frac{\lambda}{4n\pi} \left[\text{arc} \cos(\bar{I} + \Delta \bar{I}) - \text{arc} \cos(\bar{I}) \right] \tag{10}$$

$$\bar{I} = \frac{2I - (I_{max} + I_{min})}{I_{max} - I_{min}} \tag{11}$$

$$\Delta \bar{I} = \frac{2\Delta I}{I_{max} - I_{min}} \tag{12}$$

where n is the oil refractive index; λ is the wavelength of the light and it is 600 nm in normal experiment; I_{max} and I_{min} are the maximum and minimum interference intensity separately, which can be divided in 256 grades in the computer image card; ΔI is the resolution of optical interference intensity, which is one grade. The variation of the vertical resolution with respect to these factors is shown in Fig. 11. Among these factors, the wavelength is the most important in determining the vertical resolution. When the effect of all these factors is considered, the vertical resolution is about 0.5 nm when wavelength is 600 nm. The horizontal resolution depends upon the distinguishability of change coupled device (CCD) and the enlargement factor capacity of the micrometer. It is about 1 μm.

C. Thin Film Colorimetric Interferometry

The TFCI method was proposed by Hartl et al. [24,37,53]. The colorimetric interferometry technique represents an improvement of conventional chromatic interferometry in which film thickness is obtained by color matching between interferogram and color/film thickness dependence obtained from Newton rings for static contact [37,53].

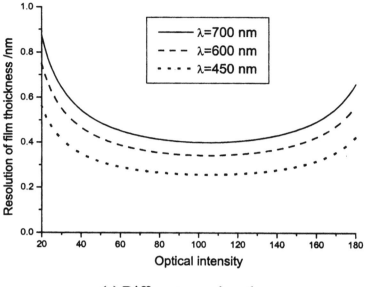

(a) Different wavelengths

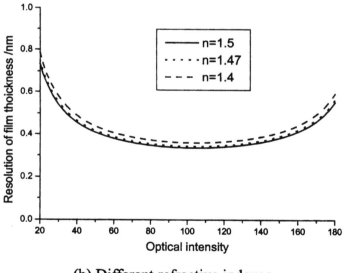

(b) Different refractive indexes

Figure 11   Resolution of film thickness vs. optical interference intensity. (a) Different wavelengths and (b) different refractive indexes. (From Ref. 2.)

The frame-grabbed interferograms with a resolution of 512 pixels × 512 lines are first transformed from RGB to CIELAB color space and then converted to the film thickness map using appropriate calibration and color matching algorithm. L^*, a^*, b^* color coordinates/film thickness calibration is created from Newton rings for flooded static contact formed between the steel ball and the glass disk coated with Cr layer. In the CIE curve shown in Fig. 12, the wavelength can be determined by the ratio of RGB separately measured by color CCD. Therefore this method has a much higher resolution than that of Gohar and Cameron's [44]

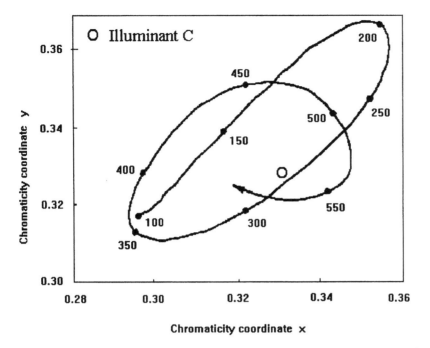

Figure 12   Interference colors in CIE x, y chromaticity diagram. (From Ref. 37.)

who measured the film thickness in terms of the colors of interference fringes observed by the eyes, and the resolution in the vertical direction is about 25 nm.

All aspects of interferogram and experimental data acquisition and optical test rig control are provided by a computer program that also performs film thickness evaluation. It is believed that the film thickness resolution of the colorimetric interferometry measurement technique is about 1 nm. The lateral resolution of a microscope imaging system used is 1.2 μm. Figure 13 shows a perspective view of the measurement system configuration. This is a conventional optical test rig equipped with a microscope imaging system and a control unit.

An optical test rig consists of a cylindrical thermal isolated chamber enclosing the concentrated contact formed between a steel spherical roller and the flat surface of a glass disk. The underside of the glass disk is coated with a thin semireflective Cr layer that is overlaid by a silicon dioxide "spacer layer," as shown in Fig. 14. The contact is loaded through the glass disk that is mounted on a pivoted lever arm with movable weight. The glass disk is driven, in nominally pure rolling, by the ball that is driven by a servomotor through flexible coupling. The test lubricant is enclosed in a chamber that is heated with the help of an external heating circulator controlled by a temperature sensor. A heat insulation lid with a hole for a microscope objective seals the chamber and helps to maintain a constant lubricant test temperature. Its stability is within ±0.2°C.

An industrial microscope with a long-working distance 20 × objective is used for the collection of the chromatic interference patterns. They are produced by the recombination of the light beams reflected at both the glass/Cr layer and lubricant/steel ball interfaces. The contact is illuminated through the objective using an episcopic microscope illuminator with a fiber optic light source. The secondary beam splitter inserted between the microscope illuminator and an eyepiece tube enables the simultaneous use of a color video camera and a fiber optic spectrometer. Both devices are externally triggered by an inductive sensor so that all measurements are carried out at the same disk position.

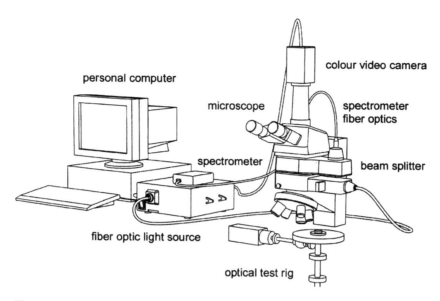

Figure 13 Experimental apparatus. (From Ref. 55.)

Spherical rollers were machined from AISI 52100 steel, hardened to a Rockwell hardness of Rc 60, and manually polished with a diamond paste to a RMS surface roughness of 5 nm. Two glass discs with a different thickness of the silica spacer layer are used. For a TFCI of about 190 nm thick, spacer layer is employed whereas FECO interferometry requires a thicker spacer layer, approximately 500 nm. In both cases the layer was deposited by reactive electron beam evaporation process and it covers the entire underside of the glass disk with the exception of a narrow radial strip. The refractive index of the spacer layer was determined by reflection spectroscopy and its value for a wavelength of 550 nm is 1.47.

III. PROPERTIES OF THIN FILM LUBRICATION

A. Relationship Between Film Thickness and Influence Factors

In EHL, the formula of film thickness was given by Hamrock and Dowson [54] as below:

$$h = 2.69 U^{0.67} G^{0.53} W^{-0.067}(1 - 0.61e^{-0.73k}) \tag{13}$$

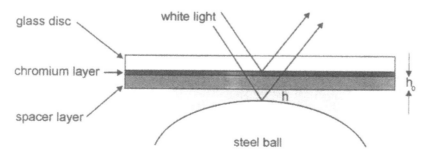

Figure 14 Schematic representation of an interferometer. (From Ref. 24.)

where U is the speed parameter, W is the load parameter, G is the materials parameter, and k is the ellipticity of the contact bodies. The speed u is very important to the film thickness in EHL. In TFL, more factors should be considered to estimate the film thickness because it is not only related to speed u, load w, initial viscosity η_0, reduced modulus of elasticity E', and radius R, but also related to molecular interactions. The relationship between the film thickness and the speed in TFL can be written as:

$$h = k_1 u^\phi \tag{14}$$

where ϕ is a speed index and k_1 is a coefficient. The speed index ϕ can be obtained by the following equation [Eq. (15)] from experimental film thicknesses at two points if the two points are close enough and other condition parameters do not change and then k_1 is constant.

$$\phi = \frac{\log\left(\frac{h_2}{h_1}\right)}{\log\left(\frac{u_1}{u_2}\right)} \tag{15}$$

1. Effect of Rolling Speed

The film thickness varies with the rolling speed as shown in Fig. 15 in which curve a is from the measured data and curve b is the measured value of thickness minus the static film thickness, i.e., the thickness of fluid film [18]. The data of curve c are calculated from the Hamrock–Dowson formula [54]. In the higher speed region (above 5 mm/sec) of Fig. 15, the film becomes thinner as speed decreases and the speed index ϕ is about 0.69 (Fig. 15, curve b), which is very close to that in Eq. (13). When the film thickness is less than 15 nm, the speed index is only about half of that in EHL and the relation between the film thickness and the speed does not obey Eq. (13) anymore. We call this point as a critical point where the slope of the film-speed curve changes significantly. This point is generally believed to be the tran-

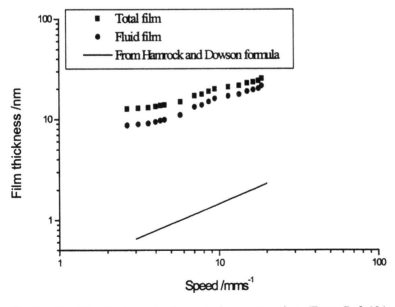

Figure 15 Film thickness in the central contact region. (From Ref. 18.)

sition point from EHL to TFL [1,2,18]. For curve a, even if the relationship between the film thickness and the speed in the logarithmic coordinates is linear in the higher speed region, the speed index is only 0.45, which is smaller than that in EHL. When the film thickness is below the critical point, the index ϕ quickly decreases to 0.18. Therefore the total film can be divided into a static film, which is a layer adsorbed on the metal surfaces, and a dynamic film, which equals the total thickness minus the thickness of the static film. These two layers play important roles in TFL.

Because the dynamic film is very thick and the thickness of the static film is very small compared to the total film thickness in EHL, the most part of the speed index is contributed by that of the dynamic fluid film, and the difference between the total and the dynamic film thickness is so small that it can be ignored. However, in the TFL region, the dynamic film only takes a smaller part of the total film and the difference between the thickness of the total film and that of the dynamic one becomes significant. Therefore if only the dynamic film is used to consider the influence of the speed, the index ϕ in the linear region will rise to 0.69, which is very close to the EHL one. While for the total film, including static and dynamic film, the index ϕ is about 0.45. In other words, using the total film, not the dynamic one, is one of the main reasons why the index of the speed in the linear region is smaller than that in EHL. However, there is also a transition point in curve b that does not contain the static film. Thus another kind of film that is similar to static film in characteristics may exist. Such film is called ordered liquid film or solidlike film [2,18].

Fig. 16 [55] shows the film thickness results for octamethylcyclotetrasiloxane (OMCTS). This fluid exhibits a deviation from linearity on a log (film thickness) vs. log (rolling speed) in the thin film region, with a thicker film than predicted from EHD theory. Because EHD film thickness is determined by the viscosity of the fluid in the contact inlet [56], it is obvious that the viscosity of OMCTS remains at the bulk value down to approximately 0.1 m/sec. However, below this speed the discretization of both central and minimum film thicknesses can be observed. The central film thickness begins to deviate from the theory at about 8 nm and the interval of the discretization is approximately 2 nm. If we take into account the molecular diameter of OMCTS that is about 1 nm, it corresponds to approximately two molecular layers.

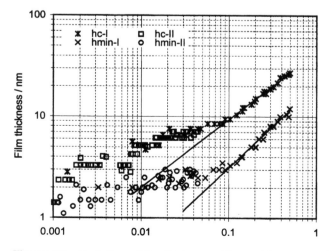

Figure 16 Variation of film thickness with speed for octamethylcyclotetrasiloxane. (From Ref. 55.)

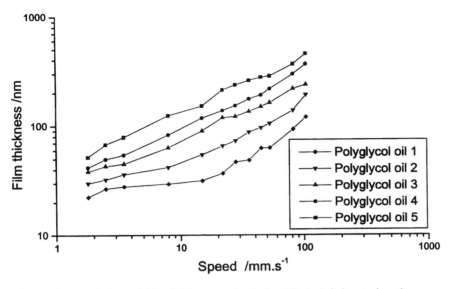

Figure 17 Variation of film thickness and relationship to lubricant viscosity.

2. Effect of Lubricant Viscosity

As shown in Fig. 17 [19], for the lubricant with higher viscosity (kinetic viscosity from 320 to 1530 mm^2/sec), the film is thick enough so that a clear EHL phenomenon can be observed, i.e., the relationship between film thickness and speed is in liner style in the logarithmic coordinates. However, for the lubricants with lower viscosity, e.g., polyglycol oils 1 and 2 with a kinetic viscosity of 47 to 145 mm^2/sec in Table 1, the transition from EHL to TFL can be seen at a speed of 8 and 23 mm/sec, i.e., the relationship between film thickness and speed becomes much weaker than that in EHL. The transition regime can be explained as follows: when the film reduces to several times thick of the molecular size, the effect of the solid surface energy on the action of molecules becomes so strong that the lubricant molecules become more ordered or solidlike. The thickness of such film is related to the lubricant viscosity or molecular size.

Table 1 Lubricants in the Experiment

Oil sample	Viscosity (mm^2/sec)/(20°C)	Index of refraction
Polyglycol oil 0	100	1.444
Polyglycol oil 1	47	1.443
Polyglycol oil 2	145	1.454
Polyglycol oil 3	329	1.456
Polyglycol oil 4	674	1.456
Polyglycol oil 5	1530	1.457
Mineral oil 13604	20.6	1.473
White oil No. 1	61	1.472
White oil No. 2	96	1.483
Paraffin liquid	25.6	1.471
Decane	7.3	1.461

3. Isoviscosity of Lubricant in Thin Film Lubrication

The isoviscosity of lubricant close to the solid surface in TFL is different from that of bulk fluid, which was discussed by Gao and Spikes [57] and Shen et al. [58]. For the homogeneous fluids, Eq. (16) predicts the elastohydrodynamic central film thickness by a power relationship of about 0.67 with velocity and viscosity, i.e.,

$$h = ku^{0.67}\eta^{0.67} \tag{16}$$

where h is the central film thickness, u is the average rolling speed, and η is the dynamic viscosity of lubricant under atmospheric pressure. The constant k is related to the geometry and modulus of tribopair and the pressure–viscosity coefficient of lubricant. Assuming that the pressure–viscosity coefficient of any boundary layer approximately equals that of the bulk fluid, the relation between isoviscosity and initial viscosity can be obtained by rearranging Eq. (16).

$$\frac{\bar{\eta}}{\eta_0} \approx \left(\frac{u_0}{u}\right)\left(\frac{h}{h_0}\right)^{1/0.67} \tag{17}$$

where $\bar{\eta}$ is the effective viscosity, u is the speed at which the film thickness is h, η_0 is the viscosity of the bulk fluid, and h_0 is the film thickness based on bulk fluid that is formed at speed u_0.

Figure 18 [58] shows the relationship of isoviscosity $\bar{\eta}$ calculated by Eq. (17) and the distance between solid surfaces. For different kinds of solid materials with different surface energy, the isoviscosity becomes very large as the film thickness becomes thinner. The isoviscosity increases about several to tens of times more than that of bulk fluid when the distance between the two solid surfaces is shorter than 3 nm. In the thick film region, the isoviscosity remains constant, which approximately equals the dynamic viscosity of bulk liquid. Therefore the isoviscosity of lubricant smoothly increases with decreasing film thick-

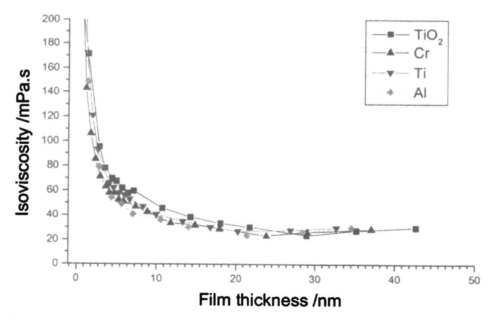

Figure 18 Isoviscosity of lubricant vs. film thickness. (From Ref. 58.)

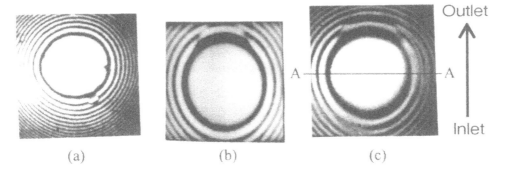

Figure 19   Interference fringes.

ness in TFL. When the film decreases down to very thin, the isoviscosity of lubricant becomes very large and a solidlike layer will be formed.

4.   Interference Fringes and Shapes of the Film Thickness Curves

Interference fringes at different speeds are shown in Fig. 19 [1,2]. In the static state, fringes are regular circles [Fig. 19(a)]. When the ball starts rolling at a speed of 3.12 mm/sec, an outlet effect appears [Fig. 19(b)]. When the speed increases further, the outlet effect becomes much stronger [Fig. 19(c)].

The shapes of film thickness along A-A as shown in Fig. 19(c) are given in Fig. 20 [1,2]. With decreasing of speed, the curve of film thickness becomes flat. The thickness at which the curve in the central cross section becomes flat is about 24 nm for the mineral oil with a viscosity of 36 mPa sec (20°C). However, in Fig. 21 [1,2] for the lubricant with a viscosity of 17.4 mPa sec, when the average thickness is about 24 nm, the curve is also bent and the curve becomes flat at a thickness of about 14 nm. This indicates that the thinner the film thickness,

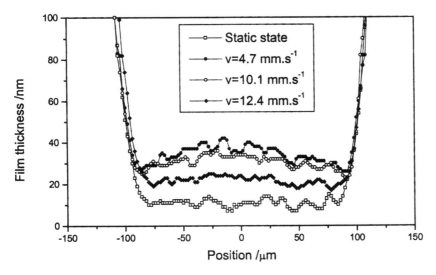

Figure 20   Film thickness in the central cross-section. Lubricant: mineral oil with viscosity of 36 mPa.s at 20°C, temperature: 25°C, diameter of ball: 20 mm, and load: 6.05 N. (From Ref. 2.)

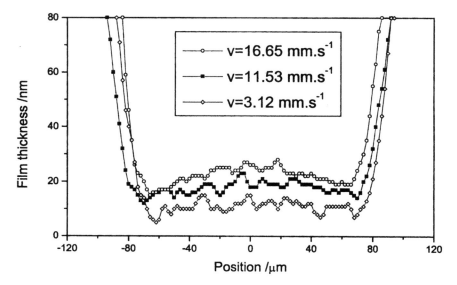

Figure 21 Film thickness in the central cross-section. Lubricant: mineral oil with viscosity of 17.4 mPa.s at 20°C, temperature: 25°C, diameter of ball: 20 mm, and load: 4 N. (From Ref. 2.)

the flatter the film in the central region. The thickness at which the shape of the film curve becomes flat is related to the critical film thickness where EHL transfers to TFL. The thicker the critical film, the thicker will the average film be at which the film curve turns flat.

Figure 22 provided by Hartl shows the isometric views and associated contour line plots of film shape for ellipticity parameter $k = 2.9$. These figures give all the characteristic features of medium-loaded point EHD contacts, i.e., nearly uniform film thickness in the central region with a horseshoe-shaped restriction. With decreasing value of k the side lobes are developed and the locations of the film thickness minima move toward the sides. The side lobes provide an effective seal to axial flow resulting in a central plateau whose extent depends on the ellipticity of the contact. Smaller values of k imply larger region of nearly uniform film thickness. Figure 23 provided by Hartl shows the variation of film thickness transverse to the direction of rolling for all three ellipticities at speed of 0.012 m/sec. This figure reveals the influence of the ellipticity on the lubricant compressibility.

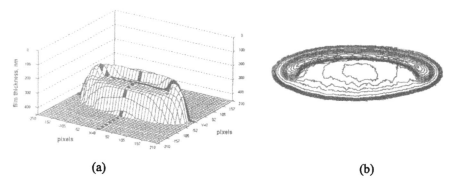

(a)   (b)

Figure 22 Isometric views and associated contour line plots of film shape for $k = 2.9$ and rolling speed of 0.021 and 0.042 ms^{-1}. (From Ref. 55.)

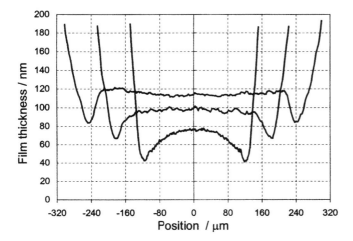

Figure 23 Variation of film thickness transverse to the direction of rolling for $k = 1$, 1.9, and 2.9 and rolling speed of 0.011 ms^{-1}. (From Ref. 55.)

B. Transition from EHL to Thin Film Lubrication

The capacity of ordered film for supporting loads is between that of the static film and that of the disordered liquid film. The orientation property of the film gradually weakens with an increase in distance from the metal surface. The transition occurs as the ordered film appears. The thickness of the ordered film is related to the initial viscosity or molecular size of lubricant, as shown in Fig. 24 [1,2,18], so that we can generally write the critical film thickness as follows:

$$h_{ct} = a + be^{c/\eta_0} \tag{18}$$

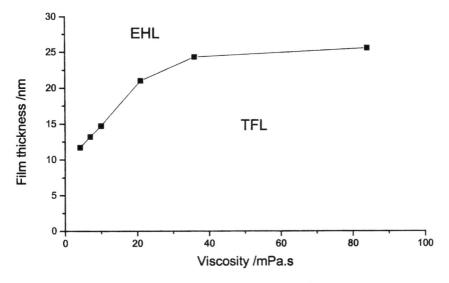

Figure 24 Critical transition film thickness vs. viscosity. Lubricant: mineral oils, load: 4 N, and ball ϕ20 mm. (From Ref. 2.)

where h_{ct} is the critical film thickness; η_0 is the initial viscosity of lubricant; a, b, and c are coefficients that are related to the molecular structure and polarity of lubricant, and the solid surface energy. These coefficients can be determined from experimental data. From Eq. (18), we can know that the thickness boundary between EHL and TFL is not a definite value, but the one varying with viscosity or molecular size, molecular polarity, solid surface energy or tension. The following equation can be used for common mineral oils without polar additives:

$$h_{ct} = 9 + 17.5e^{-8.3/\eta_0} \tag{19}$$

From Eq. (19), it can be seen that the critical film thickness changes very slowly with an increase in initial viscosity when it is close to 26 nm.

C. Effect of Solid Surface Energy

The influence of solid surface tension or surface energy [17,58] on the behavior of lubricants in TFL is shown in Fig. 25 [27]. A Cr coating, which is close to steel in surface energy, and an aluminum coating whose surface energy is much smaller than that of steel are chosen to be formed on the surfaces of the steel (E52100) balls by ion-beam-assisted deposition (IBAD) as shown in Fig. 26. The surface roughness of these balls decreases significantly in the contact region. It can be seen from Fig. 25 that the film thickness decreases as the rolling speed decreases. In the higher speed region, the film thickness changes significantly with the rolling speed. When the film thickness is less than a critical thickness (27 nm in curve Fe, 25 nm in Cr, and 18 nm in Al, Fig. 25), the film thickness decreases slightly. From the thickness difference among the three curves in Fig. 25, it can be seen that the films with the substrate of Fe and Cr are very close to each other in thickness, while they are much thicker than the film with the substrate of Al, which has a much lower surface energy than the former. Thus the larger the surface tension the thicker the total film and the critical film will be.

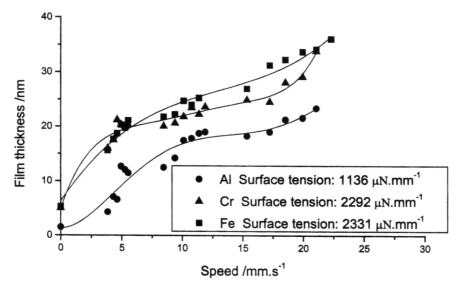

Figure 25 Film thickness with different substrate. Lubricant: mineral oil, load: 4 N, temperature: 18.5°C, and ball: φ23.5 mm. (From Ref. 17.)

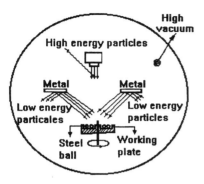

Figure 26 Coating system. (From Ref. 17.)

In addition, the film thickness enhanced with time is also affected by the surface energy of the substrate. As shown in Fig. 27 [18], the film in the contact region will become thicker with time during the running process; however, when the ball stops running, the film thickness rapidly becomes thinner and no longer changes with time. Before running, the static film thickness is 5 nm for the substrate surface of Fe and about 2 nm for Al. When the ball starts rolling, the film for Fe immediately increases to 15 nm and that for Al to 5 nm. The film thickness with the substrate surface of Fe increases by about 15 nm within 60 min but for that of Al is about 10 nm. The static film after rolling with substrate Fe is about 13 nm thick, which is much larger than 7 nm with Al coating. Therefore the higher the surface energy the larger will be the thickness of the static film and total film.

D. Friction Force of Thin Film Lubrication

In order to investigate the friction properties of lubricant film in TFL, parts C and D of the apparatus as shown in Fig. 10 were replaced by a floating device as shown in Fig. 28 [58]. The

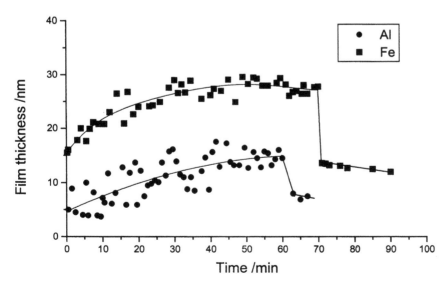

Figure 27 Film thickness with different substrates and time. Lubricant: mineral oil, load: 7 N, temperature: 18.5°C, ball: φ23.5 mm, and speed: 4.49 mm.s⁻¹. (From Ref. 17.)

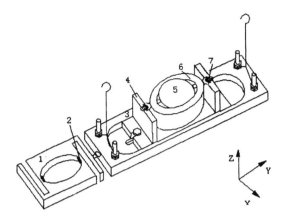

Figure 28 Floating device for friction measurement. (1) Carrier of strain gauge, (2) strain gauge, (3) beam, (4) plank, (5) steel ball, (6) oil cup, and (7) mandril. (From Ref. 58.)

steel ball is fixed so that it does not roll in the experiment and a pure sliding has been kept. The measuring system of microfriction force is composed of a straining force sensor with a resolution of 5 μN, a dynamic electric resistance strain gauge, an AD data-collecting card, and a computer.

The lower surface of the glass disk is coated with an Al, Cr, Ti, and TiO_2 layer separately by vacuum braising plating at 160 °C. These films should be plated up to a certain thickness in order to get clear interference rings. The steel ball (E52100) is 25.4 mm in diameter. The surface roughness of the glass disk and the steel ball is about 7 nm and Young's modulus is 54.88 and 205.8 GPa, respectively. When the film was in nanometer scale, the roughness in the contact area is much smaller than that of the free surface because of the elastic deformation [17]. Table 2 lists the surface energy γ of metal coatings at 25°C [59]. Table 3 shows the characteristics of three test liquids and EP.

Figure 29 [58] shows the friction coefficient vs. the sliding speed. It can be seen that different substrates make a little difference in the friction coefficients in the higher sliding speed region. The friction coefficient increases with the decrease of sliding speed. The friction coefficient with the substrate of TiO_2 is the highest, those with Cr and Ti are close to each other and that with Al is the lowest. When the sliding speed further decreases to a certain value, the friction coefficient increases sharply. It indicates that in TFL, the viscosity increases with the decrease of the sliding speed. While at a very slow speed, the lubrication regime changed into boundary regime or mixed lubrication and the friction coefficient changes from 0.15 to 0.3, which is approximately equal to that in the dry rubbing state. Furthermore, the friction is the force required to break interfacial bonds during sliding, from which it follows that the surface with the lower surface energy will exhibit the lower adhesion force, and therefore the lower friction coefficient. Therefore, the friction coefficient is closely related to the characteristics of TFL, boundary lubrication, or dry contact. The rules of

Table 2 Surface Energy of Metal Substrates

Substrate	Al	Ti	Cr	Fe
γ (mN · m^{-1})	1136	2082	2292	2331

Table 3 Characteristics of the Lubricants

Lubricant	Viscosity (mPas) (22°C)	Surface tension (mN · m^{-1}) (24°C)	Refractive index
Paraffin liquid	30	29.4	1.4612
Base oil 13604	17.4	29.57	1.4711
13604 + 10%EP	15.4	30.5	1.4675
EP	4.9	27.5	—

substrate surface energy's effect on friction are the same as that on film thickness. Higher energy leads to high friction coefficient.

Figure 30 shows that the friction coefficients with different substrates decrease slightly at first and then increase as the load increases. It can also be seen that the higher the surface energy is, the larger the friction coefficient is. A slight decrease in the friction coefficient with the increase of load is in accordance with the usual rule. But when the load increases further, the fluid film becomes thin and converts to more viscous or solidlike layer [2,39,40], which inevitably gives a great friction coefficient.

E. Effect of Slide Ratio on Thin Film Lubrication

The relation between the slide ratio ($s = 2(u_1 - u_2)/u_1 - u_2$) and the film thickness is shown in Figs. 31 to 33 [35]. The film thickness hardly changes with the slide ratio but is closely related to the average speed of the two solid surfaces ($u_1 + u_2$) for white oil No. 1 (with 5%

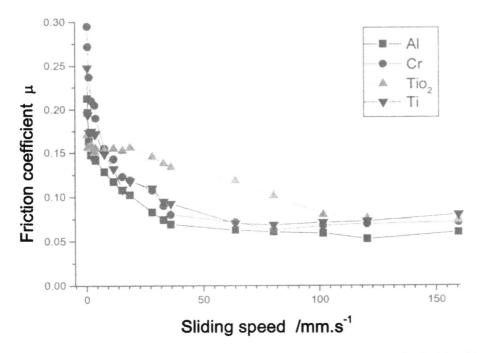

Figure 29 Friction coefficient with different substrates. Lubrication: paraffin liquid and load: 2 N. (From Ref. 58.)

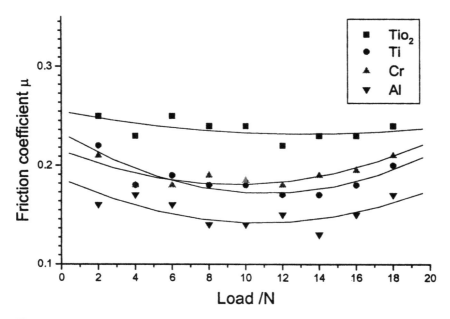

Figure 30 Friction coefficient and load on different substrates. Velocity: 15.6mm/s and lubricant: paraffin liquid. (From Ref. 58.)

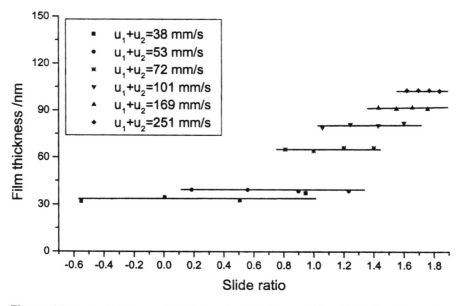

Figure 31 Film thickness with slide ratio. Lubricant: white oil No.1 + 5% nonionic acid, temperature: 20°C, relative humidity: 76%, ball diameter: 25.4 mm, and load: 2 N.

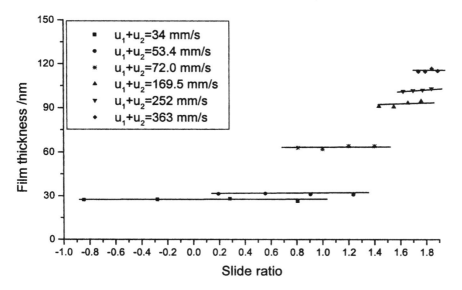

Figure 32 Film thickness with slide ratio. Lubricant: white oil No. 1 + 5% nonionic acid, temperature: 20°C, relative humidity: 76%, ball diameter: 25.4 mm, and load: 4 N.

nonionic acid in volume as additive) under a load of 2 N (Fig. 31). When the load increases to 4 N, the thickness does not change either with the slide ratio S in the range from -0.9 to 1.9 except the film thickness has a slight decrease as shown in Fig. 32. A similar phenomenon can be seen for white oil No. 2 with higher viscosity in Fig. 33. Therefore when the film thickness is beyond about 30 nm, the film is related to the total speed of the two solid surfaces but not the slide ratio. Such phenomenon is very similar to the traditional EHL theory. However, if a pure sliding condition occurs, the film will collapse easily at this speed range

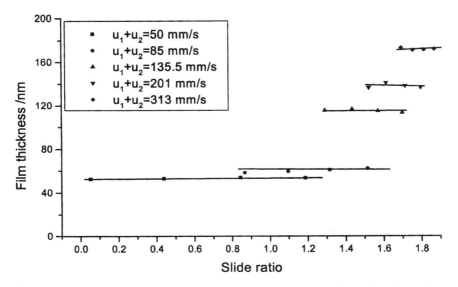

Figure 33 Film thickness with slide ratio. Lubricant: white oil No. 2 + 5% nonionic acid, temperature: 20°C, relative humidity: 76%, ball diameter: 25.4 mm, and load: 4 N.

and wear can be observed. This phenomenon may be related to the increased temperature in the contact region.

F. Effect of Electric Field

Electric voltage also has influence on the film thickness in TFL. Shen et al. [60] used hexadecane with addition of cholesteryl liquid crystals (LCs) in chemically pure lubricant to check the variation of its film thickness through applying an external DC voltage. With the technique of ROII [1,2], the effects of LCs' polarity and concentration on film thickness were investigated, and the effects of lubricant molecules on film-forming mechanism were probed.

1. Types of Liquid Crystal

Liquid crystals are organic liquids. With long-range ordered structures, they have anisotropic optical and physical behaviors, making them similar to crystal. They can be characterized by the long-range order of their molecular orientation. According to the shape and molecular direction, LCs can be classified into four types: nematic LC, smectic LC, cholesteric LC, and discotic LC. Figure 34 shows their ideal models [61].

    The sample LCs used in experiments were cholesteryl esters with low molecular weight: cholesteryl acetate (CA), cholesteryl pelargonate (CP), cholesteryl benzoate (CB), cholesteryl acrylate (CAL). In their molecular structures listed in Table 4, four tightly conjoining rings are composed of 17 carbon atoms (three rings consisting of six carbon atoms, respectively, the fourth of five carbon atoms). Such rings form comparatively linear and quasi-rigid rodlike structure.

    Liquid crystals exhibit many good lubricating properties as lubricant or additives and have attracted much attention from tribologists [62–64]. When the external electrical field is strong enough, LC molecules rearrange in the direction where the lubricant viscosity is the maximum, and an electroviscous effect occurs [62]. Kimura et al. [63] studied the active control of friction coefficient by applying a DC voltage of 30 V to the lubricants of nematic LCs and observed a 25% reduction of friction coefficient. Once the voltage was turned off, the reduction disappeared. Using nematic LCs and smectic LCs as lubricants, Mori and Iwata [64] reported that the tribological properties depended closely on their structures and their friction coefficient decreased to about 0.04. However, many other results showed that LCs used as additives exhibited better properties than as pure lubricants due to the higher stability and adaptation to temperature [65].

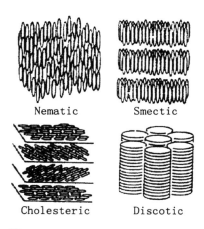

Nematic   Smectic

Cholesteric   Discotic

Figure 34 The ideal models of LCs. (From Ref. 64.)

Table 4 Molecular Structures of LCs

R		
	CA	CH_3COO-
	CP	$CH_3(CH_2)_7COO-$
	CB	$CH_3 -\!\!\bigcirc\!\!- COO-$
	CAL	$CH_3 = CH-COO-$

2. Experiment

The ROII technique [2] is shown in Fig. 9. In order to apply a DC voltage on the oil film, the lower surface of the glass disk is coated with a thin semireflecting layer of Cr, on top of which is an insulative layer of SiO_2 of about 300 nm coated at a temperature of 160°C to enhance the adhesion between the above layers and substrates as shown in Fig. 14. The lubricant film is formed between the surface of the glass disk and the surface of superpolished steel ball (E52100, 25.4 mm in diameter). The main shaft that is in contact with the Cr layer and the steel ball with a welded shaft were used as two electrodes. A pure rolling and a temperature of 30±0.5°C were kept in the test to ensure the LCs in the liquid crystalline phase. The external DC voltage varies from 0 to 1000 mV, corresponding to an electrical field intensity of 0 to 35.2 V/mm. Table 5 shows the characteristics of the tested lubricants.

3. Film Thickness of Liquid Crystals

Figure 35 [60] shows the relationship between the film thickness of hexadecane with addition of cholesteryl LCs and rolling speed under different pressures, where the straight line is the theoretic film thickness calculated from the Hamrock–Dowson formula based on the bulk viscosity under a pressure of 0.174 GPa. It can be seen that for all lubricants, when speed is high, it is in EHL regime, traditional EHL phenomena occur and a speed index ϕ of about 0.67 is produced. When the rolling speed decreases and film thickness falls to about 30 nm, the static adsorption film and ordered fluid film cannot be negligible, and the gradient reduces to less than 0.67 and the transition from EHL to TFL occurs. For pure hexadecane,

Table 5 Characteristics of the Tested Lubricants

Liquid crystal additive	Concentration	Surface tension (mN · m^{-1}) (25°C)	Apparent dynamic viscosity (mPas) (22°C)
—	—	27.1 (base oil)	2.297 (base oil)
CA	2%	25.42	2.3
CA	3%	26.96	2.579
CA	5%	27.25	2.988
CP	2%	25.49	2.583
CB	2%	25.59	2.356
CAL	2%	25.18	2.684
CAL	3%	25.83	2.384
CAL	5%	26.22	3.068

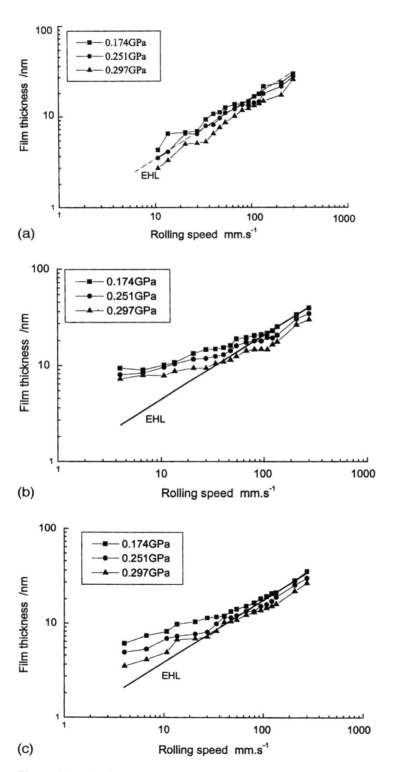

Figure 35   The film thickness with rolling speed. (a) Hexadecane, (b) 2% CB + hexadecane, (c) 2% CAL + hexadecane, (d) 2% CA + hexadecane, and (e) 2% CP + hexadecane. (From Ref. 60.)

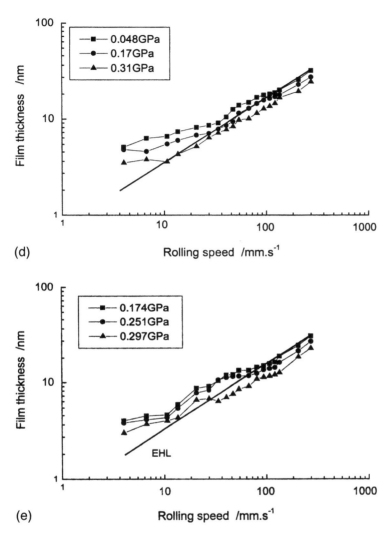

Figure 35 Continued.

because of the weak interaction between hexadecane molecules and metal surfaces, the static and ordered films are very thin. The EHL theory can maintain down to very low speed. However, for oil mixtures metal surfaces prefer to adsorb polar LC molecules and form a stronger interaction and surface force field. This leads to a thicker ordered film. Therefore, the transition from EHL to TFL occurs at high speed, i.e., the system enters into TFL when speed decreases to 101 mm/sec for CB, 98.6 mm/sec for CAL, 89.1 mm/sec for CA, and 80 mm/sec for CP, and the film thickness correspondingly reduces to 23 nm for CB, 21 nm for CAL, 18 nm for CA, and 15 nm for CP. Moreover, their practical film thickness is larger than that expected from EHL theory. A lower speed produces a greater difference between them, and the former is 3–5 times as large as the latter. Therefore the addition of LC is beneficial in the formation of the ordered film.

It can also be seen in Fig. 35 that the polarity of LC has great effects on the film thickness in TFL. For the tested cholesteryl LCs, they have the same cholesteryl structures and but different R groups. The R group in CB is benzene ring and has the highest polarity. CAL includes C=C bond and its polarity is next to CB. CA and CP belong to alkyl chain compound, their polarity is less than the former two. Moreover, less polarity results from more carbon number in alkyl chain. Thus at a speed of less than 80 mm/sec (for 0.174 GPa), the sequence of film thickness of hexadecane with 2 wt.% LCs is CB > CAL > CA > CP (according to LC additives).

Figure 36 shows the effect of the amounts of CA in hexadecane on the film thickness. It is clear that higher concentration of LC results in thicker film in BL or TFL. While in EHL (beyond 70 mm/s), for hexadecane with different LCs, weak film thickness is related to the concentration of LC. When the concentration is high, more LC molecules are adsorbed by metal surfaces to form a thicker static film and an ordered film in the low speed region because of weak fluid washing effect. However, with increasing speed, the fluid washing effect becomes stronger so that LC molecules are difficult to be adsorbed. The fluid film with bulk viscosity plays a main role and bulk viscosity becomes the key factor to determine the film thickness. When the concentration is less than 5%, as shown in Table 5, there is only a little enhancement in the viscosities of hexadecane with LCs.

In order to probe the relationship between the electric voltage and the film thickness, the properties of molecules close to solid surfaces has been investigated through applying a certain external voltage. Before each measurement, the adsorbed film left on the solid surfaces is cleared to eliminate the time effect. In addition, a low speed of 68 mm/sec is chosen to ensure the system in TFL and to reduce the fluid washing effect.

The profile of film thickness of hexadecane with LC vs. the external DC voltage is shown in Fig. 37. When the voltage is zero, the film is from 14 to 18 nm thick for the four kinds of tested lubricants. As the voltage rises and electrical field becomes stronger, the LC molecules possibly rearrange in the direction of maximum viscosity, causing an enhancement in film thickness. When the voltage increases further to 500 mV, the film thickness reaches a maximum value of about 30 nm and the intensity of the electrical field is 17 kV/mm or so. Then the film thickness hardly changes with increasing voltage.

It was explained that because a higher voltage or a stronger electric field makes LC molecules trending to rearrange in the direction of maximum viscosity, it gives rise to a greater electricoviscous effect to form a thicker film. However, when the electric field

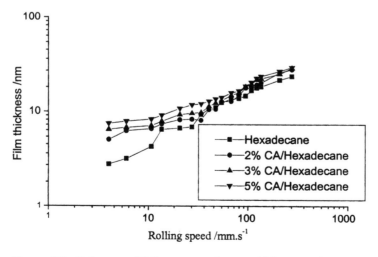

Figure 36 Influence of LC concentration on thickness under 0.174 GPa. (From Ref. 60.)

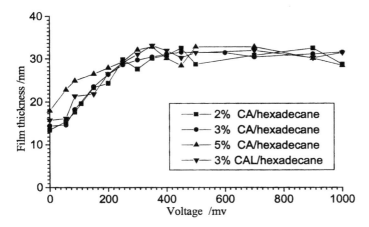

Figure 37 Influence of DC voltage on thickness. Load: 0.174 GPa and rolling speed: 68 mm.s^{-1}. (From Ref. 60.)

becomes strong to some value (depending on the type of LCs), LC molecules rearrange to the largest extent. The viscosity and film thickness also reach their maximum respectively, so that the film thickness hardly changes anymore with further increase in the voltage.

4. Isoviscosity of Liquid Crystal Films Close to Solid Surface

The relationship between the isoviscosity $\bar{\eta}$ and film thickness is shown in Fig. 38, which is calculated according to Eq. (17) from data in Fig. 35. When the film is thicker than 25 nm, the isoviscosities of hexadecane with or without LC remain a constant that approximately equals the bulk viscosity. As the film thickness decreases, the isoviscosity increases by different extent for different additives. Then, when the film thickness is about 7 nm, the

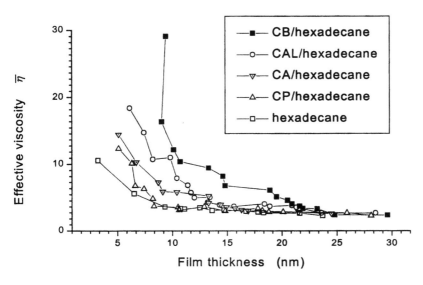

Figure 38 Isoviscosity with film thickness without voltage. Concentration: 2 wt%, and load: 0.174 GPa. (From Ref. 60.)

isoviscosity of pure hexadecane is about two times of bulk viscosity, about 3 times for CP, 4 times for CA, 6 times for CAL, and more than 10 times for CB. Thus it can also be concluded that the addition of polar compound into base oil is beneficial to the formation of thicker solidlike layer.

The isoviscosity is not only related to film thickness but also to the external voltage, as shown in Fig. 39, where the relationship between isoviscosity and external voltage is from the data in Fig. 37. When the voltage increases, the electricoviscous effect occurs and the isoviscosity rises. When the voltage rises to 500 mV, the electricoviscous effect and viscosity enhance to their largest extent. Then, with farther increasing external voltage, the isoviscosity hardly changes further. It can also be seen that a high concentration or polarity of LCs gives rise to a greater maximum of isoviscosity. Therefore it can be concluded that the stronger molecular polarity is beneficial to the formation of a thicker film in the nanoscale.

G. Time Effect of Thin Film

In the TFL regime, the film thickness varies with the running time under certain conditions [18]. Factors such as load, rolling speed, and viscosity of lubricant significantly affect the relationship between film thickness and rolling time.

The relationship between film thickness of paraffin liquid in the central contact region and running time is shown in Fig. 40. The enhancement of the film thickness produced in the running process with the running time is defined as the thickness of a time-dependent film. Before running, the thickness of the static film is 5 nm under a load of 4 N (curves A and C, Fig. 40) and 4.2 nm under 7 N (curve B, Fig. 40). When the ball starts rolling at a speed of 4.49 mm/sec (curve A), the film thickness immediately increases to 16 nm and then turns thicker with the running time. Within the 80 min of running, the thickness increases from 16 nm to about 22 nm. However, as soon as the ball stops rolling, the film collapses suddenly and the thickness decreases down to about 8 nm, which is 3 nm thicker than the static film before rolling. When the load is raised from 4 to 7 N as shown in curve B, the film thickness increases intensively with time and keeps this increasing tendency for about 40 min and then hardly changes. The increment of the total film thickness within the 70 min of running is about 13 nm. When the system stops running, the film drops down to about 10 nm thick. However, when the speed

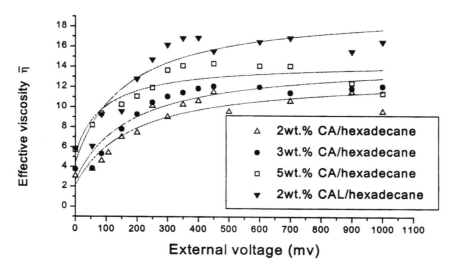

Figure 39 Isoviscosity under external voltage. Load: 0.174GPa and rolling speed: 68 mm.s^{-1}.

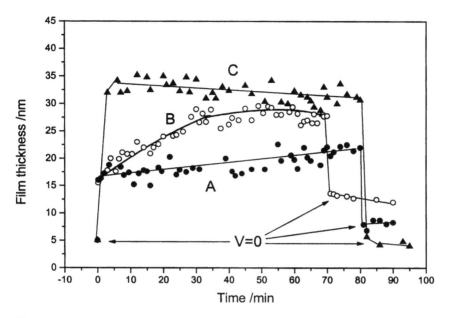

Figure 40 Film thickness vs. time. Lubricant: paraffin liquid, Temperature: 18°C, and ball: ϕ23.5 mm. (A) Speed v = 4.49 mm/s, load: w = 4 N; (B) Speed v = 4.49 mm/s, load: w = 7 N; and (C) Speed v = 17.28 mm/s, load: w = 6 N. (From Ref. 18.)

increases to 17.28 mm/sec under a load of 6 N, the film is up to about 33 nm thick, which is usually in the EHL region and it hardly varies with the running time.

The time-dependent phenomenon can also be seen in Fig. 41. The static film of decane added with 3% palimitic acid (curve A in Fig. 41) starts at 5 nm and then subsides by about 2.5 nm for nearly 15 min, and it returns to 5 nm and remains constant for more than 2 hr. But when the system starts pure rolling at a speed of 3.12 mm/sec as shown in Fig. 41 (curve B),

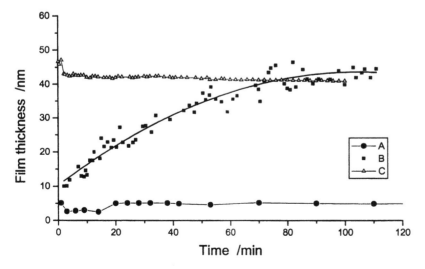

Figure 41 Film thickness with time. (A) Decane + 3% palimitic, load: 4 N, T = 30°C, v = 0 mm/s, and ball: ϕ25.4 mm; (B) Decane + 3% palimitic, load: 4 N, T = 30°C, v = 3.12 mm/s, and ball: ϕ25.4 mm; and (C) White oil No.1, load: 20 N, T = 20°C, v = 54.5 mm/s, and ball: ϕ25.4 mm. (From Ref. 18.)

the film keeps the increasing tendency for about 110 min and the thickness changes from 10 nm up to 45 nm, and then it hardly changes with time. However, when the higher viscosity of lubricant is used at a higher speed, as shown in Fig. 41 (curve C), the film thickness is 45 nm at the beginning and it reduced slightly with time.

The time-dependent film is also related to running history as shown in Fig. 42. Before the system starts running, the static film is 5 nm thick. After 60 min of running at a speed of 3.12 mm/sec, the system stops running for 10 min and the film drops down to 9 nm. When the system restarts running, the film rapidly recovers. This phenomenon indicates the films on the surface of substrates will adhere to the surface for a period of time when the rolling action has stopped, and the film thickness can pick up its value quickly once the rolling is restarted.

1. Mechanism of Time Effect

Figure 43 shows the shape of film curve in the cross section of Hertz region. Although the average film thickness (about 30 nm) of curve (a) in the central contact region is nearly the same as that of curve (b), the shapes of the two thickness curves are quite different from each other. Curve (a) is obtained at the beginning of the running at a higher speed. It is archlike in the range from -60 to 60 µm, i.e., the film thickness near the edge of the contact region is much thinner than that in the central region. This phenomenon is due to the hydrodynamic effect and the side leakage of pressure [66]. However, the shape of curve (b) obtained after 40 min of running under the condition of lower speed is very flat in the range from -60 to 60 µm. The mechanism of time effect was explained as follows [18].

The orienting force, the absorption potential of the solid surfaces, as well as the shear stress will convert lubricant molecules near the solid–liquid interface into an ordered state or solidlike in the running process. Such film will move with the ball together. Therefore the film on the surface of the ball will be getting thicker with the running time, which is similar to a snow ball getting larger when it is rolling on a snowy surface. Using this idea, the influence of pressure on the time effect can be understood. With the increase of pressure, more molecules will transfer into glassy or ordered liquid state and the static film moving simultaneously with the surface of the ball will become thicker during the running process. The isoviscosity will also be near the solid surface. Thus the total film becomes thicker with

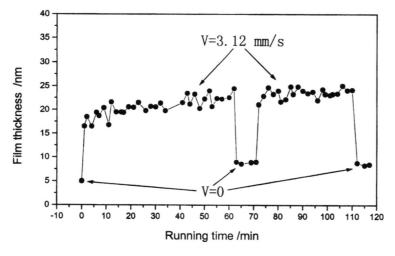

Figure 42 Film thickness with time. Lubricant: paraffin liquid, temperature: 18°C, ball: φ23.5 mm, and load: 4 N. (From Ref. 18.)

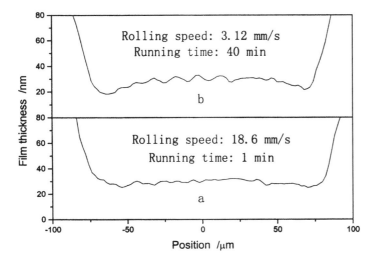

Figure 43 Film in cross section of Hertz region. Lubricant: paraffin liquid, temperature: 30°C, ball: φ23.5 mm, load: 4 N. (From Ref. 18.)

time. When the ordered film or solidlike film becomes thick enough, the number of molecules in the ordered state will balance that of the molecules washed away by fluid, and therefore the film thickness will hardly change further with time. That the thickness of ordered layer increases with pressure has also been proved by Hu et al. [67] using molecular dynamic simulation. Their results showed that the ordered structure originated from the wall–fluid interface and increased toward the middle of the film as the system pressure increases as shown in Fig. 44.

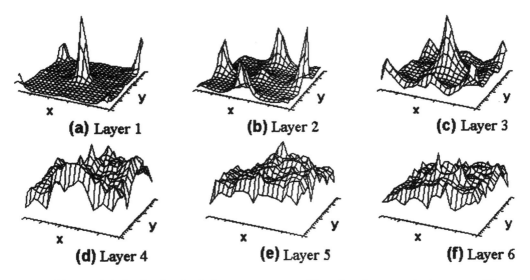

Figure 44 Spatial distribution in six layers nearest a solid wall under a constant pressure $P^*_{zz} = 3.0$. Temperature: 140 K, $\varepsilon_{wf}/\varepsilon = 3.5$, where ε_{wf} is the ratio of characteristics energy of wall-fluid interaction, ε is characteristics energy of molecular interaction. (From Ref. 67.)

H. Properties of Thin Film Lubrication with Nanoparticles

In the 1990s, much progress in the preparation and application of spherical or surface modified nanoparticles has been made, especially of nanometer-sized spherical particles, such as ultrafine diamond powder (UDP) and C60. Ultrafine diamond powder has been reported to be successfully synthesized by the detonation of explosives with negative oxygen balance [68]. As it is formed at very high velocity, high pressure, and high temperature, UDP has many unusual physical and chemical properties, such as spherical shape, nanometer sizes, large specific surface area, high surface activity, high defect density, and so on. The behavior of base oil with UDP nanoparticles was investigated by Shen et al. [69] using the technique of ROII [1,2] and a high-precision force measuring system.

Ultrafine diamond powder was prepared by the detonation of explosives [68]. Figure 45 represents the images of transmission electron microscopy (TEM) and electron diffraction [ED, in the right-hand upper corner of Fig. 45(a)]. In the ED image, the three rings clearly correspond to (111), (220), and (311) crystal surfaces, respectively, suggesting that UDP particles have complete crystal face and the structure of cubic diamond. The diameters of UFD range from 1 to 15 nm, and the average size is 5 nm. In general, they agglomerate to larger particles in the dry state as shown in the SEM image [Fig. 45(b)] because of their strong chemical activity and adsorption affinity. Table 6 lists some physical parameters of UDP.

A dispersant of polyoxyethylene nonylphenol ether by 1 wt.% and UDP with different weight ratios were added into base oils such as polyethylene glycol (PEG) or polyester (PE). Then, the mixtures were placed in an ultrasonic cleaner for 15 min to obtain a uniform mixture oil, in which hardly any sedimentation of the particles even after 2 weeks was observed. The viscosity and optical refractive index of base oils and mixture oils are given in Table 7.

In order to explore the effect of UDP on the tribological properties of lubricant oils, especially on friction force, a glass disk driven by a motor and a steel ball fixed on a mount lubricated by PE + 0.5% UDP or pure PE were rubbed against each other for 30 min under a load of 4 N and speed of 2 mm/sec. Then, the SEM images of the wear scar, which is about 0.3 mm in diameter on the steel ball, were observed (Fig. 46). It can be seen that the surface of steel ball without rubbing [Fig. 46(a)] is very smooth and when lubricated by PE + 0.5% UDP it [Fig. 46(b)] exhibits some smooth microgrooves due to hard UDP particles' plowing. However, the rubbing surface lubricated by pure PE [Fig. 46(c)] is much rougher than the former and many pits or spalls can be observed due to contact fatigue and adhesive fatigue [69,70]. (Fig. 47).

The logarithmic profiles of film thickness vs. rolling speed for the mixture oils with different concentrations of UDP are shown in Fig. 48. Under a pressure of 174 MPa shown in Fig. 48a, the film of pure PE liquid obeys the EHL theory in the higher speed region. When the film thickness decreases to about 26 nm with decreasing speed, it deviates from EHL theory and is greater than that calculated by EHL theory due to the strong adsorption between the metal surface and the functional group (–COOR) in PE. As a result, even at very low speed, the film maintains a certain thickness and load-carrying capability. When UDP is added into PE, the variation trend of film thickness with rolling speed is similar to that of pure PE liquid. However, in the low speed region, the addition of UDP causes a greater deviation in film thickness than that of pure PE because of the enhancement in viscosity resulting from the inlet viscosity effect and UDP nanoparticle adsorption effect [70,71]. In addition, as more UDP is added, a thicker film is produced, e.g., at a speed of 0.1 mm/sec, pure PE forms the thinnest film of about 6 nm, while PEs with 0.05%, 0.3%, and 0.5% UDP produce a film of 8.5, 11, and 11.5 nm, respectively.

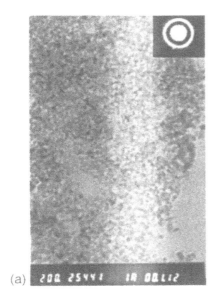

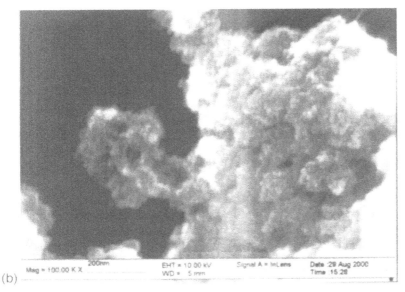

Figure 45 Images of UPD. (a) The TEM and ED image of UDP (200,000×) and (b) the SEM image of UDP.

The cases under a pressure of 297 MPa as shown in Fig. 48(b) is similar to that under 174 MPa, except that the films are a little thinner than those under the latter pressure. Furthermore, when speed is less than 1 mm/sec, the film thickness of pure PE under higher pressure decreases by a larger slope with rolling speed, and there is a little difference in film thickness of UDP-containing PEs under different loads.

The relationship between friction coefficient and sliding distance at different sliding speeds is shown in Fig. 48. It can be seen that at the beginning, friction coefficient of PEG + 0.3 wt.% UDP is almost the same for different sliding speeds, then decreases with

Table 6 Main Parameters of Synthesized UDP by Detonation of Explosives

Technical parameters	Description
Crystal structure	Single crystal and polycrystal, single crystal appearing as cubic crystal structure, and lattice constant is 0.3559–0.3570 nm
Mineralogical properties	1b style diamond
Size of crystal	1–50 nm (distribution); 5–6 nm (average)
Specific surface area	200–420 m^2/g
Specific gravity	>3 g/cm^3
Initial oxidation temperature	700 K (in air)
Productivity of diamond	Up to 10% of the detonator weight
Chemical component	86–90% C, 0.1–1% H, 2.5–3.7% N, 8.9–11.4% O

sliding distance, and at last trends to a constant. At a sliding speed of 4 mm/sec, the friction coefficient reduces to the lowest stable value of 0.055, that at 0.4 mm/sec falls off to 0.095, and that at 0.2 mm/sec to 0.125. Therefore with the same sliding distance, higher sliding speed produces a lower friction coefficient. This indicates that except for the abrasive wear effect of hard UDP particles, another new friction force-decreasing mechanism takes effect. With the increase of speed, the film becomes thicker and then the collision probability between the asperities of two surfaces and that between UDP particles and surface asperities will be decreased. Therefore the friction force at higher speed is smaller than that at lower speed.

The concentration of UDP also affects the friction coefficient as shown in Fig. 49. It is discovered that the friction coefficient of pure PEG also decreases gradually and reaches a somewhat reduced value due to the time effect of film thickness [16,18]. At a speed of 2 mm/sec and pressure of 174 GPa, the friction coefficient of pure PEG is the highest. That for PEG + 0.5% UDP ranks second. Those for PEG + 0.1% UDP and PEG + 0.3% UDP are almost the same and have the lowest friction coefficient among all tested oils. Therefore there is good concentration effect of UDP in the basic oil. If the concentration is too low, the effect of UDP on friction becomes less.

The reason why the friction coefficient decreases greatly with sliding distance when UDP is added into the basic oil is proposed in Ref. 69. Ultrafine diamond powder particles have the tendency to run in the central region as tiny ball bearing when the microgrooves have been formed as shown in Fig. 46(b). At the beginning, the hard UDP particles plow or polish the surfaces under the load, and friction force is mainly composed of mechanical plowing or micro-cutting force. With the increase of sliding distance, the microgrooves are formed and become smoother, and then more spherical UDP particles are brought into the contact area rolling as tiny ball bearing so that the sliding friction force decreases gradually.

Table 7 Parameters of Lubricants

Lubricants	η_0/mPas	n	Lubricant	η_0/mPas	n
PEG	98 (28°C)	1.458	PE	102.9 (23°C)	1.471
—	—	—	PE + 0.05 wt.% UDP	104.1	1.472
PEG + 0.1 wt.% UDP	110.6	1.459	PE + 0.1 wt.% UDP	112.3	1.473
PEG + 0.3 wt.% UDP	123.5	1.460	PE + 0.3 wt.% UDP	125.6	1.475
PEG + 0.5 wt.% UDP	134.9	1.462	PE + 0.5 wt.% UDP	136.8	1.476

PEG is polyethylene glycol and PE is polyester. n is refractive index.

Figure 46 The SEM picture of scar on steel ball. Load: 4N and rubbing time: 30 min. (a) Without rubbing, (b) PE + 0.5%UDP, and (c) PE. (From Ref. 69.)

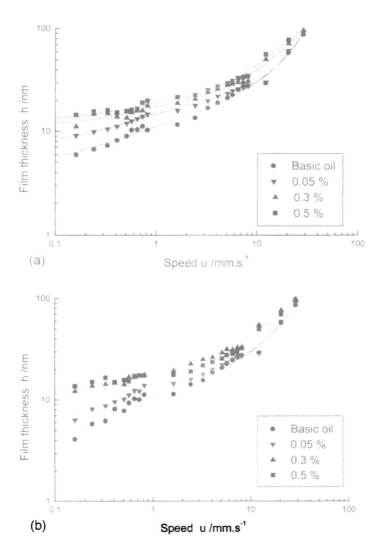

Figure 47 Film thickness for different rolling speed. Base oil: PE and temperature: 20°C (a) 174 MPa and (b) 297 MPa. (From Ref. 69.)

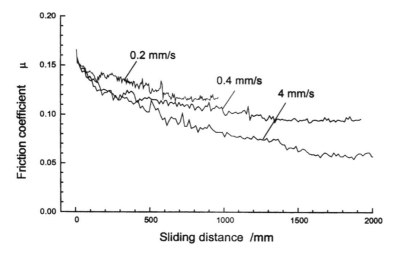

Figure 48 Friction coefficient at different sliding speed. Load: 2 N, concentration of UDP: 0.3%, and base oil: PEG. (From Ref. 69.)

Crack and pits in solid surface [Fig. 4(b),(c)] were avoided. However, it is very difficult to observe the particles' rolling directly because there is no instrument with sufficiently high resolution in the horizontal direction. In addition, a greater amount of UDP results in an enhancement in viscosity and then friction. We think that such three mechanisms may work together in the running process.

Figure 50 shows the relationship between friction coefficient and sliding speed for PEGs with different concentrations of UDP. It is clear that the friction coefficient of all

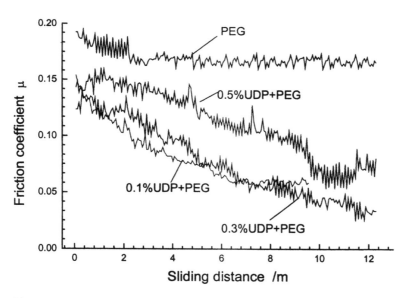

Figure 49 Friction coefficient for different concentrations. Load: 2 N, base oil: PEG, and sliding speed: 2 mm.s^{-1}. (From Ref. 69.)

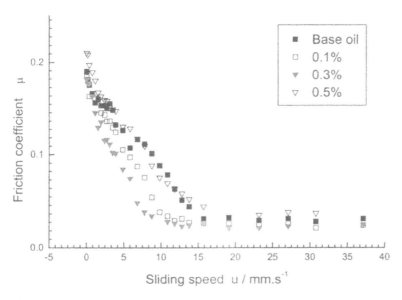

Figure 50 Friction coefficient vs. sliding speed. Base oil: PEG, temperature: 20°C, and maximum Hertz pressure: 219 MPa. (From Ref. 69.)

lubricants is about 0.22 at a speed of about 0.1 mm/sec. With increasing sliding speed, it drops sharply in the speed range from 0.1 to 15 mm/sec, and then maintains to about 0.03 when the speed is more than 15 mm/sec.

It can be concluded that the spherical diamond nanoparticles in base oil exhibit a viscosity-increasing effect making thicker film in the pure rolling state in TFL regime, and good antifriction property. Friction coefficient of UDP-containing oils decreases with sliding distance. Higher sliding speed leads to a larger decreasing slope of friction coefficient with sliding distance and a smaller stable friction coefficient. There is an optimal concentration range of UDP from 0.1% to 0.3% causing the lowest friction force. With increasing speed, the friction coefficient first drops to about 0.03 and then becomes independent of speed.

IV. THE FAILURE OF LUBRICANT FILM

The failure of the fluid film at nanoscale and the relation between the failure point and pressure, velocity and viscosity have been investigated by Luo et al. [1,19,53]. The lubricants used in experiments are given in Table 1.

According to the Reynolds theory, film thickness, pressure, speed, and lubricant viscosity obey a balance equation. The variation of film thickness of polyglycol oil No. 1 with the velocity appears in different forms under different pressures as shown in Fig. 51. Under pressures lower than 0.250 GPa, an EHL phenomenon is observed in the higher speed range (>20 mm/sec). In such case, the film thickness increases linearly with the velocity in logarithmic coordinates. However, when the film reduces to a critical thickness (about 30 nm), the further decrease in velocity can only lead to small drops in thickness and this point is usually known as the transition from EHL to TFL because the EHL rules are no longer applicable, and the critical points are related to the molecular size, solid surface energy, etc.

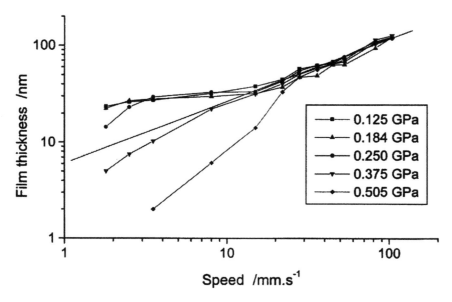

Figure 51 Film thickness with pressure. Lubricant: polyglycol oil 1, and temperature: 28°C. (From Ref. 19.)

[2,18]. When the speed decreases sufficiently, another point (at a speed of about 2.5 mm/sec) is reached at which the film thickness of polyglycol oil No. 1 under a pressure of 0.25 GPa rapidly drops from 25 to 12 nm. Because the surface roughness R_a is about 2 nm in the contact region and much smaller than that of the free surface due to the elastic deformation of the ball, the film thickness is three times thicker than the surface roughness and a full fluid film has been formed in the contact region. Thus the sudden drop in film thickness indicates that the hydrodynamic effect becomes too small to bring liquid molecules into the contact region and the fluid film cannot support the applied pressure, and then the failure of the fluid film has taken place. When the pressure changes from 0.125 to 0.505 GPa, the failure point occurs at different velocities for different pressures as shown in Fig. 51. The failure tends to occur at higher velocity values under higher applied pressures. In the other words, a larger hydrodynamic effect is required to sustain the enhanced pressure if the fluid film is to be guaranteed in the contact region.

Figure 52 shows the failure of different viscosities of lubricants. Under a pressure of 0.505 GPa, there is no film failure that occurred either when the viscosity is above 674 mm²/ sec (polyglycol oils 4 and 5). However, when velocity is reduced to about 2 mm/sec for the lubricant with a viscosity of 329 mm²/sec (polyglycol oil 3), the fluid film collapsed. For the lower viscosity lubricants (polyglycol oils 0, 1, and 2), the failure happens at higher velocities. Therefore the lower the lubricant viscosity is the higher the velocity is required to maintain a successful fluid film in the contact region.

The failure of lubricant film with the viscosity of lubricants, or the length of the molecular chains is also shown in Fig. 53 [52]. The hexadecane ($C_{16}H_{34}$) film did not fail under a pressure of 0.185 GPa in the speed ranging from 10 to 500 mm/sec. However, for the lubricants of shorter carbon chains ($C_{12}H_{26}$, $C_{10}H_{22}$, C_8H_{18}), the failure points can be clearly seen, and the shorter the carbon chain is, the higher the speed can be found at which the failure point appears. Comparing Figs. 52 and 53, it can be concluded that a lower viscosity lubricant needs higher speed to keep full oil film.

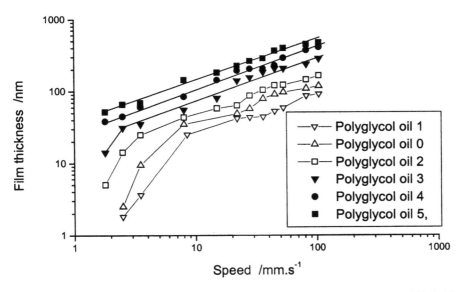

Figure 52 Film thickness vs. viscosity. Pressure: 0.505 GPa, and temperature: 28°C. (From Ref. 19.)

The failure of lubricant film is also related to the characteristics of lubricants as shown in Fig. 54. Hexadecanethiol ($C_{16}H_{34}S$) film changes from EHL to TFL at a speed of 120 mm/sec and no failure point appears in the speed range from 7 to 553 mm/sec under a heavy pressure of 0.407 GPa. However, the failure point of hexadecane film appears at a speed of 21 mm/sec under a pressure of 0.294 GPa. This indicates that the polar molecules, e.g., hexadecanethiol, form a TFL film that is difficult to collapse.

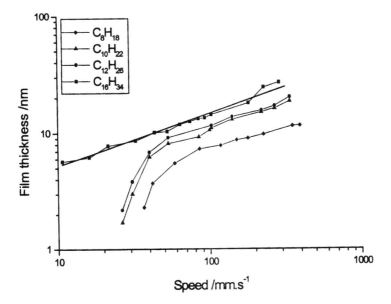

Figure 53 Film thickness with different lubricant. Pressure: 0.185 GPa and temperature: 25°C. (From Ref. 52.)

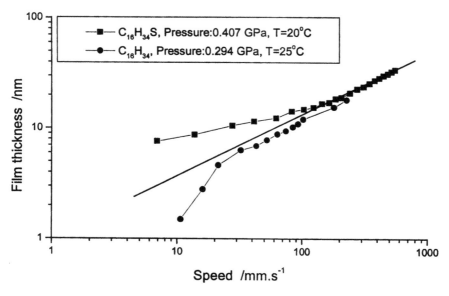

Figure 54 Film thickness with speed. (From Ref. 35.)

In order to set up the relationship among the pressure, speed, and viscosity of lubricants without polar additives, we define the fluid factor L as:

$$L = \eta_k u^{0.69} \qquad (20)$$

where η_k is the kinetic viscosity of lubricant in mm^2/sec and u is the velocity in mm/sec. The velocity index 0.69 is determined from the experimental data according to the rule that the same failure fluid factor L_f with different failure velocities and failure viscosity should meet the same pressure where the fluid film failure has taken place. The curve in Fig. 55 is called the failure curve where the lubricants are alkyl without any polar additives. If the fluid

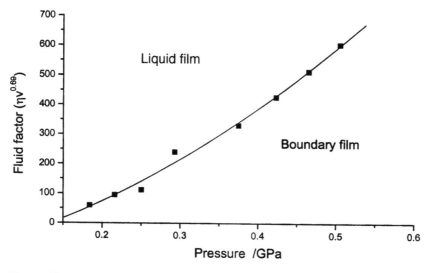

Figure 55 Failure of liquid film. (From Ref. 19.)

factor L ($\eta_k u^{0.69}$) in the point contact region of a pure rolling system is in the region above the curve in Fig. 55, the fluid film will be formed and the lubrication is in TFL or EHL regime. Whereas if it is in the region below the curve, the fluid film will fail and the lubrication regime becomes boundary lubrication or mixed lubrication. Therefore when the film thickness is thicker than three times the surface roughness R_a, the relation between failure fluid factor L_f and failure pressure P_f at which the failure of the fluid film takes place is as follows:

$$\eta_k u^{0.69} - kP_f^2 = 0 \tag{21}$$

The unit of P_f here is MPa; η_k is the kinetic viscosity of lubricant in mm^2/sec; u is the velocity in mm/sec, for the oil without polar additives; k is 23.5×10^4. If the tribopairs need to be lubricated with the fluid film in the TFL and EHL regime, the lubricant and the rolling speed should be chosen according to the pressure applied as in Eq. (2) so as to make the liquid factor L larger than the failure fluid factor L_f. Otherwise, the liquid film cannot be maintained under the pressure added.

The mechanism of the film failure is due to the limiting shear stress of the fluid film in the nanoscale [35,52]. According to the Newtonian fluid theory:

$$\tau = \eta \frac{\partial u}{\partial z} \tag{22}$$

when the relative speed u and the lubricant viscosity η are constant, the shear stress τ will increase as the film thickness decreases. If the thickness is close to zero, shear rate $\partial u / \partial z$ and the stress τ will become infinitively large according to Eq. (22). However, the maximum shear stress, τ_{max}, cannot be larger than the maximum interforce between lubricant molecules, and therefore a critical thickness must exist. If film thickness is approaching the critical value, τ is close to the limiting shear stress [72], and it will not increase anymore when the shear rate increases further. When there is an increase in pressure and the film is thinner than the critical thickness, the decrease in film thickness just causes the increase in shear rate, but not the increase in shear stress. Therefore the liquid film cannot support the extra pressure, and it will collapse. Thus the Newtonian fluid theory is no longer suitable for such conditions.

V. SUMMARY

In thin film lubrication (TFL), the film thickness is related to the surface energy of substrates, the index of effective viscosity, molecular characteristics, rolling time, speed, load, DC voltage applied, and so on. It is a lubrication regime between boundary lubrication and elastohydrodynamic lubrication (EHL). The critical film thickness, at which the transition from EHL to TFL takes place, is related to the lubricant viscosity η_0 under the atmospheric pressure, surface energy of substrate, molecular polarity, etc. The typical kind of film response to TFL regime is recognized as the ordered liquid layer or solidlike layer which becomes stronger as it is close to the solid surface.

REFERENCES

1. Luo, J.B. Study on the measurement and experiments of thin film lubrication. Ph.D thesis, Tsinghua University, Beijing, China, 1994.
2. Luo, J.B.; Wen, S.Z.; Huang, P. Thin film lubrication: Part I. The transition between EHL and thin film lubrication. Wear 1996, *194*, 107–115.

3. Bowden, F.P.; Tabor, D. *The Friction and Lubrication of Solid*; Oxford Univ. Press, 1954; 233–250 pp.

4. Adamson, A.W. *The Physical Chemistry of Surfaces,* 3rd Ed.; Interscience: New York, 1976; 447–448 pp.

5. Kingsbury, E.P. Some aspects of the thermal of a boundary lubrication. J. Appl. Phys. 1958, *29*, 888–891.

6. Cammera, A. A theory of boundary lubrication. ASLE Trans. 1959, *2*, 195–198.

7. Homola, A.M.; Israelachvili, J.N. Fundamental studies in tribology: the transition from interfacial friction of undamaged molecularly smooth surfaces to 'normal' friction with wear. Proceedings of The 5th International Congress on Tribology; Finland, 1989; 28–49 pp.

8. Dowson, D.; Higginson, G.R. A numerical solution to the elasto-hydro-dynamic problem. J. Mech. Eng. Sci. 1959, *1* (1), 6.

9. Hamrock, B.J.; Dowson, D. Isothermal elastohydrodynamic lubrication of point contact: Part I—theoretical formulation. ASME J. Lubr. Technol. 1976, *98*, 375–383.

10. Archard, J.F.; Cowking, E.W. Elastohydrodynamic lubrication at point contacts. Proc. Inst. Mech. Eng. 1965–1966, *180*(Part 3B).

11. Cheng, H.S.; Sternlicht, B. A numerical solution for the pressure, temperature and film thickness between two infinitely long, lubricated rolling and sliding cylinders, under heavy loads. ASME, J. Basic Eng. 1965, *87*, 695–707.

12. Yang, P.R.; Wen, S.Z. A generalized Reynolds equation based on non-Newtonian thermal elastohydronynamic lubrication. ASME Trans., J. Tribol. 1990, *112*, 631–639.

13. Hu, Y.Z.; Granick, S. Microscopic study of thin film lubrication and its contributions to macroscopic tribology. Tribol. Lett. 1998, *5*, 81–88.

14. Johnston, G.J.; Wayte, R.; Spikes, H.A. The measurement and study of very thin lubricant films in concentrate contact. STLE Tribol. Trans. 1991, *34*, 187–194.

15. Wen, S.Z. On thin film lubrication. Proc. of 1st Inter. Symp. on Trib.; International Academic Publisher: Beijing, 1993; 30–37 pp.

16. Luo, J.B.; Wen, S.Z. Study on the mechanism and characteristics of thin film lubrication at nanometer scale. Sci. China, Ser. A 1996, *35* (12), 1312–1322.

17. Luo, J.B.; Wen, S.Z.; Li, K.Y. The effect of substrate energy on the film thickness at nanometer scale. Lubr. Sci. 1998, *10*, 23–29.

18. Luo, J.B.; Huang, P.; Wen, S.Z. Characteristics of liquid lubricant films at the nano-scale. ASME Trans., J. Tribol. 1999, *121* (4), 872–878.

19. Luo, J.B.; Qian, L.M.; Lui, S. The failure of liquid film at nano-scale. STLE Tribol. Trans. 1999, *42* (4), 912–916.

20. Tichy, J.A. Modeling of thin film lubrication. STLE Tribol. Trans. 1995, *38*, 108–118.

21. Tichy, J.A. A surface layer model for thin film lubrication. STLE Tribol. Trans. 1995, *38*, 577–582.

22. Tichy, J.A. A porous media model of thin film lubrication. ASME J. Tribol. 1995, *117*, 16–21.

23. Matsuoka, H.; Kato, T. An ultrathin liquid film lubrication theory—calculation method of solvation pressure and its application to the EHL problem. ASME Trans., J. Tribol. 1997, *119*, 217–226.

24. Hartl, M.; Krupka, I.; Poliscuk, R.; Molimard, J.; Querry, M.; Vergne, P. Thin film colorimetric interferometry. STLE Tribol. Trans. 2001, *44* (2), 270–276.

25. Gao, G.T.; Spikes, H.A. Boundary film formation by lubricant base fluid. Tribol. Trans. 1996, *39* (2), 448–454.

26. Iliuc, I. Tribology of thin layers; Elsevier Scientific Publishing Company: New York, 1980, (Translated from the Romanian).

27. Allen, C.M.; Drauglis, E. Boundary layer lubrication: monolayer or multilayer. Wear 1969, *14*, 363–384.

28. Israelachvili, J.N.; Tabor, D. The measurement of Van Der Waals dispersion force in the range of 1.5 nm to 130 nm. Proc. R. Soc., London 1972, *A331*, 19–38.

29. Israelachvili, J. Intermolecular and surface force, 2nd Ed.; Academic Press, 1992.
30. Alsten, J.V.; Granick, S. Friction measured with a surface forces apparatus. Tribol. Trans. 1989, *32*, 246–250.
31. Granick, S. Motions and relaxation of confined liquid. Science 1991, *253*, 1374–1379.
32. Luengo, G.; Schmitt, F.; Hill, R. Thin film rheology and tribology of confined polymer melts: contrasts with bulk properties. Macromolecules 1997, *30* (8), 2482–2494.
33. Luo, J.B.; Yian, C.N. Fuzzy view point in lubricating theory. Lubr. Eng. 1989, *4*, 1–4. *in Chinese*.
34. Gupta, A.; Sharma, M.M. Stability of thin aqueous film on solid surfaces. J. Colloid Interface Sci. 1991, *149*, 392–424.
35. Luo, J.B.; Lu, X.C.; Wen, S.Z. Developments and unsolved problems in nano-lubrication. Progr. Nat. Sci. 2001, *11* (3), 173–183.
36. Gao, G.; Spikes, H.A. The control of friction by molecular fractionation of base fluid mixtures at metal surface. STLE Tribol. Trans. 1997, *40* (3), 461–469.
37. Hartl, M.; Krupka, I.; Poliscuk, R.; Liska, M. An automatic system for real-time evaluation of EHD film thickness and shape based on the colorimetric interferometry. STLE Tribol. Trans. 1999, *42* (2), 303–309.
38. Thompson, P.A.; Grest, G.S.; Robbin, M.O. Phase transitions and universal in confined films. Phys. Rev. Lett. 1992, *68* (23), 3448–3451.
39. Hu, Y.Z.; Granick, S. Microscopic study of thin film lubrication and its contributions to macroscopic tribology. Tribol. Lett. 1998, *5*, 81–88.
40. Luo, J.B.; Wen, S.Z. Progress and problems in nano-tribology. Chin. Sci. Bull. 1998, *43* (7), 369–378.
41. Jost, H.P. Tribology: the first 25 years and beyond—achievements, shortcomings and future tasks. *Tribology 2000*, 8th International Colloquium; Technische Akademic: Esslingen, 1992.
42. Hardy, W.B.; Doubleday, I. Boundary lubrication—the paraffin series. Proc. R. Soc. 1922, *A100*, 550–574.
43. Wen, S.Z. Principle of tribology; Tsinghua University Press: Beijing, 1991.
44. Gohar, R.; Cameron, A. The mapping of eslastohydrodynamic contact. ASLE Trans. 1967, *10*, 215–225.
45. Homola, A.M.; Israelachvili, J.N. Fundamental studies in tribology: the transition from interfacial friction of undamaged molecularly smooth surfaces to 'normal' friction with wear. Proceedings of The 5th International Congress on Tribology; Finland, 1989; 28–49 pp.
46. Westlake, F.J. An interferometric study of ultra-thin fluid films. Ph.D. Thesis, Univ. of London, 1970.
47. Guangteng, G.; Spikes, H.A. Properties of ultra-thin lubricating films using wedged spacer layer optical interferometry. Proc. 14th Leeds–Lyon Symp. on Trib.; Interface Dynamics: Leeds, 1988; 275–279 .
48. Ratoi, M.; Anghel, V.; Bovington, C.; Spikes, H.A. Mechanisms of oiliness additives. Tribol. Int. 2000, *33*, 241–247.
49. Wedeven, L. Optical measurements in elastohydrodynamic rolling contacts. Ph.D. Thesis, Univ. of London, 1970.
50. Kauffman, A.M. A simple immersion method to determine the refractive index of thin silica films. Thin Solid Films 1967, *1*, 131–136.
51. Born, M.; Wolf, E. Principle of optics, 5th Ed. Pergamon Press, 1975; 275–270.
52. Luo, J.B.; Liu, S.; Shi, B.; Wen, S.Z. The failure of nano-scale liquid film. Int. J. Nonlinear Sci. Numer. Simul. 2000, *1*, 419–423.
53. Hartl, M.; Krupka, I.; Liska, M. Differential colorimetry: tool for evaluation of chromatic interference patterns. Opt. Eng. 1997, *36* (9), 2384–2391.
54. Hamrock, B.J.; Dowson, D. Isothermal elastohydrodynamic lubrication of point contacts: Part III—Full flooded results. J. Lubr. Technol., ASME Trans. 1977, *99* (2), 264–276.
55. Hartl, M.; Křupka, I.; Liška, M. Experimental study of boundary layers formation by thin film colorimetric interferometry. Sci. China, Ser. A 2001, *44* (suppl), 412–417.

56. Wedeven, L.D.; Evans, D.; Cameron, A. Optical analysis of ball bearing starvation. Trans. ASME J. Lubr. Technol. 1971, *93*, 349–363.

57. Gao, G.T.; Spikes, H. Fractionation of liquid lubricants at solid surfaces. Wear 1996, *200* (1–2), 336–345.

58. Shen, M.W.; Luo, J.B.; Wen, S.Z. Effects of surface physicochemical properties on the tribological properties of liquid paraffin film in the nanoscale. Surf. Interface Anal. 2001, *32*, 286–288.

59. Yao, Y.B.; Xie, T.; Gao, Y.M. Handbook of physicochemistry; Science and Technology Press of Shanghai: Shanghai, 1985.

60. Shen, M.W.; Luo, J.B.; Wen, S.Z.; Yao, J.B. Nano-tribological properties and mechanisms of the liquid crystal as an additive. Chin. Sci. Bull. 2001, *46* (14), 1227–1232.

61. Demus, D.; Goodby, J.; Gray, G.W. Handbook of liquid crystals; Wiley-VCH: New York, 1998.

62. Morishita, S.; Nakano, K.; Kimura, Y. Eletroviscous effect of nematic liquid crystal. Tribol. Int. 1993, *26* (6), 399–403.

63. Kimura, Y.; Nakano, K.; Kato, T. Control of friction coefficient by applying electric fields across liquid crystal boundary films. Wear 1994, *175*, 143–149.

64. Mori, S.; Iwata, H. Relationship between tribological performance of liquid crystals and their molecular structure. Tribol. Int. 1996, *29* (1), 35–39.

65. Bermudez, M.D.; Gines, M.N.; Vilches, C.; Jose, F. Tribological properties of liquid crystals as lubricant additives. Wear 1997, *212* (2), 188–194.

66. Wen, S.Z.; Yang, P.R. Elastohydrodynamic lubrication; Tsinghua University Publisher: Beijing, 1992; 261–288.

67. Hu, Y.Z.; Wang, H.; Guo, Y.; Zheng, L.Q. Simulation of lubricant rheology in thin film lubrication: Part I. Simulation of poiseuille flow. Wear 1996, *196*, 243–248.

68. Chen, P.W.; Yun, S.R.; Huang, F.L. The properties and application of ultrafine diamond synthesized by detonation. Ultrahard Mater. Eng. 1997, *3*, 1–5.

69. Shen, M.W.; Luo, J.B.; Wen, S.Z. The tribological properties of oils added with diamond nano-particles. STLE Tribol. Trans. 2001, *44* (3), 494–498.

70. Liu, J.J.; Cheng, Y.Q.; Chen, Y. The generation of wear debris of different morphology in the running-in process of iron and steels. Wear 1982, *154*, 259–267.

71. Ryason, P.R.; Chan, I.Y.; Gilmore, J.T. Polishing wear by soot. Wear 1990, *137*, 15–24.

72. Bair, S.; Khonsari, M.; Winer, W.O. High-pressure rheology of lubricants and limitations of the Reynold's equation. Tribol. Int. 1998, *31* (10), 573–586.

9
Tribology of Metal Cutting

Viktor P. Astakhov
Astakhov Tool Service, Rochester Hills, Michigan, U.S.A.

I. INTRODUCTION

Tribology plays a pivotal role in materials processing, particularly in metal cutting operations. The tribological conditions in these operations—real area of contact, stress distribution along contact areas, interfacial temperature fields, and highly active and freshly generated (nascent) surfaces—are more severe than in other applications.

Because the economic and technical feasibility of a process or product can be dictated by wear, tribological knowledge can help strengthen the competitiveness of the manufacturing industries and minimize the energy and resources they consume. Consider, for example, machining in the automotive industry, where the cost of the tool itself is insignificant, while the downtime and direct cost to change the tool each time it is worn or it fails may be many times greater because such a change, if unscheduled, requires usually to shut down the entire production line. Or consider the cost of an expensive aircraft engine part rejected (at best) or failed (at worst) due to premature fracture of a cutting tool during machining or due to high residual stress generated in machining by the worn tool.

Current demands for tribological advances to support improved productivity coincide with additional challenges to tribology posed by the increased utilization of engineered materials. Some of these materials are useful as tools and dies and can perform optimally due to their improved properties. Others are difficult-to-machine work materials and create tribological problems during machining such as severe tool wear, great residual stresses in the machined surface, metallurgical and structural changes of the machined surface, and many others.

Direct effects of tribology in metal cutting, such as wear of the tools and surface quality of finished products, are obvious. Indirect effects, less readily evident, are equally important. It is a known fact, for example, that a product's tribological history during manufacturing may later determine such characteristics as reliability, corrosion and irradiation resistance (important in nuclear power industry), fatigue tolerance, and frictional properties. The tribological problems may inhibit or impede the introduction of new or advanced machining processes such as high-speed machining, combined machining, etc.

Figures for the economics of tribology in metal cutting are difficult to obtain due to the diversity and pervasiveness of the field. The replacement of a prematurely worn $2 tool insert or a $80 broken gundrill may hold production of a $1 million machine or assembly line. But how much of the associated cost can be attributed to tribology? Should the costs of replacement, downtime, capital invested in less-than-optimally efficient equipment, missed opportunities, etc. be included? Unfortunately, many researches have concentrated on the energy and direct costs aspects of tribology in metal cutting—energy consumption and energy savings due to tribology, saving on cutting tool consumption, and quality improvement that could accrue from advances in tribology.

The ASME Research Committee on Lubrication has studied [1] the role of tribology in energy conservation. It concluded that about 5.5% of U.S. energy consumption is used in primary metals and metal-processing industries and that 0.5% can be saved through advances in tribology of metal removal and forming processes, achievable through relatively modest research expenditures and effort. The combined potential savings in manufacturing alone of 1.8% of the national energy consumption totaled about $21.5 billion per year [1]. Among the economic activities surveyed, the manufacturing sector was estimated to provide the greatest potential savings per dollar spent on tribology research.

Tribological conditions encountered in machining are severe [2–5]. The contact stress at the tool–chip interface is very high resulting in high shear stresses along the tool–chip contact area. The real area of contact is near the apparent area at the plastic part of the tool–chip contact where the shear stress may be much higher than the yield shear stress of the original work material. The chip surface sliding over the tool face is a virgin and thus chemically active. The mechanical properties of the chip contact layer are different (usually much superior) than those of the work material. The contact temperature may reach more than 1000°C in machining difficult-to-machine materials. As a result, chemical interactions between the tool, the work material, and the environment are crucial in machining. Similarly, abrasion, adhesion, seizure, diffusion, or their complex combination may occur between the tool and the chip.

Although there are a great number of research papers and book on the contact conditions on the tool rake face published in the last 50 years, we are very far from a clear understanding of the nature and complicity of tribological phenomena in this region. Simple force diagram is still in use for determining "an average friction coefficient" on the tool rake face based on assumptions of equality and colinearity of the resultant force acting on the shear plane and tool face, although it is well known that a coefficient of friction is inadequate to characterize the sliding between chip and tool [6]. On the other hand, the friction and normal forces and shear and normal stresses on the tool rake face could be obtained experimentally by conducting orthogonal machining tests and measuring cutting and trust components of the cutting force parallel and normal to the tool rake face. The friction coefficient thus obtained, unfortunately, does not match with common experience [1]. The same can be said about the shear and normal stresses distributions [5].

Similar processes take place at the tool flank face(s) where the tool is in contact with the work material primarily due to work material spring back [2]. The nature and importance of this spring back are not fully understood and thus appreciated, although the wear of the flank face often defines tool life. When material is cut by the cutting edge, it first deforms elastically and then plastically. When this edge passes over the deformed region, this region springs back due to the reversibility of elastic deformation. The heavier the cut, the greater volume of the work material deforms causing larger spring back. Besides, this work material has been plastically deformed so that its properties are different from those of the original material. As a result, the contact stresses at the flank face(s) are much higher than might be expected from just spring back [7].

Although the temperatures at the tool flank(s)–work material contact area are much smaller than those at the tool–chip interface, the properties of the work material in this contact are modified by its strain hardening. Moreover, the sliding speed at the tool flank–

work material interface is much greater than that at the tool–chip interface (2–10 folds). Due to this high sliding speed, a great amount of heat generated at this interface.

The contact temperatures at the tool flank(s)–work material interface, however, are much smaller than those at the tool–chip interface because the significant amount of the heat energy generated dissipates into the workpiece usually having a significant mass. It may not be the case while machining work materials having low thermal conductivity. As such, much smaller amount of heat energy generated is carried out by the chip and the workpiece so that the tool temperatures become much higher concentrating in the regions to the cutting edge.

There are three principal ways to reduce the severity of the contact processes in metal cutting and thus reduce the tool wear: cooling and lubricating of the machining zone [8–14], coatings on the cutting tools [15–17], and modification of the workpiece chemical composition [18–28]. Usually, these are used in their combination, although the compatibility of a particular combination of the cutting fluid, tool coating, and workpiece chemical composition is practically ignored.

It follows from the foregoing consideration that the understanding of the tribology of metal cutting is of great importance. This should provide clear guidelines in the selection of parameters of the metal cutting system maintaining its coherence, high productivity, and efficiency. Therefore this chapter aims to present an entirely new insight into the nature of contact processes at the tool–chip and the tool–workpiece interfaces accounting for their relative motions and the cyclic nature of the cutting process.

II. METAL CUTTING PROCESS

In modern test books (for example, Ref. 29), machining is identified as a deforming process that forms and shapes metals and alloys. What completely ignored is that a principal difference exists between machining and all other metal forming processes. In machining, the physical separation of the layer being removed from the rest of the workpiece must occur. As a result, one solid workpiece should be separated into at least two items, the part and the chip. Such separation process is known as fracture [29] and thus should be so regarded. To achieve fracture, the stress in the chip formation zone should exceed the strength of the work material (under a given state of stress imposed by the cutting tool), whereas other forming processes are carried out applying stress sufficient to achieve the well-known shear flow stress in the deformation zone trying to avoid fracture.

The ultimate objective of machining is to separate the layer being removed from the rest of the workpiece with minimum possible plastic deformation and thus energy. Therefore the energy spent on the plastic deformation in machining must be considered as wasted. On the other hand, any other metal-forming process, especially involving high strains (deep drawing and extrusion), uses plastic deformation to accomplish the process. Parts are formed into useful shapes such as tubes, rods, and sheets by displacing the material from one location to another [30]. Therefore the better material, from the viewpoint of metal forming, should exhibit a higher strain before fracture occurs. It is understood that this is not the case in metal cutting, as the best material should exhibit no strain at fracture. Unfortunately, this does not follow from the traditional metal cutting theory [2–4] which normally utilizes the shear strength or, at best, the shear flow stress (the term was specially invented for metal cutting to cover discrepancies between the theoretical and experimental results) to calculate the process parameters (cutting force, temperatures, and contact characteristics), although everyday machining practice confirms that these parameters are lower in cutting brittle materials having higher shear strength.

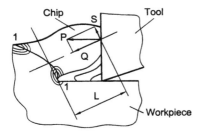

Figure 1   Single shear plane model.

Astakhov [5] suggested and proved that the cutting process takes place in the cutting system composing of the following components: the workpiece, the chip, and the tool. Various dynamic interactions of the system components define the cutting process. Fig. 1 shows the model of chip formation. As seen, the application of penetration force P results in compressive force Q and bending moment $M = SL$ in the deformation zone. As such, the chip forms under the combined action of the compressive and bending stresses. The maximum combined stress acts along surface 1–1, and thus chip sliding occurs along this surface. If it is so, there should be a way to optimize the performance of such a system using the physical properties of its components rather than the system output parameters (cutting forces, tool wear, integrity of the machined surface, productivity, etc.). As such, a distinctive criterion for the system performance should be established.

III. ENERGY FLOWS IN THE CUTTING SYSTEM

Generalization of the theoretical and experimental results on the performance of the cutting system allows the graphical representation of energy flows and conversions in the cutting system as shown in Fig. 2. The total energy entering the cutting system, U_{cs}, is a part of the energy produced by the drive motor, U_{em}. Obviously,

$$U_{cs} = U_{em}\eta_{pt} \tag{1}$$

where η_{pt} is the efficiency of the power train that connects the cutting system and the drive motor of the machine tool.

Ideally, if there is no energy loss in the cutting system, then the energy transmitted through the cutting tool having the cutting tool energy U_T to the chip having the chip energy U_c and then to the chip formation zone having the deformation energy W_f is equal to the energy to fracture the layer being removed, U_f. As such, $U_{cs} = U_T = U_c = W_f = U_f$. In real cutting systems, however, the energy losses occur due to elastic and plastic deformations of their components as well as friction losses in the interactions of these components. These losses do not correlate directly with the failure of the layer being removed. These energy losses are converted into heat, which, in turn, affect these losses even further.

A. Cutting Tool

According to the conservation law, the work done over compression, W_{Tc}, and its bending, W_{Tb}, of the cutting tool transforms into its potential energy, U_T, and is partially spent on internal friction during deformations, which results in heat generation, Q_l. Because the

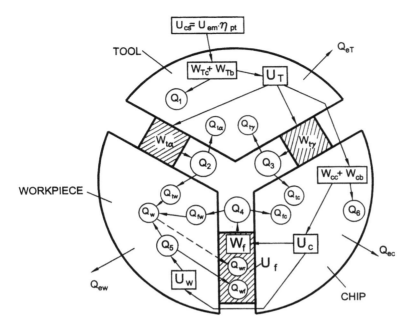

Figure 2 Graphical representation of energy flows in the metal cutting system.

cutting tool is subjected normally only to elastic deformation, Q_1 is small. However, when the cutting tool is not rigid and/or when it vibrates with an appreciable amplitude during machining, the share of this heat energy may become significant.

The potential energy U_T is then spent as the work done by the frictional forces on the tool flanks and rakes, $W_{t\alpha}$ and $W_{t\gamma}$, respectively, and on the work done over the chip. The works of the frictional forces, $W_{t\alpha}$ and $W_{t\gamma}$, convert into heat, Q_2 and Q_3, respectively. Heat Q_3 generated at the tool–chip interface then conducted into the tool, $Q_{t\gamma}$, and into the chip, Q_{tc}. The amounts of heat conducted into the tool and the chip are in the reverse proportion to their thermal resistances. Similarly, the thermal energy generated at the tool flank(s)– workpiece interface, Q_2, is then distributed between the tool, $Q_{t\alpha}$, and the workpiece, Q_{tw}, in the reverse proportion to their thermal resistances. In the current consideration, the notion of the thermal resistance is used as more general compared to simple thermal conductivity because the cutting system is a dynamic system, so a number of different system parameters contribute to the thermal resistances of its components, which are in relative motion with respect to each other.

Because the discussed directions of heat flow are as shown in Fig. 2, the tool contact surfaces (the tool–chip and tool–workpiece interfaces) always have higher temperatures than the rest of the cutting tool. The temperature gradient decreases with the distance from the tool rake face. As a result, the heat flows in the direction of the flank surface until the temperature gradient becomes zero due to the heat flow from the flank surface. The same can be said about heat flow from the tool flank contact surface. The exception is a small region adjacent to the cutting edge. Here, with decreasing distance between tool rake and flank surfaces, it can be assumed that the power of one heat source exceeds that of the other. As such, heat would flow from the tool into the workpiece or the chip whichever heat source is weaker. In reality, however, it does not occur because the friction conditions on the tool rake and flank surfaces as well as their temperatures in the region adjacent to the cutting edge are balanced so that the temperature gradient here is zero. Therefore it can be concluded that the heat flows only into the cutting tool from the tool–chip and tool–workpiece interfaces. The other surfaces of the cutting tool dissipate the heat energy into the environment, $Q_{\text{en-T}}$.

Hereafter, the working part of the cutting tool is referred to as the cutting wedge, which is enclosed between the tool rake and flank contact surfaces meeting to form the cutting edge. This cutting wedge is under the action of stresses applied on the tool–chip and tool–workpiece contact surfaces. Moreover, due to heat that flows into the cutting tool, this wedge has high temperature. As a result, the gradual exhaustion of the resource of the tool material takes place as well explained in Ref. 31.

B. Chip

The potential energy of the cutting tool is spent on the deformation of the chip. The chip serves as a lever to transmit the applied load into the chip formation zone (Fig. 1). A part of the work done by the compressive force and the bending moment, $W_{\text{cc}} + W_{\text{cb}}$ over the chip, is dissipated in the chip converting into heat Q_6. The other part is the potential energy of the chip U_{c} that makes its contribution W_{f} to the formation of the fracture energy U_{f}. As discussed above, a part Q_{tc} of the heat Q_3 generated at the tool–chip interface is also conducted into the chip. Besides, the chip also receives a certain part Q_{fc} of the heat generated in the chip formation zone due to plastic deformation of the layer being removed and heat due to fracture, Q_4.

Heat Q_{tc} from the tool–chip interface flows into the chip primarily due to thermal conductivity and partially due to mass transfer because the chip moves over the tool–chip interface. Heat Q_{fc} enters the chip due to mass transfer because the velocity of chip is normally higher than that of heat conduction in the chip [5]. In other words, the heat generated in the conversion of the layer being removed into the chip is transferred by the mass of the moving chip from the deformation zone, not due to its thermal conductivity. As a result, the moving mass of the layer being removed is converted into the chip and the moving chip changes its energy along the tool–chip interface. This is true for each chip formation cycle. The chip temperature results in its higher plasticity and thus the energy loss in the transmission of the compressive force and bending moment into the chip formation zone increases and, in turn, increases Q_6.

Additionally, the heat $Q_{\text{en-c}}$ is released to the environment.

C. Workpiece

It is demonstrated in Ref. 5 that the surface of the maximum combined stress, which eventually becomes the surface of fracture or chip separation, changes its position within each chip formation cycle. Due to this fact, the energy of fracture of the layer being removed is distributed over a certain volume of the work material causing plastic deformation of this region which normally extends below the surface that conditionally separates the layer being removed and the rest of the workpiece. Obviously, some part of the potential energy transferred into the chip formation zone is spent on the plastic deformation of this region, U_{w}. It is confirmed experimentally and manifests itself by the cold working of the machined surface and results in the machining residual stress [32]. As such, the heat Q_5 is released.

Besides heat Q_5, heat Q_{tw} from the tool flank–workpiece contact and heat Q_{fw} from the chip formation zone are also supplied into the workpiece forming the total thermal energy

$Q_w = Q_s + Q_{tw} + Q_{fw}$. As discussed earlier [33], this thermal energy cannot elevate the energy level in the fracturing of the layer being removed in the direction of the cutting speed because the velocity of heat conduction is much lower than the cutting speed. However, its part Q_{wr} transferred in the direction of the feed motion can change this energy level [33].

A special case takes place in the machining of a workpiece having a small diameter. As such, heat Q_w, having been reflected many times by the outer surface of the workpiece, affects the energy level of its entire cross section so that much higher part Q_{wr} than Q_{wf} is added to the energy of fracture U_f.

Additionally, some heat Q_{en-w} is released to the environment through convection and conduction.

IV. EFFICIENCY OF THE CUTTING SYSTEM

It is obvious that not all the energy required by the cutting system U_{cs} is spent for the fracture of the layer being removed, i.e., for performing useful work U_f. As discussed above, a part of the energy spent in the cutting system is dissipated in the components of this system changing their properties and in the environment. As a result, the cutting system consumes more energy than needed for the fracture of the layer being removed. It is clear that the better the components of the cutting system are organized, the smaller the difference is between these two energies. Therefore it appears to be reasonable to introduce the notion of the efficiency of the cutting system.

The word *efficiency* is a term used virtually everyday in a myriad of circumstances; thus it needs to be clearly defined for any specific usage. For metal cutting systems, it is the ratio (expressed as a percentage) of the actual energy required for the fracture of the layer being removed and the total energy spent in the cutting system

$$\eta_{cs} = \frac{U_f}{U_{cs}} \tag{2}$$

The introduced energy efficiency of the cutting system makes sense if and only if the energy U_f in Eq. (2) is constant which could be regarded as the benchmark so that the amount of the actual energy spent for machining can be compared with this target. The amount of energy needed to fracture a unit volume of the work material would depend on many factors. However, the state of stress, strain rate, and temperature are of prime importance. However, for a given cutting system, the system parameters, including system geometry and regime, uniquely define the stress and strain at fracture and thus U_f [5].

In the design of practical cutting systems, achieving maximum efficiency is not the ultimate goal. Rather, its optimization should be considered. This is because one needs to assure the required time period of existence of the cutting system. In other words, for a given tool material, the cutting wedge should be so shaped and the cutting regime should be so selected that the tool wear rate is in the required limits. The latter condition may be at odds with the requirement of maximum efficiency. Moreover, a particular value of the cutting system efficiency may be further corrected accounting on the required quality, including integrity of the machined surface, productivity, and other practical constraints.

The specific energy of fracture of the layer being removed is determined under the state of stress in this layer as

$$U_f = \int_0^{\varepsilon_f} \sigma(\varepsilon) d\varepsilon \tag{3}$$

where ε_f is the stain at fracture [5]. The specific cutting energy required by the cutting system can be determined as

$$U_{cs} = \frac{P_c \tau_c}{V_c} \qquad (4)$$

where P_c is the cutting power, τ_c is the machining time, and V_c is the volume of the work material cut during time τ_c.

Because $V_c = fdv\tau_c$, where f is the cutting feed (m/rev), d is the depth of cut (m), and v is the cutting speed (m/sec), and $P_c = F_z v$, where F_z is the power component of the cutting force (N), the final expression for the efficiency of the cutting system can be written as

$$\eta_{cs} = \frac{fd}{F_z} \int_0^{\varepsilon_f} \sigma(\varepsilon) d\varepsilon \qquad (5)$$

A series of turning tests were carried out to reveal the influence of various parameters of the cutting system on its efficiency. The test setup, methodology, and conditions were the same as discussed earlier in Ref. 33. Four general-purpose triangular cutting inserts having shape SNMM1200404M and made of P20 carbide produced by different carbide suppliers were selected for the test to avoid bias due to possible variations of tool material properties. The following characteristics of the carbide inserts were compared: transverse rupture strength (TRS), compressive strength, density, fracture toughness, grain size, hardness, lattice parameters, magnetic susceptibility, mean thermal expansion coefficient, melting point, thermal conductivity, specific heat, Poisson's ratio, Young's modulus, and wear resistance. Special precautions were taken to avoid possible influence of the workpiece diameter, the static and dynamic rigidity of the setup, built up edge (BUE), etc. For each supply of work materials, the composition and element limits were requested from the steel supplier and then were compared with the actual results obtained using LECO® SA-2000 Discharge-Optical Emission Spectrometer. The hardness of the work materials has been determined over the complete cross section, and the cutting tests were conducted only on the bars where the hardness was within the limits $\pm 10\%$. Special metallurgical parameters such as the microstructure, grain size, inclusions count, etc. were inspected using quantitative metallography.

Figure 3 shows the influence of the rake angle. As seen, the efficiency of the cutting system increases with the rake angle; this result was anticipated because it follows from everyday machining practice. More pronounced effect of the rake angle is observed when the depth of cut, cutting speed, and feed are increased. This influence is greater for difficult-to-machine materials (as an example, the results for steel AISI 52100 are presented). What was not anticipated is that the chip compression ratio and tool–chip contact area remained practically the same in the discussed tests; that is, these two important parameters of the cutting system were not affected by the tool rake angle. In other words, the following contradiction should be resolved: the cutting force and thus the energy spent in cutting depend on the rake angle, while the chip deformation caused by this force and thus tool–chip contact length do not.

First of all, it should be shown that the obtained result is not application-specific; that is, it is of general nature and thus is independent on the particular cutting regime. To do this, the chip compression ratio and the tool–chip contact length should be determined as a function of the Peclet criterion defined as [5]

$$Pe = \frac{vt_1}{w_w} \qquad (6)$$

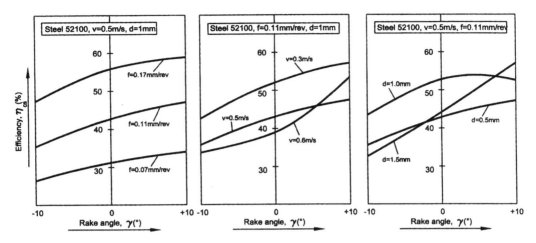

Figure 3   Influence of the rake angle on the efficiency of the cutting system. Work material—steel AISI 52100.

where w_w is the thermal diffusivity of workpiece material (m^2/sec),

$$w_w = \frac{k_w}{(c\rho)_w} \tag{7}$$

where k_w is the thermal conductivity of workpiece material [J/(m sec °C)] and $(c\rho)_w$ is the volume specific heat of workpiece material [J/(m^3 °C)].

Figure 4 shows the chip compression ratio and the relative tool chip–contact area as the functions of the Peclet criterion for different rake angles. These results can be used to explain why cutting with negative rake angles requires more energy. As seen from Fig. 1, if the rake angle decreases and the chip cross section remains unchanged, the applied penetration force P must be increased to keep the same bending moment on the surface of maximum combined stress because chip formation occurs due to this moment [5]. This increased penetration force multiplied on the cutting speed yields the increased required energy.

Another important conclusion follows from these results. Because the tool–chip contact area does not change while cutting force increases, the stress at the tool–chip interface increases. This increase could be significant. However, it does not affect the de-

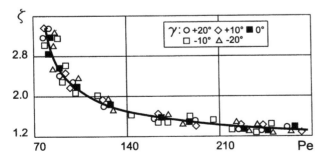

Figure 4   Chip compressing ratio (ζ) and the relative tool chip contact area (c/a where c is the tool–chip contact length and a is the uncut chip thickness) as functions of the Peclet criterion (Pe). Work material—steel AISI 52100.

formation of the chip because the chip compression ratio does not change. This proves the conclusion made in Ref. 5 that there is no a so-called secondary shearing in metal cutting. The observed chip contact layer, which is in the known publications (for example, Ref. 2), is used as an evidence of the existence of such a zone, forms, in reality, in the chip formation zone [5], and then just spreads over the tool–chip interface.

The cutting feed has marked effects on the system efficiency for difficult-to-machine and brittle materials, while the influence of cutting speed and the depth of cut are more noticeable for easy-to-machine materials. Figure 3 also confirms the well-known practical finding that one should use as high a feed as allowed by the strength of the tool, quality of machining, and other constraints in order to increase the efficiency of the cutting system. Although the influence of the cutting speed seems to be at odds with the existing perception of machining, it is correct because an increase in the cutting speed causes higher heat generation on the tool–chip and tool–workpiece interfaces. Moreover, because the chip compression ratio decreases with the cutting speed (Fig. 4), the relative tool–chip velocity increases even further causing the intensification of heat generation at the tool–chip interface.

Figure 5 shows the influence of cutting speed on the efficiency of the cutting system. As seen, although this influence depends on many cutting parameters, the influence of the work material is of prime importance. This is natural because a great part of the energy "wasted" in the cutting process is spent on the plastic deformation of the chip and the workpiece

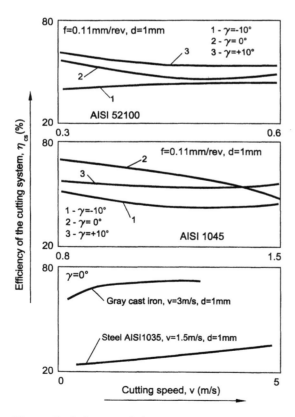

Figure 5 Influence of the cutting speed on the efficiency of the cutting system for different work materials.

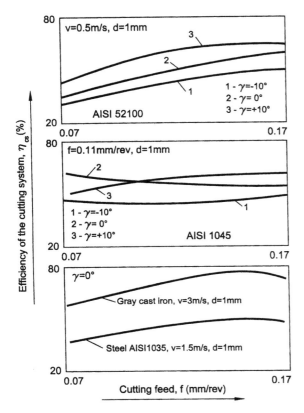

Figure 6 Influence of the cutting feed on the efficiency of the cutting system for different work materials.

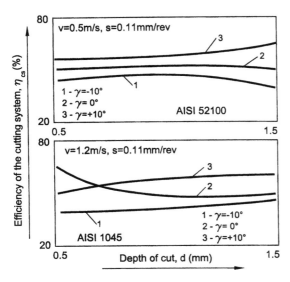

Figure 7 Influence of the depth of cut on the efficiency of the cutting system for different work materials.

surface. Therefore machining of more brittle work materials is more efficient than that of ductile ones. Although this fact is well known in machining practice, the existing metal cutting theory, operating with shear strength and flow shear stress, cannot explain it. However, when one accepts the discussed approach, this problem disappears (as well as many other problems in understanding and implementation of the metal cutting theory).

Figure 6 shows the influence of the cutting feed. As with the cutting speed, the work material influence is the strongest. The influence of the rake angle varies with the cutting feed depending upon the tool rake angle in nonlinear and nonproportional manner. The same can be said on the influence of the depth of cut shown in Fig. 7.

V. CUTTING TOOL WEAR

In plastic deformation processes, concern over the wear is often overshadowed by considerations of forces or material flow. Except for hot extrusion, die life is measured in hours and days or in thousands of parts. In metal cutting, however, tool wear is the dominant concern because process conditions are chosen to give maximum productivity or economy, often resulting in tool life in minutes. Central to the problem are the following: high contact temperatures at the tool–chip–workpiece interfaces which lead to softening of the tool material and promote diffusion and chemical (oxidation) wear; high contact pressures at these interfaces and sliding of fresh, not previously encountered work material layers promote abrasive and adhesion wear; and cyclic nature of the chip formation process leads to the temperature and cutting force fluctuations at the tool–chip–workpiece interfaces, which can cause cracking due to thermal fatigue. The mentioned wear mechanisms (well discussed in Ref. 34) may be active alone or, more frequently, in combination.

The nature of tool wear, unfortunately, is not clear enough yet in spite of numerous investigations. Although various theories have been introduced hitherto to explain the wear mechanism, the complicity of the processes in the cutting zone hampers formulation of a sound theory of cutting tool wear. Cutting tool wear is a result of complicated physical, chemical, and thermomechanical phenomena. Because different "simple" mechanisms of wear (adhesion, abrasion, diffusion, oxidation, etc.) act simultaneously with predominant influence of one or more of them in different situations, identification of the dominant mechanism is far from simple, and most interpretations are subject to controversy [34]. As a result, experimental or postprocess methods are still dominant in the studies of tool, and only topological or simply geometrical parameters of tool wear are selected and thus reported in tool-life consideration. Therefore proper measurements and reporting should be of prime concern in tool wear considerations.

A. Types of Tool Wear

The principal types of tool wear, classified according to the regions of the tool they affect, are the following:

- Rake face or crater wear (Fig. 8(a)) produces a wear crater on the tool rake face. According to Standard ANSI/ASME B94.55M-1985 [35], the depth of the crater KT is selected as the tool-life criterion for sintered carbide tools. The other two parameters, namely, the crater width KB and the crater center distance KM, are important if the tool undergoes resharpening.

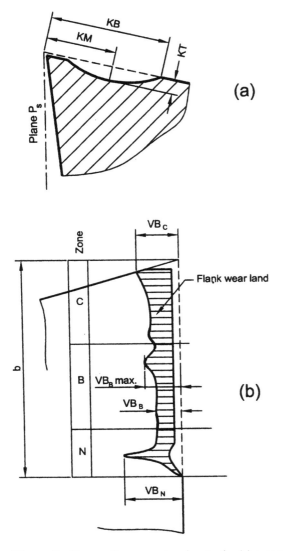

Figure 8 Types of wear on turning tools: (a) crater wear; (b) flank wear.

- Relief face or flank wear (Fig. 8(b)) results in the formation of a wear land. For the purpose of wear measurement [35], the major cutting edge is considered to be divided into the following three zones: (a) Zone C is the curved part of the cutting edge at the tool corner; (b) Zone N is the quarter of the worn cutting edge length b farthest away from the tool corner; and (c) Zone B is the remaining straight part of the cutting edge between Zones C and N. The maximum VB_{Bmax} and the average VB_B width of the flank wear are measured in Zone B, the notch wear VB_N is measured in Zone N, and the tool corner wear VB_C is measured in Zone C.

B. Characterization of Tool Wear

The proper assessment of tool wear requires some characteristics. The selection of these characteristics depends upon a particular objective of the tool wear study. Most often, dimensional accuracy dictates this selection, i.e., the need to manufacture parts within the tolerance limits assigned for tool wear. As such, tool life defined by this criterion may be referred to as dimensional tool life. Dimensional tool life can be characterized by the time within which the tool works without adjustment or replacement (T_{c-1}), by the number of parts produced (N_{p-1}), by the length of the tool path (L_{c-1}), by the area of the machined surface (A_{c-1}), and by the linear relative wear (h_{1-r}). All the listed characteristics are particular and thus, in general, do not allow the optimal control of cutting operations, comparison of different cutting regimes, the assessment of different tool materials, etc. For example, dimensional tool life of little help if one needs to compare cutting tools that worked at different cutting speeds and feeds and/or when the flank wear lands are not the same. The dimensional wear rate, the relative surface wear, and the specific dimensional tool life are much more general characteristics [2] to be used in metal cutting tests conducted everywhere from the research laboratory to the shop floor level.

The dimensional wear rate is the rate of shortening of the cutting tip in the direction perpendicular to the machine surface taken within the normal wear period, i.e.,

$$v_h = \frac{dv_r}{dT} = \frac{h_r - h_{r-i}}{T - T_i} = \frac{vh_{l-r}}{1000} = \frac{vfh_s}{100} \quad (\mu m/min) \tag{8}$$

where h_r and h_{r-i} are the current and initial radial wear, respectively, T and T_i are the total and initial operational time, respectively, and h_s is the relative surface wear.

As seen, the dimensional wear rate is reverse proportional to tool life and does not depend, as tool life, on the selected wear criterion (a particular width of the flank wear land, for example).

The relative surface wear is the radial wear per 1000 sm^2 of the machined surface

$$h_{vs} = \frac{dh_r}{dS} = \frac{(h_r - h_{r-i})100}{(l - l_i)f} \quad (\mu m/10^3 \, sm^2) \tag{9}$$

where h_{r-i} and l_i are the initial radial wear and initial machined length, respectively, and l is the total machined length.

As seen, the relative surface wear is reverse proportional to the overall machined surface and, in contrast to it, does not depend on the selected wear criterion.

The specific dimensional tool life is the area of the workpiece machined by the tool per 1 μm of its radial wear

$$T_{UD} = \frac{dS}{dh_r} = \frac{1}{h_s} = \frac{(l - l_i)f}{(h_r - h_{r-i})100} \quad (10^3 \, sm^2/\mu m) \tag{10}$$

The relative surface wear and the unit dimensional tool life are versatile tool wear characteristics because they allow comparison of different tool materials for different combinations of the cutting speed and feed for different selected criteria for tool life. Table 1 presents a comparison of different characteristics of tool life.

It is possible to use the width of the wear land at the tool point (nose) (current VB_C and initial VB_{C-i}) instead of the radial wear (Fig. 9), i.e.,

$$h_{vs} = \frac{(VB_C - VB_{C-i})100}{(l - l_i)f} \quad (\mu m/10^3 \, sm^2) \tag{11}$$

Table 1 Comparison of Different Characterizations of Tool Life

Characteristic	Designation/ equation	Restriction factors				Possibility to use in calculations of dimension accuracy
		Cutting speed, v (m/min)	Cutting feed, f (mm/rev)	Dimensions of the machined part (surface)	Tool wear VB_m or WB_r	
Machining time without adjustment or replacement of the tool (min)	$T_{c\text{-}l}$	+	+	−	+	No
Number of part produced without adjustment or replacement of the tool	$N_{p\text{-}l}$	−	−	+	+	No
Length of the tool path	$L_{c\text{-}l} = vT_{c\text{-}l}$	−	+	−	+	No
Area of the machined surface	$A_{c\text{-}l} = 10vT_{c\text{-}l}f$	−	−	−	+	No
Linear relative wear	$h_{l\text{-}r} = \dfrac{(h_r - h_{r\text{-}i})1000}{(l - l_i)}$	−	+	−	−	Yes
Dimensional wear rate	$v_h = \dfrac{vh_{l\text{-}r}}{1000}$	+	+	−	−	Yes
Relative surface wear	$h_{vs} = \dfrac{(h_r - h_{r\text{-}i})100}{(l - l_i)f}$	−	−	−	−	Yes
Specific dimensional tool life	$T_{UD} = \dfrac{(l - l_i)f}{(h_r - h_{r\text{-}i})100}$	−	−	−	−	Yes

Note that "+" means that the restrictive factors should be kept the same when using this characteristic for the comparison cutting tools and regimes.

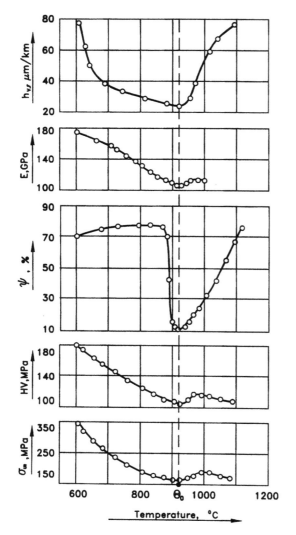

Figure 9 Effect of the temperature on the mechanical properties (ultimate stress, σ_{UTS}, elongation ψ, and hardness HV) of the work material and tool volumetric wear per length of cutting path (h_{vF}). Work material: pure iron; tool material: carbide P10 (89% WC, 15% TiC, 6% Co). (From Ref. 5.)

Although such a substitution is correct, it is difficult to correlate VB_C with the dimensional accuracy of machining. This is due to the plastic lowering of the cutting edge.

   It is possible to use the width of the wear land at the tool point (nose) (current WB_r and initial WB_{r-i}) instead of the radial wear, i.e.,

$$h_{vs-f} = \frac{(h_r - h_{r-i})100}{(l - l_i)f} \quad (\mu m/10^3 sm^2) \tag{12}$$

Although such a substitution is correct, it is difficult to correlate WB_r with the dimensional accuracy of machining. This is due to the plastic lowering of the cutting edge.

Makarow [36] proposed a new characteristic for assessment of tool wear. This is called the instant dimensional wear, h_{ins}, and calculates as the ratio of the dimensional wear rate and the cutting speed, i.e.,

$$h_{ins} = \frac{v_h}{v} = \frac{\dfrac{vh_{l\text{-}r}}{1000}}{v} = \frac{h_{l\text{-}r}}{1000} \quad (\mu m) \tag{13}$$

C. Contact Stresses at the Flank Contact Surface

The greatest advances in the consideration of the flank forces in the machining were made by Makarow [36] who probably, for the first time, correlated the processes on the rake face and on the flank contact surfaces considering various interrelationships among cutting phenomena. On the basis of his findings, he formulated the law which was presented as the first metal cutting law (the Makarow's law) by Astakhov [5]: For given combination of the tool and workpiece materials, there is the cutting temperature, referred to as the optimum cutting temperature, at which the combination of minimum tool wear, minimum stabilized cutting force, and highest quality of the machined surface is achieved. This temperature is invariant to the way it has been achieved. Because there is a very strong correlation between the cutting temperature and the cutting speed, we can say that the cutting speed at which the optimum cutting temperature is achieved can be referred to as the optimum cutting speed, v_{opt}. In this law, the cutting temperature is determined as the average integral temperature on the tool–chip interface so that it can be measured by the tool–work thermocouple technique [5].

Figure 9 illustrates the physics behind OCT. As seen, there are a number of important workpiece properties that, at this temperature, become favorable in terms of metal cutting.

A cutting regime corresponding to the optimum cutting temperature (OCT) is referred to as the optimum cutting regime and, as the optimum temperature is invariant to the way it has been achieved, there is an infinite number of possible optimum regimes. Therefore independence of OCT on the geometry of the cutting tool, the coolant used in machining, cutting speed and cutting feed, etc. makes it possible to optimize the cutting process using a physical criterion of optimization [5]. Any departure from OCT (to higher or lower cutting temperatures) leads to an increase in tool wear. This explains the known fact that the application of the cutting fluid does not necessarily reduce the tool wear.

Optimum cutting temperature can be defined using a tool-life test where the volumetric tool wear per unit length of the tool path is established as a function of the cutting temperature. The temperature corresponding to the minimum of tool wear is considered as the OCT. However, a complete tool-life test is expensive and time-consuming, although it has to be carried out only once for a given combination of tool and workpiece materials. Therefore it has been suggested to conduct a test to determine OCT at constant depth of cut and feed rate varying the cutting speed and measuring the tool wear, main component of the cutting speed, and roughness of the machined surface simultaneously [5]. Figure 10 shows an example of the results where OCT is determined to be 875°C. As seen, OCT can be defined as corresponding the minimum stabilized cutting force. OCT can also be found without an actual cutting experiment [5]. If the microhardness of tool and work materials are considered as functions of temperature, OCT can be thought of as the temperature corresponding to the maximum difference of these microhardnesses as shown in Fig. 11.

In the author's opinion, the optimum cutting speed v_{opt} defined this way is the only technically sound cutting speed because the other known optimum cutting speed are determined using highest productivity, minimum cost per part, etc. and thus are associated with operating rather than technical conditions. The introduced parameter seems to be the

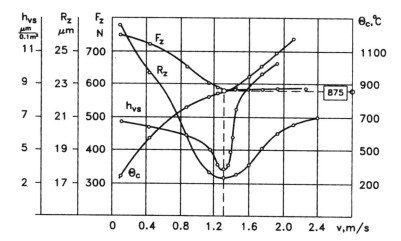

Figure 10 Experimental determination of OCS using longitudinal turning of 4340 steel. Tool material: carbide P10. Cutting tool regime: feed $f = 0.15$ mm/rev, depth of cut $t_1 = 1$ mm. Tool geometry: normal rake angle $\gamma = 10°$, normal flank angle $\alpha = 8°$, cutting edge angle $\kappa_r = 45°$, minor cutting edge angle $\kappa_{r1} = 25°$, nose radius $r_n = 1$ mm. (From Ref. 5.)

only proper technical parameter to compare different grade of carbides, different cutting regimes, and machinability of different work materials. There is no other way to compare different tool materials available today. For example, the answer to a simple question of how to compare different carbide grades does not exist: should one compare them using the same cutting regime which can be suitable for one grade and completely unacceptable for the other? If the comparison is carried out using different cutting regime, then one should know how to compare the results.

To obtain an idea about process at the tool flank, the forces acting on this flank should be considered. The normal force is the result of the work material resistance to tool

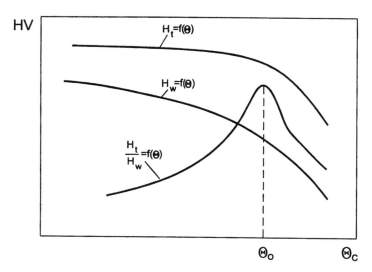

Figure 11 Method of OCT determination based on hardness measurements. (From Ref. 5.)

penetration into the workpiece. This force depends upon compression yield strength of the surface layer of the work material, tool–workpiece contact area, and curvature of the edges of the flank wear land. The normal force defines the normal stress at the tool–workpiece interface and through the apparent friction coefficient, the tangential force and shear stress distribution of this interface.

The properties of the work and tool materials, tool geometry, as well as the cutting regime determine the contact phenomena of the tool–workpiece interface. As such, the cutting speed has the strongest influence [5]. However, this influence is not obvious. To clarify the issue, a series of turning tests were carried out. The test setup, methodology, and conditions were the same as discussed earlier in Ref. 33. The experimental results obtained using different groups of work materials show that the influence of the cutting speed on the contact characteristic at the flank–workpiece interface cannot be generalized because it differs considerably from one work material to another. To illustrate this point, Fig. 12 shows the influence of the cutting speed on the normal, q_{e-f}, and shear, q_{s-f}, stresses at the flank contact area and the apparent friction coefficient at the tool–chip interface, μ_{ff}, in machining of high-carbon tool steel AISI W5, while Fig. 13 shows these stresses for heat-resistant nickel-based high alloy Inconel 718. As seen, the normal and shear stress distributions do not follow the similar trend for these materials. However, regardless of the differences in the values and trend of the normal and shear stresses, two important similarities can be observed:

- The minimum tool wear occurs at the optimum cutting speed v_{opt}.
- The apparent friction coefficient reaches its lowest value at this speed v_{opt}.

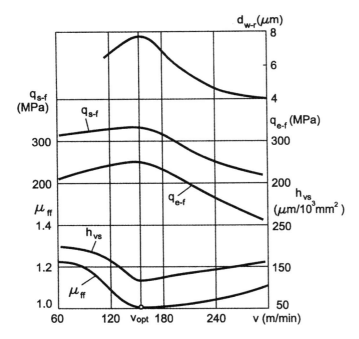

Figure 12 Influence of the cutting speed on the normal, q_{e-f}, and shear, q_{s-f}, stresses at the flank contact area, the apparent friction coefficient, μ_{ff}, and depth of recutting, d_{w-r}, in machining of high-carbon tool steel AISI W5.

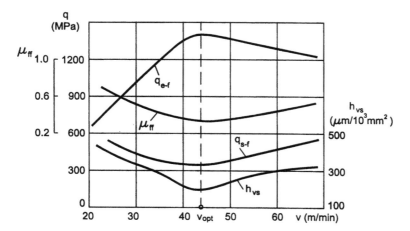

Figure 13 Influence of the cutting speed on the normal, q_{e-f}, and shear, q_{s-f}, stresses at the flank contact area and the apparent friction coefficient, μ_{ff}, in machining of Inconel 718.

The observed phenomena have to be explained. As such, the known works on tool wear which try to suit the Taylor's tool-life equation are of little help simply because they do not consider the physics of metal cutting, trying to substitute this physics by phenomenological observations.

To explain the discussed phenomena, one should recognize that metal cutting is the purposeful fracture of the work material as defined by Astakhov [5]. According to Atkins and May [37] and Komarovsky and Astakhov [31], there is a marked increase in tensile strain to fracture and also in the work of fracture, at about 0.0–0.25 of the melting point (T_m); similar changes occur in other measures of ductility such as Charpy values (CVN) as shown in Fig. 14. It explains a number of "strange" results obtained by Zorev [2] in his tests at low cutting speeds. This phenomenon also explains the great size of the zone of plastic deformation observed at low cutting speeds and incorporated in the model discussed by Astakhov [5]. The known built up edge is the result of the discussed high plasticity region.

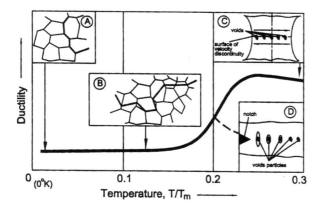

Figure 14 Changes in ductility and typical associated mechanisms of fracture for bss material in the temperature range $< 0.3T_m$. (A) Low-temperature intergranular cracks; (B) twinning or slip leading to cleavage; (C) shear fracture at particles; (D) low-energy shear at particles.

Exceptions are certain fcc metals and alloys (Al, Cu, Ni, and Pb) that do not normally cleave. As such, there is no transition in values, which merely gradually rise with temperature.

The increase in ductility over the "transition temperature range" is followed by a gradual drop beyond about $0.35T_m$ (Fig. 15). It is believed that it happens due to the continuing fall in the Pierls–Nabarro stress which opposes dislocation movement, coupled with the emergence of cross-slip (as opposed to Frank–Read sources) as a dislocation generator as the temperature is raised [37]. In the author's opinion, the cause is in dilations–compressions reactions as explained in Ref. 31.

At high temperatures, grain boundaries become significant. Below about $0.45T_m$, grain boundaries act principally as barriers inhibiting cleavage and causing dislocation pileups. At higher temperatures, the regions of intense deformation, which are contained within the grains at lower temperatures, now shift to the grain boundaries themselves. Voids are nucleated and cracks then develop on the grain boundaries. Shear stresses on the boundaries cause relative sliding of the grains, and voids are reduced in region of stress concentrations (see Fig. 15, position D). Therefore around this temperature region can be referred as the ductility valley. Experiments showed [5] that the reduction of plasticity may reach twofold and even more for high alloys. The presence of this valley is the physical cause of the existence of the optimum cutting temperature.

At temperatures $(0.5–0.6)T_m$, recovery and recrystallization process set in (recovery relates to a redistribution of dislocation sources so that dislocation movement is easier, and in recrystallization, the energy of dislocations generated during prior deformation is used to nucleate and grow new grains, thus effecting an annealed structures over long times). The net effect is increased ductility causing a bump shown in Fig. 15.

The high contact stresses at the tool flank–workpiece interface cause the elastic recovery (sometime referred to as spring back) of the machined surface. This can be observed through the so-called recutting. Recutting is observed when cutting using a fresh-sharpened tool with the minor cutting edge having cutting edge angle $\kappa_{r1}=0°$ as

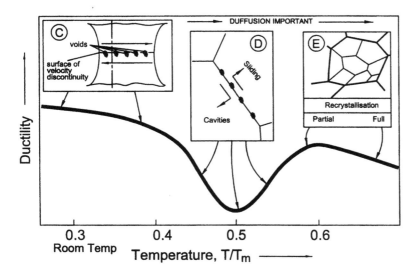

Figure 15 Sequence of events leading to fracture by growth and linkage of grain boundary cavities. Changes in ductility with temperature for fcc, bcc, and hcp alloys at temperatures $> 0.3T_m$. (C) Shear at particles, (D) cavities along grain faces, (E) recrystallization suppresses cavitation.

shown in Fig. 16. As such, the minor cutting edge of length l_m does not simply slide over the machined surface. Rather, it cuts some layer having depth $d_{w\text{-}r}$.

The depth of recutting, $d_{w\text{-}r}$, depends upon the work material hardness and the cutting speed, while it does not practically depend on the uncut chip cross-sectional area (the product of the depth of cut and the cutting feed) [38]. The maximum on the curve $d_{w\text{-}r} = f(v)$ (Fig. 8) is observed at the same cutting speed as the maximum of $q_{e\text{-}f}$. This speed coincides with the optimum cutting speed corresponding to the minimum tool wear. Makarow [36] showed that for carbon steels (tool material—P20, $l_m = 1$ mm, $f = 0.78$ mm/rev), the depth of recutting, $d_{w\text{-}r}$, calculates as

$$d_{w\text{-}r} = 0.11\sigma_{uts} \quad (\mu m) \tag{14}$$

while for hardened tool carbon alloys (tool material—P30, $l_m = 2.88$ mm, $f = 1.59$ mm/rev, $\theta_{opt} = 750°C$),

$$d_{w\text{-}r} = 0.19 HRC \quad (\mu m) \tag{15}$$

D. Plastic Lowering of the Cutting Edge

Normally, abrasion, adhesion, diffusion, and oxidation types of cutting tool wear are discussed in literature on metal cutting. However, in machining of difficult-to-machine materials and in high-speed machining, the plastic lowering of the cutting edge is the predominant cause of tool premature breakage. This is due to the plastic deformation of the cutting wedge (a part of the tool between the rake and the flank contact surfaces). Figure 17 shows the result of plastic deformation of the cutting wedge which is known as its plastic lowering. As seen, plastic lowering is characterized by actual h_f and apparent h_y wear lands. As such, the tool rake and flank angle change and these changes are characterized by the following parameters: l_α, h_α, and l_y as shown in Fig. 17. As known [5], when temperatures at the tool–chip interface reach 1000–1200°C, the cutting wedge deforms plastically. The cause of this deformation is high-temperature creep.

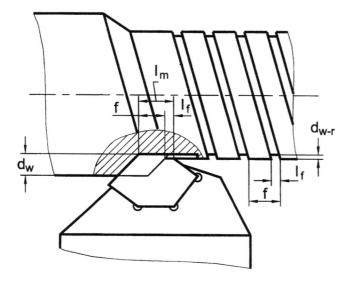

Figure 16   Formation of the machined surface under recutting.

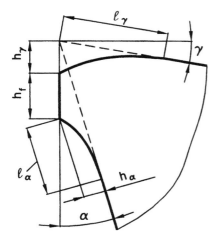

Figure 17   Plastic lowering of the cutting edge.

Creep is progressive deformation of a material at constant stress. The engineering creep curve shown in Fig. 18 represents the dependence of plastic deformation of a metal when constant load and temperature are applied. As seen, upon loading of a preheated specimen, deformation increases rapidly from zero to a certain value ε_0 known as the initial rapid elongation [31]. There is no need for additional energy for this deformation because it occurs due to the thermal energy that already exists in the specimen, so the work done by the internal forces begins with the level of energy that has already been achieved. In other words, if the temperature is a characteristic of the thermal energy and deformation and stress characterize the work done by the external forces, then the critical amount of energy accumulates in the material as the result of their summation.

Among the phases, normally present in a tool carbide, the plastic deformation is greater in the cobalt phase as it follows from Fig. 19. This plastic deformation results in tearing off of the grains of carbide from deforming cobalt layers, "ploughing" this deforming layer by hard inclusions contacting in the work material, and "spreading" of the tool material on the chip and workpiece contact surfaces. If the temperature increases further, a

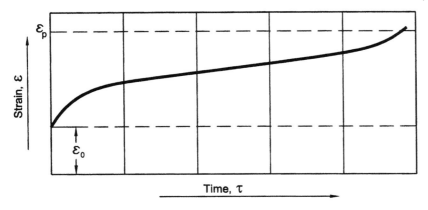

Figure 18   Engineering creep curve.

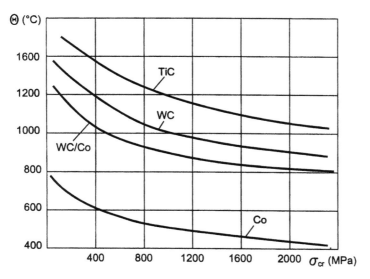

Figure 19   Creep resistance of different phases of a carbide.

liquid layer forms between tool and workpiece, which is quickly removed [36]. This liquid layer is due to diffusion leading to the formation of low-melting-point compound Fe_2W having melting temperature $T_M = 1130°C$.

The introduced cause for the plastic lowering of the cutting edge at high temperatures has an excellent experimental confirmation known (but not explained) from everyday machining practice. Although fine grain carbides are superior over those having coarse grains in terms of wear resistance, coarse grain carbides, having higher creep resistance, exhibit much smaller plastic lowering of the cutting edge under the same cutting conditions.

E. Resource of the Cutting Wedge

As follows from the foregoing analysis, the energy flows to the zone of the fracture of the layer being removed through the cutting wedge (Fig. 2). Out of three components of the cutting system, the only component that has an invariable mass of material and which is continuously loaded during the process is the cutting wedge. As such, the overall amount of energy, which can be transmitted through this wedge, is entirely determined by the physical and mechanical properties of the tool material.

On the contrary, the material of the chip is not subjected to the same external force because the chip is an ever-grooving component; that is, new and new sections are added to the chip during each cycle of chip formation [5], while "old" sections move out of the tool–chip contact with the cutting wedge and thus do not experience the external load. The same can be said about the workpiece in which volume and thus mass change during the cutting process as well as the area of load application imposed by the cutting tool.

When the cutting wedge looses its cutting ability due to wear or plastic lowering of the cutting edge (creep, as considered above, Fig. 17), the work done by the external forces that causes such a failure is regarded as the critical work. For a given cutting wedge, this work (or energy) is a constant value. The resource of the cutting wedge therefore can be represented by this critical work.

According to the principle of physical theory of reliability [31], each component of a system has its resource spent during operation time at certain rate depending on the operating conditions. This principle is valid for a wide variety of operating conditions providing that changes from one operating regime to another do not lead to any structural changes in materials properties (reaching the critical temperatures, limiting loads, chemical transformations, etc.). As such, the resource of a cutting tool, r_{ct}, can be considered as a constant, which does not depend on a particular way of its consumption, i.e.,

$$r_{ct} = \int_0^{\tau_1} f(\tau, R_1) d\tau = \int_0^{\tau_2} f(\tau, R_2) d\tau \qquad (16)$$

where τ_1 and τ_2 are the total operating time on the operating regime R_1 and R_2, respectively, till the resource of the cutting tool is exhausted.

If the initial recourse of the cutting tool is represented by the above-discussed critical energy, the flow of energy through the cutting tool exhausts this resource. The amount of energy that flows through the cutting tool depends on the energy to fracture, U_f, of the layer being removed, which, in turn, defined the total energy, U_{cs} (Fig. 2), required by the cutting system to exist. Therefore there should be a strong correlation between a parameter (or metric) characterizing the resource of the cutting tool (for example, flank wear VB) and the total energy, U_{cs}.

A series of cutting tests were carried out to prove this hypothesis experimentally. The test setup, methodology, and conditions were the same as discussed earlier in Ref. 33. The work material was steel AISI 52100 and the cutting tool material was carbide P10 (cutting inserts SNMM120404). The tool wear was measured using a toolmaker's microscope equipped with a digital video system.

The experimental results are shown in Table 2. As follows from this table, there is a very strong correlation, which does not depend on a particular cutting regime, cutting time, and other parameters of the cutting process, between the total work required by the cutting system and the flank wear. Figure 20 shows the correlation curve.

The discovered correlation between the energy passed through the cutting wedge and its wear can be used for the prediction of tool life and optimum cutting speed allowing avoiding expensive and time-consuming tool-life tests. Moreover, the multiple experimental results, obtained in machining of different work materials using different cutting tools, proved that this correlation holds regardless of the particular manner the resource of the tool was spent. This finding allows to introduce a new method for the determination of tool life. The essence of the method can be described as follows.

The energy required by the cutting system during the time period corresponding to tool life T_{ct} can be represented as

$$U_{cs} = W_{cs} T_{ct} \qquad (17)$$

where W_{cs} is the power required by the cutting system (W). Note that U_{cs}, when selected for a given tool material using the correlation curve similar to that given in Fig. 20 for the accepted tool-life criterion, is a sole characteristic of the tool material, i.e., its resource, and thus can be used for calculating tool life in cutting different work materials.

The power of the cutting system, W_{cs}, is determined as a product of the power component of the cutting force (often referred to as the cutting force), P_z, and the cutting speed, v, i.e.,

$$W_{cs} = P_z v \qquad (18)$$

Table 2 Conditions of Tests and Experimental Results

Test no.	Feed (mm/rev)	Depth of cut (mm)	Operating time (sec)	Flank wear VB (mm)	Energy of the cutting system (kJ)
1	0.07	0.1	8540	0.45	0.88
2	0.07	0.1	6680	0.41	0.63
3	0.07	0.1	4980	0.39	0.52
4	0.07	0.1	1640	0.29	0.25
5	0.07	0.1	9120	0.45	0.91
6	0.07	0.1	7660	0.42	0.68
7	0.07	0.1	6260	0.41	0.55
8	0.07	0.1	4900	0.37	0.41
9	0.07	0.1	3450	0.35	0.47
10	0.07	0.1	5380	0.38	0.65
11	0.07	0.1	4240	0.34	0.35
12	0.07	0.1	3150	0.30	0.33
13	0.07	0.1	2075	0.26	0.24
14	0.07	0.1	1036	0.20	0.15
15	0.07	0.1	2980	0.37	0.37
16	0.07	0.1	1940	0.32	0.30
17	0.07	0.1	1190	0.27	0.17
18	0.09	0.1	938	0.15	0.05
19	0.09	0.1	1874	0.18	0.10
20	0.09	0.1	2840	0.22	0.22
21	0.09	0.1	3820	0.25	0.17
22	0.09	0.1	4810	0.31	0.39
23	0.12	0.1	775	0.20	0.08
24	0.12	0.1	1520	0.23	0.18
25	0.12	0.1	2350	0.26	0.29
26	0.12	0.1	3220	0.27	0.30
27	0.14	0.1	675	0.20	0.18
28	0.14	0.1	1315	0.21	0.12
29	0.07	0.2	1295	0.21	0.13
30	0.07	0.2	2610	0.27	0.30
31	0.07	0.2	5420	0.44	0.82
32	0.07	0.8	1316	0.20	0.19

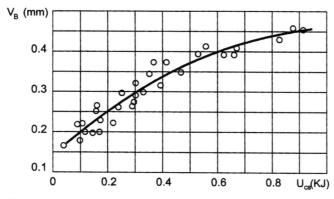

Figure 20 Correlation curve.

In turn, the power components of the cutting force can be determined experimentally depending upon the cutting parameters as [33]

$$P_z = C_{P_z} d^{n_z} f^{m_z} v^{k_z} \tag{19}$$

where C_{P_z} is the constant of the work material and n_z, m_z, and k_z are exponents.

Substituting Eqs. (18) and (19) into Eq. (17) and expressing tool life, one can obtain an equation which determines tool life for a given cutting regime

$$T_{ct} = \frac{U_{cs}}{C_{P_z} d^{n_z} f^{m_z} v^{k_z+1}} \tag{20}$$

If it is necessary to know the cutting speed corresponding desired tool life, then Eq. (20) can be expressed as

$$v = \left(\frac{U_{cs}}{C_{P_z} d^{n_z} f^{m_z} T_{ct}} \right)^{\frac{1}{k_2+1}} \tag{21}$$

It is obvious that U_{cs} is selected depending on the tool flank wear and depends not only on the properties of the tool material, but also on the tool geometry. Therefore the correlation curve $U_{cs} = f(V_B)$ should be corrected accounting on the particular tool geometry. As a result, there are countless numbers of possible combinations "cutting tool material–tool geometry" to account for the influence of the tool geometry. To avoid the influence of tool geometry, the volume of worn tool material, V_W, can be used.

The results of foregoing analysis suggest that a prospective way to achieve repeatability of cutting tools with inserts is the certification of cutting inserts of standard shapes. The number of standard shapes of cutting inserts (including their geometry) is relatively small so each insert producer should be able to provide a correlation curve $U = f(VB)$ for each shape and tool material. Table 3 shows an example of such correlations for different shapes and tool materials of inserts. The correlation curves in this table were obtained for $VB \le 0.4$ mm. The data presented in Table 3 are valid under the condition that the tool material does not loose its cutting properties due to excessive temperature.

F. Wear Curves

Tool wear curves illustrate the relationship between the amount of wear of the tool flank (rake) face and time of cutting τ_m or the overall length of cutting path L. These curves are

Table 3 Correlation Curves for Some Tool Materials

Tool material	ISO code of the shape	Correlation curve	Critical temperature (°C)
TS332 (Al_2O_3, 2300HV)	SNMN 120404M	$U = \exp(9.6VB^2)$	1200
VOK60 ($Al_2O_3 + TiC$ 94 HRA)	SNMN 120404M	$U = \exp(10.91VB)$	1200
Silinit-P ($Si_3N_4 + Al_2O_3$, 96 HRA)	SNMN 120404M	$U = 573VB^2$	1200
TN20 (75%TiC,15%Ni, 10%Co, 90HRA)	SNMN 120404M	$U = 434.46 \cdot 10^{-3}VB^2$	780
Kiborit (96% cBN, KNH 32–36 GPa)	RNMM 1200404M	$U = 50VB^{1/2}$	1400

represented in linear coordinate systems using the results of cutting tests where flank wear VB_{Bmax} is measured after certain periods of time (Fig. 21(a)) or after certain length of the cutting path (Fig. 21(b)). Such curves often have three general regions. The first region (I in Fig. 21(b)) is the region of primary or initial wear. Relatively high rate of tool wear in this region is explained by accelerated wear of the tool layers damaged during its sharpening (regrinding). The second region (II in Fig. 21(b)) is the region of steady-state wear. This is the normal operating region for the tool. The third region (III in Fig. 21(b)) is known as the tertiary or accelerated wear region. Accelerated tool wear in this region is usually accompanied by high cutting forces, temperatures, and sever tool vibrations. Normally, the tool should not be used within this region.

Tool wear depends not only on the cutting time or the length of the cutting path, but also on the parameters of tool geometry (rake, flank, inclination angles, etc.), cutting regimes (cutting speed, feed, depth of cut), and properties of the work material (hardness, toughness, structure, etc.). In practice, however, the cutting speed is of prime concern in the considerations of tool wear. As such, tool wear curves are constructed for different cutting speeds keeping other machining parameters invariable. In Fig. 22, three characteristic tool wear curves (average values) are shown for three different cutting speeds, v_1 through v_3. v_3 is the fastest cutting speed and therefore corresponds to the fastest wear rates. When the amount of wear reaches the permissible tool wear VB_{Bc}, the tool is said to be worn out.

VB_{Bc} is typically 0.15 to 1 mm. It is often selected on the ground of process efficiency and often called the criterion of tool life. In Fig. 22, T_1 is tool life when the cutting speed v_1 is

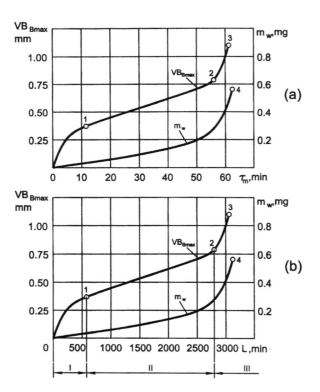

Figure 21 Typical tool wear curves for flank wear: (a) as a function of time; (b) as a function of cutting path.

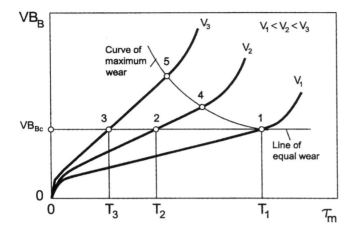

Figure 22 Flank wear vs. cutting time at various speeds.

used, T_2 when v_2 is used, and T_3 when v_3 is used. When the integrity of the machined surface permits, the curve of maximum wear instead of the line of equal wear should be used (Fig. 22). As such, the spread in tool life between lower and higher cutting speed becomes less significant. As a result, a higher productivity rate can be achieved which is particularly important when high-speed CNC machines are used.

VI. METHODS OF INCREASING TOOL LIFE

Among many known methods of increasing tool life, the coating of cutting tools and cutting fluid (coolant) applications are of prime concern. These two are briefly discussed in this section.

A. Coatings

Some of the newest methods to increase the life of cutting tools are the thin film hard coatings and the thermal diffusion processes. These methods find ever-increasing applications and brought significant advantages to their users (Fig. 23). The thin film hard coatings are nitride or carbide-based ceramics, with a thickness of 3–10 μm. They are applied by three types of techniques:

- Chemical vapor deposition (CVD), in which the components of the coating (e.g., titanium and nitrogen) are supplied in gaseous form, and the thermochemical reaction to form the coating is produced on the surface of the tool, heated to approximately 1000°C.
- Physical vapor deposition (PVD), in which the metal component of the coating is produced from solid, in a high-vacuum environment. The generation of the metal atoms is accomplished by evaporation or ion bombardment methods, at temperature of approximately 500°C.
- The thermal diffusion (TD) is applied in a molten borax bath, with addition of vanadium, at approximately 1000°C. The resultant vanadium carbide coating has very good results in numerous applications.

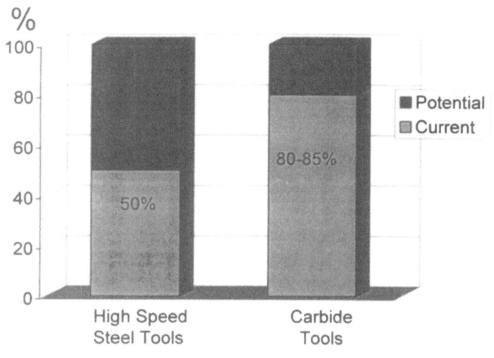

Figure 23   Coated tool usage in the United States.

Chemical vapor deposition coatings have been commercially available for about 30 years, and the fact that more than half of the inserts sold are CVD-coated testifies to the effectiveness of these coatings. However, the high temperatures (about 1000°C) involved in the CVD process create an embrittlement called "eta phase" at the coating/substrate interface. Depending on its extent, the embrittlement can affect performance in operations involving interruptions of cut and inconsistency of workpiece microstructure such as found in some nodular irons. Recently developed medium temperature CVD (MTCVD) coatings have shown a reduced tendency to the formation of eta phase. MTCVD-coated tools offer increased resistance to thermal shock and edge chipping compared to conventional CVD-coated tools. The result is greater tool life as well as increased toughness compared to high-temperature CVD coatings [15].

Physical vapor deposition (PVD) coatings also offer advantages over CVD coatings in certain operations and/or workpiece materials. Commercialized in the mid-1980s, the PVD coating process involves relatively low deposition temperatures (approximately 500°C) and permits coating of sharp insert edges. (CVD-coated insert edges are usually honed before coating to minimize the effect of eta phase.) Sharp, strong insert edges are essential in operations such as broaching, gear shaving, milling, drilling, threading and cutoff, and for effective cutting of the so-called "long-chip" materials such as low-carbon steels. In fact, a wide range of "problem" materials—such as titanium, nickel-based high alloys, and nonferrous materials—can be productively machined with PVD-coated tools. From a workpiece structure point of view, sharp edges reduce cutting forces, so PVD-coated tools can offer a true advantage when machining thin-wall components or when the machining residual stresses are the issue.

The first PVD coatings were titanium nitride (TiN), but more recently developed PVD technologies include titanium carbonitride (TiCN) and titanium aluminum nitride (TiAlN), which offer higher hardness, increased toughness, and improved wear resistance. TiAlN tools, in particular, through their higher chemical stability, offer increased resistance to chemical wear and thereby increased capability for higher speeds. Recent developments in PVD coatings include "soft" coatings such as molybdenum disulfide (MoS_2) for dry drilling applications. Combination of soft/hard coatings, such as MoS_2 over a PVD TiN or TiAlN, also demonstrated great potential, as the hard (TiN or TiAlN) coating provides wear resistance, while the softer, more lubricious outer layer expedites chip flow [15–17]. The basic PVD coating are listed in Table 4 and their properties are shown in Table 5. Effectiveness of various coatings on cermet cutting tools is discussed in Ref. 39.

It should be pointed out, however, that a great variety of coatings available in the market make the selection of the most suitable one for a particular application very cumbersome. The trial-and-error method is widely used in such a selection simply because the coating properties are poorly correlated with cutting conditions. There could be many controversial opinions on the same coating used in "similar" application (sometimes, it works very well; sometimes, it does not help at all).

Table 4 Basic PVD Coatings

Titanium nitride TiN	The gold-colored coating offers excellent wear resistance with a wide range of materials and allows the use of higher feeds and speeds. Forming operations can expect a decrease in galling and welding of workpiece material with a corresponding improvement in surface finish of the formed part. A conservative estimate of tool life increase is 200% to 300%, although some applications see as high as 800%.
Titanium carbonitride TiN(C,N)	Bronze-colored Ti(C,N) offers improved wear resistance with abrasive, adhesive, or difficult-to-machine materials such as cast iron, alloys, tool steels, copper and its alloys, Inconel, and titanium alloys. As with TiN, feeds and speeds can be increased and tool life can improve by as much as 800%. Forming operations with abrasive materials should see improvements beyond those experienced with TiN.
Titanium aluminum nitride (Ti,Al)N	Purple/black in color, (Ti,Al)N is a high-performance coating which excels at machining of abrasive and difficult-to-machine materials such as cast iron, aluminum alloys, tool steels, and nickel alloys. (Ti,Al)N's improved ductility makes it an excellent choice for interrupted operations, while its superior oxidation resistance provides unparalleled performance in high-temperature machining.
Chromium nitride CrN	Silver in color CrN offers high thermal stability, which in turn helps in the aluminum die casting and deep draw applications. It can also reduce edge buildup commonly associated with machining titanium alloys with Ti-based coatings.

Table 5 Basic Physical Properties of PVD Coatings

Property	Titanium nitride TiN	Titanium carbonitride TiN(C,N)	Titanium aluminum nitride (Ti,Al)N	Chromium nitride CrN
Color	Gold	Bronze	Purple/black	Silver
Hardness	2800HV	3000HV	2800HV	2000HV to 2200HV
Coating thickness (μm)	2–4	2–4	2–4	3–5
Thermal stability	550°C (1000°F)	400°C (750°F)	750°C (1350°F)	800°C (1470°F)
Lubricity TiN/steel	0.4–0.55	0.5–0.6	0.5–0.6	0.55–0.65
Deposition temperature	500°C (930°F)	500°C (930°F)	500°C (930°F)	350°C (660°F)
Cost comparison	Base	1.5 × Base	2 × Base	1.75 × Base

It is worthwhile to mention here the following.

- A specific coating would be beneficial if and only if it is properly used. If this is the case, tool life of the coated tool increases 2–3 times compared to that of the uncoated tool. In some cases, increases of 10% to 50% in productivity have been demonstrated in some applications.
- Coatings will typically not solve tooling problems and thus efforts in this direction are fruitless. In other words, the selection of the proper cutting tool and machining regime for the cutting operation should be accomplished as a prerequisite for coating applications.
- Coating usually adds 0.6–1 μm of surface roughness. Postcoat polishing is possible; however, no data on this process are available [15].
- Coating does change the dimensions of the cutting tool. Change depends on the coating, its specified thickness, and the coating process. Typically, PVD is recommended for high tolerance tools and CVD for loose tolerance tools. Most PVD coatings add 2–3 μm per side to a tool or component. CVD and PVD CrN are thicker and can add 10 μm or more in some cases. Processing temperatures may grow or shrink some substrate materials. CVD temperatures, in particular, affect the heat treatment conditions of tools and components and can cause dimensional changes.

B. Cutting Fluids (Coolants)

1. General

Cooling and lubrication are important in reducing the severity of the contact processes at the cutting tool–workpiece interfaces. Historically, more than 100 years ago, water was used mainly as a coolant due to its high thermal capacity and availability [8,9]. Corrosion of parts and machines and poor lubrication were the drawbacks of such a coolant. Oils were also used at this time as these have much higher lubricity, but the lower cooling ability and high costs restricted this use to low cutting speed machining operations. Finally, it was found that oil added to the water (with a suitable emulsifier) gives good lubrication properties with the good cooling and these became known as the soluble oils. Other substances are also added to these to control problems such as foaming, bacteria, and fungi. Oils as lubricants for machining were also developed by adding extreme pressure (EP) additives. Today, these two types of cutting fluids (coolants) are known as water-emulsifiable oils and straight cutting oils. Additionally, semisynthetic and synthetic cutting fluids were also developed to improve the performance of many machining operations [10,11].

Although the significance of cutting fluids in machining is widely recognized, cooling lubricants are often regarded as supporting media that are necessary but not important [12]. In many cases, the design or selection of the cutting fluid supply system is based on the assumption that the greater the amount of lubricant used, the better the support for the cutting process. As a result, the contact zone between the workpiece and the tool is often flooded by the cutting fluid without taking into account the requirements of a specific process. Moreover, the selection of the type of the cutting fluid for a particular machining operation is often based upon recommendations of sales representatives of cutting fluid suppliers without clearly understanding the nature of this operation and the clear objectives of cutting fluid application. The brochures and web sites of cutting fluid suppliers are of little help in such selection. The techniques of cutting tool application, which include the cutting fluid pressure, flow rate, nozzles' design, and location with respect to the machining zone, filtration, temperature, etc., are often left to the machine tool designers. Moreover, the machine operators are often those who decide the point of application and flow rate of the cutting fluid for each particular cutting operation.

On the other hand, it was pointed out that the cutting fluids also represent a significant part of manufacturing costs. Just two decades ago, cutting fluids accounted for less than 3% of the cost of most machining processes. These fluids were so cheap that few machine shops gave them much thought. Times have changed, and today, cutting fluids account for up to 15% of a shop production cost [40]. Figure 24 illustrates the cost of production of camshafts in the European automotive industry [12–14]. The conspicuous high share of the costs for cooling lubrication technology reaches 16.9% of the total manufacturing costs. As seen from Fig. 19, the costs of purchase, care, and disposal of cutting fluids are more than two folds higher than tool costs, although the main attention of researchers, engineers, and managers is focused on the improvement of cutting tools. Moreover, cutting fluids, especially those containing oil, have become a huge liability. Not only does the Environmental Protection Agency regulate the disposal of such mixtures, but many states and localities also have classified them as hazardous wastes.

At present, many efforts are being undertaken to develop advanced machining processes using less or no coolants [7,9,41]. Promising alternatives to conventional flood coolant applications are the minimum quantity lubrication (known as MQL) and dry machining technologies. It was pointed out, however, that the use of MSQ will only be acceptable if the main tasks of the cutting fluid [42] (heat removal—cooling, heat and wear reduction—lubrication, chip removal, corrosion protection) in the cutting process are successfully replaced [7]. As such, the understanding of the metal cutting tribology plays a vital role.

2. The Action of Cutting Fluids

A still open question in metal cutting regards the action of cutting fluids. When cutting fluids are applied, the existence of high contact pressure between chip and tool, particularly along the plastic part of the tool–chip contact length, should apparently preclude any fluid access to the rake face. In spite of this, to explain the marked influence which cutting fluids have on

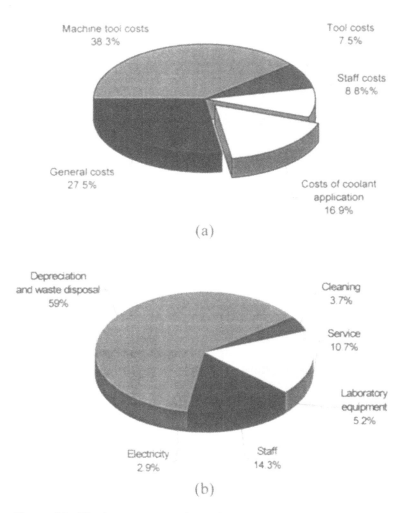

(a)

(b)

Figure 24 Pie-chart representations of: (a) manufacturing cost at the German automotive industry; (b) structure of coolant cost [13,14].

the cutting process outputs (cutting force and temperature, surface finish and residual stresses in the machined surface, and tool wear), the theory considering these fluids as boundary lubricants is still leading [43].

Despite a relatively great number of publications on cutting fluids, only very few of them have been aimed to understand the role of a cutting fluid in the complex mechanics of the cutting process [44–51]. To account for cutting tool penetration to the rake face, four basic mechanisms of cutting fluid access to the rake face have been suggested, namely, access through capillarity network between chip and tool, access through voids connected with buildup edge formation, access into the gap created by tool vibration, and propagation from the chip blackface through distorted lattice structure. However, no conclusive experimental evidences are available to support these suggestions. It was observed that cutting fluids reduce (sometimes) the tool–chip contact length.

To understand the action of cutting fluids, the above-discussed system-based model should be considered as shown in Fig. 25. At the beginning of a chip formation cycle (Phase 1), the actions of the cutting fluid are as follows:

1. Contamination of the rake face at A providing lubricating between the chip and the rake face
2. Contamination of the two chip elements sliding over each other
3. Cooling the zone of plastic deformation at C and thus limiting the flow shear stress in this zone
4. Lubrication and cooling of the flank–workpiece interface at D

At the middle of the cycle (Phase 2), the actions are as follows:

1. Contamination of the rake face at A providing lubricating between the chip and the rake face.
2. Cooling of the free surface of the partially formatted chip at B.
3. Cooling the zones of plastic deformation at C and E that increases the flow shear stress of the work material in these zones [52]. As such, the stress at fracture of the work material is achieved with less plastic deformation that promotes the formation of cracks [5,37] along the surface of the maximum combined stress.
4. Lubrication and cooling of the flank–workpiece interface at D.

At the end of a chip formation cycle (Phase 3), the actions are as follows:

1. Contamination of the rake face at A providing lubricating between the chip and the rake face.
2. Cooling of the free surface of the partially formatted chip at B reducing its plastic deformation and thus the chip compression ratio [2,5].
3. Cooling the zones of plastic deformation at C and E promoting propagation of cracks. When the cutting fluid penetrates into the crack formed in the chip-free surface (Fig. 1), it suppresses the above-discussed healing of these cracks. Our multiple analyses of the chip structures obtained in cutting with and without cutting fluid prove this fact [53].
4. Lubrication and cooling of the flank–workpiece interface at D.

Multiple experimental results are available to prove adequacy of the proposed model [5,52–58].

When the cutting fluid is applied by simple flooding of the machining zone, the weakest actions of the cutting fluid is observed at A and D. The application of a high-pressure cutting fluid jet significantly increases tool life and lowers the cutting forces [59,60].

The relative influence of the cutting fluid actions at A–D significantly depends on the frequency of chip formation and thus on the cutting speed [5]. The higher the cutting speed, the lower the viscosity of the cutting fluid should be in order to penetrate into the above-discussed cracks formed on the chip-free surface. This explains why soluble oils of low viscosity are more efficient at high cutting speeds compared to straight oils.

3. Types of Cutting Fluids

There are five major types of the cutting fluids available today.

1. Straight cutting oils. These are oil-based materials, which generally contain what are called extreme pressure or antiweld additives. These additives react under pressure and heat to give the oil better lubricating characteristics. These straight cutting oils are most

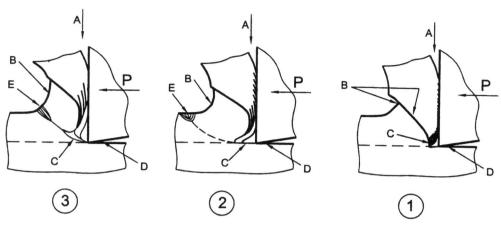

Figure 25 Cutting fluid action at different stages of a chip formation cycle.

often used undiluted. Occasionally, they are diluted with mineral oil, kerosene, or mineral seal oil to either reduce the viscosity or the cost. They will not mix with water and will not form an emulsion with water. The advantages of straight cutting oils are good lubricity, effective antiseizure qualities, good rust and corrosion protection, and stability. Disadvantages are poor cooling, mist and smoke formation at high cutting speeds, and high initial and disposal cost. Straight oils perform best in heavy-duty machining operations and very critical grinding operations where lubricity is very important. These are generally slow-speed operations where the cut is extremely heavy. Some examples would include broaching, threading, gear hobbing, gear cutting, tapping, deep-hole drilling, and gear grinding. Straight oils do not work well in high-speed cutting operations because they do not dissipate heat effectively. Because they are not diluted with water and the carryout rate on parts is high, these oils are costly to use and therefore only used when other types of cutting fluid are not applicable.

2. Water-emulsifiable oils. They are more commonly referred to as soluble oils. This, however, is a misnomer because they are not really soluble in water but rather form an emulsion when added to water. These emulsifiable oils are oil-based concentrates, which contain emulsifiers that allow them to mix with water and form a milky white emulsion. Emulsifiable oils also contain additives similar to those found in straight cutting oils to improve their lubricating properties. They contain rust and corrosion inhibitors and a biocide to help control rancidity problems. Advantages of water-emulsifiable oils are good cooling, low viscosity and thus adequate wetting abilities, nonflammable and nontoxic, easy to clean from small chips and wear particles using standard filters, and relatively low initial and disposal cost. Disadvantages are low lubricity, rancidity, misting, low stability (components have different degradation levels), and, in mass production, requires everyday expensive maintenance in order to keep the required composition. Water-emulsifiable oils are the most popular cutting fluids in use today. Because they combine the lubricating qualities of oil with the cooling properties of water, they can be used in a wide range of both machining and grinding operations.

3. Synthetic fluids. Sometimes referred to as chemical fluids, these synthetic cutting fluids are water-based concentrates, which form a clear or translucent solution when added to water. These fluids contain synthetic water-soluble lubricants, which give them the necessary lubricating properties. In addition, these synthetic fluids contain rust and corrosion

inhibitors, biocides, surfactants, and defoamers. Synthetic cutting fluids do not contain any oil. Advantages of synthetic cutting fluids are resistance to rancidity, low viscosity and thus good cooling and wetting, good rust protection, little misting problems, nontoxic, completely nonflammable and nonsmoking, good filtration with standard filters, and biodegradable. Disadvantages are insufficient lubricity for heavy-duty applications, reaction with nonmetal parts, and residue is often a problem. As disposal problems become an ever increasing problem with the advent of the Resource Conservation and Recovery Act, synthetic fluids, because they present less of a disposal problem than emulsifiable oils, become more popular because synthetics are easier to treat than emulsifiable oils before they can be disposed. Synthetics are most definitely the products of the future. A very large percentage of the development work on cutting fluids is devoted to improving the synthetic fluid technology. However, there are still some problems and still some machining and grinding operations that for one reason or another cannot be done using a synthetic fluid. The major problem is that lubrication has always been the big problem for synthetic coolants. Another problem caused by synthetics is the sticky and gummy residue that is sometimes left when water evaporates from the solution mix. Metal safety on nonferrous metals is a problem with some synthetics because of their relatively high pH (8.5 to 10.0) and the lack of oil to act as an inhibitor.

4. Semisynthetic fluids. These are synthetic fluids, which have up to 25% of oil added to the concentrate. When diluted with water, they form a very fine emulsion that looks very much like a solution, but in fact, is an emulsion. The oil is added to improve lubricity. When synthetic fluids were in their early stages, lubricity was a big problem, so the semisynthetics were introduced. However, with the technology in synthetic lubricants improving, lubricity is not the problem it once was for synthetic fluids and therefore the semisynthetic is becoming less popular.

5. Liquid nitrogen. Liquid nitrogen (having temperature 196°C) is used as a cutting fluid for cutting difficult-to-machine materials where chip formation and chip breaking present a significant problem [61,62]. Liquid nitrogen is used to cool workpiece (for example, internally supplied under pressure in case of tubular-shaped workpieces), to cool the tool (which has the internal channels through which liquid nitrogen is supplied under pressure), or by flooding general cutting area.

Although the required properties of the cutting fluid should be formulated for each particular machining operations, some of the qualities required in a good cutting fluid could be listed as follows: (a) good lubricating qualities to reduce friction and heat generation; (b) good cooling action to dissipate the heat effectively that is generated during machining; (c) effective antiadhesion qualities to prevent metal seizure between the chip and the rake face; (d) good wetting characteristics which allow the fluid to penetrate better into the contact areas as well as in the cracks; (e) should not cause rust and corrosion of the machine components; (f) relatively low viscosity fluids to allow metal chips and dirt to settle out; (g) resistance to rancidity and to formation of a sticky or gummy residue on parts or machines; (h) stable solution or emulsion [provide safety work environment (nonmisting, nontoxic, and nonflammable (smoking)]; and (i) should be economical in use, filter, and dispose.

If there was one product that met all the particular requirements to the cutting fluid, the selection of a cutting fluid would be easy. But there just is not such a product. Moreover, many of the above-listed properties often cannot be guaranteed without actual testing of a particular cutting tool in a production environment. Such a testing, however, is expensive and time-consuming. Therefore a method to compare different cutting fluids for a particular machining operation should be beneficial. Although there are a number of attempts to develop such a method (for example, Refs. 48 and 63–66), no method has been developed for qualifying and comparing the performance of one cutting fluid to another [57].

REFERENCES

1. Komanduri, R.; Larsen-Basse, J. Tribology: The cutting edge. ASME Mech. Eng Jan. 1989, *110*, 74–78.
2. Zorev, N.N. *Metal Cutting Mechanics*; Pergamon Press: Oxford, 1966.
3. Shaw, M.C. *Metal Cutting Principles*; Clarendon Press: Oxford, 1984.
4. Oxley, P.L.B. *Mechanics of Machining: An Analytical Approach to Assessing Machinability*; John Wiley & Sons: New York, 1989.
5. Astakhov, V.P. *Metal Cutting Mechanics*; CRC Press: Boca Raton, 1998/1999.
6. Finnie, I.; Shaw, M.C. The friction process in metal cutting. Trans. ASME 1956, *78*, 1649–1657.
7. Chen, N.N.; Pun, W.K. Stresses at the cutting tool wear land. Int. J. Mach. Tools Manuf. 1988, *28*, 79–92.
8. McCoy, J.S. Introduction: Tracing the historical development of metalworking fluids. Metalworking Fluids; Byers, J.P., Marcel Dekker: New York, 1994; 1–23 pp.
9. Machado, A.R.; Wallbank, J. The effect of extremely low lubricant volumes in machining. Wear 1997, *210*, 76–82.
10. Gunderson, R.C.; Hard, A.W. *Synthetic Lubricants*; Reinhold: New York, 1962.
11. Schey, J.A. *Metal Deformation Processes: Friction and Lubrication*; Marcel Dekker Inc.: New York, 1970.
12. Brinksmeier, E.; Walter, A.; Janssen, R.; Diersen, P. Aspects of cooling application reduction in machining advanced materials. Proc. Inst. Mech. Eng. 1999, *213*, 769–778.
13. Brinksmeier, E.; Walter, A.; Janssen, R.; Diersen, P. Aspects of cooling lubrication reduction in machining advanced materials. IMechE, Part B 1999, *213*, 769–778.
14. Nasir, A. General comments on ecological and dry machining. In Network Proceedings "Technical Solutions to Decrease Consumption of Cutting Fluids"; Sobotin-Sumperk: Czech Republic, 1998; 10–14 pp.
15. Arnold, D.B. Trends that drive cutting tool development. Metalworking Technology Guide 2000, Modern Machine Shop online, 2000 http://www.mmsonline.com/articles/mtg0003.html.
16. Segal, L.; Tovbin, R. Hard coatings for heavy duty stamping tools. SAE, Paper Number 1999-01-3230, http://www.sputtek.com/paper1999013230a.htm.
17. Flood, P.D. Thin film coatings and the cutting tool industry. In http://www.multi-arc.com/presentations/cutting/sld001.htm.
18. Optiz, H. Tool wear and tool life. Proceedings of the International Conference "Research in Production," Carnegie Institute of Technology, Pittsburgh; ASME: New York, 1963; 107–113 pp.
19. Okushima, K.; Hoshi, T.; Narataki, N. Behaviour of oxide-layer adhered on tool face when machining Ca-deoxidized steel. J. Jpn. Soc. Precis. Eng. 1968, *34*, 478–485.
20. Yamada, H.; Yoshida, S.; Kimura, A. On the appearance of inclusion in Ca-free cutting steel and its machinability. J. Iron Steel Inst., Jpn. 1971, *57* (13), 2110–2118.
21. Tipnis, V.A. Influence of metallurgy on machining-free machining steels. In: *On the Art of Cutting Metals—75 Years Later*; Kops, L., Ramalingam, S., Ed.; ASME: New York, 1982; 119–132.
22. Kissling, R. *Non-Metallic Inclusion in Steels*; The Institute of Metals Press: London, 1989.
23. Nordgren, A.; Melander, A. Tool wear and inclusion behaviour during turning of a calcium-treated quenched and tempered steel using coated cemented carbide tools. Wear 1982, *139*, 209–223.
24. Fang, X.D.; Zhang, D. An investigation of adhering layer formation during tool wear progression in turning of free-cutting stainless steel. Wear 1996, *197*, 169–178.
25. Qi, H.S.; Mills, B. On the formation mechanism of adherent layers on a cutting tool. Wear 1996, *198*, 192–196.
26. Mills, B.; Hao, C.S.; Qi, H.S. Formation of an adherent layer on a cutting tool studied by micro-machining and finite element analysis. Wear 1997, *208*, 61–66.

27. Tieu, A.K.; Fang, X.D.; Zang, D. FE analysis of cutting tool temperature field with adhering layer formation. Wear 1998, *214*, 252–258.
28. Qi, H.S.; Mills, B. Formation of a transfer layer at the tool–chip interface during machining. Wear 2000, *245*, 136–147.
29. Dieter, G.E. *Mechanical Metallurgy,* 3rd Ed.; McGraw-Hill: New York, 1986; 541–542 pp.
30. DeGarmo, E.P.; Black, J.T.; Kohser, R.A. *Materials and Processes in Manufacturing*; Prentice Hall: Upper Saddle River, NJ, 1997.
31. Komarovsky, A.A.; Astakhov, V.P. *Physics of Strength and Fracture Control: Fundamentals of the Adaptation of Engineering Materials and Structures*; CRC Press: Boca Raton, 2002.
32. Outeiro, J.C.; Dias, A.M.; Lebrun, J.L.; Astakhov, V.P. Machining residual stresses in AISI 3161 steel and their correlation with the cutting parameters. Int. J. Mach. Sci. Technol. 2002, *6*, 251–270.
33. Astakhov, V.P.; Shvets, S.V. A novel approach to operating force evaluation in high strain rate metal deforming technological processes. J. Mater. Process. Technol. 2001, *117*, 226–237.
34. Schey, J.A. *Tribology in Metalworking*; American Society for Metals: Metals Park, Ohio, 1983.
35. American National Standard. *"Tool Life Testing With Single-Point Turning Tools" ANSI/ ASME B94.55M-1985*; ASME: New York, 1985.
36. Makarow, A.D. *Optimization of Cutting Processes*; Mashinostroenie: Moscow, 1976. *in Russian.*
37. Atkins, A.G.; May, Y.W. *Elastic and Plastic Fracture. Metals, Polymers, Ceramics, Composites, Biological Materials*; John Wiley & Sons: New York, 1985.
38. Makarow, A.D. *Machining of Difficult-to-Machine Materials*; Mashinostroenie: Moscow, 1972.
39. Rahman, M.; Seah, K.H.W.; Goh, T.N.; Lee, C.H. Effectiveness of various coatings on cermet cutting tools. J. Mater. Process. Technol. 1996, *58*, 368–373.
40. Graham, D. Dry out. Cutt. Tool Eng. 2000, *52*, 1–8.
41. Sreejith, P.S.; Ngoi, B.K.A. Dry machining: machining of the future. J. Mater. Process. Technol. 2000, *101*, 287–291.
42. De Chiffre, L. Function of cutting fluids in machining. Lubr. Eng. 1988, *44*, 514–518.
43. Bailey, J.A. Friction in metal machining-mechanical aspect. Wear 1975, *31*, 243–253.
44. Williams, J.A.; Tabor, D. The role of lubricants in machining. Wear 1977, *43*, 275–292.
45. Williams, J.A. The action of lubricants in metal cutting. J. Mech. Eng. Sci. 1977, *19*, 202–212.
46. Doyle, E.D.; Horne, J.G.; Tabor, D. Frictional interactions between chip and rake face in continuous chip formation. Proc. R. Soc. Lond. 1979, *A366*, 173–183.
47. De Chiffre, L. Mechanics of metal cutting and cutting fluid action. Int. J. Mach. Tool Des. Res. 1977, *17*, 225–234.
48. De Chiffre, L. Mechanical testing and selection of cutting fluids. Lubr. Eng. 1980, *36*, 33–39.
49. De Chiffre, L. Frequency analysis of surfaces machined using different lubricants. ASLE Trans. 1984, *27*, 220–226.
50. De Chiffre, L. What can we do about chip formation mechanics? CIRP Ann. 1985, *34*, 129–132.
51. De Chiffre, L. Function of cutting fluids in machining. Lubr. Eng. 1988, *44*, 514–518.
52. Brown, R.H.; Luong, H.S. The influence of microstructure discontinuities on chip formation. CIRP Ann. 1976, *25*, 49–52.
53. Astakhov, V.P. A treatise on material characterization in the metal cutting process. Part I: A novel approach and experimental verification. J. Mater. Process. Technol. 1999, *96*, 22–23.
54. Lindberg, B.; Lindstrom, B. Measurements of the segmentation frequency in the chip formation process. CIRP Ann. 1983, *32*, 17–20.
55. Sidjanin, L.; Kovac, P. Fracture mechanisms in chip formation processes. Mater. Sci. Technol. 1997, *13*, 439–444.
56. Itava, K.; Uoda, K. The significance of dynamic crack behavior in chip formation. CIRP Ann. 1976, *25*, 65–70.
57. Kovac, P.; Sidjanin, L. Investigation of chip formation during milling. Int. J. Prod. Econ. 1997, *51*, 149–153.

58. Tonshoff, H.K.; Amor, R.B.; Andrae, P. Chip formation in high speed cutting (HSC). SME Technical Paper MR99-253, 1999.
59. Crafood, R.; Kamenski, J.; Lagerberg, S.; Ljungkrona, O.; Wretland, A. Chip Control in tube turning using a high-pressure water jet. Proc. Inst. Mech. Eng. Part B 1999, *213*, 761–767.
60. Li, X. Study of the jet-flow rate of cooling in machining. Part 2: simulation study. J. Mater. Process. Technol. 1996, *62*, 157–165.
61. Hong, S.Y.; Ding, Y.; Ekkens, R.G. Improving low carbon steel chip breakability by cryogenic cooling. Int. J. Mach. Tools Manuf. 1999, *39*, 1065–1085.
62. Wang, Z.Y.; Rajurkar, K.P. Cryogenic machining of hard-to-cut materials. Wear 2000, *238*, 169–175.
63. Corvell, L.V. A method for studying the behavior of cutting fluids in wear of tool materials. Trans. ASME 1958, *80*, 1054–1059.
64. De Chiffre, L. Testing the overall performance of cutting fluids. Lubr. Eng. 1978, *34*, 244–251.
65. Medaska, M.K.; Nowag, L.; Liang, S.V. Simultaneous measurement of the thermal and tribological effects of cutting fluid. Mach. Sci. Technol. 1999, *3*, 221–237.
66. Gugger, M. Putting fluids to the test. Cutt. Tool Eng. 1999, *51*, 1–6.

10
Tribology in Metal Forming

Emile van der Heide
TNO Industrial Technology, Eindhoven, The Netherlands

Dirk Jan Schipper
University of Twente, Enschede, The Netherlands

I. INTRODUCTION

The field of metal-forming processes is large and covers processes such as wire drawing, rolling, deep drawing, forging, extrusion, hydroforming, and rubber pad forming. Metal-forming processes are all designed to mechanically deform metal into a shape without material removal, in contrast with manufacturing processes such as machining or punching. Each forming process has its own characteristic features in terms of the material flow and in terms of the basic operating variables (Sec. II.A), as shown in Fig. 1. A common classification method is to distinguish bulk-forming processes, where material is plastically deformed in three dimensions, and sheet metal forming (SMF), where sheet material is plastically deformed in two dimensions (the thickness of the sheet is more or less constant). As an example, the setup for a commonly applied SMF process is shown in Fig. 2. Thin, initially flat, sheet material is clamped between a blankholder and a forming die. The punch moves in the direction indicated by the arrow and forces the sheet material to flow into the forming die. In this way, the metal sheet is formed into a cup. Usually the sheet material is precoated with a metal-forming lubricant to control friction and wear that arises because of sliding between the metal sheet and the blankholder and the forming die. The application of tribology—the science and technology of interacting surfaces in relative motion—to metal-forming processes often raises questions such as: how can a micro-oriented approach solve problems related to a 2800-ton transfer press 33 m long with metal-forming tools up to 30 tons, making millions of pressing operations each year? An answer to this question is given by looking at a general aim of the manufacturing industry, i.e., to make products of constant (high) quality in an increasingly competitive way. Control of friction and wear contributes to this aim by eliminating uncertainty in production and by reduction of the amount of waste. In general, it can be stated that control of friction and wear in the interface between tooling and processed material will make production equipment more stable, which in turn guarantees the required constant process output. Tool wear, in particular, has a direct negative influence on the product's dimensions and on the surface quality. Because metal-forming tools represent high economical value and because change of tooling causes

Metalworking process	Basic operating variables		
	pressure ($10^7 N/m^2$)	speed (m/s)	temperature (°C)
Drawing 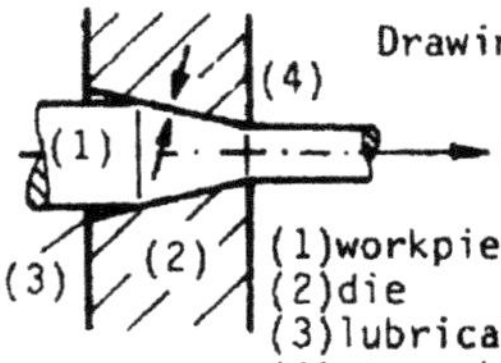 (1)workpiece (2)die (3)lubricant (4)atmosphere	...20	0.0005 (bar and tube) ...1 (heavier wire) ...15-30 (fine wire)	cold: room temperature + ΔT=200-300 warm: 400-600 hot : 800 for Mo,W
Pressing 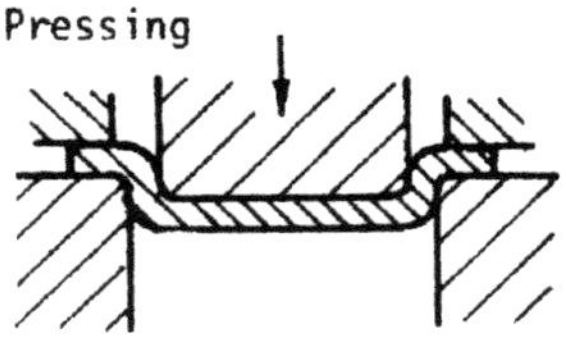	...80	0.02-0.5	room or elevated temperature (e.g. 500-800 for Ti)
Rolling 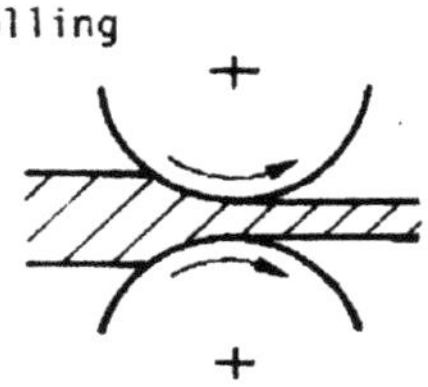	20-150	5-25 (cold rolling) 0.1-0.5 (hot rolling)	room temperature or elevated temperature (400-1250, hot rolling)
Forging 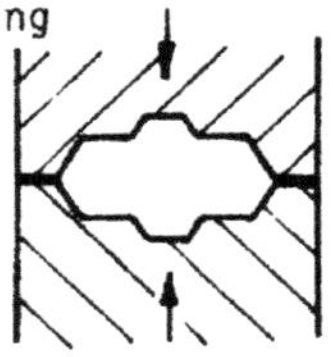	30-130	0.02-0.2 (hydr. presses) ...10 (hammers)	100-200 (cold) 500-600 (warm) ...1200 (hot forging)
Extrusion 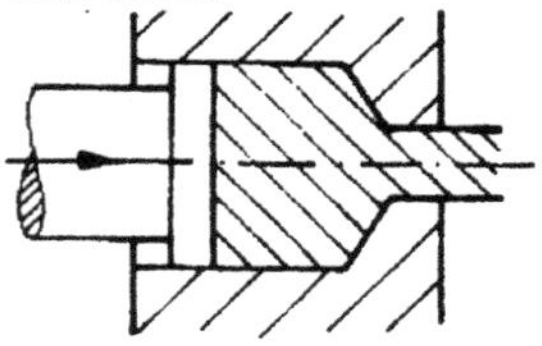	100-200	0.05 ("hot shot") 0.5-5	room or elevated temperature (1000-1200 hot extrusion) 600 non-ferrous

Figure 1 Basic operating variables for metal forming processes. (Adapted from Ref. 1.)

standstill in production, it is clear that improvement and prediction of tool life is of high industrial importance.

From the point of view of tribology, metal-forming processes differ strongly from classical applications such as roller bearings, bushings, and gears, i.e., with respect to the type of motion and the type of deformation. Classical tribology considers the situation in which surfaces deform elastically and "see" each other after each turn or stroke, thus allowing for run-in of both surfaces. Running in, a crucial aspect in tribological design, is a phase of initial high friction and/or high wear, in which small geometrical defects like surface

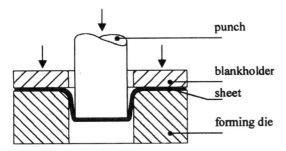

Figure 2 Deep drawing.

roughness are adjusted. After run-in, wear and friction will stabilize to a low level, assuming a well-chosen combination of materials. In metal-forming processes only the tool material deforms elastically and is subjected to run-in. The processed material deforms plastically and is always "fresh." This discrepancy with classical contact situations is also reflected in tribological testing where reciprocating machines and pinring types of devices dominate. Because the predicting capacity of these existing methods is poor for contact situations that occur in metal forming, it is clear that new methods are considered.

In this work, the effect of tool coatings on the performance of forming processes is shown for SMF. First the "system approach" is introduced in Sec. II and applied to lubricated SMF processes. It provides a framework for describing the effect of tool coatings in well-established tribological terms. Section III shows which aspects of coatings are important in SMF. A framework for successful tool treatment is constructed in this section and projected on a range of available coating and surface-treatment techniques. In Sec. IV, the friction and wear behavior of coated tools in metal-forming situations is given, measured at the laboratory scale. Section V confirms the positive effect of tool coatings by means of results from industrial trails. In Sec. VI, the main conclusions are summarized.

II. TRIBOLOGICAL SYSTEM

A. Systems Approach

Analysis of tribological contact situations is generally done based on the so-called systems approach [1]. Basically, this means that a tribological contact situation is separated from the application studied by using a hypothetical system envelope. The contact situation separated by this envelope is regarded as a system, i.e., a set of elements, interconnected by structure and function. The structure of contact situations in SMF operations can be reduced to the interaction between the forming tool surface and the sheet material surface, respectively, in the presence of a lubricant and surrounded by the environment. It should be noted that the sheet material is always fresh in the contact, meaning that one of the contacting surfaces is not able to run-in. A schematic drawing of the tribological system is shown in Fig. 3.

The connections between the system and the rest of the application can be reduced to input, the operating variables, and output: friction and wear. Section II.B describes the range of the operating variables and their effect on friction in lubricated SMF. The system components, sheet material, tool material, and lubricants are discussed in detail in Secs. II.C, II.D, and II.E, respectively.

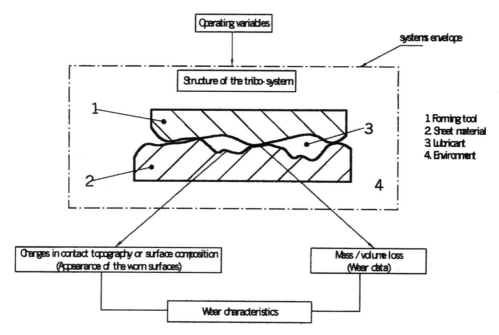

Figure 3 Tribological system in sheet metal forming. (Based on Ref. 1.)

B. Lubrication Regimes and Friction

The subject of friction and friction modeling in metal forming was studied by several authors [2,3]. Similar to many other lubricated tribosystems, it is shown that lubricated SMF contacts obey a Stribeck type of behavior [2]. This means that the coefficient of friction f, defined as the ratio of the friction force and the applied normal force, can be plotted in a diagram as a function of the dimensionless lubrication number $L = \eta v/(p_{\mathrm{m}} R_{\mathrm{a}})$, introduced by Schipper [4]. The lubrication number represents the main operational variables of study in lubricated contacts: lubricant inlet viscosity (η), velocity (v), mean contact pressure (p_{m}), and the combined center line average roughness (R_{a}). The generalized form of the Stribeck curve is given in Fig. 4.

Three lubrication modes can be distinguished:

• (Elasto) hydrodynamic lubrication (E)HL. No physical contact between the asperities of the interacting surfaces occurs; the velocity difference v between the surfaces is accommodated by shear in the lubricant film, which results in relative low friction forces and, consequently, a low coefficient of friction. In this mode, the bulk viscosity of the lubricant determines the properties of the lubricant film. Typically, the value of f is of the order of 0.01.

• Boundary lubrication (BL). In this lubrication mode there is physical contact between the asperities of the interacting surfaces. All the load is carried by the interacting asperities. The velocity difference is accommodated by shearing of boundary layers, the properties of which do not depend on the bulk viscosity of the lubricant. Typically, values for f are in the range of $0.1 < f < 0.15$.

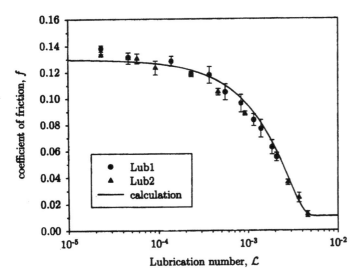

Figure 4   Generalized Stribeck curve. (Adapted from Ref. 2.)

- Mixed lubrication (ML). This mode represents the intermediate regime between BL and (E)HL where a part of the load is carried by hydrodynamic action and the remaining part of the load is carried by the interacting asperities. The coefficient of friction ranges from EHL to BL values.

An important engineering aspect of this diagram is the possibility to predict the transition values for L [4]. For SMF systems it is shown [2] that BL or a part of ML will be the operational lubrication mode. The performance of boundary layers is therefore of prime importance in tribological contacts in sheet metal forming processes.

C. Sheet Materials

The range of commonly used sheet materials consists of four basic groups: unalloyed low-carbon steel, stainless steel, aluminum alloys, and copper alloys. Within these groups a wide spectrum of possibilities is commercially available to fulfill the specific needs of products, ranging from automotive side panels to brass instruments. Table 1 summarizes some properties of the sheet materials from the abovementioned basic groups [5]. In this chapter only unalloyed low-carbon steel and stainless steel are taken into account.

Unalloyed low-carbon steel sheet material (<0.05 wt.% C, <1 wt.% alloying elements) is produced by hot rolling of casted steel into a thin strip with a minimum thickness of 1.5 mm. This hot-rolled sheet material is used for products that require only relatively low surface quality. Application of hot-rolled steel with its extremely hard and brittle skin could lead to high abrasive wear of the tools during forming processes. Although the importance of hot-rolled steel increases, it has not replaced the traditional cold-rolled steel. Cold rolling is used to further reduce the thickness of hot-rolled sheet material to a representative thickness of 0.7 mm, reeled into coils. By using pretextured rolls it is possible to create a high-quality sheet surface or produce specific roughness textures such as electrodischarge

Table 1 Survey of Representative Sheet Materials

Material	Typical example	Yield stress (MPa)	Tensile strength (MPa)	Brinell hardness (HB)
Carbon steel	DC 06	180	270–350	80 HB
	HSLA Docol 500 YP	500–620	570–710	—
Stainless steel	AISI 304 / X5 CrNi 18 10	210	520–720	150–190 HB
	AISI 316 / X5 CrNiMo 17 12 2	220	520–670	150–190 HB
	AISI 409 / X2CrTi12	220	380–560	150–180 HB
	AISI 430 / X6Cr17	260	430–630	150–190 HB
Al alloys	Al99.5 (O)	20	65	20 HB
	AlMn1 (O)	30	90	27 HB
	AlMg4 (O)	100	240	65 HB
	AlCu4Mg1 (T4)	275	425	120 HB
Cu and Cu alloys	SW-Cu (F22)	max 140	220–260	40–65 HB
	CuZn 30 (F27)	max 160	270–350	55–85 HB
	CuSn8 (F33)	max 190	330–380	65–95 HB

Source: Ref. 5.

texturing (EDT) and electron beam texturing (EBT) at the sheet's surface. Typical yield strength of these (mild) steels is about 150–300 MPa.

The increasing demand for light and strong structures has led to the development of carbon steel sheet material with high(er) strength, allowing the application of thinner cross sections in mechanical designs. Typical yield strength of these materials ranges from 500 to 700 MPa. These sheet materials require higher press forces during SMF-processes. Therefore, contact pressures in tribocontacts with high-strength steel are higher, which in turn could result in increased volume wear of the tools, assuming a constant specific wear rate for the tool materials.

A large volume of carbon steel is zinc coated to improve the lifetime of products that are sensitive to corrosion. Zinc layers are deposited by electrolytic or thermal processes, resulting in layers of 5–20 μm with hardness of about 60 HV 5gr–350 HV 5gr. In using these sheet materials, zinc layer delamination should not occur during forming. Other coatings often used for packaging steel are tin and chromium.

Ferritic stainless steel, a specific class of corrosion-resistant, tough steel results from alloying with at least 12 wt.% Cr (max. of 1.2 wt.% C). It is mainly used in domestic (washing, refrigeration, and cooking equipment) and high-temperature applications (up to 1100°C). The weldability of sheets of this steel is limited. Commonly used ferritic stainless steels are AISI 409 and 430. Another group of stainless steels contains 18% Cr and at least 8% Ni to increase the corrosion resistance of the sheet material (austenitic stainless steel). The combination of mechanical properties, corrosion resistance, and high energy absorption of austenitic stainless steel sheet makes it possible to design car components with improved crash safety. Commonly used austenitic stainless steels are AISI 304 and 316. Stainless steel sheet material is known to be sensitive to galling [6].

D. Tool Materials

Iron–carbon alloys are probably the most widely used tool materials in metal-forming applications. When relatively cheap material is needed, e.g., for large dies or low loaded applications, cast Fe–C alloys are chosen. Two groups can be distinguished: cast iron and cast steel. Cast iron, an Fe–C alloy with 2.5–4 wt.% carbon and essential alloying elements such as silicon and phosphorous, is used for large car body dies. For low loaded conditions gray cast iron GG 25 can be used, although present automotive demands require at least ductile cast iron grade GGG 60, but preferably GGG 70 L. Cast steel, applied for tools with smaller dimensions, is characterized by a maximum carbon content of 0.45% and small amounts of alloying elements such as manganese, molybdenum, and chromium. Alloying is done to improve the hardenability and the wear resistance of the material. Typical cast steel tool materials are 16MnCr5 and 42CrMo4.

Metal-forming tools subjected to more demanding loading conditions need higher alloyed, wrought Fe–C alloys, usually called tool steel. Although tool steels represent only 0.1% of the total steel production [7], considerable effort is put into the development of alloys with improved lifetime in high volume forming applications. Easy machinability in the soft annealed state and dimensional stability and polishability in the hardened state must be combined with toughness, compressive strength, and wear resistance during the life of the tool. This combination of properties has resulted in a wide range of alloys, generally standardized into cold work tool steels, hot work tools, and plastic mold steels. Cold-forming tool steels can be applied to a maximum temperature of about 200°C and have a typical carbon content between 0.45 and 3 wt.%. Alloying elements are usually carbide-forming elements such as Cr, Mo, V, and W. Hot-forming tool steels can withstand incidental temperatures of 800°C and possess improved thermal fatigue resistance. The carbon content of these steels range from 0.3 to 0.6 wt.%. Commonly used tool steel grades are given in Table 2. Alloying with carbide-forming materials such as Cr, Mo, W, and V (at the same time raising the carbon content) is generally done to improve the volume percent of carbides, which could significantly reduce volume wear of tools. The maximum amount of carbides is about 25 vol.% for conventional hot-formable tool steels. Increasing the amount of carbides above 25% requires powder metallurgy (PM), a process that starts with metal powder. A large variety of powder metallurgy routes exists to produce PM steels. One of the routes is to fill a capsule with metal powders, weld it gastight, and use hot isostatic pressing to densify the material to an ingot. Next, hot forging is used to wrought the material into billets of tool steel. The use of PM results in a considerably better combination of ductility and abrasive wear resistance [8] compared with conventional high-C, high-Cr tool steels. By

Table 2 Composition of Typical Tool Steels

Group	Steel no.	C	Si	Mn	Cr	Mo	W	V
Hot forming	WN 1.2344	0.40	1.0	0.40	5.10	1.30		1.00
	WN 1.2365	0.31	0.30	0.35	2.90	2.80		0.50
Cold forming	WN 1.2363	1.00	0.30	0.55	5.30	1.10	—	0.20
	WN 1.2379	1.55	0.30	0.30	11.50	0.70	—	1.00
	WN 1.3344[PM]	1.3	0.6	0.3	4.2	5.0	6.4	3.1
	VANADIS 4[PM]	1.5	1.0	0.4	8	1.5	—	4.0
	VANADIS 10[PM]	3.0	1.0	0.5	8	1.5	—	9.8

Values are in weight percent. PM indicates powder metallurgy.

changing the carbide type, size, and amount, the properties of the steel can be altered to increase either the ductility or the resistance to abrasive wear. The composition of two PM steels, VANADIS 4 (higher ductility, lower resistance to abrasive wear) and VANADIS 10 (lower ductility, higher resistance to abrasive wear), is given as an example in Table 2 [9].

A second group of materials that can be applied for high-demand loading situations is formed by cemented carbides, also produced by PM technology. A commonly applied cemented carbide composition, used for forming tools, consists of tungsten carbides (WC) bonded together by cobalt. A tool material with a high resistance to impact can be realized by increasing the amount of cobalt up to 30%, combined with a coarse grain sized WC (30 μm). Resistance to abrasive wear requires a lower Co content (6 wt.% GT6) and WC of a fine grain size (1–10 μm). Another type of cemented carbides combines a Cr–Mo metal binder with up to 45 wt.% TiC (Ferro-titanit). Both cemented carbide types are made using either cold isostatic pressing and subsequent sintering or hot isostatic pressing [10].

Alternative tool materials are developed for specific cases. A particularly interesting group of materials is technical ceramics, such as Al_2O_3, SiC, ZrO_2, and Si_3N_4, which have been used for decades in all sorts of tribological systems. The combination of high resistance to corrosion, high hardness and favorable "dry" running properties is used to make sliding contacts in all sorts of mechanical equipment. Especially, Si_3N_4 is a promising material for forming applications with limited or no lubrication [11], although the poor impact strength of the material limits its application. A second group of interest is formed by soft tooling materials, which have the advantage of fast machining and low development time, and which are used for small- to medium-volume production. Typical materials are Zn alloys such as ZnA14Cu3, aluminum bronzes (drawing of stainless steel), and engineering plastics, such as epoxy resin with metal fillers to improve the mechanical properties. For some aluminum sheet materials, wooden tools (resin-impregnated laminated and condensed wood) can be applied. The wear resistance of these materials is much lower than that of cast iron in the same application [12].

E. Forming Lubricants

The main function of lubricants during SMF processes is to control the friction forces that arise because of the sliding contact between sheet material and tool material. Second, lubricants are used to avoid or delay galling problems by protecting the tooling from direct contact with the sheet material. These functions can be met under mixed or hydrodynamically lubricated conditions (Sec. II.B) by typical lubricant base fluids of mineral (petroleum) or synthetic origin, mixed with additives, related to viscosity regulation. Friction and wear control at boundary lubricated conditions (Sec. II.B) requires specific additives that form boundary layers at the contacting surfaces. The performance of these lubricants is mainly determined by the thermal stability of the formed boundary layer [13,14], which in turn may depend on the reactivity of the sheet's surface.

The performance of boundary layers [15] is generally classified based on the mechanism of layer formation, i.e., physical adsorption, chemical adsorption, and chemical reaction. Boundary layers formed by the first mechanism typically consist of clustered, long-chain hydrocarbons with a polar "head." The polar group adheres to the contacting surface by physical adsorption, forming high-viscosity hydrocarbon layers. Typical additives are long-chain alcohols and fatty acids. This group of layers is meant to reduce friction and wear under mild loading conditions. It should be noted that fatty acids will only adhere by physical adsorption if unreactive contacting surfaces (e.g., platinum or chromium) are applied. For reactive metal surfaces the second mechanism will be more likely to occur, i.e., chemical adsorption. This mechanism combines physical adsorption with chemical reaction with the surfaces to form a metal soap. Stearic acid ($C_{17}H_{35}COOH$), e.g., will chemically react with the metal surface to form iron stearate ($C_{17}H_{35}COOFe$), which is firmly linked to the surface. Additives that form boundary layers by chemical adsorption are usually referred to as friction modifiers [16]. Typical additives that belong to this group are fatty acids, long-chain fatty amides, and esters. This group of layers is more resistant to increased contact temperatures and therefore used for wear and friction reduction at moderate loading conditions. This is in contrast to the third group of layers that is formed by a chemical reaction of lubricant constituents and the contacting surfaces. Usually, complex compounds based on P, S, B, and Cl are used to form metal salts with high temperature stability. The latter makes these additives suitable for wear protection at severe loading conditions were extreme pressure (EP additives) causes high contact temperatures. Especially, B and Cl are suspected to harm the environment and—without protective measures—human health.

In selecting lubricants for SMF processes, two important steps in production must be considered: compatibility with the oil applied in the mill and compatibility with the coating-process. Cold-rolled steel has to be protected from corrosion during transport of the coils and during storage at the press shop; therefore a preservation agent is applied at the sheet surface. After the forming process, a coating is often applied at the surface for decorative reasons or for prevention of corrosion. In both cases, expensive cleaning of the steel is necessary if there are compatibility problems with the selected forming lubricant. If cleaning is inevitable, chemical reaction layers must be avoided because environmentally unfriendly solvents are needed to remove the layers from the formed product. From the point of view of cleaning and the environment, it is obvious that adsorption is preferred as the mechanism for boundary layer formation.

Because of the complexity of lubrication, both in terms of costs (selecting, buying, applying, and cleaning) and in terms of ecotoxicity of the waste, the choice of lubricant is often a subject of discussion. As a result, many alternative technologies are developed. Examples are preservation oils with improved performance during forming (so called prelubes), dry permanent "lubricants," and precoated sheets.

III. IMPROVING TOOL LIFE BY SURFACE TREATMENT AND COATING TECHNOLOGY

A. Classification of Wear in Sheet Metal Forming

Wear, defined as "the progressive loss of substance from the operating surface of a body occurring as a result of relative motion at the surface," [17] is an important source of failure in SMF practice. Wear influences the stability of the equipment, the production speed, and the quality of the products and the product's surface. It is also recognized in practice that wear strongly depends on the materials processed and on the complexity of the product's shape, both illustrating the system dependence of wear.

A classification scheme that fits to the general practice of SMF is found in Ref. 18. The latter combines the structure of the tribosystem, the type of motion, and the type of work with four main *wear mechanisms*: adhesion, abrasion, surface fatigue, and tribochemical or corrosive wear.

    1.   Adhesive wear: wear by transference of material from one surface to another during relative motion due to a process of solid phase welding. Particles that are

removed from one surface are either temporarily or permanently attached to the other surface [17].

2. Abrasive wear: wear by displacement of material caused by hard particles or hard protuberances [17].
3. Surface fatigue or fatigue wear: removal of particles detached by fatigue arising from cyclic stress variations [17].
4. Tribochemical wear: removal of chemical reaction products formed as a result of chemical interactions between the elements of a tribosystem initiated by tribological action [19].

These four basic mechanisms or any combination of them are involved in wear processes. The type of motion is used to distinguish the wear processes: sliding, oscillation, erosive, impact, and rolling wear (Fig. 5).

Typical contact types occurring in SMF are examined in Ref. 2. The application of the classification scheme shows that "sliding wear" is expected to be the dominant wear process and that all four basic wear mechanisms could play a role in this process. Rolling, oscillation, and impact wear will also occur in SMF; however, the contribution of these wear processes to the total amount of wear can be neglected when considering SMF in general. Erosive wear is less likely to occur in a press shop.

A closer look at wear-related failure in SMF shows two important types of sliding wear: volume loss of the tool surface and galling [20]. The first wear type is characterized by a damaged tool surface. Reshaping of the surface is necessary to avoid products with unacceptable dimensions. The rule of thumb to minimize this type of sliding wear is to increase the hardness of the tool surface (less abrasion) or to apply a surface with low affinity/solubility with the counter surface (less adhesion) (see, e.g., Ref. 21). The second wear type deserves more attention, perfectly illustrated by the International Research Group on Wear of Engineering Materials IRG-OECD (Organization for Economic Co-operation and Development) definition of galling: "a severe form of scuffing associated with gross damage to the surfaces or failure. Note: The use of this term should be avoided.

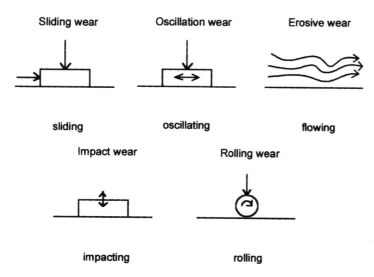

Figure 5   Wear processes. (Adapted from Ref. 18.)

More detailed description of the damage is desirable" [17]. Although many definitions are constructed for galling, it is clear that material transfer plays an important role in all worn systems denoted by this term [22]. Galling, in the context of metal forming, is associated with the tendency for lubricant film breakdown resulting in pick-up of sheet material by the tool surface and subsequent scoring (= severe scratching) of the work piece surface [6].

Three phases can be recognized in SMF galling processes [23–25]:

- Initiation
- Lump growth
- Severe scratching or seizure

Initiation of material transfer is related to local lubricant failure and occurs at surface defects such as grinding marks or carbides [23,24]; in fact, it can initiate at any high roughness peak that interacts with the sheet material. Once initiation has occurred, lumps start to grow as a function of the amount of products formed. The transferred lumps of sheet material adhering to the tool surface now interact with the sheet, creating scratches at the product's surface. The amount and the depth of scratches increase with production time. Polishing of the tool surface is necessary to avoid an unacceptable product surface and extreme friction forces during forming. Unfortunately, no rule of thumb exists for avoiding or delaying this type of sliding wear, other than applying mirror finish in tool polishing.

Wear resistance of forming tools, although important, is only one of the controlling parameters during the development phase of tooling. Machinability is also very important to minimize the development time of the tools, not to forget the ability to withstand the mechanical loading of the application. This combination of properties is hard to be met within one material. Because surface treatment of tool material enables the use of advanced materials locally, it is considered a cost-effective way to meet the conflicting demands of tool design.

As stated, wear-related failure in SMF processes is generally due to sliding wear. The selection and application of coating technology/surface treatment should therefore be focused on minimizing sliding wear or, specifically, on minimizing volume loss of the tool surface and on avoiding galling.

B. Conventional Surface Treatments—Proven Technology

Surface treatment of forming tools is traditionally limited to heat treatment (hardening), nitriding, and hard chromium plating. These three types of treatment are considered "proven technology" for automotive forming applications and are used extensively for all kinds of tooling. In fact, these treatments are classical examples of minimizing sliding wear in conventional lubricated tribological contacts as gears (case hardening, nitriding) and hydraulic cylinders (hard chromium plating).

Heat treatment of cast iron or low alloyed cast steel is usually done by induction or flame hardening, creating a case of hard material around a tough core of unhardened steel or cast iron. Typical hardness values range from 500 to 750 HV; typical case depths range from 250 to 5000 μm. Heat treatment of high-C, high-Cr tool steels and PM tool steels is an essential step in toolmaking. Vacuum hardening and tempering at elevated temperatures result into a tough hardened material with hardness ranging from 450 up to 900 HV. Maximum tempering temperature is about 520–550°C. The overall hardness of the material reduces if the tempering temperature is chosen at higher levels.

Nitriding of steel and some cast irons [26] is done by either gas (process temperature 480–590°C) or plasma treatment (process temperature 340–565°C), aiming at increasing the nitrogen content of the surface. Because of this treatment, a diffusion zone is created with a top layer of Fe–N alloys (white layer). The hardness of the diffusion zone decreases as a function of depth, relative to the steel surface. Depending on the substrate, hardness values of 500–1600 HV can be reached. A typical range of total case depth is 20–500 μm. The nitriding process can be combined with carburizing (carbonitriding or nitrocarburizing), which may offer advantages for some steel alloys.

Hard chromium plating has been applied to forming tools since 1965 [27] and has the advantage of a relatively low deposition temperature (40–70°C), in combination with the possibility to treat tools with large dimensions. The chromium coating is produced using an electrochemical process that involves the use of an electric current and a chromium bath [28,29]. Hard chromium layers can be deposited with a thickness from 3 to 500 μm (typical layer thickness, 40 μm), resulting in a hardness of 850–1100 HV. The hardness of chromium layers decreases when temperatures above 200°C are applied.

The application of alternative treatments or coating technologies is relatively unknown for forming tools, although many cases are reported in the literature where diffusion or vapor deposition technology showed promising results. The reason for this can be understood by considering the long time and the high costs involved before a technology becomes "proven technology" and by looking at the specific and general demands that should be met to successfully apply surface treatment and coating technology in SMF processes.

C. Framework for Successful Tool Treatment

A framework of design and practical rules for successful application of surface treatment and coating technology to SMF tools is difficult to construct because of the system dependence of wear. Nevertheless, the following seven rules will generally increase the chances of success.

1. Coating equipment should be large enough to handle heavy and large substrates because forming tools have considerable dimensions and are generally made from solid steel.
2. The coating process temperature must not exceed the tempering temperature of the tool steel because of loss of substrate hardness and dimensional changes. In some cases, the hardening and tempering process can be combined in one surface-treatment process.
3. Repair of the tool coating must be possible.
4. The surface finish of the coating must be excellent; in many cases it is necessary to have an Ra roughness of less than 0.1 μm. Local misprints are not allowed since they will be copied on the counter face. Because polishing or grinding of hard layers are both expensive and difficult, it is clear that the chosen treatment should produce a layer with a finish comparable to that of the untreated substrate.
5. The adhesion of the layer to the substrate must be excellent to withstand the variations in stress situations.
6. Possible wear of the layer should cause gradual volume loss. Local detachment of the layer (spalling) should not occur.
7. Application of a layer should not result in fluctuating friction forces in the contact between sheet and tool. A stabile forming process requires a constant friction level rather than a low friction level [30].

Besides these seven rules it also important to look at future demands for tool coatings. Three general trends can be detected:

1. Tool surfaces should be able to perform under poor or nonlubricated conditions.
2. The wear resistance of tool surfaces should be good enough to process relatively hard materials such as rough hot-rolled steel and high-strength steel.
3. Tool coatings should be applicable to soft tool materials like zinc alloys in order to upgrade low-volume tools for use in medium- and high-volume production.

Heat treatment, nitriding, and hard chromium plating do meet the seven general rules. With these processes, it is possible to handle large tools in a cost-effective way and produce surfaces of high quality. Wear of conventionally treated tools is generally gradual and, especially, hard chromium plating is suited for local repair. It is sure, however, that the conventional type of treatment will not sufficiently meet the demands of the near future especially with respect to the demand of poorly lubricated or nonlubricated forming. Thus, it is important to consider and evaluate alternative hard coatings.

D. Alternative Coatings and Surface Treatments

The application of thick, hard coatings is generally done by a hard facing technique. From this group of techniques, fusion welding and laser cladding are of especial interest because of the possibility to deposited Co- and Ni-based layers with high resistance to sliding wear [31]. Local surface melting assures a good bonding of the layer to the substrate. Examples of such coatings are Stellite 6 (28 wt.% Cr, 3 wt.% N, 4 wt.% Ni, balance Co) or Tribaloy 800 (18 wt.% Cr, 28 wt.% Mo, 4 wt.% Si, balance Co). Typical hardness ranges from 400 to 800 HV, layer thickness from 0.10 to 5 mm. A third, more experimental technique is laser surface alloying of tool steels. This thermochemical process involves melting of the substrate material by means of a laser beam and alloying with high-melting-point, hard particles, such as titanium-or vanadium carbides. The resulting layer consists of a metallic matrix with the added hard particles embedded in the matrix [32]. Until now, the application of hard facing techniques for tooling is limited because of the necessary mechanical posttreatment of the layer for reasons of geometry and surface finish. A major advantage of hard facing techniques is the possibility of local deposition and local repair of thick layers.

A commonly applied alternative for hard chromium plating is electroless Ni. The nickel layer is deposited by submerging the tool in an electroless Ni bath. An autocatalytic chemical reduction process (no electric current required) results in a Ni alloy, mostly Ni–P and Ni–B. The use of composite coatings based on Ni–P with a fine dispersion of hard particles (SiC) or soft particles (PTFE) can offer advantages in poorly lubricated tribosystems. The coating thickness ranges from 0.2 to 200 μm, although typically 2–30 μm is applied. The process temperature depends on the Ni-bath used, ranging from 85° to 100°C. Additional heat treatment at about 400°C is required to obtain coatings with a hardness between 500 and 1200 HV (Ni–B).

Vapor deposition [33] is used for production of thin (1–5 μm), hard, ceramic coatings and is well established in industrial applications, such as the production of cutting and grinding tools. Important features of chemical (CVD) or physical (PVD) vapor deposition processes is the use of a vacuum chamber. In CVD, a reactive gas mixture at temperatures of about 1000°C is introduced. In PVD, a physical deposition process such as sputtering or evaporation operating in a plasma is applied. The latter process is usually performed at a typical process temperature of 450°C. Combinations of PVD and CVD are also possible. Coatings that are produced by PVD, CVD, or combinations of both processes are often titanium-and chromium nitrides or carbides. The exact composition of the coating depends strongly on the manufacturing route. Hardness values can range from 1000 to 4000 HV, depending on the composition of the coating. Limiting features of vapor deposition techniques for application in forming tools are the process temperature, which could easily result in hardness loss of the substrate, and the use of a vacuum chamber.

A specific vapor-deposited coating of interest for SMF applications of the near future is usually called diamond-like carbon (DLC). This type of hard, mostly amorphous, carbon film can be described as a "network of graphite clusters, linked into islands by sp3 bonds" [34]. Depending on the deposition process and conditions, the films may contain varying amounts of hydrogen. Alloying with metals or silicon improves in some cases the adhesion to the substrate [35] or the tribological performance [36]. Although the adhesion strength to tool steel has still to be optimized, promising results are reported for various applications where non- or poorly lubricated conditions are required [37]. The tribological performance of a DLC coating deteriorates at elevated temperatures [38].

Chemical diffusion techniques involve the modification of the chemical composition of a surface by diffusion through the surface [39]. This type of processes is usually only applied on cast iron and steel. *Chromizing* consists of diffusion of chromium into high-carbon steel by heating at 950–1300°C in chromium-rich media, often in the solid state (pack chromizing). The resulting layer consists of complex chromium carbide. The hardness of the layer can be 1200–1300 HV; the diffusion zone is about 100–200 μm thick. *Boronizing*, the diffusion of boron into steel, can be done in solid, liquid, or gas media at temperatures between 900° and 1100°C, and results in the formation of iron borides. Hardness values range from 900 to 2300 HV. Relatively thick layers can be produced (5–250 μm). The fatigue strength of the coating is weak because of high porosity and consequent crack initiation. *Vanadizing*, the diffusion of vanadium into steel, is usually done by immersion in a salt bath [40] at high temperature, creating a carbide coating of mainly VC. The total layer thickness can range from 2 to 15 μm; the resulting hardness can be as high as 3200–3800 HV. The process temperature, 850–1050°C, requires post heat treatment to restore the hardness of the steel substrate (this is also necessary after chromizing and boronizing).

IV. MEASURING FRICTION AND WEAR AT LABORATORY SCALE

A. Introduction

The effect of surface treatment on tool life or product quality should preferably be measured by actually performing tests at an industrial scale. Practical reasons such as the unavailability of a trail press or lack of time and money to perform extensive trails has created the need for tribological screening tests. In general, tribological testing is done assuming that industrial applications can be reduced to contact situations with input and output such as that illustrated with the system approach in Sec. II. The extracted system is then simulated at a laboratory scale using the same system components and input. Since structure and input are similar, it is assumed that the output of the experiment correlates with tribological performance in industrial practice.

Wear tests related to SMF differ from conventional wear tests because of the specific system components involved and the required type of motion. Actually, the following requirements should be met [41]:

1. Since sheet–tool interaction has to be simulated, it is essential that a well defined, reproducible contact between sheet and tool material is maintained during the test.
2. It is essential that the sheet material in contact is always fresh.

Naturally, the general requirements for a tribotest should be obeyed, i.e.,

3. The operational variables should be applied as in industrial practice.
4. Wear and friction should be measured directly.
5. Friction should be measured independently of the normal force.

Because the method should also be economically feasible, final requirements are the following:

6. The test pieces used should be "easy" to make.
7. The test method should allow for sliding distances in the kilometer range.

B. Short Sliding Length Laboratory Tests

As a result of the first two requirements, it is clear that modified conventional tribotesters such as rotational and reciprocating testing devices could only generate data linked to SMF for one turn or stroke. This type of testing is often used to measure the effect of sheet roughness and lubricant viscosity on friction as a function of sliding speed and normal pressure; see, e.g., Ref. 30 and Sec. II (Stribeck diagram).

The effects of surface treatment on friction and wear are reported only for continuously rotating or reciprocating sliding. Especially thin hard coatings made by either chemical or physical vacuum deposition are frequently used in sliding contacts on laboratory scale. Franklin and Beuger [42], e.g., show sufficiently low wear rates without using lubricants in a pin-disk test device, when using sliding surfaces coated with CVD or PVD layers such as TiN, TiC, Cr_xC_y, and a DLC (combined low wear and low friction). The counter surface in such testing is usually ball bearing steel 100Cr6 or tool steel WN 1.2379.

The most frequently used tribotests for SMF conditions make use of strip material drawn over a simple radius. These types of tests have the advantage of an easy to manufacture test piece geometry; however, a major disadvantage is that the friction force in the contact cannot be measured directly and independent of the normal force. Example tribometers are the strip stretching test (Fig. 6) where a strip is pulled over a radius, applying a constant normal force and pulling velocity at one end of the strip. The other end of the strip is fixed. This test is also performed allowing for linear displacement of both ends of the strip (radial strip drawing test). The back-pull force can be used to increase the normal force acting on the radius [3]. Test results show that coatings are effective in avoiding or delaying galling mechanisms [3]. The draw bead test can be regarded as a variant of this type of testing. Strip material is drawn through a set of three cylinders, as shown in Fig. 7, by imposing a horizontal displacement at the end of the strip while suppressing vertical displacement at the other end of the strip [3]. Tests are performed with cylinders that can rotate freely or with fixed cylinders. The draw bead simulator can also be used to measure the effect of a hard weld in prewelded blanks on volume loss of tool material [43]. Simulations

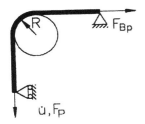

Figure 6 Strip stretch test. (Adapted from Ref. 3.)

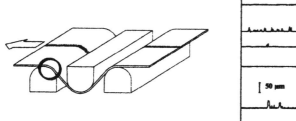
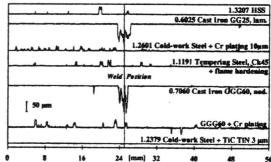

Figure 7 Measurement principle and 2-D profile of a draw bead after the experiment, indicating the position of the weld during sliding contact. Conditions: speed = 0.07 m/s, radius of the draw bead cylinders = 6 mm, 25.4 mm apart, lubricated with a prelube 5 gr/m², sheet material St 14, shot blast $R_a = 1.1$ μm. (Adapted from Ref. 43.)

indicated that a relatively soft tool material like cast iron is extremely sensitive to abrasive wear caused by interaction with the hard weld. The application of (coated) tool steels showed promising results (Fig. 7).

A cylinder on strip method (Fig. 8) to measure friction and wear in SMF-like contacts that fit the framework of requirements is presented in Ref. 2. With this method, friction can be measured directly and independent of the normal force, using a well-defined contact between tool material and strip material. The sliding tool is used for friction measurements, the rotating tool for support during the test. By using a tensile tester for clamping the strip material, experiments can be conducted under conditions of controlled (plastic) deformation of the strip.

De Rooij [25] performed unlubricated sliding tests with the cylinder-on-strip tribometer for 0.5 m on Al 6061 T4 strip material, with several coated cylinders made of WN 1.2379. Results clearly showed that a nonmetal containing DLC prevented galling to occur; only details from the highest roughness peaks are removed due to the sliding actions. PVD TiN, TiC, CrN, and TiCN did not prevent galling or produced friction coefficients lager than 1. As can be seen from Fig. 9, polishing has a large effect on the frictional behavior, indicating

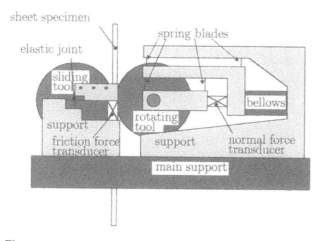

Figure 8 Friction device for the cylinder on strip tribometer. (Adapted from Ref. 25.)

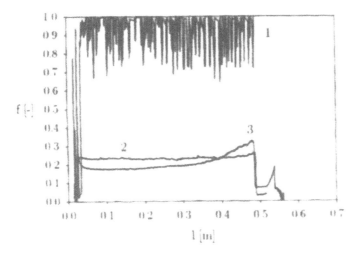

Figure 9 Coefficient of friction for CrN (1), polished CrN (2) and CrN + MoS2 topcoat (3) in sliding contact with A1 6016 T4, no lubrication, v = 0.005 m/s, Fn = 350 N. (Adapted from Ref. 25.)

the importance of tool roughness. A top coating of MoS_2, which could offer benefits in reciprocating sliding contacts, wears off within 0.5 m sliding distance, leaving a CrN coating as surface.

C. Long Sliding Length Laboratory Tests

1. Lubricated Sliding

A well-accepted intermediate stage between practice and laboratory testing uses a rather complex test piece geometry involving bending over a radius [44] or drawing a strip between

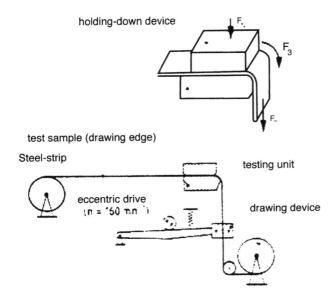

Figure 10 PtU model wear test, TU-Darmstadt. (Adapted from Ref. 46.)

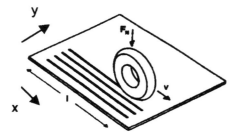

Figure 11   Schematic representation of the TNO slider—on—sheet test.

two stationary test pieces, respectively simulating the sheet-tool contact at the die radius and the blankholder contact during deep drawing (Fig. 10).

Secondly, long sliding length experiments can be performed with the Netherlands Organisation for Applied Scientific Research (TNO) slider-on-sheet tribometer (Fig. 11). This tribometer consists of a sliding contact between a ring, made of the tool material of interest, and sheet material used in the application. Each track is made next to the previous track, in the same direction with sliding speed v and under a normal load of F_n, thus assuring virgin sheet material in the contact. If the tracks are made 1 mm apart from each other, it is possible to realize 1-km sliding distance on 1-m^2 sheet material.

Results with the PtU model wear test indicated that long tool lives are reached when material transfer from the sheet to the tool is avoided. The ability to avoid galling is expressed in terms of roughness of the strip material before and after sliding contact. Measurements of the maximum scratch depth on the (lubricated) cold-rolled steel strip showed that hard chromium plating, boronizing, and vanadizing improved the lifetime of the die radius between 5 and 10 times, compared with uncoated, hardened die radii. The application of nitriding, CVD TiC, CVD TiN, and CVD TiC + TiN, however, prevented scratch formation within the measured test length. Consequently no lifetime is measured and therefore no ranking was possible. For lubricated stainless steel strip, the results differed slightly. Nitriding, hard chromium plating, boronizing, and CVD Cr_xC_y increased tool life slightly, but in all cases deep scratches were formed at the stainless steel strip surface. The use of aluminum bronze, CVD TiC, CVD TiN, and vanadizing prevented scratch formation within the tested sliding distance [44]. Experiments with PVD TiN-coated cast steel WN

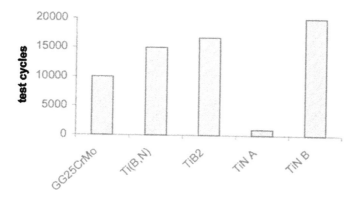

Figure 12   Lifetimes of TiN coated GG 25CrMo, with (TiN B) and without Ti intermediate layer (TiN A) compared to Ti(B,N), TiB$_2$ coated and uncoated grey cast iron. (Adapted from Ref. 46.)

Table 3 Lubricant Properties

Code	Kin. viscosity (mm^2/sec, 40°C)	Density (kg/m^3)	Chlorine content (wt.%)	Sulfur content (wt.%)
Lub1	160	0.900	0	0
Lub2	228	0.910	<0.05	≤0.3
Lub3	2.5	0.809	<0.01	0.26

1.0443 (185 HV) showed a decrease in material transfer in combination with aluminum- and zinc-coated steel, compared with uncoated steel [45]. An intermediate Ti layer is used to increase the bonding of the layer to the substrate. SEM analyses indicated plastic deformation of the tool steel due to the sliding contact, which resulted in local coating detachment. In Ref. 46, it is shown that gray cast iron GG 25 CrMo is also suited for thin film deposition, be it in combination with an intermediate layer, which prevents delamination initiation at the interface due to the presence of graphite lamellae (Fig. 12).

Similar results are found using the TNO slider-on-sheet tribometer with stainless steel sheet AISI 304. Three lubricants designed for light duty use with stainless steel are selected (Table 3) to investigate the potential of PVD TiN-coated PM tool steel (Vanadis 23, hardened at 1150°C and tempered three times at 550°C to a resulting hardness of 64 HRc). One slider is coated with a commercially available PVD TiN layer. The coated and the uncoated rings have similar roughness patterns created by polishing (R_a of about 0.05 μm parallel and perpendicular to the sliding direction). Typical friction and wear data in sliding contact with uncoated Vanadis 23 are presented in Fig. 13. After 60 m sliding distance, friction increased to a relatively high level with large deviations within one track ($f = 0.22$, SD = 0.05). The resulting depth of the tracks in this regime is also high ($R_y = 22$–24 μm). A detailed view of the slider's wear scar shows a transferred layer of stainless steel (Fig. 14). Because friction is closely related to the scratch depth it is possible to define the lifetime of system by taking the sliding distance at which f becomes larger than x. The latter value depends on the actual application [41].

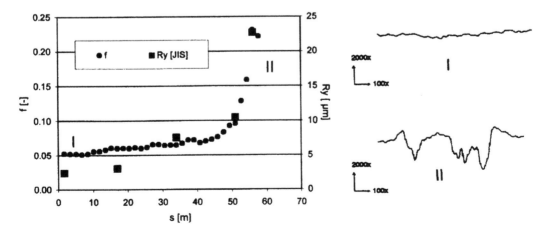

Figure 13 Typical friction and wear data for galling experiment. The profile measurements of the deformation tracks on the sheet correspond with the positions in the graph indicated by I and II (Lub2, $F_n = 100$ N, $v = 0.5$ m/s).

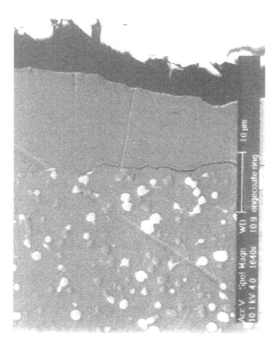

Figure 14 Transferred stainless steel on top of the PM tool steel.

Performing the experiments with undoped or less doped lubricants showed similar results; however, galling occurs at shorter sliding distances. The latter lubricants have significant practical advantages, i.e., easy to clean and environment-friendly formulation. Experiments performed with the PVD TiN-coated slider clearly demonstrated the potential of this surface-treatment technique by enhancing the lifetime of the easy-to-clean lubricants to more than 3 km of sliding distance.

Another set of experiments is performed with sliders made out of conventional tool steel WN 1.2379. A selection of commercially available surface treatments is applied to this set (see Table 4). Not all surface treatments gave the same results in combination with stainless steel under the conditions tested. Figure 15 compares the $L_{f > 0.15}$ values for a set of surface treatments, perfectly showing the superior behavior of PVD TiCN, CVD TiN, and TiC in preventing sliding wear, compared with conventional nitrocarburizing and hard chromium plating.

2. Nonlubricated Sliding

The effect of removing the lubricant from the sliding contact can be demonstrated by sliding tests on aluminum or stainless steel sheets. A comparison of the friction behavior of three systems is given in Fig. 16. High friction in this graph is accompanied by severe material transfer from the sheet to the tool. Ceramic coatings such as PVD CrN, TiN, TiCN, etc. cannot prevent severe material transfer from the sheet to the tool material; therefore galling occurs in a relative short sliding length, which is similar to the uncoated case. This behavior is also found for sliding on stainless steel sheet. Diamond-like carbons are considered suitable to prevent material transfer, although these types of layers are consumed during the sliding contact with the sheet material.

Table 4 Tool Coatings

Substrate	Surface treatment	R_a roughness (µm)		Layer thickness (µm)	Substrate hardness HV10
		//	⊥		
WN 1.2379	Trough hardened	0.02	0.04	—	783
	Nitrocarburizing	0.22	0.16	80	524
	Chromizing	0.05	0.06	19	722
	Chromium plating	0.06	0.04	58	785
	Electroless nickel L1	0.03	0.02	78	785
	Electroless nickel dispersion layer N	0.06	0.09	44	785
	Electroless nickel dispersion layer L2	0.08	0.06	20	785
	CVD Cr_xC_y	0.10	0.09	7	773
	CVD TiC	0.05	0.06	2	751
	CVD TiN	0.04	0.04	4	772
	PVD TiCN	0.07	0.08	5	505

3. Soft Tooling

Volume loss of alternative soft tooling materials in sliding contact with zinc-coated steel sheet is an important aspect for low-volume tools (prototyping, special series) [12]. Research done with 0.7-mm zinc-coated, cold-rolled deep draw steel DC 04 showed a large effect of the zinc layer type. Three zinc coating types are used: hot dip galvanized (GI), electrogalvanized (EZ), and galvannealed (GA); see Table 5 for characteristic roughness and hardness data. Slider-on-sheet tests are done with sliders made from the selected alternative tooling

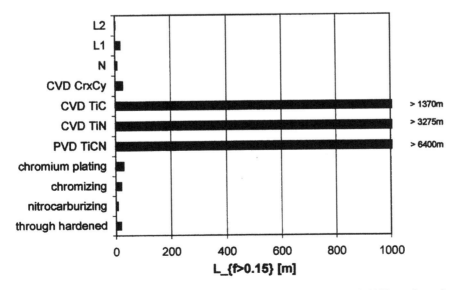

Figure 15 Slider-on-sheet test results for surface treated WN 1.2379 tool steel on lubricated stainless steel. Lub1, F_N = 200 N, v = 0.5 m/s.

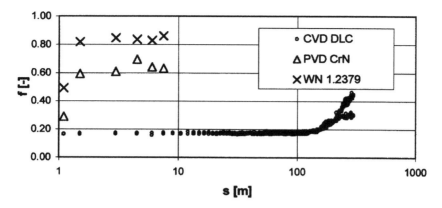

Figure 16 Coefficient of friction as a function of the sliding distance for coated and uncoated WN 1.2379, in sliding contact with AI 5754 O, rinsed with acetone, no additional lubricant. $F_N = 50$ N, $v = 0.5$ m/s.

materials. Each material is tested on the three sheet materials. Table 6 summarizes the results. The combinations indicated by the letter "g" in Table 6 were prone to galling, and, consequently, no k values can be given. Results show that the wear rate of a soft tool material can change two orders of magnitude as a result of the zinc-layer type used. Furthermore, it is shown that the relative performance of alternative tool materials is strongly related to the hardness of the (tooling and sheet) materials. Industrial forming tests with a selection of alternative tooling materials confirmed the model wear test results.

V. INDUSTRIAL FORMING TESTS WITH COATED TOOLS

A. Introduction

The effect of tool coatings on wear is reported in a variety of papers, accompanied by many leaflets from coating companies claiming good results with their specific layers. From this set a small selection is taken in this section, illustrating the possibilities of thin hard layers in lubricated and dry forming tests. The presented forming tests are generally in good agreement with the presented test results of Sec. IV, although a prediction is hard to give of the amount of products that can be made in practice.

B. Lubricated Forming Tests

Vetter et al. [47] investigated the effect of replacing uncoated tool by PVD CrN-coated tools for a set of combined stamping and drawing applications. In most cases, an increase in tool

Table 5   Characteristic Roughness and Hardness Data Sheet Material

| Sheet | R_a (μm) | | Hardness zinc layer |
	// rolling direction	⊥ rolling direction	
GI	1.07	1.08	54–59 HV0.001
EZ	1.10	1.09	52–53 HV0.001
GA	1.14	1.23	232–319 HV0.005

Table 6 Wear Coefficients of Alternative Tool Materials in Sliding Contact with Zinc-Coated Steel

k tool material 10^{-6} (mm^3 N^{-1} m^{-1})	Epoxy resin	Laminated wood	ZnA14Cu3	Cast iron		
				GG-25	GG-30	GGG-60
GI	18.7	26.5	2.41	g	g	g
EZ	56.1	40.2	3.67	g	g	g
GA	1450	664	89.5	0.716	0.472	0.161

$F_n = 100$ N, $v = 0.37$ m/sec.

life is found in the range from 3 to more than 100 times compared with the uncoated situation. Second, it was possible to reduce the amount of lubricant in some cases when replacing uncoated tool by PVD CrN-coated tools. According to Vetter et al. [47], low friction and avoidance of material transfer can be explained by the generation of Cr_2O_3 at the coated surface.

An extensive report of the application of PVD CrN applied by various processes is given in Ref. 48. In certain forming applications a typical increase in tool life of 3–6 times is reported using low-temperature-deposited PVD CrN (200°C, 5 μm) on conventional tool steels. CrN was also successfully used in cutting of Cu, Al, Ni, and Ti alloys and in pressure die casting of Zn and Al alloys.

Austenitic stainless steel WN 1.4301 (AISI 304) is known for its tendency to gall. Severely damaged products are reported when applying a simple axisymmetrical deep drawing process and nonchlorinated lubricants in combination with uncoated tool steel WN 1.2379 [6,49]. The use of PVD TiN-coated WN 1.2379 prevented galling within the used series [49]. The same results are found by [50] using a different product geometry based on bending. The effect of ceramic coatings applied by PVD on WN 1.2379 becomes especially pronounced when using nonchlorinated lubricants (see Table 7). The drawing radius of the aluminum bronze tool used within this research was subjected to volume wear.

A U-bending test designed to rank the galling resistance of tool materials is used to compare coated and uncoated tools with respect to Al 5052 and Al 1050, using a lubricant with 5 wt.% Cl [51]. Results showed that uncoated tool steel WN 1.2379 is sensitive to galling, taking into account the scratched area on the sheet and the depth of the scratches. Applying one of the tool coatings, CVD TiC, PVD CrN, or PVD TiN, clearly reduced this sensitivity, regardless of the coating used. Best results are reported for tailored cermet

Table 7 Results of Hat Drawing Tests with Coated Tools and Different Lubricants

	Treatment	Lub. A, 71 cSt	Lub. B, 190 cSt	Lub. C, 160 cSt	Lub. D, 80 cSt	Lub. AE, 36 cSt
WN 1.2379	Hardened	20,000	2,400	2,000	500	500
WN 1.2379	CVD TiN	20,000	4,300	20,000	19,000	2,000
WN 1.2379	PVD TiCN	20,000	20,000	20,000	2,000	200

The maximum test run was 20,000 hats. Sheet material AISI 304 (EDT-) skin passed.
Lub. A, Cl-containing reference lubricant.
Lub. B–D, alternative lubricants (no Cl).
Lub. E, preservation oil for DC type of sheet.
Source: Adapted from Ref. 50.

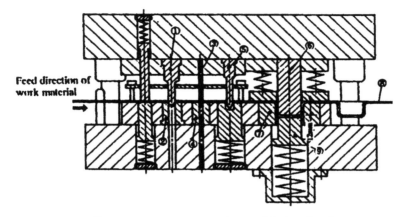

① Punch for piercing a hole to guide the pilot pin,
② Die for piercing the hole to guide the pilot pin,
③ Punch for piercing the blank, ④ Die for piercing
the blank, ⑤ Pilot pin, ⑥ Deep drawing pinch,
⑦ Deep drawing die, ⑧ Work material (aluminum
sheet), ⑨ Stripper

Figure 17 Progressive die used in nonlubricated drawing tests with Al sheet material. (Adapted from Ref. 52.)

consisting of Ni and MoB (1020 HV). A nondefined cemented carbide performed in between the coated tool steel and the MoB cermet.

C. Nonlubricated Forming Tests

A demonstration of the potential of DLC coatings in poorly lubricated forming process with Al sheet material is given in Ref. 52. A progressive die, shown in Fig. 17, used for cup drawing of a 1-mm-thick aluminum 5000 alloy, is coated with various coatings. Results with a poor-performing lubricant (water-soluble) again pointed out that replacing today's highly doped forming lubricants by relatively harmless, environmentally friendlier alternatives increases the speed at which galling dominates the forming process. Table 8 summarizes the results of the test.

Table 8 Effect of Lubrication on Galling for a Cup Drawing Test Using the Tooling of Fig. 17

	Tool life	
Coating	Water-soluble lubricant	No lubricant
None	233	0
TiCN	2237	10
DLC	1216	1500
WC/C	>7000	15
HVOF sintered carbide (70 μm) + DLC	—	6190

The PVD coatings are applied on tool steel WN 1.2379. Sheet material is Al 5000.
Source: Adapted from Ref. [52].

In this case, the metal containing hard carbon coating PVD WC/C enables the production of more than 7000 products. The amorphous carbon coating suffered from partial delamination after 1216 products. An uncoated tool could only produce 233 products. Although not reported, it is less likely that the other two conventional surface treatments, chromium plating or nitriding, would improve this behavior. The same tests repeated without lubrication clearly underlines the demand for new surface-treatment technology, incorporating near-zero wear in dry sliding conditions. Only with the DLC-coated tool products could this be made.

Similar results for axisymmetrical deep drawing 1.5-mm-thick Al 1200-0 are found by Bolt [53]. Without lubrication the conventional heat-treated WN 1.2379 tool was not able to make products because of rupture of the sheet material (high friction caused by severe scratching). Drawing with PVD CrN tools did not improve the performance of the tooling. By using a DLC-coated tool, 4500 products could be made. However, microscopy revealed local delamination of the coating.

The application of DLC and metal-containing DLC films in dry forming applications with stainless steel is still in its development stage, although positive results are reported when using very hard a-C:H coatings ($H > 30$ GPa) [37]. The main problem is delamination of the coating at highly stressed positions at the tool.

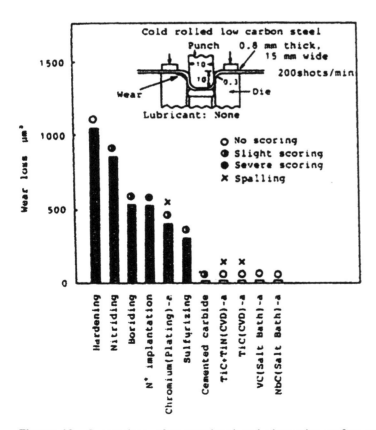

Figure 18 Comparison of conventional and alternative surface treatments under dry forming conditions. (Adapted from Ref. 54.)

Less difficult to form without lubricant is cold-rolled low-carbon steel. Axisymmetrical drawing tests performed in 1983 by Arai [54] showed that CVD TiC, CVD TiC + TiN, TD-VC, and TD-NbC are perfectly capable of avoiding galling, although spalling occurred in some cases. From Fig. 18 it can also be seen that conventional surface treatments cannot protect the tool from wear loss (hardening) or adherence of sheet material (nitriding/chromium plating).

VI. CONCLUSIONS

The application of tribology to SMF processes contributes to the general industrial aim to make products of high quality at an increasingly competitive way by increasing the tool life and realizing a constant level of friction during forming. Lubricant technology in combination with advanced friction models enables the designer of metal-forming processes to estimate the coefficient of friction accurately. An example of this is given Fig. 4, where theory and experiments are in agreement. Prediction of wear in SMF is less developed as friction modeling. A close look at wear-related failure in metal forming showed that sliding wear, and more specifically volume loss of tool material, and galling are the dominant wear types. The selection and application of coated tools should therefore be focused on minimizing these wear types. A proven way to handle sliding wear is to use heat-treated, nitrided, or hard chromium-plated tools. Nonlubricated forming conditions combined with the introduction of relatively hard sheet materials enhances, however, the need for alternative thin or thick hard layers.

Results from laboratory-scale tribotests, suited for simulation of contact situations that arise during SMF processes, indicate the positive effect of vapor-deposited layers in lubricated sliding experiments. Long sliding test results show that DLC coatings are potentially suited for poorly lubricated processes even for sheet materials known for their sensitivity for galling such as stainless steel. Forming tests at industrial scale have proven the positive effect of especially vapor-deposited layers for lubricated and nonlubricated forming.

REFERENCES

1. Czichos, H. *Tribology: A Systems Approach to the Science and Technology of Friction, Wear and Lubrication*; Tribology Series, 1; Elsevier Scientific Publishing Company: Amsterdam, 1978.
2. ter Haar, R. *Friction in Sheet Metal Forming: The Influence of (Local) Contact Conditions and Deformation*; Ph.D. thesis, University of Twente, The Netherlands, 1996.
3. Sniekers, R.J.J.M. *Friction in Deep Drawing*; Ph.D. thesis, Eindhoven Technical University, The Netherlands, 1996.
4. Schipper, D.J. Transitions in the Lubrication of Concentrated Contacts; Ph.D. thesis, University of Twente, The Netherlands, 1988.
5. Bolt, P.J. *Materialen-vormgeven van dunne metaalplaat*; FME-CWM VM 111: Zoetermeer, NL, 1996; 20–42.
6. Andreasen, J.L.; Eriksen, M.; Bay, N. Major process parameters affecting limits of lubrication in deep drawing of stainless steel. In Proceedings 1st International Conference on Tribology in Manufacturing Processes '97; Dohda, K., Nakamura, T., Wilson, W.R.D., Eds.; Gifu University: Gifu, Japan, 1997; 122–127.
7. Consemüller, K.; Hribernik, B.; Schneider, G. Tool steels present and future. In Proceedings of

the 4th International Conference on Tooling; Berns, H., Hinz, H.-F., Hucklenbroich, I.-M., Eds.; Verlag Schürmann and Klagges KG: Bochum, Germany, 1996; 3–16.

8. Sandberg, O.; Hillskog, T. Dimensional stability in service and properties of tool steels for high precision tooling, Proceedings Advances in Powder Metallurgy & Particular Materials-1999; Metal Powder Industries Federation: Princeton, NJ, 1999; Vol. 9, 79–92.

9. Steels for cold work tooling, Uddeholm Tooling; Hagfors, Sweden, 2001.

10. Santham, A.T.; Tierney, P.; Hunt, J.L. Cemented carbides. *ASM Handbook*, Properties and Selection: Nonferrous Alloys and Special-Purpose Materials, Section Special-Purpose Materials; ASM: Materials Park, OH, October 1, 1999; Vol. 2.

11. Doege, E.; Dröder, K. Einsatz von Keramik als Werkzeugwerkstoff in der Blechumformung. Bänder, Bleche, Rohre 1997, *12*, 16–21.

12. van der Heide, E.; Burlat, M.; Bolt, P.J. Wear of alternative forming tool materials due to sliding contact with zinc coated steel sheet in press operations. EUROMAT 2001, AIM: Rimini, Milano, Italy, 2001.

13. Dizdar, S. Formation and Failure of Chemireacted Boundary Layers in Lubricated Steel Contacts, Ph.D. thesis, KTH, Stockholm, Sweden, 1999.

14. Blok, H. The postulate about the constancy of scoring temperature. In *Interdisciplinary approach to friction and wear, Symposium, Troy*; Ku, P.M., Ed.; NASA SP-237: New York, 1969; 153–248.

15. Bowden, F.P.; Tabor, D. *The Friction and Lubrication of Solid*; Clarendon Press: Oxford, 1950.

16. Kenbeek, D.; Buenemann, T.F.; Rieffe, H. Review of organic friction modifiers—contribution to fuel efficiency? SAE 2000-01-1792: Warrendale, PA, 2000.

17. de Gee, A.W.J., Rowe, G.W., Eds. Glossary of Terms and Definitions in the Field of Friction, Wear and Lubrication—Tribology, Technical Report, IRG-OECD; OECD Publications: Paris, France, 1969.

18. DIN 50320: Verschleiß-Begriffe, Systemanalyse von Verschleißvorgängen, Gliederung des Verschleißgebietes. Beuth Verlag: Berlin, 1979.

19. Zum Gahr, K.-H. *Microstructure and Wear of Materials*, Tribology Series 10; Elsevier Science Publishers: Amsterdam, 1987.

20. Christiansen, S.; De Chiffre, L. Topographic characterization of progressive wear on deep drawing dies. Tribol. Trans. 1997, *40* (2), 346–352.

21. Hogmark, S.; Hedenqvist, P.; Jacobson, S. Tribological properties of thin hard coatings: demands and evaluation. Surf. Coat. Technol. 1997, *90*, 247–257.

22. Rigney, D.A.; Chen, L.H.; Sawa, M. Transfer and its effects during unlubricated sliding. In Proceedings of Symposium on Metal Transfer and Galling in Metallic Systems, TMS; Merchant, H.D., Bhansali, K.J., Eds.; TMS: Orlando, FL, 1986; 87–102.

23. Schedin, E.; Lehtinen, B. Galling mechanisms in lubricated systems: a study of sheet metal forming. Wear 1993, *170*, 119–130.

24. Schedin, E. Galling mechanisms in sheet forming operations. Wear 1994, *179*, 123–128.

25. de Rooij, M. B. Tribological Aspects of Unlubricated Deepdrawing Processes, Ph.D. thesis, University of Twente, The Netherlands, 1998.

26. Nicoletto, G.; Tucci, A.; Esposito, L. Sliding wear behaviour of nitrided and nitrocarburized cast irons. Wear 1996, *197*, 38–44.

27. Sugimoto, Y.; Arai, T. Present situation of application of hard coatings onto tooling in Japan. Proceedings ITC, Nagoya; Japanese Society of Tribologists: Tokyo, Japan, 1990; 1981–1986.

28. Bushan, B.; Gupta, B.K. *Handbook of Tribology—Materials, Coatings and Surface Treatments*; 1st Ed.; McGraw-Hill: New York, 1991; 10.22–10.40.

29. Newby, K.R. Industrial (hard) chromium plating. *ASM Handbook Surface Engineering*; ASM: Materials Park, OH, 1994; Vol. 5, 177–191.

30. Emmens, W.C. Tribology of Flat Contacts and Its Application in Deep Drawing, Ph.D. thesis, University of Twente, The Netherlands, 1997.

31. Frenk, A.; Kurz, W. Microstructural effects on the sliding wear resistance of a cobalt-based alloy. Wear 1994, *174*, 81–91.

32. Kron, P.; Bleck, W.; Dahl, W. Laser surface alloying of tool steels-characterization in respect to microstructure and wear. In Proceedings 4th International Conference on Tooling; Berns, H., Hinz, H-Fr., Hucklenbroich, M., Eds.; Verlag Schürmann and Klagges KG: Bochum, Germany, 1996; 411–419.

33. Driessen, J.P.A.M. Low-Temperature Chemical Vapour Deposition of Titanium Nitride, Ph.D. thesis, TU-Delft, The Netherlands, 1999.

34. Grill, A. Review of the tribology of diamond-like carbon. Wear 1993, *168*, 143–153.

35. Bewilogua, K.; Dimigen, H. Preparation of W-C:H coatings by reactive magnetron sputtering. Surf. Coat. Technol. 1993, *61*, 1344–150.

36. Meneve, J., et al. Low friction and wear resistant a-C:H/a-Si$_{1-x}$C$_x$:H multilayer coatings. Surf. Coat. Technol. 1996, *86–87*, 617–621.

37. Taube, K. Carbon-based coatings for dry sheet-metal working. Surf. Coat. Technol. 1998, *98*, 976–984.

38. Vanhulsel, A., et al. Study of the wear behaviour of diamond-like coatings at elevated temperatures. Surf. Coat. Technol. 1998, *98*, 1047–1052.

39. Bushan, B.; Gupta, B.K. *Handbook of Tribology—Materials, Coatings and Surface treatments*; 1st ed.; McGraw-Hill: New York, 1991; 11.36–11.45.

40. TRD method. Proc. ITC, Nagoya; Japanese Society of Tribologists: Tokyo, Japan, 1990; 1965–1972.

41. van der Heide, E.; Huis in 't Veld, A.J.; Schipper, D.J. The effect of lubricant selection on galling in a model wear test. Wear 2001, *251/1–12*, 973–979.

42. Franklin, S.E.; Beuger, J. A comparison of the tribological behaviour of several wear-resistant coatings. Surf. Coat. Technol. 1992, *54/55*, 459–465.

43. Vermeulen, M.; Scheers, J. Effect of pre-welded blanks on the wear of deep drawing tools, presentation themadag tribologie VPT, Apeldoorn; TNO Industrial Technology: Eindhoven, Netherlands, 1996.

44. Woska, R. Einfluß ausgewählter Oberflächenschichten auf das Reib-und Verschleißverhalten beim Tiefziehen. Ph.D. thesis, TU Darmstadt, Germany, 1982.

45. Schulz, A.; Stock, H.R.; Mayer, P.; Staeves, J.; Schmoeckel, D. Deposition of TiN PVD coatings on cast steel forming tools. Surf. Coat. Technol. 1997, *94–95*, 446–450.

46. Matthes, B., et al. Tribological properties and wear behaviour of sputtered titanium-based hard coatings under sheet-metal-forming conditions. Mater. Sci. Eng. 1991, *A1 40*, 593–601.

47. Vetter, J., et al. Hard coatings for lubrication reduction in metal forming. Surf. Coat. Technol. 1996, *86–87*, 739–747.

48. Navinšek, B.; Panjan, P.; Milošev, I. Industrial applications of CrN (PVD) coatings, deposited at high and low temperatures. Surf. Coat. Technol. 1997, *97*, 182–191.

49. Foller, M.; Schruff, I.; Uhlig, G. Improved deep drawing properties of austenitic stainless steels due to PVD-coated tools. In Proceedings 5th International Conference on Tooling; Jeglitsch, F., Ebner, R., Leitner, H., Eds.; Montauniversität Leoben: Leoben, Austria, 1999; 327–334.

50. Jordan, F.; Heidbuchel, P. Friction and wear of stainless steel, IDDRG Working Group Meetings, Birmingham, UK, June 1999.

51. Sato, T.; Besshi, T. Anti-galling evaluation in aluminum sheet forming. J. Process. Technol. 1998, *83*, 185–191.

52. Murakawa, M.; Koga, N.; Takeuchi, S. Utility of diamondlike carbon-coated dies as applied to deep drawing of aluminum sheets. In Proceedings of the 1st International Conference on Tribology in Manufacturing Processes '97; Dohda, K., Nakamura, T., Wilson, W.R.D., Eds.; Gifu University: Gifu, Japan, 1997; 322–327.

53. Bolt, P.J. TNO report 99MI-5116, "Thin—precoated—aluminum sheet," Eindhoven, The Netherlands, 1999.

54. Arai, T. Technical report, Toyota Central R&D Lab, 1983. In *Tribological Properties of Thin Hard Coatings Used in Metal Forming*, Proceedings ITC, Nagoya; Dohda, K., Ed.; Japanese Society of Tribologists: Tokyo, Japan, 1990; 1973–1980.

11

Tribology in Textile Manufacturing and Use

Stephen Michielsen
College of Textiles, North Carolina State University, Raleigh, North Carolina, U.S.A.

I. INTRODUCTION

Since earliest times, fibers have played a key role in the survival of the human species, from use in food storage (baskets), weapons (strings), shelter (skins sown together), and clothing. Today, textiles are used in a wide variety of industries, including defense (composites in aircraft), civil engineering (roofing, soil stabilization, reinforced concrete), transportation (synthetic tire cords, tarps, straps), medicine (artificial arteries and tendons), as well as the traditional carpet, apparel, and home furnishings. As demand for these products has increased, so has the fiber processing speeds. Initially, natural oils and waxes were sufficient to satisfy the tribological criteria. Today, synthetic fibers are spun at speeds in excess of 6000 m/min, 24 hr a day, 7 days a week. Without proper lubrication, both the fibers and the processing equipment are quickly damaged. In processing short staple fibers (e.g., cotton, polyester staple), rotor speeds in excess of 100,000 rpm are used. In these cases, the friction cannot be too high or the fibers will abraid or even melt, thus disrupting this high-speed process. Perhaps surprisingly, the friction must also not be too low or the fibers will slide off the rollers in the process and cause process upsets. In staple fiber processing, there must also be substantial fiber-to-fiber cohesion or the fibers will simply pull apart during processing. The fibers also are contacted by many different surfaces, e.g., chromed steel rollers, ceramic guides, steel or ceramic travelers, and rubber rollers. At each point in these complex processes, the friction must be within the correct range or the process will be disrupted. Typically, a fiber "finish" is applied to control the friction and the static charge dissipation. (Note: For the purpose of this chapter, fibers and filaments will be used interchangeably. Strictly speaking, filaments are long, continuous narrow stands, while fibers are short, narrow strands.)

Once the filaments have been made, there are many more processing steps. These include cutting them into short lengths, called staple fibers, spinning these fibers into yarns (the cotton process), heat setting the yarns to stabilize their structure, weaving, knitting, and scouring, to name a few. In each process, machine parts rub against the fibers. This results in yet another problem, triboelectric charging of the fibers. The charged fibers repel each other, making processing impossible. In addition, large voltages can lead to arcing, which can, in turn, ignite flammable vapors from the lubricants. In a computer room, triboelectric charging of the operator's shoes from walking across the carpet can lead to destruction of very expensive equipment and data. Considerable research has gone into developing fiber finishes that provide the proper lubricity and sufficient antistatic characteristics. Without these finishes, textile manufacturing could not exist as we know it today.

A third area where lubricants are often required is in the end use. For some applications, high friction is desirable, others need very low friction, and still others need an intermediate friction, but it must be of the correct level. For example, in polyaramid body armor, if the interfiber friction is too low, the fibers spread apart when the projectile hits the garment, allowing it to penetrate easily. If the interfiber friction is too high, the fabric is extreme stiff and terribly uncomfortable. Fibers used to reinforce rubber goods (tires, belts) must adhere to the rubber. Frequently, the adhesion is imparted via a surface adhesion-promoting layer. Carpet fibers must not pick up oil (soil) and must dissipate triboelectric charges caused by walking across the surface. Apparel fibers must continue to have reduced friction to impart softness, yet not pick up body oils (ring around the collar). Obviously, this is a daunting task.

To solve these problems, considerable effort has been spent in developing topical finishes. These finishes must perform the several functions mentioned above. They must act as a lubricant, an antistatic agent, and must provide cohesion to the yarn. Although the finish must reduce friction, it must not lower it too much. During processing, fibers experience many different conditions. In some regions, they travel over metal or ceramic guides, at other times they experience high tension or high pressures on metal rolls, while at other times, fiber–fiber contact is the dominant friction. At each step of the process, the friction must be correct.

At the same time, environmental issues have become important. Some estimates of the scale of this problem are that about 50% of the effluent from textile mills is attributable to the finishes. In other words, U.S. textile mills emit more than 0.5 billion lb of finish oil to the environment annually. Thus, there has been a push to replace these fugitive finishes with permanent ones through surface modifications.

This chapter will not describe the tribological aspects of the machinery for processing fibers, yarns, and fabrics. Rather, it will entirely concentrate on synthetic and natural organic polymer-based fibers including polyester, nylon, cotton, and wool, as well as others. The first section will deal with traditional approaches to fiber finishes. Next, conventional methods for measuring the tribological characteristics are described. The third section will address surface modifications performed on fibers, specifically to modify their tribological behavior. This will include changes in the surface texture as well as permanent chemical modifications. Finally, areas of research that are critical for future surface modification for tribological purposes will be mentioned.

II. TRADITIONAL FIBER FINISHES

As described above, finishes are applied to fibers to lubricate them, to provide antistatic protection, and to provide cohesion of the yarn bundle. These finishes are applied to yarns by two major methods: metering the finish onto the yarn and using a kiss roll, which is usually a ceramic roll that rotates in the finish solution/dispersion and transfers the finish to the yarn on contact. In this section, we will describe the lubricants first, and the antistatic agents second. The last topic in this section will consider some of the physicochemical issues involved in the design of fiber finishes.

A. Lubrication

Two types of friction are critical to fiber manufacturing, boundary friction, and hydro-dynamic friction. Boundary friction is the dominant friction at low speeds and high pressures. This occurs within the yarn or fabric and whenever the fibers travel on rolling surfaces, such as godets. In both cases, the relative speed is nearly zero. In this region, stick–slip behavior is observed where initially no movement occurs between the surfaces even when stress is applied to the yarn. At some stress level, determined by the static friction coefficient and the normal load, the yarn or fibers begin to slip past the counter surface. This continues until the tension drops below a certain value determined by the kinetic friction, whereupon the fibers once again stick to the counter surface. The main purposes of boundary lubricants are to keep the fiber surface from being damaged by the machine parts (slip) and to provide cohesion (stick) within the yarn bundle. They accomplish this by forming a thin surface layer with low shear strength. Typical chemicals used in textile applications for boundary lubricants are fatty-based molecules such as alcohols, alkanes, acids, amines, and amides. Other good boundary lubricants are polyethylene or poly (ethylene glycol) waxes [1]. As the chain length increases, the lubricant becomes solid, resulting in a significant decrease in the boundary friction [2]. In addition, the boundary friction decreases when the molecule is adsorbed onto one or both surfaces and as the chain-end polarity increases, providing that this increases the adsorption of the lubricant onto the surfaces through the polar chain-end.

   Hydrodynamic lubricants provide protection to the fiber at high speeds, such as when the fibers are being pulled across stationary guides and draw pins (used to stretch the fiber to increase its strength). The hydrodynamic lubricant keeps the fiber surface off the machine parts to reduce wear by providing a thin liquid film between the surfaces. The viscosity of the lubricant and the speed at which the yarn passes the machine surface maintains the lubricant film. As the viscosity is increased, the friction increases, but if the viscosity is too low, the film cannot be maintained and the surfaces come into contact, leading to excessive wear. Typical chemicals used for hydrodynamic lubricants are mineral and vegetable oils or their derivatives. Recently, synthetic oils and fatty acid esters have been used because they allow better control of the viscosity and have improved thermal stability [3].

B. Antistatic Agents

Fibers both during processing and during use come into contact with a large number of surfaces, e.g., shoes, furniture, processing rolls, cards, weaving, and knitting machinery components. The constant contacting/loss of contact transitions lead to contact or tribo-electric charging. Large potentials are readily generated (thousands of volts). Unless these charges are dissipated, they can lead to painful electrical discharges or even sparks that can ignite organic vapors. Rolfe [4] lists two types of problems associated with triboelectric charging of textile substrates, nuisance and hazard. Under the hazard category, she includes cling (apparel), dust pickup (carpets, upholstery, curtains, and industrial outerwear), shocks (carpets, car seats, sweaters, laundry belts), machinery disruption (laundry belts, paper-maker felts), and electrical interference (carpets/clothing affecting computers, electronic components, and photographic film). The major hazard is a spark near an inflammable or explosive material. To alleviate this problem, the fibers are treated with antistatic agents during processing. These agents include organic salts, quaternary ammonium salts, alkyl phosphate salts, or polyols. However, after processing, these agents are removed by washing. New antistats can be applied during laundering, but this is not practical for carpets or for clothing that may be exposed to water many times between launderings.

C. Physicochemical Considerations

The spin finish must have the proper boundary and hydrodynamic friction for each surface that the fiber comes into contact with, and provide adequate antistatic protection. It must also be able to be uniformly applied to the fibers, survive the elevated temperatures used in postformation processing, have low vapor pressure, have low surface tension, and be nonhazardous to the operator and the consumer. The finish ingredients should be low-smoke and not lead to varnishing on hot surfaces, and the vapors and their oxidized derivatives should not be harmful to humans. At the same time, any ingredients that wash off the fibers must not pose a health risk in water. Boundary lubricants should adsorb onto the fiber and/or the metal parts of the machinery in a monolayer film. Hydrodynamic lubricants should have the lowest viscosity compatible with processing. The antistatic agent should provide a means to conduct the electric charge away. In addition, these ingredients should not absorb into the fiber or into rubber rollers used in some processes. The spin finish should have a low surface tension to aid in spreading the finish onto/into the yarn, but should not foam. When possible, the spin finish should also perform well over a wide range of processing conditions and in many different processes. Otherwise, the finish may have to be removed and additional finishes applied. This adds both to the cost of manufacturing and to the environmental impact. This is a daunting task, often employing complex mixtures of finish ingredients. Slight changes in the performance can lead to major production losses. Slade [5] has provided a recent review of traditional fiber finish technology.

III. TRIBOLOGICAL MEASUREMENTS IN FIBERS AND TEXTILES

The friction and the ability of the finish to dissipate charge are both functions of the relative humidity and temperature, which must be carefully controlled during the measurements. Several devices have been developed to measure the friction, the charge-generating potential and the resistivity of the fiber finish. The main ones are discussed below.

A. Fiber Friction

There are several methods used to measure the friction between fibers and various other materials, most are based on the capstan equation

$$T_2/T_1 = \exp(\mu\theta) \tag{1}$$

where T_1 is the tension on the fiber or yarn on the low tension side of the circular substrate against which the fiber friction is being measured (see Fig. 1), T_2 is the tension on the high tension side, θ is the wrap angle on the substrate, and μ is the friction coefficient. The main difference in these instruments is the speed at which they measure friction. The substrate can be a single filament, a yarn package, metal parts designed to simulate machine parts, plastics, or other substrate of interest.

A relatively simple device for measuring single fiber to single fiber friction is shown schematically in Fig. 2 (Rame-Hart). In the author's laboratory, this type of tribometer has been interfaced to a computer so that all data collection and analysis is automatically

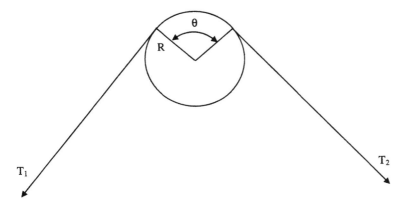

Figure 1 Capstan configuration for measuring friction of fibers. A fiber or yarn with tension T_1 is pulled around a pin of radius R through an angle θ by a tension T_2.

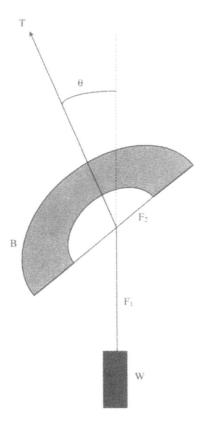

Figure 2 Crossed cylinder geometry for measuring single fiber friction against another single fiber or a metal wire. A fiber or metal wire, F_2, is mounted on a metal bow, B. A second fiber, F_1, is suspended from a load cell, T, and a weight, W, is hung from the other end. Fiber F_1 is deflected from the vertical by an angle θ by moving the bow in the horizontal plane. The friction force is given by $T-W$ as the bow is moved slowly in the vertical plane.

performed. However, only very low speeds are available with this instrument because of the difficulty of handling long lengths of single fibers.

Both commercial and custom-built instruments have been developed to measure the friction of yarns. There are two major designs, as shown in Fig. 3. The main difference is again the speed ranges available with these instruments. The design shown in Fig. 3(a) is used for speeds < 80 m/min, while the design shown in Fig. 3(b) is used for higher speeds. The instrument shown in Fig. 3(a) uses a fixed weight for tension T_1, tension T_2 is measured by the load transducer, and $\theta = 180°$. The substrate, S, can be a metal rod or a small package of yarn. A yarn is attached to the load cell, draped over S, and the fixed load attached to the other end. A motor rotates the substrate at various speeds and the tension T_2 is recorded. The output from T_2 can be sent to a strip chart recorder or to a computer for recording and subsequent analysis. The device shown in Fig. 3(b) is generally used to measure fiber to metal friction at moderate to high speeds. The Rothschild F meter is one such instrument. In this case, the yarn is pulled through a tension gate to control the input tension, which is measured by a tension transducer to obtain T_1. The yarn is wrapped around a pin through a known angle, usually 180°. The pin is often chrome. The yarn then passes through another tension

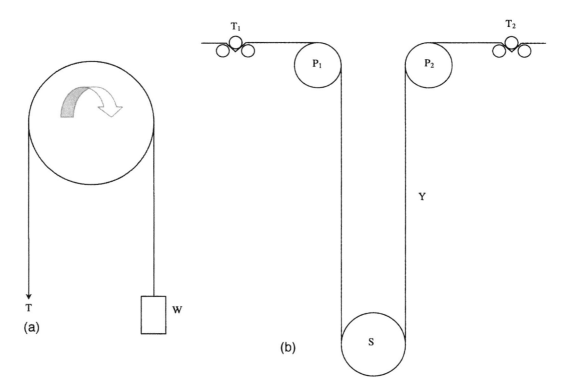

Figure 3 Higher-speed tribometers. (a) Moderate-speed tribometer. A yarn is attached to a load cell, T, and placed around a rotating cylinder. A weight, W, is hung from the other end. The cylinder can be a metal or a carefully wrapped yarn package. The friction is measured according to the capstan equation. (b) High-speed tribometer. Yarn Y passes through tension measuring device T_1, across a very low friction pulley, P_1, around the test substrate, S, across another very low friction pulley, P_2, and finally through a second tension measuring device, T_2. The yarn can be pulled through this system at speeds of a few thousand meters per minute.

transducer to obtain T_2, and around a drive shaft that pulls the yarn through the instrument. Finally, it is fed into a waste container. Typical speeds in this device are hundreds to a few thousand meters per minute.

B. Fabric Friction

Figure 4 shows the basic procedure for measuring fabric friction against other fabrics or process equipment. The fabric is attached to a sled and the sled is placed on the other substrate (other fabric or materials similar to the processing equipment). A weight is placed on the sled and the sled is pulled in the horizontal plane at a constant speed. The force required to move the sled is recorded and the friction coefficient, μ, is calculated:

$$\mu = F/W \tag{2}$$

F is the force required to pull the sled and W is the weight of the sled, the additional weight and the fabric sample. Because of their simplicity, these devices are usually custom-built and attached to conventional tensile testing equipment, which are designed to move the sled at a constant speed and to measure the force to pull the sled.

C. Triboelectrical Charging and Dissipation

Typically, triboelectric charging of the yarn is measured by pulling the yarn across an insulated pin and measuring the charge deposited on the pin. The Rothschild F meter is easily adapted for these measurements. This is not a simple measure of the triboelectric charging because some of the charge is dissipated by conduction along the fiber to a grounded metal portion of the instrument via the antistatic agents and through moisture in the air. The main purpose of this test is to ensure that dangerous voltages do not develop during fiber processing.

There are a number of methods for measuring the charge dissipation. The simplest of these techniques is to simply wrap the yarn onto two metal pins and measure the resistance to the flow of current at a known voltage. From the number of filaments, N, in the yarn, their

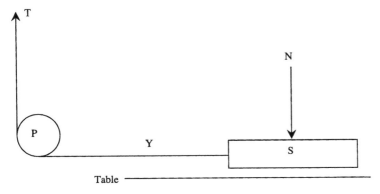

Figure 4 Fabric friction tribometer. The test fabric is attached to the table or the sled, S, or both. A normal load, N, is applied (comprising the weight of the sled and any additional weights). The sled is pulled by a yarn, Y, which wraps around a low friction pulley and is attached to the load cell, T, on a tensile testing instrument's crosshead.

circumference, C, the number of wraps, W, and the distance between the electrodes, D, the surface resistivity, σ_s, can be determined:

$$\sigma_s = \frac{R_m}{D} NWC \qquad (3)$$

where R_m is the measured resistance. The value of σ_s for fibers during processing must be below 10^{10} Ω to prevent dangerous voltages from developing, and for use in computer rooms, it is required that σ_s should be less than 10^5 Ω.

IV. APPROACHES TO SURFACE MODIFICATION

There have been many approaches to modifying the surface of fibers to modify their tribological properties as well as other surface properties. The most common methods are treatments with reactive chemical species in a corona, in a plasma, and under ultraviolet (UV) illumination. Additional treatments include laser ablation, mechanical abrasion, and through chemical activation and subsequent reactions including surface activated polymerization. Recently, surface grafting of polymeric chains has also been explored. Each of these approaches has its own strengths and weaknesses. The most obvious difficulty is the high production speed of fibers. For example, at processing speeds of 6000 m/min, the yarn travels 10 cm in only 1 msec. If the treatment zone is only 1 m long, the surface modification must be complete in less than 10 msec. For many treatments, a 1-m-long treatment zone would be prohibitively expensive, yet often treatment times in excess of several seconds are required. Furthermore, the permanent change in the surface must provide the desired benefit without degrading the other critical properties. It is not surprising that until recently, there have been few commercially acceptable surface treatments except for altering the surface tension (wetting) and perhaps the extent of triboelectric charging. This section will attempt to review the approaches used to date, their strengths, and their shortcomings. It will end with a brief discussion of future work that is needed to expand the development of permanent surface modification.

A. Surface Free Energy

Improving the spreading of the lubricant on the fiber surface improves the performance of hydrodynamic lubricants. Various additives are used for this purpose. Surface free energy (surface tension) controls spreading of fluids on the surface of fibers. The difference between the surface free energy of the fiber and the other materials it contacts also controls the triboelectric charging. Changing the surface free energy is perhaps the easiest surface modification. To lower the surface tension, small alkane, silicone, or fluorochemical molecules are grafted through plasma, UV, or chemical activation. To raise the surface tension to increase wetting, oxygen or air plasmas may be used to increase the surface polarity and thus, the surface tension. In fact, virtually any surface modification will result in a change in the surface tension. Changing the surface tension will also change the tribological characteristics of the fibers. Subsequent sections will provide specific examples of these individual techniques in more detail.

B. Friction

Because hydrodynamic friction depends on the flow of a fluid, it seems unlikely that surface modification can directly imparted hydrodynamic lubrication. However, as mentioned above, improved wetting of the lubricant can improve the performance of the hydrodynamic lubricant. On the other hand, boundary friction only requires that the lubricant be able to prevent direct contact of the two friction surfaces and that the layer between the two surfaces has low shear strength. With a judicious choice of materials, it should be possible to design a system in which the two substrates do not strongly interact to provide low boundary friction. On the other hand, it is also possible to increase the boundary friction by choosing surface materials that adhere to each other. A subsequent section discusses several examples of these approaches.

C. Triboelectric Charging/Charge Dissipation

Because all large-scale commercial fibers are insulators with bulk resistivities of $>10^{12}$ Ω/cm, any charge deposited on the fibers will remain there until they come into contact with a conductor. Thus, when you walk across a carpet in shoes during periods of low humidity and touch a doorknob, a spark jumps between your hand and the doorknob. This unpleasant shock is a result of the triboelectric charging of your shoes and hence of your body.

Arridge [6] showed that contact electrification of nylon 6,6 was directly proportional to the contact potential of the material to which the nylon 6,6 was contacted. Using a contact potential for gold of 4.7 V, he found that nylon 6,6 had a contact potential of 4.2–4.4 V. When contacted by metals with a lower contact potential, the nylon 6,6 picked up a negative charge, while contact with a higher contact potential metal lead to a positive charge. Follows et al. [7] found that absorbed dyes did not affect the contact potential. However, small islands of silicone oil would consistently change the contact potential. After cleaning, the contact potentials of the polyamides returned to their normal value. Davies [8] measured the contact electrification for several additional polymers. His results are shown in Table 1. His value for nylon differs slightly from that of Arridge, but they used different work functions for gold (Arridge used 4.7 eV, while Davies used 4.6 eV.) Thus, rubbing nylon 6,6 with a poly(vinyl chloride) (PVC) comb results in very large charges being generated while rubbing poly(ethylene terephthalate) (PET) fibers with Teflon® or polytetra fluoroethylene (PTFE) results in almost no charge generation.

This suggests that if nylon carpet comes into contact with only one dominate material, the triboelectric charge can be minimized by modifying the nylon surface to have the same contact potential as that of its contacting surface, e.g., leather. However, if two or more materials with significantly different contact potentials commonly come into contact with the carpet fibers, triboelectic charging cannot be eliminated. In this case, a better approach is to permit triboelectric charging, but provide a means to dissipate the charge rapidly.

Table 1 Contact Electrification of Polymer Surfaces

Polymer	Contact potential (eV)
PVC	4.85
Polyimide	4.36
Polycarbonate	4.26
PTFE	4.26
PET	4.25
Polystyrene	4.22
Nylon 6,6	4.08

Brennan et al. [9] applied thin fluoropolymer coatings to nylon films. They measured the charge pickup as a function of the fluoropolymer thickness and found that the charge transfer was confined to a layer less than 3 nm thick. The charge transfer depended approximately on the logarithm of the fluoropolymer thickness up to a thickness of 3 nm. Thereafter, it did not change. This indicates that thin coatings significantly alter triboelectric charging. However, no benefit was seen for thicknesses larger than a few nanometers. Thus, it seems that surface modifcation greatly affected triboelectric charging.

Several approaches have been used to create permanently antistatic garments and carpets. The most common approach has been to incorporate metal or other conductive fibers in the fabric or carpet. Many carpets today contain "antistatic" fibers, often containing carbon particles, to dissipate the charge. DuPont manufactures a carbon black filled core of a core/shell fiber (Negastat™). Another approach has been to modify the surface of the fibers to reduce triboelectric charging or to provide a conductive coating. Rolfe [4] describes a polyester fiber in which a core/shell fiber is spun with a poly(ethylene terephthalate) core and a lower melting temperature copolyester shell. She further modified the copolyester surface by heating the fiber to soften the shell and embedding carbon particles into the surface. This greatly lowers the resistance and helps dissipate the tribo-electric charge rapidly. However, this approach limits the color range of the carpets using these fibers. In addition, this approach is not useful in many garment applications. Thus, a method for making the fiber surface more conductive without imparting color has been sought for a long time.

V. SURFACE MODIFICATION

One way to alter the friction of fibers is to change their surface texture. This forms the basis of both the chemical etching techniques and laser ablation, as described below.

A. Chemical Etching

One method of modifying the surface of fibers is to etch the surface chemically. Early attempts to shrink-proof wool used this approach to modify wool's surface. Scales cover the surface of wool fibers (see Fig. 5.) The friction coefficient depends on the direction of sliding.

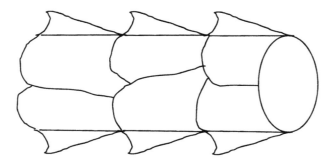

Figure 5 Schematic of the surface of wool. The cortex of the fiber is surrounded by cuticle cells that form a sawtooth structure on the surface of the fibers. The cuticle cells are irregularly shaped and cover the entire surface of the fiber. See Ref. 12 for scanning electron microscopy images of the surface of wool.

When these scales slide against each other, if the scales on both fibers point in the same direction, the friction is low. If they point in opposite directions as the scales slide toward each other, the friction is high. The high friction can be explained by the ratcheting effect [10].

Bradbury [11] treated wool by using several shrinkproofing chemical techniques and measured the changes in the friction coefficients and the amount of surface modification. These treatments included acid chlorination, permanganated treatment, permonosulfuric acid treatment, and the deposition of polyglycine. Each of these treatments affected only the cuticle or surface of the wool fibers. In each case, the difference between the "antiscale" friction coefficient, μ_1, and the "with scale" friction coefficient, μ_2, decreased as the surface modification increased. In particular, they found that polyglycine deposited on the surface of the wool fibers, reducing the surface roughness [12]. Because the roughness decreased, they found that the difference between the "antiscale" friction coefficient, μ_1, and the "with scale" friction coefficient, μ_2, decreased as the coverage increased. Thus, as the degree of surface modification increases, $\mu_1 - \mu_2$, decreases. Although $\mu_1 - \mu_2$ decreased, both μ_1 and μ_2 increased. Indeed, in further studies, Bradbury et al. [13] found that shrinkproofing correlated well with both the degree of surface modification and the difference in the with scale and antiscale friction coefficients. It did not matter whether both friction coefficients increased or decreased, only that the difference decreased.

B. Chemical Etching—PE

Silverstein et al. [14] studied the effect of several strong oxidizers on the surface of polyethylene fibers. They found that chromic acid, potassium permanganate, and hydrogen peroxide all removed oxidized finish oils from the surface of the fibers, but only chromic acid significantly altered the surface texture. It also greatly increased the amount of oxygen species on the surface of the fibers and enhanced wetting. All of these effects are expected to affect the frictional characteristics of polyethylene, but they did not measure the friction. However, the enhanced wetting will improve the performance of topically applied hydrodynamic lubricants.

C. Chemical Etching—PET

Achwal [15] found that chemically etching poly(ethylene terephthalate) (PET) with NaOH increased the softness (related to a decrease in friction) and greatly reduced the static charging. Static charging of the control sample resulted to a potential of 0.4 kV, while static charging of the chemically etched samples resulted in a potential of only 0.1 kV. Charge dissipation times were also reduced from 60 sec in the control to 10 sec in the etched sample. The greatest reductions occurred in etching solutions of ethanolic NaOH and of aqueous NaOH, to which stearyltrimethylammonium chloride was added as a catalyst. In both solutions, the static charging and charge dissipation times decreased the most after the fiber lost 15% of its weight (from the surface).

D. Laser Modification

Bahners et al. [16] found that upon intense UV irradiation, PET developed large surface rolls. The periodicity and the depth of the rolls depended on the intensity and the number of laser pulses. They found that the rolls formed because of extremely high surface temperatures and high thermal gradients at the fiber surface (which are attributed to the high absorptivity of PET at these UV wavelengths). In polymers with lower absorptivity [polyethylene and polypropylene (PP)], no rolls were formed at these exposures. They also found that the roll structure depended on how much the fibers had been permanently stretched during processing, i.e., the draw ratio of the fibers. High draw ratios result in high levels of molecular orientation within the polymer chains. Upon melting, the polymer chains attempt to return to their preferred random coil conformations. This results in substantial contraction, which forms the roll structures.

Schollmeyer and coworkers also examined the effect of irradiating other polymers with high intensity UV eximer lasers. Knitter and Schollmeyer [17] modified the surface roughness of various types of fibers by laser ablation and localized melting with subsequent surface contraction. Although they did not discuss the effects of this surface modification on the tribological properties, they found increased adhesion and increased wetting of the fibers. The increased wetting should enhance the performance of traditional hydrodynamic fiber finishes while the increased surface roughness should alter the frictional characteristics of these fibers.

Wong et al. [18] found differences in the surface morphology based on the intensity of the laser irradiation. At high intensity, material was ablated from the surface, there was localized melting, and carbon-rich particles deposited on the surface resulting in a rough surface with features of the order of a few microns in size. These features would be expected to alter the frictional characteristics while the debris is expected to increase wear of both the fibers and processing equipment. At lower intensities, they found no ablation. However, within a narrow intensity range, surface roughness on a size scale of 100 nm was observed, although there was no surface debris. Although the authors did not report any friction data, these two different structures could be expected to have different frictional characteristics.

E. Core/Shell

One approach to modifying the fiber surface is to spin a bicomponent fiber with a different surface polymer (shell) than the bulk polymer (core) as shown in Fig. 6. For these fibers, the

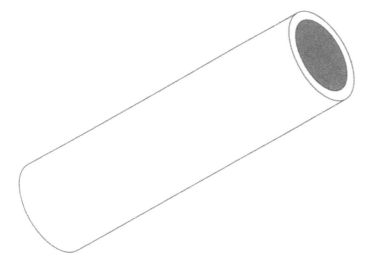

Figure 6   Core/shell fiber. The core is shaded gray while the shell is white.

tribology is determined by the shell polymer, while the core can provide color, stiffness, or conductivity. There are several commercial spinning machines that are capable of making such fibers. Several companies use this method for incorporating carbon particles in the core or in the shell of bicomponent fibers to produce antistatic fibers. Originally, these bicomponent fibers were developed to provide a conductive core to dissipate triboelectrically generated charges. For example, DuPont manufactured a carbon black filled core of a core/shell fiber (Negastat™). Another approach has been to modify the surface of the fibers to reduce triboelectric charging or to provide a conductive coating. Rolfe [4] describes a polyester fiber in which a core/shell fiber is spun with a poly(ethylene terephthalate) core and a lower melting temperature copolyester shell. The surface is modified by heating the fiber to soften the shell and embedding carbon particles into the surface. This greatly lowers the electrical resistance and helps dissipate the triboelectric charge. However, this approach limits the color range of the fibers. In addition, this approach is not useful in many garment applications. Thus, a method for making the fiber surface more conductive without imparting color has been sought for a long time.

The main disadvantages of this approach are that (1) only relatively thick shells (>1 µm) can be made, (2) the equipment is expensive and more difficult to operate, and (3) the surface polymer will affect other critical fiber properties, such as dyeing, hand (how the fabric feels), bending rigidity, strength, and so forth. With the exception of the cases described above, this approach is currently being used to make premium fibers, but without concern for their tribological characteristics.

F. Corona

Yarns and fabrics can be corona-treated by pulling the yarn or fabric between high-frequency, high-voltage electrodes of a corona discharge device. The air around the high voltage electrodes is ionized and the ions impinge on the yarn or fabric. After corona treating polypropylene fibers in air, Hautojärvi and Laaksonen [19] applied varying amounts of lubricant and antistatic agents. They found that corona treatment made the antistatic agent much more effective, reducing the surface resistance by more than 1 order of magnitude, thus reducing the amount of antistatic agent needed. They attributed this enhanced activity to a more uniform spreading of the antistatic liquid because of the increased surface tension of the treated fibers. They also found that triboelectric charging of the fibers, when processed in a card (a device made from steel and used processing of staple fibers), can be reduced by approximately 50%. Finally, they found the fiber to stainless steel friction coefficient to be reduced from 0.57 to 0.36 at a lubricant add on level of only 0.12 wt.%. Thus, the effectiveness of both the antistatic agents and the lubricants were greatly enhanced by increasing the wetting uniformity of the spin finish solutions on PP fibers. This improvement was presumably the result of an increase in the surface tension of the fibers probably caused by the reaction of ionized oxygen species (e.g., ionized ozone from oxygen, hydroxyl ions from water vapor) with the polypropylene surface. Untreated polypropylene has a low surface tension because it is a hydrocarbon. The oxidized surface is expected to have a much higher surface tension thus enhancing wetting.

By addition of other chemicals to the corona device, it is possible to introduce other surface species. Seto et al. [20] used a corona discharge to initiate polymerization of poly(acrylic acid) (PAAc) onto the surface of polyethylene and nylon 6 nonwoven fabrics. They propose that the corona discharge creates peroxides and radicals from oxygen, carbon dioxide, and water vapor in the air. These radicals and peroxides form on the surface of the fibers. These highly reactive species then initiate polymerization resulting in grafted poly(acrylic acid). After treatment, they measured the rate at which an electric charge was dissipated. At 20°C and 40% relative humidity, the untreated fabrics required more than 5 min to lose 1/2 their initial charge, while after treating, the polyethylene-g-polyacrylic acid (PE-g-PAAc) fabric and nylon 6-g-PAAc fabric lost one-half of their initial charges in 8 and in 12 sec, respectively. They attributed these substantial increases in the rate of charge dissipation to the moisture uptake of the PAAc coating. The antistatic properties of these fabrics were greatly improved by this treatment. This is important for the dissipation of triboelectric charges developed during processing.

G. Plasma

Many authors have modified the surface of fibers by using radio frequency (RF) plasma treatments. In this approach, the fibers are placed in a vacuum chamber, the air is pumped out, the desired treatment gas introduced at pressures of about 0.1–3 Torr, and a radio frequency (RF) field turned on. The RF field strips electrons from the gas, which then attacks the surface. The plasma gas may abstract a proton, leaving behind a free radical, or it may directly react with the surface. The radicals can then react with other gases present (such as monomers, thus initiating polymerization) or, after removing the fibers from the chamber, the radicals can react with moisture in the air, forming acidic and alcoholic functionality on the surface. If polymerizable monomers are included in the plasma chamber, polymers can be formed. Plasma polymerized coatings are (usually) highly crosslinked.

For example, Wakida et al. [21] treated wool and PET for 30 sec in an atmospheric argon/acetone and in an argon/helium plasma. In both cases, substantial increases in the amount of oxygen species were found on the surface. They attributed the increase in oxygen containing species to the formation of radicals on the surface of the fibers. After removing the fibers from the chamber, these radicals react with the most reactive species in air. These tend to be the oxygen containing species. The increase in oxygen containing species increases the surface tension, which leads to increased wetting by water of the treated fibers. Although they did not measure the effects of these treatments on friction, increased wetting should enhance the uniformity of finish oils on the surface of the fibers.

Gao and Zeng [22] used an oxygen plasma to modify the surface of ultrahigh molecular weight poly(ethylene) (UHMW-PE) to improve adhesion between the fibers and an epoxy matrix. They found that the surface tension increased from 34 to 60 mJ/m^2 [23]. They also found the surface groups to be COH, CO, COOH, and COO. They did not measure the tribological properties, but it is anticipated that the boundary friction coefficient for the contact of this fiber with most other materials would be increased because of the increased polarity of the surface and thus increased adhesion to the counter surface.

Wang et al. [24] also used an oxygen plasma to attach a polymerization catalyst to the surface of Kevlar™. They subsequently polymerized polyethylene onto the surface. This treatment greatly enhanced the adhesion of the fibers to a polyethylene matrix. Although they did not report frictional characteristics, this treatment should greatly reduce the boundary friction for the contact of this fiber with most other materials because polyethylene behaves similarly to hydrocarbon waxes.

Griesser et al. [25] used RF plasma to polymerize hexamethyldisiloxane onto elastomeric sutures for heart surgery. The unmodified sutures exhibited substantial stick–slip friction, which resulted in oscillation in the suture during surgery and even suture breaks. After modification, the dynamic friction remained unchanged, but the static friction decreased until the static friction was the same as the dynamic friction and there were no stick–slip transitions. They attributed this change to the lack of polar structures in the surface grafted polyhexamethyldisiloxane. They postulated that by eliminating the polar structures, the adhesion between the suture and the heart tissue should be reduced, thus reducing the static friction.

Negulescu et al. [26] used a radio frequency plasma in tetrachlorosilane to modify the surface of poly(ethylene terephthalate) (PET). After plasma treatment, they exposed the fabrics to air. The grafted chlorosilane was rapidly converted to hydroxylsilanes, resulting in a high concentration of –OH groups on the surface. Both the surface tension and the roughness of the fibers markedly increased. They measured an increase in the friction coefficient of the PET against stainless steel wire from 0.21 for the untreated fabric to 0.26 for the treated fabric. The strong adhesion of –OH groups to most metals is the likely cause of the increased friction coefficient.

Cohen et al. [27] used radio frequency plasma to treat Spectra 900 polyethylene fibers (Honeywell, Petersburg, VA) with ammonia, carbon dioxide, and acrylic acid. They subsequently treated the ammonia treated yarns with toluene 2,4-diisocyanate (TDI) or with TDI and poly(propylene glycol) (PPG). Some of the CO_2-treated yarns were subsequently treated with thionyl chloride, $SOCl_2$ and lithium aluminum hydride ($LiAlH_4$) to convert the acid groups to hydroxyl groups. Next, these yarns were reacted with TDI, or with TDI and PPG. The results of these surface modifications and their friction results are shown in Table 2. They were able to increase the friction of PE from 0.13 for the untreated yarn to 0.29 by using an ammonia plasma with subsequent treatment with TDI and a low-molecular-weight PPG. Other treatments gave intermediate friction coefficients.

Although RF plasma treatment can readily change the surface properties of fibers, a major limitation to traditional plasma treating of fibers is that all of the air must be removed from the fibers and the plasma treating gas allowed to react with the surface (unless the treatment gas is air). Typical treatment times last seconds to minutes. At commercial processing speeds, this is a formidable problem. Atmospheric plasmas have been developed but they only allow plasma treatments with air, unless some method is provided to remove the air and replace it with some other gas. For example, Tsai et al. [28] describe an atmospheric plasma chamber that allows plasma treatment in a variety of gases. However, they cannot remove air from the chamber and thus they plasma treated with carbon dioxide and oxygen. This results in modifying the surface with various oxygen containing species. The increase in surface concentration of oxygen containing species increases the surface tension. They also observed significant surface roughening after plasma treatment. Although both of these effects should affect the tribological characteristics of the treated fabrics, they did report any tribological data.

Table 2 Effect of Plasma Surface Modification on the Friction of Spectra 900 PE Fibers

Treatment	Friction coefficient
None	0.13
NH_3 plasma	0.16
Acrylic acid plasma	0.17
CO_2 plasma	0.20
CO_2 plasma + TDI + PPG (2007 amu)	0.20
CO_2 plasma + TDI + PPG (2976 amu)	0.23
NH_3 plasma + TDI + PPG (1012 amu)	0.29

Even when atmospheric plasma treatment imparts beneficial tribological character-istics, there is still a major hurdle to be overcome. As mentioned in the Introduction of this chapter, fiber-processing speeds are extremely high. However, most plasma treatment times are a few minutes long. Thus, thousands of meters of yarn must be in the plasma simultaneously! Unless the treatment times are greatly reduced, the author believes that plasma treatment will not be a viable process for tribological modification of large volume fibers and yarns.

H. Chemical Treatments

A number of different chemical treatments have been used to modify the surfaces of fibers. These include chemical etching as described above, as well as chemical grafting, polymer-izing a different polymer onto the surface, and encapsulating the fibers with another polymer. Mares and Oxenrider [29,30] used materials in which one end of the molecule was miscible with the fiber polymer and the other end was immiscible. These materials partition themselves at the fiber–air interface. They used fluorocarbon chains as the immiscible segment, which then formed the surface of the fiber. Although they did not measure the effect of this treatment on friction, they observed a substantial increase in the oil repellency. These materials should also lower the friction. This approach is similar to that used in many synthetic boundary lubricants for metals where one end of the lubricant molecule adsorbs strongly onto the metal while the other end is only weakly adsorbing, and forms the lubricious portion of the lubricant.

I. Polymerizing

Two general approaches have been used to modify the surface of fibers by polymerizing another polymer onto the fiber surface. In one approach, the fiber is encapsulated in the new polymer, but there are no chemical bonds between the fiber surface and the encapsulating polymer. In the second approach, the new polymer grows from polymerization initiation sites on the fiber surface, and thus forms direct chemical links to the surface. Often, the actual material is a combination of the two approaches.

Ono et al. [31] reacted a hydroxy-terminated poly(dimethyl siloxane) and an epoxy-modified poly(dimethyl siloxane). The epoxy groups on one silicone react with the hydroxyl groups on the other silicone to form a durable crosslinked coating. This method encapsu-lates the fibers. They were able to reduce the static friction of untreated polyester yarns against hard chrome plating from 0.24 down to less than 0.16 for the treated fibers. In addition, the treatment could not be removed with multiple dry cleanings.

Using a similar approach, Totten and Pines [32] applied various alkoxysilylalkyl functional silicones to the surfaces of fibers. These materials readily crosslink and thus encapsulate the fibers in a low friction coating. They defined the staple pad friction coefficient as the sliding force of a 4540-g sled divided by the sled weight. They did not specify the sled material. They pulled the sled across a 1.0-g fiber sample that was placed on a table, which covered with emery paper. They found the staple pad friction coefficients for the treated fibers as low as 0.17, while the untreated fibers had a staple pad friction coefficient of 0.35–0.45. They also reported that this reduction was durable to washing. Greene and Daniele [33] used a similar approach but incorporated epoxy-containing silicones to form the crosslinks. They were able to reduce the staple pad friction coefficient from over 0.45 with no treatment to less then 0.25 after treatment and curing the surface coating. Again, these coatings did not wash off.

J. UV Photoinitiated Grafting

Ultraviolet grafting is another commonly used method to polymerize materials onto the surface and to alter the surface properties of fibers. In UV grafting, small molecules, usually monomers containing a vinyl group, and a photoinitiator are applied to the fiber. When shining UV light onto the fiber, the photoinitiator abstracts a proton from the surface leaving behind a free radical. This free radical then can initiate polymerization of the monomers. The surface polymer is usually highly crosslinked. The coating thickness depends on the duration of the UV exposure. It can reach several micrometers, essentially encapsulating the fiber. This approach has the advantage in that the fibers do not have to be placed in an evacuated chamber. However, they do need to be exposed to intense UV light, typically for several seconds to minutes. As with plasma treatment, the time duration is often prohibitive for all but the most expensive fiber applications.

Ranby [34] illuminated polyethylene, polypropylene, fibers polystyrene, and poly (ethylene terephthalate) fibers, films, and sheets with UV light in the presence of benzophenone or benzophenone derivatives. During the exposure, acrylate or methacrylate monomers were also present. The excited benzophenone molecules abstracted hydrogen atoms from the surface of the polymeric substrate, leaving free radicals attached to the surface. These radicals initiated polymerization of the monomers. By using monomers containing additional reactive groups (e.g., glycidyl acrylate), subsequent grafting to other materials could be performed. By first immersing the fibers immersed in a liquid bath containing both the initiator and the monomers, UV exposure times as short as a few seconds were sufficient to form the polymeric coating. For example, on UV exposure of PP in the presence of a photoinitiator and acrylamide, they formed a poly(acrylamide) coating on the fibers. There was a significant increase in the surface tension and in the adhesion of the fibers to Araldite resin (used in composites). More than 50% of the surface polymer formed a chemical graft to the surface. Although this approach can enhance the adhesion of fibers to composite resins, its effect on friction is poorly understood.

The UV-photoinititated polymerization of various polymers onto the surface of fibers suffers from the same incompatibility between the processing speeds of current commercial processes and the required duration of treatment. Even a few seconds requires uniform UV exposure for hundreds of meters of yarn. This renders the process impractical for all but the slowest processes. Another potential problem with photopolymerization is that they fibers within a yarn can be fused together by the photopolymer. This causes the yarn to become a stiff composite in which the matrix polymer is just the photopolymer. This approach may have its own commercial benefits, but is beyond the scope of this work.

K. Chemical Grafting of Polymers

Rather than polymerizing a new polymer onto the surface of a fiber, Michielsen [35] used a different approach. He grafted preformed polymers to the surface of nylon 6,6 fibers. He grafted poly(dimethyl siloxanes) (PDMS) of various molecular weights. The PDMS molecules contained either epoxy or carboxyl groups, which react readily with the amino end-groups of nylon 6,6. By varying the molecular weight of the PDMS, he was able to control the surface coverage. He showed that the boundary friction coefficient decreased linearly with the molecular weight (M_W) up to ~70 kDa. He argued that tethered chains attached to the surface and those that have no specific interactions with the surface should deposit as "mushrooms" [36]. When two surfaces covered with these tethered chains come into contact, the mushrooms will keep the substrate surfaces apart. Furthermore, because of the excluded volume effect, tethered chains on opposing surfaces should repel each other and not entangle [37]. If the mushrooms readily deform under shear, the friction coefficient should be low. However, if the surface has large bare regions, the mushrooms will not be able to keep the surfaces apart. The friction coefficient will then depend on the fraction of the load-bearing surface that is bare. Michielsen illustrated this by graphing the boundary friction coefficient against the reduced area of the mushrooms. He defined the reduced area of the mushrooms as

$$A_{\text{red}} = \frac{\pi R_{\text{g}}^2}{\pi R_{\text{n}}^2} \tag{4}$$

where R_{g} is the radius of gyration of the lubricant molecule and R_{n} is one-half the distance between adjacent amino end-groups on the nylon, assuming that they form a hexagonal close-packed array on the surface.

$$R_{\text{g}}^2 = \frac{nl^2}{6} C_{\infty} \tag{5}$$

where n is the number of repeat units in the polymer chain, l is the root mean square length of a backbone bond, and C_{∞} is the molecular weight $M = nM_0$, where M_0 is the molecular weight of the repeat unit. Because R_{g}^2 and M both depend linearly on n, the reduced area was found to depend linearly on M_{W}:

$$\mu = (1 - \phi)\mu_{\text{s}} + \phi\mu_{\text{p}} \tag{6}$$

where μ is the friction coefficient of the modified surface, ϕ is the area fraction of the grafted polymer, μ_{p} is the friction coefficient associated with the grafted polymer, and μ_{s} is the friction coefficient of the substrate polymer.

Using this approach, he was able to reduce the boundary friction coefficient from 0.32 for bare nylon to < 0.08 for a reduced area of 0.95. In addition, the friction coefficient could be adjusted to any desired value between these two limits by simply changing the molecular weight of the bound lubricant and hence changing ϕ. His results are shown in Fig. 7. He found similar trends when he used a maleic anhydride containing polyethylene although the

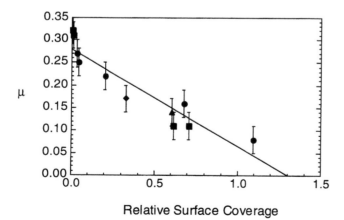

Figure 7 Friction coefficient for several grafted polymeric lubricants. Line is linear regression for all points as represented in Eq. (6). ● $\pi R_{\text{g}}^2/90$ nm^2 and ▲ $\pi R_{\text{g}}^2/270$ nm^2 for PDMS oils, ■ $\pi R_{\text{g}}^2/90$ nm^2 and ◆ $\pi R_{\text{g}}^2/270$ nm^2 for PE lubricants. (From Ref. 35, ©John Wiley & Sons, 1999.)

range of available molecular weights of the modified polyethylene was limited. The maleic anhydride moiety reacts with the amine ends of nylon 6,6. Thus the same analysis worked equally well for two quite different surface grafted polymers. He confirmed the analysis given in Eq. (6) by multiple-angle X-ray photoelectron spectroscopy, which provides both the relative coverage and the thickness of the coating. Surprisingly, coatings only 5–10 nm thick were sufficient to reduce the friction coefficient to less than 0.08.

Although very little is known about this approach beyond what is given above, Eq. (4) indicates that the optimum molecular weight of the grafted polymer depends on the concentration of reactive sites on the substrate surface. For a higher concentration of reactive sites, the optimum molecular weight of the grafted polymer is lower. However, this also results in a thinner surface coating. It is not clear how the friction coefficient depends on the thickness.

VI. CONCLUSIONS

Surface modification of fibers can be a cost-effective means to tailor the surface properties on fibers while maintaining the bulk properties of the fiber. To alter the surface tension, only very thin coatings are needed, perhaps as thin as a few atomic layers. To limit (or enhance) triboelectric charging, the surface electron binding energy can be modified by grafting or polymerizing a material with the desired binding energy. In this application, coatings of ~ 30 nm are sufficient.

To reduce the friction of the surface, a layer with low shear strength whose thickness is sufficient to separate the bare surfaces is desired. For typical melt spun fibers, this layer can be as thin as 5–10 nm. To increase the friction, surface modifications that increase the surface polarity can be used. Because these layers can be so thin, very expensive materials can be used without incurring large cost penalties.

However, most methods that have been used to modify the surface of fibers requires several seconds or longer. As fiber-processing speeds are often in excess of 3000 m/min, these methods have not found widespread use. However, when batch processing of the fibers is possible (i.e., shrinkproofing of wool), or when the value of the surface modification is high enough to justify slow processing speeds (i.e., coronary sutures), these methods have been used. Only the bicomponent core/shell fibers are made on a large scale because of the perceived value of reducing static generation in carpets. The author believes that this scenario will continue until treatment methods are developed that can be completed in a few milliseconds and the value of the surface modification justify the added cost of manufacture.

VII. FURTHER WORK

Although considerable research has been performed to modify the surface of fibers, very little work has been carried out to characterize the effect of the modification on the fibers' tribology. Only rare examples have been found where the friction or the triboelectric charge generation or charge dissipation has been reported. Considerable opportunity exists to enhance the performance of fibers while reducing the environmental impact of the traditional fiber finishes. However, faster methods of surface modification are needed in which treatment times of less than 10 msec are sufficient and in which air does not need to be excluded.

In addition, it is not known how thick a coating is needed. We do not know whether complete or partial coverage is needed. Little is known about the tribological characteristics of molecules when they are bound to a surface. For example, poly(dimethyl siloxane) has the reputation of being a poor boundary lubricant for nylon fibers. Yet, in studies by Ono et al. [31], Totten and Pines [32], Greene and Daniele [33], and Michielsen [35], various silicones exhibited excellent boundary lubrication. Likewise, almost nothing is known about how substrate properties and morphology affect the tribological properties of the modified fibers. Considerable research needs to be carried out in these areas.

REFERENCES

1. Patterson, H.T.; Proffitt, T.J., Jr. *Fatty Acids in Industry*; Johnson, R.W. Fritz, E., Eds.; Marcel Dekker, Inc.: NY, 1989; 503 pp.
2. Proffitt, T.J., Jr.; Patterson, H.T. Oleochemical surfactants and lubricants in the textile industry. JAOCS 1988, 65, 1682.
3. Lee, D.E. Int. Fiber J. Surfactants in Spin Finishes. Aug. 1991, 94 pp.
4. Rolfe, Sue. Problem solving with specialist antistatic fibers. Textiles 1993, 22 (2), 9–11.
5. Slade, P.E. *Handbook of Fiber Finish Technology*; Marcel Dekker: New York, 1998.
6. Arridge, R.G.C. The static electrification of nylon 6,6. Br. J. Appl. Phys. 1967, 18, 1311–1316.
7. Follows, G.W.; Lowell, J.; Wilson, M.P.W. Contact electrification of polyamides. J. Electrost. 1991, 26, 261–273.
8. Davies, D.K. Charge generation on dielectric surfaces. Br. J. Appl. Phys. 1969, 2 (2), 1533–1537.
9. Brennan, W.J.; Lowell, J.; O'Neill, M.C.; Wilson, M.P.W. Contact electrification: the charge penetration depth. J. Phys. D, Appl. Phys. 1992, 25, 1513–1517.
10. Lindberg, J.; Gralén, N. Measurement of friction between single fibers: II. Frictional properties of wool fibers measured by the fiber twist method. Text. Res. J. 1948, 18, 287.
11. Bradbury, J.H. The theory of shrinkproofing wool: Part II. Chemical modification of the fiber surface and its effect on felting shrinkage, friction, and microscopic appearance. Text. Res. J. 1961, 31, 735–743.
12. Bradbury, J.H.; Rodgers, G.E. The theory of shrinkproofing of wool: Part IV. Electron and light microscopy of polyglycine on the fibers. Text. Res. J. 1963, 33, 454–458.
13. Bradbury, J.H.; Rogers, G.E.; Filshie, B.K. The theory of shrinkproofing of wool: Part V. Electron and light microscopy of wool fibers after chemical treatment. Text. Res. J. 1963, 33, 617–630.
14. Silverstein, M.S.; Breuer, O.; Dodiuk, H. Surface modification of UHMWPE fibers. J. Appl. Polym. Sci. 1994, 52, 1785–1795.
15. Achwal, W.B. A comparative study of the surface action of caustic soda on polyester fabrics under different conditions. Man-Made Text. India 1984, 27, 185, 187–190, 216.
16. Bahners, Thomas; Kesting, Wolfgang; Schollmeyer, Eckhard. Designing surface properties of textile fibers by UV-laser irradiation. Appl. Surf. Sci. 1993, 69, 12–15.
17. Knitter, Dierk; Schollmeyer, Echard. Surface structuring of synthetic polymers by UV-laser irradiation: Part IV. Applications of excimer laser induced surface modification of textile materials. Polym. Int. 1998, 45, 110–117.
18. Wong, Wilson; Chan, Kwong; Yeung, Kwok Wing; Lau, Kai Shui. Chemical surface modification of poly(ethylene terephthalate) by excimer irradiation of high and low intensities. Mater. Res. Innovat. 2001, 4, 344–349.
19. Hautojärvi, Joni; Laaksonen, Sanna. On-line surface modification of polypropylene fibers by corona treatment during melt-spinning. Text. Res. J. 2000, 70, 391–396.
20. Seto, Fusako; Muraoka, Yoichiro; Sakamoto, Nobuyuki; Kishida, Akio; Akashi, Mitsuru. Surface modification of synthetic fiber nonwoven fabrics with poly(acrylic acid) chains prepared by corona discharge induced grafting. Angew. Makromol. Chem. 1999, 266, 56–62.

21. Wakida, Tomji; Tokino, Seiji; Niu, Shouhua; Kawamura, Haruo; Sato, Yukihiro; Lee, Munceul; Uchiyamam, Hiroshi; Inagaki, Hideo. Surface characteristics of wool and poly (ethylene terephthalate) fabrics and film treated with low-temperature plasma under atmospheric pressure. Text. Res. J. 1993, *63*, 433–438.

22. Gao, S.; Zeng, Y. J. Surface modification of ultrahigh molecular weight polyethylene fibers by plasma treatment. I. Improving surface adhesion. Appl. Polym. Sci. 1993, *47*, 2065–2071.

23. Gao, S.; Zeng, Y. J. Surface modification of ultrahigh molecular weight polyethylene fibers by plasma treatment. II. Mechanism of surface modification. Appl. Polym. Sci. 1993, *47*, 2093–2101.

24. Wang, Q.; Kaliaguine, S.; Ait-Kadi, A. Catalytic grafting: A new technique for polymer fiber composites. III Polyethylene plasma treated Kevlar TM fibers composites: Analysis of the fiber surface. J. Appl. Polym. Sci. 1993, *48*, 121–136.

25. Griesser, Hans J.; Chatelier, Ronald C.; Martin, Chris; Vasic, Zoran R.; Gengenbach, Thomas R.; Jessup, George. Elimination of stick–slip of elastomeric sutures by radiofrequency glow discharge deposited coatings. J. Biomed. Mater. Res. 2000, *53*, 235–243.

26. Negulescu, Ioan I.; Despa, Simona; Chen, Jonathan; Collier, Billie J.; Despa, Mircea; Denes, Agnes; Sarmadi, Majid; Denes, Frank S. Characterizing polyester fabrics treated in electrical discharges of radio-frequency plasma. Text. Res. J 2000, *70*, 1–7.

27. Cohen, Samuel H.; Schreuder-Gibson, Heidi; Stapler, John T. Study of surface modified poly (ethylene) yarns. Scanning Microsc. 1992, *6*, 997–1008.

28. Tsai, Peter P.; Wadsworth, Larry C.; Reece Roth, J. Surface modification of fabrics using a one-atmosphere glow discharge plasma to improve fabric wettability. Text. Res. J. 1997, *67*, 359–369.

29. Mares, F.; Oxenrider, B.C. Modifications of fiber surfaces by monomeric additives: Part I. Extrusion techniques. Text. Res. J. 1977, *47*, 551–561.

30. Mares, F.; Oxenrider, B.C. Modifications of fiber surfaces by monomeric additives: Part II. Absorption of fluorocarbon additives by polyethylene terephthalate. Text. Res. J. 1978, *48*, 218–229.

31. Ono, Michikaze; Ito, Toshio; Oshima, Masharu; Ogawa, Nobuyoshi; Nakamura, Yukimasa. A method for manufacturing durable synthetic fibers which have good softness and smoothness, JP Application 58-58473, 1983.

32. Totten, George E.; Pines, Arthur N. US Patent 4579964, 1986.

33. Greene, George H.; Daniele, Robert A. Jr. US Patent 4661405, 1987.

34. Ranby, Bengt. Surface modification of polymers by photoinitiatied graft polymerization. Makromol. Chem., Macromol. Symp. 1992, *63*, 55–67.

35. Michielsen, Stephen. The effect of grafted polymeric lubricant molecular weight on the frictional characteristics of nylon 6,6 fibers. J. Appl. Polym. Sci. 1999, *73*, 129–136.

36. de Gennes, P.G. Scaling theory of polymer adsorption. J. Phys. (Paris) 1976, *37*, 1445–1452.

37. Williams, D.R.M. Cylindrical and spherical polymer brushes: moduli and force laws. Macromolecules 1993, *26*, 6667–6669.

12
Biotribology

Hong Liang and Bing Shi
University of Alaska, Fairbanks, Fairbanks, Alaska, U.S.A.

Tribology is the knowledge of lubrication, friction, and wear. Tribology of biomaterials is the study of the performance and function of biomaterials in tribological applications. Tribology plays an important role in improving the design and making successful biomaterials for medical and clinical purposes. Joints of the human body, such as the hip, knee, jaw, dental parts, etc., all need to consider the wear and lubrication properties. In this chapter we will give a general introduction of biotribology and provide an overview of recent progress in biotribological research.

I. INTRODUCTION

Tribology is concerned with friction, wear, and lubrication. It is the science of interacting surfaces in relative motion. In fact, tribological problems are very common everywhere, even in the human body. When tribology is used to study the biological system, it is called biotribology, which is one of the newest fields to emerge in the discipline of tribology. It describes the study of friction, wear, and lubrication of biological systems, mainly synovial joints such as human hips and knees.

Artificial tribosystems have been systematically investigated for more than 150 years. Natural tribosystems have served as models for the artificial ones. The earliest successful implants were bone plates, introduced in the early 1900s to stabilize bone fractures and accelerate their healing. As early as in the late 1950s, the first total hip replacement was introduced using polytetrafluoroethylene (PTFE) as the cup-bearing surface [1]. However, PTFE undergoes aseptic loosening. Research and later caveats from researchers testing the interpositional implant (IPI) showed that PTFE was not an appropriate material to use as a load-bearing surface in the body.

In the late 1960s, tribology was introduced into living and artificial human joints [2]. Sir John Charnley, a British orthopedic surgeon, was knighted for his development of joint replacement [2]. He developed the fundamental principles of the artificial hip and designed an artificial hip in the mid and late 1960s that still sees widespread use today. Frank Gunston developed one of the first artificial knee joints in 1969. Since then, joint replacement surgery has become one of the most successful orthopedic treatments. The development has been seen in many areas. For example, the tribological behavior of synovial joints and their replacement, wear of human dental tissues, the lubrication by plasma of red blood cells in

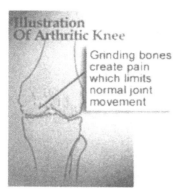

Figure 1   Knee joint deteriorates. (From Ref. 5.)

narrow capillaries, the wear of replacement heart valves, the wear of screws and plates in and on bone in fracture repair, the tribology of skin and the friction of hair, biosensors, artificial fingers, etc., are all topics of study.

In the late 1970s, a temporomandibular joint (TMJ) replacement using PTFE as the bearing counterface was invented [1]. In 1983, the PTFE implant, which was called the interpositional implant, was approved for the market. Because PTFE exhibits a low coefficient of friction and has been used extensively as a bearing surface in other engineering applications, it would seem an appropriate choice for an implant material in theory [1]. However, of the more than 25,000 PTFE TMJ implants received by patients, most failed. With the development of the research on IPI, evidence did exist that PTFE was not an appropriate implant material [1].

The number of hip replacements done in the world per year numbers between 500,000 and 1 million. The total number of knee replacements done in the world per year is less, but probably still between 250,000 and 500,000. It is estimated that 85% to 95% of approximately 600,000 total joint replacements performed each year in the United States (more than 120,000 artificial hip joints are being implanted annually) will still be functioning after 10 years, based on research reported in orthopedic medical journals [3,4]. Artificial joint replacement is performed when the cartilage that lines the joint deteriorates, resulting in bones grinding against each other (shown in Fig. 1) [5], if nonsurgical treatments, such as anti-inflammatory drugs, walking aids, and support braces do not offer relief.

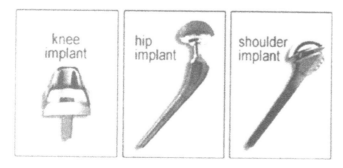

Figure 2   Joint replacements. (From Ref. 5.)

Joint replacement surgery is most commonly performed for hips, knees, and shoulders, as shown in Fig. 2, although toes, fingers, and elbows have been successfully replaced [5]. Most artificial joint replacements are designed to remove the diseased areas of the joint and replace them with metal-and-plastic implants designed specifically to restore that joint's function and stability.

II. MATERIALS IN BIOTRIBOLOGY

A. Introduction

As advances have been made in the medical sciences, the aging population has become important. More organs, joints, and other critical body parts will wear out and must be replaced to maintain good quality of life for elderly people. Biomaterials play a major role in replacing or improving the function of every major body system (skeletal, circulatory, nervous, etc.). Some common implants include orthopedic devices such as total knee and hip joint replacements, spinal implants, and bone fixators; cardiac implants such as artificial heart valves and pacemakers; soft tissue implants such as breast implants and injectable collagen for soft tissue augmentation; and dental implants to replace teeth/root systems and bony tissue in the oral cavity.

Artificial organs and joints are foreign bodies of the living system. Properties to be considered when choosing biomaterials are strength, biocompatibility, infection, rejection, wear, friction, fatigue, and corrosion of the materials. Requirements for materials used as implants in the living system are listed in Table 1 [6].

B. Materials Used in Biotribology

The properties of biomaterials (such as materials to replace bone, cartilage, muscle, and connective tissue) are related to organized macromolecular subassemblies and their functions. Despite the success of surgical implants such as artificial hip and knee joints,

Table 1 Requirements of Implants

Compatibility	Mechanical properties	Manufacturing properties
Tissue reactions	Elasticity	Fabrication methods
Changes and properties	Yield strength	Consistency and conformity to all requirements
Mechanical	Ductility	Quality of raw materials
Physical	Toughness	Superior techniques to obtain excellent surface finish or texture
Chemical	Time-dependent deformation	Capability of materials to get safe and efficient sterilization
Degradation leads to local deleterious changes	Creep	Cost product
Harmful systematic effects	Ultimate strength Fatigue strength	

the materials used still do not meet the requirements. Many potential causes of failure for total hip arthroplasty are, e.g., [7]

1. Deficiencies in design (size and shape) of the device for a particular patient (e.g., an undersized noncemented stem)
2. Surgical problems (e.g., problematic orientation or problems in wound healing)
3. Host abnormalities or diseases (e.g., osteopenia)
4. Infection
5. Material fracture, wear, and corrosion

Synovial joints, including the knee joints, are typically found at the ends of long bones and permit a wide range of motion [7]. This type of joint is surrounded by a fibrous articular capsule that is lined by a thin synovial membrane. It is important to note that the surfaces of each bone are not in direct contact, but are covered by special articular cartilages, which resemble the hyaline cartilages found elsewhere in the body.

Synthetic biomaterials, such as stainless steel, titanium alloy, polymers, and ceramic composite, undergo degradation through fatigue and corrosive wear because of load bearing and the salty environment of the human body. At the same time, deposits of inorganic salts can scratch weight-bearing surfaces, making artificial joints stiff and awkward. The increasing life expectancy of the aging population and the need to surgically treat arthritis in increasing numbers of young patients are placing greater demands on the durability and expected clinical lifetime of artificial joints. At present, well-designed artificial joints such as hip joints have a lifetime of at most 10 to 15 years.

For biomaterial design, engineers must consider the physiological loads to be placed on the implants to obtain sufficient structural integrity. Material choices also must take into account biocompatibility with surrounding tissues, the environment and corrosion issues, friction and wear of the articulating surfaces, and implant fixation either through osseointegration (the degree to which bone will grow next to or integrate into the implant) or bone cement. One of the major problems plaguing these devices is purely material related: wear of the polymer cup in total joint replacements. A summary of tribobiomaterials is given in Table 2. These conventional materials fit into the specific needs of bioapplications. The specific properties are listed in Table 3.

1. Metals and Alloys

Replacement arthroplasty made important advancements during the 1950s and 1960s through the contributions of G. K. McKee and Sir John Charnley. McKee introduced metal-on-metal hip prosthesis in which components were originally made of stainless steel, which was rapidly changed to cobalt–chromium–molybdenum alloy to mitigate the excessive friction and rapid loosening of the stainless steel pair [12].

Applications of titanium alloys as biomaterials are increasing due to the lower modulus, superior biocompatibility, and enhanced corrosion resistance when compared to more conventional stainless steel and cobalt-based alloys. Figure 3 shows the calcification of the surface of a porous titanium disk in the presence of different concentrations of a novel surface coating in an in vitro model.

Recently, new titanium alloy compositions have been developed. These first orthopedic alloys included Ti6Al7Nb, Ti5Al2.5Fe, Ti12Mo6Zr2Fe, Ti15Mo5Zr3Al, Ti15Mo3Nb3O, Ti15Zr4Nb2Ta0.2Pd, Ti15Sn4Nb2Ta0.2Pd, Ti13Nb13Zr, and Ti35Nb5Ta7Zr [12].

Table 2 Summary of Tribobiomaterials

Material	Application	Major properties, description
Metal: brass, stainless steel, nickel plated, nickel-plated steel, zinc-plated steel	Inserts	
Alloy: titanium alloys [6], titanium aluminum vanadium alloy [8], cobalt chromium alloy [9], cobalt chromium molybdenum alloy [10]	Total joint replacement	Wear and corrosion resistance
Inorganic: diamond-like carbon	Biocompatible coatings	Reduced friction and increased wear resistance
Ceramics [11]: Al_2O_3, ZrO_2, Si_3N_4, SiC, B_4C, quartz, bioglass (Na_2O–CaO–SiO_2–P_2O_5), sintered hydroxyapatite $[Ca_{10}(PO_4)_6(OH)_2]$	Bone joint coating	Wear and corrosion resistance
Polymers: ultrahigh molecular weight polyethylene (UHMWPE),	Joint socket	Wear, abrasion, and corrosion resistance
Polytetrafluoroethylene (PTFE)	Interpositional implant temporomandibular joint (jaw)	Low coefficient of friction
Polyglycolic acid	Joint bone	Elastic with less wear
Polyurethane	Leaflet heart valve	Highly biocompatible, high strength, and dynamic ranges of breathability
Composites: specialized silicone polymers	Bone joint	Wear, corrosion, and fatigue resistance

Table 3 Other Material Analysis Methods

Analytical information	Techniques used
Elemental analysis, surface contaminations	ESCA (XPS), TOF-SIMS
Chemical bonding analysis, functionalities, oxidation states	TOF-SIMS, ESCA, PM-IRRAS
Molecular analysis, traces (contaminants, additives, oligomers, low molecular mass products), molecular structure (unsaturation, end groups, tacticity, branching)	TOF-SIMS, ESCA
Quantitative analysis, stoichiometry, chemical composition	ESCA (XPS), TOF-SIMS
Chemical mapping, depth profiling	ESCA (XPS), TOF-SIMS
Morphology, roughness, nanoindentation, viscoelastic moduli	AFM, multimode nanoindentation
Wettability, surface forces, adhesion energy	AFM, chemical tips contact angle

Figure 3 The calcification (white deposit) of the surface of a porous titanium disk (gray) in the presence of different concentrations of a novel surface coating in an in vitro model. (From Ref. 6.)

2. Polymers

Several types of polymers, such as ultrahigh molecular weight polyethylene (UHMWPE), PTFE, urethane, etc., have been used in biological applications such as implants of the knee, hip, jaw, heart valves, etc.

The initial metal-on-metal form of replacements introduced by McKee in the 1950s was replaced in the 1960s and 1970s by metal-on-polymer form of replacements. In the 1960s, Sir John Charnley developed the concept of low-friction arthroplasty. The early Charnley joint consisted of a stainless steel femoral head and a PTFE acetabular cup. However, PTFE had unacceptable wear properties, and its replacement with UHMWPE, which had a much lower wear rate, could solve this problem.

Research has been done to investigate the physical, chemical, and clinical properties of UHMWPE, the service time, and the friction coefficient. Especially, wear tests were done to evaluate wear rate and the size and morphology of the polyethylene wear debris.

3. Composite and Surface Engineering in Biological Materials

Metallic materials have been widely used as implants. Some polymeric biomaterials such as polyethylene, polycarbonate, and polyetheretherketone are biodegradable. However, biocompatibility is not often found in these materials. The design of materials suitable for use as replacements for damaged or diseased organs in the human body has long been a major objective of the biomedical engineer. There are two major ways of improving the biocompatibility of materials: coating techniques and surface modification.

a. Coating Techniques. Coating for biological materials is the physical absorption of a biocompatible surface onto a medical device to improve its performance, especially for long-term implant applications.

An example is the diamond-deposition technique [13]. This technique is becoming increasingly practical for coating materials with hard, smooth films of diamond and other hard carbon films. Diamond thin films have been used to coat titanium alloy (Ti6Al4V) and cobalt chrome (Co26Cr6Mo) alloy, which are used to make the load-bearing components of artificial human joints. The alloys were chosen because of their high bulk modulus, relatively good biocompatibility, and excellent fatigue resistance. However, surface wear of these artificial joints often lead to corrosion. The wear debris causes inflammation and eventual aseptic loosening of the prosthesis. Coatings on the surface of the artificial joints are able to increase the service lifetime. Diamond coatings and diamond-like (DLC) coatings are chemically inert and hard. Research on DLC coatings shows that the friction coefficient and wear are obviously small [13].

Nanotube-reinforced tetrahedral amorphous carbon films are being developed as wear-resistant coatings for artificial joints, as shown in Fig. 4 [15a]. The first step is to grow

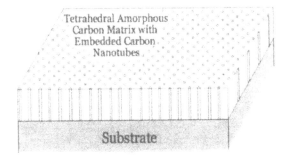

Figure 4 Schematic image of nanotube reinforced tetrahedral amorphous carbon films. (From Ref. 13.)

aligned arrays of carbon nanotubes (CNTs) on a metal substrate. Then a matrix of tetrahedral amorphous carbon (TAC) is grown to complete the film. Similar to the diamond and DLC coatings, this carbon nanotube coating also has high biocompatibility. There are several advantages of carbon nanotube films over diamond and DLC films. Nanotubes grown by microwave plasma chemical vapor deposition (MPCVD) tend to align themselves perpendicularly to the sample surface despite the shape of the substrate. A CNT/TAC film grows easily over curved surfaces. Thus, it remains well adhered to the substrate. Disadvantages of CNT/TAC films are the following: the nanotubes might be dissolved during TAC deposition and CNTs may allow corrosion of the substrate by acting as ion channels. Figure 5 shows CTN coatings on an acetabular cup implant.

Besides carbon coatings, glass coatings for artificial joints are often used. A biologically active glass that enables metal implants to bond with bone could significantly extend the lifetime of artificial hips, knees, and other medical-reconstructive devices. A bioactive silicate glass has been developed. The thickness of the coating is 20–200 μm. The glass layer can be composited at the metal–glass and glass–bone interfaces so that the coating binds with both bone and metals.

b. Surface-Modification Techniques. Surface-modification techniques for biomaterials are based on the functionalization of the surface of medical devices through covalent attachment to bioactive molecules. The modification of the surface of polymers and metals will enhance the biocompability of the biomaterials.

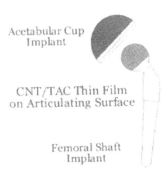

Acetabular Cup
Implant

CNT/TAC Thin Film
on Articulating Surface

Femoral Shaft
Implant

Figure 5 CTN coatings on acetabular cup implant. (From Ref. 13.)

4. Biopolymers

Biomaterials used to date, as mentioned earlier, are metals, ceramics, polymers, composites, and biopolymers. The first four classes are conventional materials adapted from traditional materials processing. Biopolymers, on the other hand, involve materials synthesis and bioengineering. There are three basic approaches: DNA-linked material building, tissue culture, and biopolymerization.

Recently, scientists developed a method to link DNA to gold, a noble metal [14]. There are biochemical approaches to molecular structures of precisely defined dimensions ranging from 1 nm to 10 cm in length. Conversely, the synthesis of precisely defined unnatural molecular architectures beyond 25 nm in length is often unattainable due to solubility, material throughput, and characterization constraints [15b]. Therefore, as nanotechnological needs advance, syntheses could rely on self-assembling strategies using natural scaffolds as templates for the construction of synthetic nanostructures [16]. Most studies involving DNA self-assembly have focused on duplex interactions between complementary DNA strands, the insertion of synthetic units through covalent tethering to specific nucleotides involving the assembly of materials through electrostatic interactions on the phosphate groups along a DNA backbone [17].

The goal of tissue culture is to develop a synthetic alternative to orthopedic tissues such as bone, ligament, and cartilage. Once implanted, the presence of cells and growth factors will initiate bone regeneration throughout the 3-D pore network. As regeneration continues, the matrix is slowly resorbed by the body. Upon complete degradation, the implant site is filled with newly regenerated bone and is free of any residual polymer. Examples are given of tissue-engineered bone (Fig. 6), tissue-engineered ligament (Fig. 7), tissue-engineered cartilage (Fig 8), and cell culture (Fig. 9) [18]. In studies examining cell growth on these matrices, osteoblast cell lines have been used as well as bone cells isolated from rat calvaria to create model systems for cell growth on bioerodable materials. This image shows osteoblasts growing on the surface of the 3-D microsphere matrix. It is interesting to note that the cells are growing in a circumferential pattern due to the structure of the matrix.

Biopolymerization has been used to generate intelligent and smart materials. Very often one can find them in sensoring and electronic devices.

Biomaterials are the future of medicine. From tissue removal and replacement, to future regeneration, biomaterials will take part in advancement in the near future. Biomaterial and implant research will continue to concentrate on serving the needs of medical device manufacturers and recipients, as well as medical professionals. Development

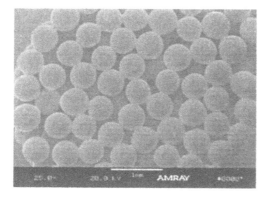

Figure 6   Tissue-engineered bone. (From Ref. 18.)

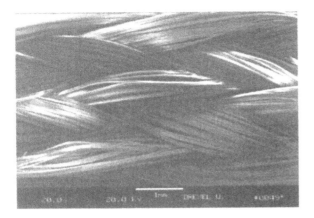

Figure 7 Tissue-engineered ligament. (From Ref. 18.)

of technologies will meet those needs by designing materials with improved strength, shape, function, and behavior. Future biomaterials will incorporate biological factors (such as bone growth) directly into an implant's surface to improve biocompatibility and bioactivity. New projects will be directed at materials development for improved mechanical integrity, corrosion resistance, and biocompatibility. The understanding of the nature of the human body parts such as knee, jaw, dental parts, and hip, etc. will lead to the application of these techniques in future materials design.

C. Surface Characterization of Biomaterials

Materials characterization methods are mechanical testing, chemical analysis, microstructural characterization, explant analysis, coating evaluation, development of special test techniques and component design and analysis including structural analysis, reliability and probabilistic modeling, engineering design optimization, and failure analysis.

Characterization is particularly difficult when the materials are used for the human body, requiring in vivo test conditions.

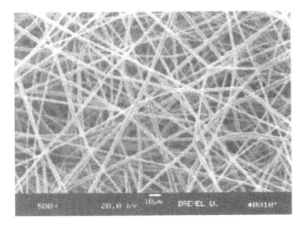

Figure 8 Tissue engineered cartilage. (From Ref. 18.)

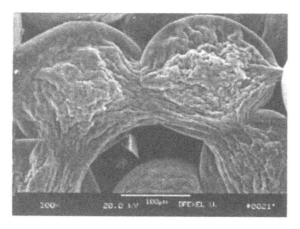

Figure 9   Cell culture. (From Ref. 18.)

The *atomic force microscope (AFM)* is widely used in materials analysis at nanometer scale [19,20]. It is also used in biomaterial research. Figure 10 is an AFM image showing the surface structure of a hydrated *Staphylococcus epidermis* biofilm formed on an implant surface [21]. The channels observed are believed to provide nourishment that sustains the biomaterial-associated infection. Using an AFM, cortical bone is imaged at ultrahigh magnification (shown in Fig. 11) [22]. Advanced in situ microscopy testing devices allow biomaterials and biological materials to be observed at high magnification with the AFM while being subjected to controlled levels of stress or deformation.

Scanning electron microscopy (SEM) is used to characterize the reaction rate and ion release imaging modes of metallic biomaterials [23]. SEM can be used to analyze micro-tribological properties. It can analyze the structure of two surfaces, which leads to the understanding of microtribological properties. The scanning potential microscope (SPM) can also be used in measuring the microcorrosion processes on metallic biomaterial surfaces. The SPM has imaged localized preferential corrosion areas on the samples. Localized corrosion potential is clearly associated with topographic features [24].

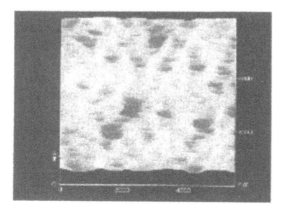

Figure 10   AFM image of a hydrated *Staphylococcus epidermis* biofilm. (From Ref. 21.)

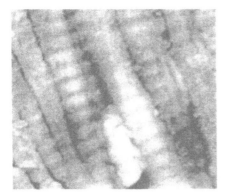

Figure 11 AFM image of bone (*Staphylococcus epidermis* biofilm). (From Ref. 22.)

Confocal laser-scanning microscopy (CLSM) is a rapidly advancing technique used to produce crisp and precise images of thick specimens in fluorescent and reflective light modes by rejection of out-of-focus light via a confocal pinhole shown in Fig. 12 [25]. This feature of confocal microscopes, known as "optical sectioning," makes it possible to scan a sample at various x–y planes corresponding to different depths and, by ordering these planes into a vertical stack, reconstruct a 3-D image of the specimen. The confocal microscope does not require physical sectioning of thick samples. It also precludes the need for extensive specimen processing. Therefore, the CLSM is one of the most efficient methods available to gain 3-D information on living biological specimens and biomaterials. Confocal images were obtained of osteoblasts grown on a nanotopography surface and stained for tubulin and actin [25].

New fluorescence techniques have been developed to identify the interactions of virtually any kind of molecule with cell membranes. These studies are important for a large number of fundamental biological processes such as membrane biogenesis and signaling.

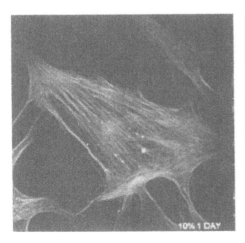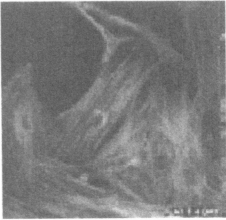

Figure 12 Confocal images of osteoblast grown on a nanotopography surface and stained for tubulin and actin. (From Ref. 25.)

Auger electron spectroscopy (AES) is then used to identify and quantify calcium and phosphate, the major inorganic components of bone. The deposited mass depends on the surface chemistry and a clear correlation between surface pretreatment and calcium/phosphate ratios is observed. This can, in some cases, also be correlated to macroscopic properties such as surface wettability [26].

Materials scientists continue to develop novel surface modifications to enhance textile performance. Wettability is one important characteristic of a surface. Wettability of textile fibers can be enhanced to assist the dyeing operation during processing; and the fibers can be made wettable for application in, e.g., materials of tissue culture. Therefore, the *surface tension meter* and *contact angle tester* are very useful in measuring the surface force, wettability, and adhesion energy.

D. Summary of Surface Characterization Techniques

Some other methods that are used to characterize the properties of biomaterials are listed in Table 3.

Other analysis methods used in biological research are Fourier transform infrared and Fourier transform Raman spectroscopy and microscopy, X-ray diffraction, X-ray fluorescence, etc.

III. FUNDAMENTALS OF BIOTRIBOLOGY

The development of biomedical engineering has made biotribology an independent research field to improve the quality of human life. Biotribology will continually be one of the most prospective areas in the near future. The fundamentals introduced in this section include the biomechanics, materials, and surface science issues.

A. Friction

Normal joints are remarkably effective, with coefficients of friction lower than those obtainable with manufactured journal bearings. Furthermore, the constant process of renewal and restoration ensures that living articular tissues have durability far superior to that of any artificial bearing [27]. No artificial joint can equal the performance of a normal human joint. Some comparative friction coefficients are shown in Tables 4 and 5.

B. Lubrication

1. Biolubrication Fundamentals

The mechanics of joint lubrication has provided a focus of investigation beginning with the unique structure of the bearing surface. Articular cartilage is elastic, fluid-filled, and backed by a relatively impervious layer of calcified cartilage and bone. This means that load-induced compression of cartilage will force interstitial fluid to flow laterally within the tissue and to surface through adjacent cartilage. As that area, in turn, becomes load bearing, it is partially protected by the newly expressed fluid above it. This is a special form of hydrodynamic lubrication, so-called because the dynamic motion of the bearing areas produces an aqueous layer that separates and protects the contact points.

Table 4 Examples of Two Surfaces Sliding Relative to Each Other

Surface in contact	Coefficient of friction (μ)
Rubber tire/dry road	1.0
Metal/metal	1.0
Glass/glass	1.0
Perspex/Perspex	0.8
Perspex/steel	0.3
Wood/wood	0.25
Ice/ice	0.05
Synovial joints	0.02
Articular cartilage with saline	0.01
Articular cartilage with synovial fluid	0.004
Articular cartilage/articular cartilage	0.02–0.001

Source: Ref. 28.

Boundary layer lubrication is the second major low-friction characteristic of normal joints. Here, the critical factor is proposed to be a small glycoprotein called lubricin. The lubricating properties of this synovium-derived molecule are highly specific and depend on its ability to bind to articular cartilage where it retains a protective layer of water molecules. Lubricin is not effective in artificial systems and thus does not lubricate artificial joints.

Other lubricating mechanisms have been proposed; some remain under investigation. Interestingly, hyaluronic acid, the molecule that makes synovial fluid viscous (synovia means "like egg white"), has largely been excluded as a lubricant of the cartilage-on-cartilage bearing. Instead, hyaluronate lubricates a quite different site of surface contact—that of synovium on cartilage. The well-vascularized, well-innervated synovium must alternately contract and then expand to cover nonloaded cartilage surfaces as each joint moves through its normal range of motion. This process must proceed freely. Were synovial tissue to be pinched, there would be immediate pain, intraarticular bleeding, and inevitable functional compromise. The rarity of these problems testifies to the effectiveness of hyaluronate-mediated synovial lubrication.

2. Synovial Fluids

Synovial fluid contributes significant stabilizing effects as an adhesive seal that freely permits sliding motion between cartilaginous surfaces while effectively resisting distracting forces. This property is most easily demonstrated in small articulations such as the metacarpophalangeal joints. The common phenomenon of "knuckle cracking" reflects the fracture of this adhesive bond. Secondary cavitation within the joint space causes a radiological obvious bubble of gas that requires up to 30 min to dissolve before the bond can be reestablished and the joint can be "cracked" again. This adhesive property depends on the normally thin film of synovial fluid between all intraarticular structures. When this film enlarges as a pathologic effusion, the stabilizing properties are lost.

In normal human joints, a thin film of synovial fluid covers the surfaces of synovium and cartilage within the joint space. The volume of this fluid increases when disease is present to provide an effusion that is clinically apparent and may be easily aspirated for study. For this reason, most knowledge of human synovial fluid comes from patients with joint disease. Because of the clinical frequency, volume, and accessibility of knee effusions, our knowledge is largely limited to findings in that joint.

Table 5 Coefficient of Friction for Dental Materials

Material couples	Dry	Wet
Amalgam on:		
Amalgam	0.19–0.35	
Bovine enamel		0.12–0.28
Composite resin		0.10–0.18
Gold alloy		0.10–0.35
Porcelain	0.06–0.12	0.07–0.15
Bone on:		
Metal (bead-coated)	0.50	
Metal (fiber-mess-coated	0.60	
Metal (smooth)	0.42	
Bovine enamel on:		
Acrylic resin	0.19–0.65	
Amalgam	0.18–0.22	
Bovine dentin	0.35–0.40	0.45–0.55
Bovine enamel	0.22–0.60	0.50–0.60
Chromium–nickel alloy	0.10–0.12	
Gold	0.12–0.20	
Porcelain		
Composite resin on:		
Amalgam	0.13–0.25	0.22–0.34
Bovine enamel	0.30–0.75	
Gold alloy on:		
Acrylic	0.6–0.8	
Amalgam	0.15–0.25	
Gold alloy	0.2–0.6	
Porcelain	0.22–0.25	0.16–0.17
Hydrogel-coated latex on:		
Hydrogel	0.054	
Latex on:		
Glass	0.470	
Hydrogel	0.095	
Metal (bead-coated) on:		
Bone	0.54	
Metal (fiber-mess-coated) on:		
Bone	0.58	
Prosthetic tooth materials	0.21	0.37
Acrylic on acrylic	0.23	0.30
Acrylic on porcelain	0.34	0.32
Porcelain on acrylic	0.14	0.51
Porcelain on porcelain		

The coefficient of friction is defined as the ratio of tangential force to normal load during a sliding process.
Source: Ref. 29.

The basic functions of synovial fluid are as follows [30]:

1. Provide lubrication to reduce frictional resistance to joint movement.
2. Provide nutrition to articular cartilage that may have a poor blood supply via vessels of the synovial membrane and the hypochondral blood vessels from the underlying marrow cavity.
3. Protect the joint structures when subjected to large compressive forces.
4. Provide a liquid environment within a narrow pH range.
5. Remove various products of metabolism.

In the synovium, as in all tissues, essential nutrients are delivered and metabolic by-products are cleared by the bloodstream perfusing the local vasculature. Synovial microvessels contain fenestrations that facilitate diffusion-based exchange between plasma and the surrounding interstitium. Free diffusion provides full equilibration of small solutes between plasma and the immediate interstitial space. Further diffusion extends this equilibration process to include all other intracapsular spaces including the synovial fluid and the interstitial fluid of cartilage. Synovial plasma flow and the narrow diffusion path between synovial lining cells provide the principal limitations on exchange rates between plasma and synovial fluid.

This process is clinically relevant to the transport of therapeutic agents in inflamed synovial joints. Many investigators have made serial observations of drug concentrations in plasma and synovial fluid after oral or intravenous administration. Predictably, plasma levels exceed those in synovial fluid during the early phases of absorption and distribution. This gradient reverses during the subsequent period of elimination when intrasynovial levels exceed those of plasma. These patterns reflect passive diffusion alone, and no therapeutic agent is known to be transported into or selectively retained within the joint space.

Metabolic evidence of ischemia provides a second instance when the delivery and removal of small solutes becomes clinically relevant. In normal joints and in most pathologic effusions, essentially full equilibration exists between plasma and synovial fluid. The gradients that drive net delivery of nutrients (glucose and oxygen) or removal of wastes (lactate and carbon dioxide) are too small to be detected. In some cases, however, the synovial microvascular supply is unable to meet local metabolic demand, and significant gradients develop. In these joints, the synovial fluid reveals a low oxygen pressure (PO_2), low glucose, low pH, high lactate, and high carbon dioxide pressure (PCO_2). Such fluids are found regularly in septic arthritis, often in rheumatoid disease, and infrequently in other kinds of synovitis. Such findings presumably reflect both the increased metabolic demand of hyperplastic tissue and impaired microvascular supply.

Consistent with this interpretation is the finding that ischemic rheumatoid joints are colder than joints containing synovial fluid in full equilibration with plasma. Like other peripheral tissues, joints normally have temperatures lower than that of the body's core. The knee, for instance, has a normal intraarticular temperature of 32°C. With acute local inflammation, articular blood flow increases and the temperature approaches 37°C. As rheumatoid synovitis persists, however, microcirculatory compromise may cause the temperature to fall as the tissues become ischemic.

The clinical implications of local ischemia remain under investigation. Decreased synovial fluid pH, for instance, was found to correlate strongly with radiographic evidence of joint damage in rheumatoid knees. Other work has shown that either joint flexion or quadriceps contraction may increase intrasynovial pressure and thereby exert a tamponade effect on the synovial vasculature. This finding suggests that normal use of swollen joints may create a cycle of ischemia and reperfusion that leads to tissue damage by toxic oxygen radicals.

Normal articular cartilage has no microvascular supply of its own and, therefore, is at risk in ischemic joints. In this tissue, the normal process of diffusion is supplemented by the convection induced by cyclic compression and release during joint usage. In immature joints, the same pumping process promotes exchange of small molecules with the interstitial fluid of underlying trabecular bone. In adults, however, this potential route of supply is considered unlikely, and all exchange of solutes may occur through synovial fluid. This means that normal chondrocytes are farther from their supporting microvasculature than are any other cells in the body. The vulnerability of this extended supply line is clearly shown in synovial ischemia.

The normal proteins of plasma also enter synovial fluid by passive diffusion. In contrast to small molecules, however, protein concentrations remain substantially less in synovial fluid than in plasma. In aspirates from normal knees, the total protein was only 1.3 g/dL, a value roughly 20% of that in normal plasma. Moreover, the distribution of intrasynovial proteins differs from that found in plasma. Large proteins such as IgM and cr2-macroglobulin are underrepresented, whereas smaller proteins are present in relatively higher concentrations. The mechanism determining this pattern is reasonably well understood. The microvascular endothelium provides the major barrier limiting the escape of plasma proteins into the surrounding synovial interstitium. The protein path across the endothelium is not yet clear; conflicting experimental evidence supports the fenestrae, intercellular junctions, and cytoplasmic vesicles as the predominant sites of plasma protein escape. What does seem clear is that the process follows diffusion kinetics. This means that smaller proteins, which have fast diffusion coefficients, will enter the joint space at rates proportionately faster than those of large proteins with relatively slow diffusion coefficients.

In contrast, proteins leave synovial fluid through lymphatic vessels, a process that is not size selective. Protein clearance may vary with joint disease. In particular, joints affected by rheumatoid arthritis (RA) experience significantly more rapid removal of proteins than do those of patients with osteoarthritis. Thus, in all joints, there is a continuing, passive transport of plasma proteins involving synovial delivery in the microvasculature, diffusion across the endothelium, and ultimate lymphatic return to plasma.

The intrasynovial concentration of any protein represents the net contributions of plasma concentration, synovial blood flow, microvascular permeability, and lymphatic removal. Specific proteins may be produced or consumed within the joint space. Thus, lubricin is normally synthesized within synovial cells and released into synovial fluid where it facilitates boundary layer lubrication of the cartilage-on-cartilage bearing. In disease, additional proteins may be synthesized, such as IgG rheumatoid factor in RA, or released by inflammatory cells, such as lysosomal enzymes. In contrast, intraarticular proteins may be depleted by local consumption, as are complement components in rheumatoid disease.

Synovial fluid protein concentrations vary little between highly inflamed rheumatoid joints and modestly involved osteoarthritic articulations. Microvascular permeability to protein, however, is more than twice as great in RA as in osteoarthritis. This marked difference in permeability leads to only a minimal increase in protein concentration, because the enhanced ingress of proteins is largely offset by a comparable rise in lymphatic egress. These findings illustrate that synovial microvascular permeability cannot be evaluated from protein concentrations unless the kinetics of delivery or removal is concurrently assessed [31].

C. Tribological Failure of Bioimplants

At present, well-designed implants, such as hip and knee replacements, have excellent clinical success rate, at 10–12 years in most patients. However, in the second decade of the prosthesis's life the failure rate and need for revision operations increase dramatically. In the last 5 years, researchers have identified osteolysis, induced by wear debris, as a major cause of long-term failure.

Temporomandibular joint disorder is estimated to affect 30 million Americans with approximately 1 million new patients diagnosed yearly. In the past decade, at least 100,000 patients have received alloplastic material in their joints with another 300,000 patients having received autogenous tissue including costochondral rib graft. These implants have become a disaster because the failure rate is approaching 80% [31].

1. Fracture

Fracture happens often in bone and orthopedic implants when load-bearing ability is important. Once fractured, biomaterials cannot be regenerated as natural bones. Therefore, structure and properties are particularly important. It is also very important to induce lubricants to improve the biomaterial properties in the same time. The inducing of the lubricants could reduce the wear, which is one of the main problems that could lead to fracture of the materials. Research on thin film lubrication can be found in listed references [32,33]. The two leading mechanisms in joint replacement wear are foreign body wear and surface fatigue.

2. Wear

When a piece of materials comes into the articulating joint surfaces, wear occurs. Wear of the articulating surfaces in artificial hip and knee joints gives rise to production of wear particles of sizes of submicrons and larger. The negative biological effects of these wear particles are considered to be one important factor that limits the long-term clinical performance of this type of devices. There is therefore an immediate need for development of novel and improved material combinations for the articulating surfaces in artificial joints. Such development is in turn dependent on an improved understanding of the wear processes involved and how these are influenced by different material properties and conditions. Equally important is to develop reliable and predictive methods for stimulating the wear processes under in vitro conditions, preferentially in an accelerated way.

In order to provide a basis for systematic development of novel material combinations and surfaces with improved wear properties and to develop reliable in vitro screening methods for assessing the wear of different candidate combinations, wear research is done to obtain an increased understanding of the mechanisms underlying wear in artificial joint prostheses. The long-term engineering goal is to develop a material (surface) combination that does not produce biologically harmful wear particles under the physiological conditions that prevail in artificial hip and knee joints.

3. Third-Body Wear

Bioenvironments are mostly wet. Wear may be increased for wet sliding wear compared with dry sliding wear for unfilled polymers [34]. The wear volume also increases in carbon fiber and glass-reinforced composites compared with unfilled polymers, thus enhancing the cutting/lowing/cracking mechanism of abrasive particles against the polymer or polymer composite material [34]. In a physiological environment, metallic, ceramic, or polymeric wear particles may be trapped between two moving surfaces, causing three-body wear, which generally causes a significantly higher wear rate than two-body wear. Other mechanisms for increased in vivo wear include environmental stress cracking, polymer degradation, microstructural imperfections, and creep.

Experiments have been conducted to demonstrate the increase of wear rates under in vivo vs. in vitro conditions for artificial joints. Dry in vitro tests of hip joints predicted an average wear rate of 0.02 mm a year; the average clinical penetration rate was found to be 0.07 mm a year over the first 9 years of use [35]. This wear may cause loosening of the prostheses by the resulting poor mechanical fit between the ball and socket of the hip. If the wear becomes extensive, the artificial hip may need to be replaced.

The wear particle is, however, of much greater concern. Generation of wear debris is an important factor both because of the potential for wear debris to migrate to distant organs, particularly the lymph nodes where accumulation of particle-containing macrophages causes enlargement and chronic lymphadenitis [36], and because of local physiological responses such as inflammatory, cytotoxid, and osteolytic reactions.

It is now apparent that large quantities of micron and submicron wear debris from most materials cause nonspecific osteolysis of local bone. For example, a major cause of late failures of artificial hips has been loosening between the hip stem and the partially resorbed surrounding bone, causing pain and requiring hip replacement. Sometimes femoral bone weakened by osteolysis can fracture under high loading conditions and require replacement. It was thought that this condition was the result of osteolysis of the bone caused by bone cement particles released by wear and fatigue microfractures [37], and it was occasionally called bone cement disease.

A new generation of artificial hips has been developed and introduced that uses bioingrowth of bone into porous surfaces of the hip stem so that cement is not required; thus it was hoped to eliminate bone cement disease. Unfortunately, the newer bioingrowth artificial hips have continued to fail by the same osteolytic mechanism, in the absence of any bone cement. Further research has shown that polyethylene [38], titanium, and ceramic wear debris can cause the same osteolytic problem associated with bone cement debris. Both microsized particulates seen in earlier histological slides, as well as submicron particulates now found using digestion and scanning electron microscopic techniques, have been implicated in osteolytic destruction of the bone-implant interface. Assuming a million steps per year and 0.03 to 0.08 mm per year wear into polyethylene acetabular cups, one recent calculation estimates that on an average as many as 470,000 polyethylene particles of 0.5-μm size could be released to the surrounding tissues at each step [39]. Other examples of nonspecific osteolysis presumed to be caused by wear debris of devices include bone resorption around failed elastomeric finger, toe, and wrist prostheses and around failed poly(tetrafluoroethylene)–carbon fiber temporomandibular joint (jaw) prostheses. Therefore, important factors in the selection of materials for use in medical implants are the quantity, size, shape, and composition of wear debris that may be released in vivo.

A total hip replacement is clinically successful with most designs that utilize a UHMWPE acetabular surface articulating with either a metallic or ceramic femoral head component [40]. Because of significant localized contact stresses at the ball/socket interface, small regions of UHMWPE tend to adhere to the metal or ceramic ball. During the reciprocating motion of normal joint use, fibrils will be drawn from the adherent regions on the polymer surface and break off to form submicrometer-sized wear debris. This adhesive wear mechanism, coupled with fatigue-related delamination of the UHMWPE (most prevalent in knee joints), results in billions of tiny polymer particles being shed into the surrounding synovial fluid and tissues. The biological interaction with small particles in the body then becomes critical. The body's immune system attempts, unsuccessfully, to digest the wear particles (as it would a bacterium or virus). Enzymes are released that eventually result in the death of adjacent bone cells, or osteolysis. Over time, sufficient bone is resorbed around the implant to cause mechanical loosening, which necessitates a costly and painful implant replacement, or revision. Aseptic loosening, often accompanied by osteolysis, continues to be a source of long-term clinical failure. It was suggested that the macrophage response to phagocytosis of particulate wear debris, occurring in interaction between the cement and bone, was an important causative factor in osteolysis, leading to eventual loosening [41]. The particulate debris has emerged as a major factor in the long-term performance of joint replacement prostheses [42]. When present in sufficient amounts, particulates generated by wear, fretting, or fragmentation induce formation of an inflammatory, foreign-body granulation tissue that has the ability to invade the bone-implant interface [43–50]. This may result in progressive periprosthetic loss of bone, a loss that threatens the fixation of prostheses inserted with or without cement [15,51,52].

Studies of wear debris extracted from actual tissue samples of patients whose implants failed as a result of aseptic loosening have generated significant information regarding wear particle size, shape, and surface morphology. Atomic force microscopy was used to produce detailed, high-resolution images of wear particles a few hundred nanometers in size. Wear debris studied sometimes exhibits a cauliflower-like surface morphology. By combining wear debris and cellular response studies, engineers and biologists will be able to better understand implant failure and to reengineer implants to prevent future problems.

The wear problem that occurs with an artificial joint implant component (socket) constructed of UHMWPE is illustrated in Fig. 13. At left is unworn UHMWPE. The sample at right has undergone a friction and wear test vs. cobalt chromium (artificial joint ball material). The fibrillation and small particles are characteristic of an adhesive wear mechanism, which can result in surrounding bone loss and the need for implant replacement [1].

4. Corrosion and Corrosive Fatigue

The second type of wear is subsurface fatigue. High contact stresses in the artificial joints will cause cracks in the biological materials and will propagate beneath the surface. Figure 14 shows a typical knee crack [53]. The following are examples of cracks in the tibial component of a total knee.

Corrosive attack in the taper crevice of modular implants made of similar metals or mixed-metal combinations [54–58]. The corrosive attack results in metal release and mechanical failure of the component [59].

Corrosive fatigue, on the other hand, is expected due to cyclic loading and the corrosive environment of the human body. The fractures occurred at the grain boundaries

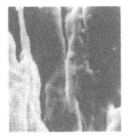

Figure 13 SEM micrographs of a socket of UHMWPE before and after wear test. (From Ref. 1.)

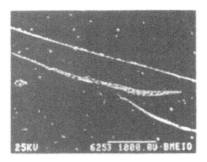

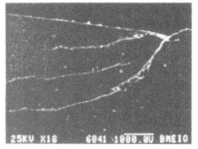

Figure 14   Knee crack. (From Ref. 41.)

of the microstructure and appeared to be the result of three factors: porosity at the grain boundaries; intergranular corrosive attack, initiated both at the head–neck taper and at free surface; and cyclic fatigue loading of stem [60].

Mechanical testing will be carried out in parallel with finite element analysis (FEA). The purpose of FEA is to determine the effect of different design features on the stress in the materials and, just as importantly, the effect of the support conditions from the bone. For instance, the components could be supported in strong rigid bone or, at the other extreme, the bone could resorb locally leaving regions of fibrous tissue.

D. Tribology of Joints

A joint is formed by the ends of two or more bones that are connected by thick tissues. The functions of joints are first, *to permit motion*, and second, *to provide stability*. The joints can be classified into two types as upper limb such as shoulder, elbow, etc., which does not need to carry the body weight during motion, and as lower limb such as hip, knee, and ankle due to the similar motion, load, and lubrication. However, the upper limb sometimes can be shown to be subjected to quite high loads under certain conditions. The joints can also be classified based on the structure of three types: fibrous, cartilaginous, and synovial joints. Joints can also be further classified according to the anatomical and physiological properties as follows:

1. Plane: essentially flat surfaces, such as the intermetacarpal joints that can perform gliding, nonaxial movement.
2. Ball and socket: ball-shaped head fits into concave socket, such as shoulder and hip joints that can do triaxial movement. When the socket is well rounded it is called cotyloid or cup shaped. When less well rounded it is called glenoid.

3. Ellipsoid: oval shaped condyle fits into elliptical cavity. This is a spheroidal joint, one of the axes being longer, such as the wrist joint that can perform movement in two planes at right angles such as flexion–extension and abduction–adduction.
4. Hinge: spool-shaped surface fits into concave surface, such as elbow, knee, and ankle that can perform uniaxial movement.
5. Saddle; Saddle shaped bone fits into socket that is concave–convex in the opposite direction, such as the first carpometacarpal joint that can perform the biaxial movement similar to ellipsoidal.
6. Condylar: one bone articulates with the other by two distinct surfaces whose movements are not dissociable. These surfaces are termed condyles, such as knee and temporomandibular joints.
7. Pivot: arch-shaped surfaces rotate about a rounded or peglike pivot, like superior radioulnar and atlanto-odontoid joints that can perform rotational and uniaxial movements.

The tribological behavior of human joints is superb. The lubrication regime results in the ability of joints to operate under a wide range of conditions ranging from high loads at low speeds to low loads at high speeds and go from a rest position to sliding motion under the most severe condition without damage to the joints. The excellence of the lubrication is reflected in the low friction coefficient, which typically lies in the range 0.003–0.015 under comparatively high-load and low-speed conditions. Wear in joints is minimal under normal circumstances and the cartilaginous surfaces last an entire lifetime of 70 years or more [61].

The bone ends of a joint are covered with a smooth layer called cartilage. Normal cartilage allows nearly frictionless and pain-free movement. However, when the cartilage is damaged or diseased by arthritis, joints become stiff and painful. Every joint is enclosed by a fibrous tissue envelope or capsule with a smooth tissue lining called synovium. The synovium produces fluid that reduces friction and wear in a joint.

Under normal healthy conditions, the human joints are fluid-film lubricated, boundary lubricated, and mixed lubricated. The synovial joints are perfect bearings with low friction and low wear and last almost all life long.

Artificial joints, on the other hand, are lubricated by a mixed regime of fluid film and solid-to-solid contact. Because of the contact, friction and wear are still problems, and one of the main problems in artificial joints is the formation of wear particles.

An arthritic or damaged joint is removed and replaced with an artificial joint called a prosthesis. The goal is to relieve the pain in the joint caused by the damage done to the cartilage. Although hip and knee replacements are most common, joint replacement can be performed on other joints, including the ankle, foot, shoulder, elbow, fingers, etc. Figure 15 shows pictures of some joints [62,63].

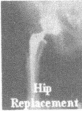

Figure 15 Pictures of some joints. (From Refs. 5 and 62.)

Main research on joint prosthesis usually concentrates on research of biomaterials used as implants; the shape of the implants for different joints according to the biomechanism; and the tribological properties of the implants including the lubrication friction and wear, which connect directly to the service time of the artificial joints.

Joints can be classified into two types: upper limb such as finger, which does not need to carry the body weight during motion, and lower limb such as hip, knee, and ankle due to the similar motion, load, and lubrication. The friction, lubrication, and wear of some of the artificial joints will be introduced in the following.

1. Knee Joints

The knee joint is formed by the lower leg bone, called the tibia or the shin bone, and the thighbone, called femur. The knee structure can be seen in Fig. 16 [53].

Artificial knee joints are developed for persons who suffer pain and disability from an arthritic knee that could not be solved by methods other than a surgical solution.

In an arthritic knee the damaged end of the bones and cartilage are replaced with metal and plastic surfaces that are shaped to restore knee movement and function (shown in Fig. 17) [53].

2. Hip Joints

The hip joint is a ball and a socket joint, formed by the upper end of the femur (the largest bone in the body), the ball, and a part of the pelvis called the acetabulum, the socket. The hip joint is one of the largest and most heavily loaded joints of the human body.

Figure 18 shows a normal hip and a hip with arthritis. In an arthritic hip, the damaged ball (the upper end of the femur) is replaced by all-metal artificial joints or a metal ball attached to a metal stem fitted into the femur, and a plastic socket is implanted into the pelvis, shown in Fig. 19 [53]. Most modern artificial hip joints combine a metal femoral component (ball) with a UHMWPE acetabular cup.

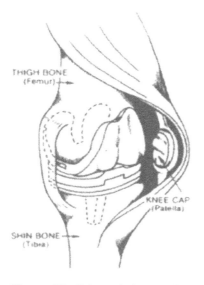

Figure 16 Schematic image of the knee. (From Ref. 53.)

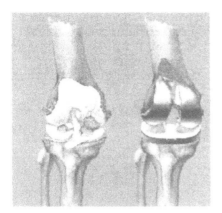

Figure 17 Artificial knee joints. (From Ref. 53.)

Primary total hip replacement (THR) is most commonly used for hip joint failure caused by osteoarthritis; other indications include, but are not limited to, rheumatoid arthritis, avascular necrosis, traumatic arthritis, certain hip fractures, benign and malignant bone tumors, arthritis associated with Paget's disease, ankylosing spondylitis, and juvenile rheumatoid arthritis. The aims of THR are relief of pain and improvement in function. The hip joint is located where the upper end of the femur meets the acetabulum. The femur, or thighbone, looks like a long stem with a ball on the end. The acetabulum is a socket or cuplike structure in the pelvis, or hipbone. This "ball-and-socket" arrangement allows a wide range of motion, including sitting, standing, walking, and other daily activities. During hip replacement, the surgeon removes the diseased bone tissue and cartilage from the hip joint. The healthy parts of the hip are left intact. Then the surgeon replaces the head of the femur (the ball) and the acetabulum (the socket) with new, artificial parts. The new hip is made of materials that allow a natural, gliding motion of the joint.

3. Temporomandibular Joints

The TMJ provides all jaw mobility and is crucial for chewing, talking, and swallowing. This joint can deteriorate from disease or trauma that, in severe cases, necessitates replacement by

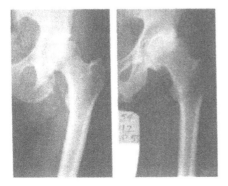

Figure 18 A normal hip (left) and a hip with arthritis (right). (From Ref. 53.)

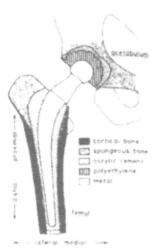

Figure 19 A metal femoral component (ball) with a UHMWPE acetabular cup. (From Ref. 53.)

an artificial joint. For many years, less than optimum technologies existed for TMJ implants. Materials engineers know that the reason PTFE exhibits such a low coefficient of friction is that a thin film of the material is continuously transferred onto the opposing bearing surface. Although this transfer film acts as a lubricant, it also, by virtue of its formation, subjects the material to an adhesive wear mechanism. In the case of PTFE TMJ implants, surrounding tissues quickly became overwhelmed by wear debris, and the immune system response resulted in osteolysis, causing massive destruction of the joint and surrounding tissues. For those people who received the implants, this was truly a tragedy; many suffered severe facial deformities, and most experienced unbearable pain and were no longer able to chew, swallow, or sleep.

IV. BIOTRIBOLOGY TESTING

A. Tribology Tests

Classical experimental techniques of practical biotribology comprise measuring of friction force and torque at first or, generally, measuring of forces. A typical laboratory setup designed at Tallinn Technical University is shown in Fig. 20 [64].

The test rig is proposed for testing joint models and joint implants. The load is registered by piezoelectric force sensor. The rotation velocity parameters of the specimen can also be changed. The tested joint state is monitored by additional sensors (temperature gauges, electrical characteristics—depending on the task of the experiment). The most important output quantity in the test system is the value of friction force or torque formed at relative motion of the rubbing surfaces. The rig is suitable for comparative testing of joint implants and experimental investigation of models of biotribosystems.

Wear test is a basic method to evaluate the durability of biomaterials. *Pin-on-disc* experiments are carried out on different material combinations using systematically varied tribological conditions (lubrication, loading, contact geometry, sliding speed). Novel candidate materials will be investigated, as well as current materials (UHMWPE) that have been modified by different methods (e.g., plasma technology, microfabrication, and polymer

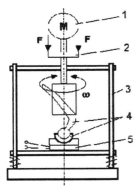

Figure 20 Functional scheme of joint prostheses and materials test rig. 1, motor; 2, loading unit; 3, frame; 4, joint prosthesis; 5, piezoelectric load and torque gauge. (From Ref. 65.)

processing). The chemical and structural properties of the tested materials and wear products are characterized by *spectroscopic and microscopic analysis techniques*. Results from *multiaxial joint simulator* experiments, as well as analysis of components from retrieved clinical implants, constitutes references for the in vitro simulation studies.

After the tribological experiments, materials characterization will be necessary for better understanding of the wear mechanism of the tribology.

Tribological study reported that abrasive wear, adhesion wear, and fatigue wear of UHMWPE have been identified as three basic mechanisms [65–69]. The UHMWPE pins sliding against thermal-oxidation-treated Ti6A14V alloy discs with different surface roughness was investigated in water by Shi et al. [70]. Evidence of microfatigue wear was found. The wear rate was reported at 1.07×10^{-4} when the average surface roughness is in 0.020–0035 µm. While the surface roughness is higher to 0.06, the wear rate increases to 7.51×10^{-4} mm^3/m [70]. Examples of microfatigue wear are shown in Fig. 21 [70]. The wear of UHMWPE is affected by the motion direction [71]. Researchers found softening or multimode wear in wear screening of UHMWPE materials.

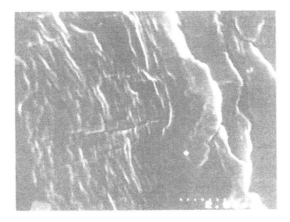

Figure 21 Fatigue wear of UHMWPE against thermal-oxidation-treated Ti6A14V. (From Ref. 70.)

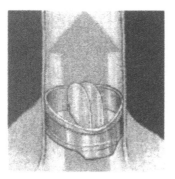

Figure 22   Artificial mechanical heart valves.

B. Electrochemical Tests

The material properties and the mechanical and electrochemical processes were observed to control the corrosion attack to the hip replacement. Some *electrochemical test* methods are the following:

1. Short-term tests: the effect of cyclic load magnitude on fretting corrosion currents
2. Long-term tests: rest tests, cyclic mechanical loading, scratch test, etc. [8–10].

Artificial mechanical heart valves are made of pyrolytic carbon to prevent complications associated with blood clotting, but this material can be subject to cyclic fatigue. An *acoustic emission-based system* is developed for detecting crack initiation and the growth of existing flaws during controlled-stress testing of artificial heart valves shown in Fig. 22.

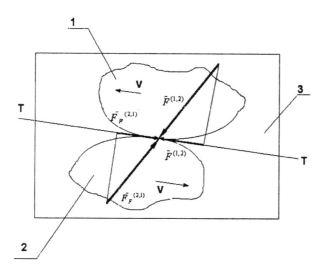

Figure 23   Frictional interaction of bodies: 1, 2, solid contacting and moving bodies; 3, medium. (From Ref. 64.)

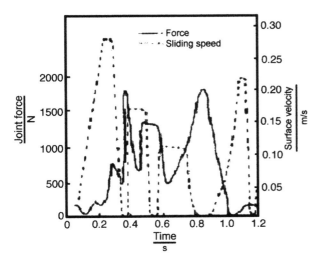

Figure 24 Knee joint forces and surface velocities at different parts of the walking cycles (after Seedhom). (From Ref. 70.)

V. BIOMECHANICAL ANALYSIS

Forces, contact stresses, and contact geometry are generally analyzed using mechanical approaches. Figure 23 shows the interactions between two contacting and relatively moving bodies [64,72–75]. In this figure, several different specific treatments and statements of tribological systems are not considered. Here V is the relative moving velocity, T is the tangential force, F is the contact force, 1 and 2 indicate the two bodies, and 3 the media.

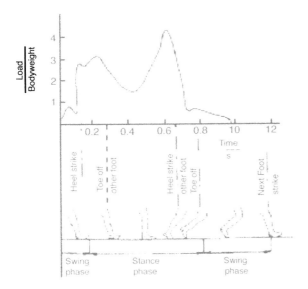

Figure 25 Hip joint force (after Paul). (From Ref. 70.)

The bonding is generated due to direct contact of the bodies through an interstitial substance (a liquid film, a gas flow, etc.) or by an effect of electric or magnet field. The effect of the interstitial substance is treated as a reaction force in mechanics. Its projection on the tangent plane of the contacting bodies is therefore the friction force. These four components are all inside an environment or medium, such as vacuum in a biological subject. The components are two moving bodies, interstitial substance (may not exist), and a medium. Therefore, the first major problem of tribology as well as biotribology is friction between the bodies—all questions connected with friction bonding. Moving bodies wear by generating debris or materials loss. Many tribosystems do not show any wear during the system's life span, but many do. Lubrication is only an instrument for controlling and mediating the major biotribological processes [73–77]. Today, there are a limited number of mathematical models for biotribosystems. The friction is generally treated as an elastohydrodynamic formula. The boundary lubrication theory is most likely the appreciable one. Such theories are applied in analyses of human and animal joints [73–77].

Loads and motions in the joints of the lower limb are different from those of the upper limb. Figure 24 shows the knee joint forces and surface velocities at different parts of the walking cycle during typical walking [70]. The cycle is about 1.2 sec per period and maximum loads can reach 1500 N when the heel strikes the ground or the toe is pushing off [70]. Figure 25 shows hip joint forces. From both figures, the highest loads correspond to the lowest velocity. When the surface velocities are the maximum, the loads are the lowest [71].

REFERENCES

1. Blanchard, C.R. Biomaterials: body parts of the future. Technol. Today, Fall 1995.
2. Dowson, D. Progress in tribology: A historical perspective, New directions in tribology, Plenary and invited papers from the First World Tribology Congress, 8–12 September, 1997. In *First World Tribology Congress*; Hutchings, I.M., Ed.; Bury St. Edmonds: Mechanical Engineering Publications for the Institution of Mechanical Engineers: London, 1997; 3–20.
3. McLaughlin, J. 8–10 year results using the Taperloc femoral component in uncemented total hip arthroplasty. Annual Meeting of the American Academy of Orthopedic Surgeons, New Orleans, February 1994.
4. Ritter, K.; Merrill, A.; Worland, Richard; Saliski, John; Helphenstine, Jill V.; Edmondson, Kristen L.; Keating, E.; Michael, Faris; Philip, M.; Meding, John B. Flat-on-flat, nonconstrained, compression molded polyethylene Paris; total knee replacement. Clin. Orthop. Relat. Res. 1995 Dec, *321*, 79–85.
5. http://www.biomet.com/patients/bestimplant.html. Accessed July 2001.
6. Long, M.; Rack, H. Review, titanium alloys in total joint replacement—a materials science perspective. Biomaterials 1998, *19*, 1621–1639.
7. Wang, A.; Essner, A.; Polineni, V.K.; Sun, D.C.; Stark, C.; Dumbleton, J.H. Lubrication and wear of ultra-high molecular weight polyethylene in total joint replacements. *New Directions in Tribology*; Hutchings, I.M., Ed.; I Mech. E., London, Sept. 8–12, 1997; 443–458.
8. Gilbert, J.L.; Jacobs, J.J. The mechanical and electrochemical processes associated with taper fretting crevice corrosion: a review. In *Modularity of Orthopedic Implants, ASTM STP 1301*; Marlowe, D.E., Parr, J.E., Mayor, M.B., Eds.; American Society for Testing and Materials: Philadelphia, 1997; 45–59.
9. Gilbert, J.L.; Buckley, C.A. Mechanical–electrochemical interactions during in vitro fretting corrosion tests of modular taper connections. In *Total Hip Revision Surgery;* Galante, J.O., Rosenberg, A.G., Callaghan, J.J., Eds.; Raven Press Ltd.: New York, 1995; 41–50.

10. Goldberg, J.R.; Gilbert, J.L. Electrochemical response of CoCrMo to high-speed fracture of its metal oxide using an electrochemical scratch test method. J. Biomed. Mater. Res. 1997, *37*, 421–431.

11. Sarikaya, M.; Fong, H.; Frech, D.; Humbert, R. Biomimetic assembly of nanostructured materials. Mat. Sci. Forum 1999, *293*, 83–98.

12. Long, M.; Rack, H.J. Titanium alloys in total joint replacement—a materials science perspective. Biomaterials 1998, *19*, 1621–1639.

13. www.dop.uab.edu/~mfries/interests.html. Accessed August 2001.

14. Shi, B.; Ajaji, O.; Fenske, G.; Erdemier, A.; Liang, H. Tribological performance of some alternative bearing materials for artificial joints. Wear 2003, *255*, 1015–1021.

15a. Conjugated Macromolecules of Precise Length and Constitution. Organic Synthesis for the Construction of Nanoarchitectures. Chemical Reviews 1996, *96* (1), 537–553.

15b. Molecular Self-Assembly and Nanochemistry: A Chemical Strategy for the Synthesis of Nanostructures. Additional Info: Engineering a Small World: From Atomic Manipulation to Microfabrication. Science November 29, 1991, *254* (5036), 1312–1319.

16a. Niemeyer, C.M. DNA als Werkstoff fur die Nanotechnologie. Angewandte Chemie. 1997, *109* (6), 603–606.

16b. A DNA-based method for rationally assembling nanoparticles into macroscopic materials. Nature 1996, *382* (6592), 607–609.

16c. Mirkin, C.A.; Letsinger, R.L.; Mucic, R.C.; Storhoff, J.J. Organization of 'nanocrystal molecules' using DNA. Nature 1996, *382* (6592), 609–611.

16d. Alivisatos, A.P.; Johnson, K.P.; Peng, X.; Wilson, T.E. The electrophoretic properties of a DNA cube and its substructure catenanes. Electrophoresis September 1991, *12* (9), 607–611.

16e. Chen, J.; Seeman, N.C. DNA components for molecular architecture. Accounts of Chemical Research 1997, *30* (9), 357–363.

16f. Seeman, N.C. Acc. Chem. Res. 1997, *30*, 357–363.

17a. Bach, D.; Miller, I.R. Interaction of deoxyribonucleic acid with poly-4-vinylpyridine. Biochim. Biophys. Acta Feb 21, 1966, *114* (2), 311–325.

17b. Bach, D.; Miller, I.R. Interaction of DNA with heavy metal ions and polybases: cooperative phenomena. Biopolymers 1968, *6* (2), 169–179.

17c. Felgner, P.L.; Gadek, T.R.; Holm, M.; Roman, R.; Chan, H.W.; Wenz, M.; Northrop, J.P.; Rigold, G.M.; Danielsen, M. A highly efficient, lipid-mediated DNA-lipofecion procedure. Proc. Natl. Acad. Sci. USA 1987, *84*, 7413–7417.

17d. Vollenweider, H.J.; Sogo, J.M.; Koller, H.H. A routine method for protein-free spreading of double- and single-stranded. Proc. Natl. Acad. Sci. USA 1975, *72* (1), 83–87.

18. http://laurencin.coe.drexel.edu/. Accessed October 2003.

19. Lu, X.C.; Shi, B.; Luo, J.B.; Wang, J.H.; Li, H.D.; Lawrence, K.Y. Investigation on micro-tribological behavior of thin films using friction force microscope. Surf. Coat. Technol. 2002;, 128–129, 341–345.

20. Lu, X.C.; Wen, S.Z., et al. A friction force microscope employing laser beam deflection for force detection. Chin. Sci. Bull. 1996, *41* (22), 1873–1876 (in English).

21. http://www.swri.org/3pubs/brochure/d06/matstruc/msaddlg.htm. Accessed October 2003.

22. http://www.swri.edu/3pubs/brochure/d10/appbio/appbiom.htm. Accessed October 2003.

23. Gilbert, J.L.; Smith, S.M.; Lautenschlager, E.P. Scanning electrochemical microscopy of metallic biomaterials: reaction rate and ion release imaging modes. J. Biomed. Mater. Res. 1993, *27*, 1357–1366.

24. Smith, S.M.; Gilbert, J.L. The scanning potential microscope: an instrument to image microcorrosion process on metallic biomaterials surfaces. Winter 1991, *2* (2), 11–16.

25. www.biomaterials-partnership.org.uk. Accessed October 2003.

26. Nygren, H.; Tengvall, P.; Lundström, I. The initial reactions of TiO_2 with blood. J. Biomed. Mater. Res. 1997, *34*, 487–492.

27a. http://www.orthop.washington.edu/arthritis/general/joints/04. Accessed October 2003.

27b. Spector, M. Biomaterial Failure. Orthop. Clin. North Am. April 1992, *23* (2), 211–217.

28a. www.uq.edu.au/~anvikippe/. Accessed October 2003.
28b. Woodruff, L.L.; Baitsell, G.A. *Foundations of Biology*; The Macmillan Company: NY, 1951.
29.   http://www.lib.umich.edu/dentlib/Dental_tables/Coeffric.html. Accessed October 2003.
30.   http://www.uq.edu.au/~anvkippe/gmc/joints.html. Accessed October 2003.
31.   Baird, D.N.; Rea, W.J. The temporomandibular joint implant controversy: a review of autogenous alloplastic materials and their complications. J. Nutr. Environ. Med. 1998, *8*, 289–300.
32.   Luo, J.B.; Liu, S.; Shi, B.; Lu, X.C.; Wen, S.Z.; Li, K.Y. The failure of nano-scale liquid film. Int. J. of Nonlinear Sci. and Numer. Simul. 2000, *1*, 419–423.
33.   Luo, J.; Qina, Linmao; Wen, Shizhu, Wen, L.; Wen, S.; Li, L. The failure of fluid film at nanoscale. STLE, Tribol. Trans. 1999, *42* (4), 912–916.
34.   Lhymn, C. Effect of environment on the two-body abrasion of polyetheretherketone (PEEK)/carbon fiber composites. ASLE Trans. 1987, *30* (3), 324–327.
35.   Dowling, J.M. Wear analysis of retrieved prostheses. *Biocompatible Polymers, Metals, and Composites*; Szycher, M., Ed.;(1983). Technomic Publ.: Lancaster, PA, 1983; 407 pp.
36.   Benz, E.; Sherburne, B.; Hayek, J.; Godleski, J.J.; Sledge, C.V.; Spector, M. Migration of polyethylene wear debris to lymph nodes and other organs in total joint replacement patients. Trans. Soc. Biomater. 1990, *XVII*, 83.
37.   Willert, H.G.; Bertram, H.; Buchhorn, G.H. Osteolysis in alloarthroplasty of the hip—the role of bone cement fragmentation. Clin. Orthop. Relat. Res. 1990, *258*, 108–121.
38.   Willert, H.G.; Bertram, H.; Buchhorn, G.H. Osteolysis in alloarthroplasty of the hip—the role of ultra-high molecular weight particles, Clin. Orthop. Relat. Res. 1990, *258*, 95–107.
39.   McKellp, H.A.; Schmalzried, T.; Park, S.H.; Campbell, P. Evidence for the generation of submicro polyethylene wear particles by micro-adhesive wear in acetabular cups. Trans. Soc. Biomater. 1993, *XVI*, 184.
40.   Wang, A.; Essner, A.; Polineni, V.K.; Sun, D.C.; Stark, C.; Dumbleton, J.H. Lubrication and wear of ultra-high molecular weight polyethylene in total joint replacement. In *New Directions in tribology;* Hutchings, I.M., Ed.; World Tribology Congress: London, 1997; 443–458.
41.   Willert, H.G.; Semlitsch, M. Reactions of the articular capsule to wear products of artificial joint prostheses. J. Biomed. Mater. Res. 1977, *11*, 157–164.
42.   Urban, R.M.; Jacobs, J.J.; Gilbert, J.L.; Galante, J.O. Migration of corrosion products from modular hip prostheses. J. Bone Jt. Surg. Sep. 1994, *76-A* (9), 1345–1359.
43.   Amstutz, H.C.; Campbell, P.; Kossovsky, N.; Clark, I.C. Mechanisms of clinical significance of wear debris-induced osteolysis. Clin. Orthop. 1992, *276*, 7–18.
44.   Boss, J.H.; Shajrawi, I.; Soundry, M.; Mendes, D.G. Histomorphological reaction patterns of the bone to diverse particulate implant materials in man and experimental animals. In *Particulate Debris from medicak Implants: Mechanisms of Formation and Biological Consequences, ASTM STP 1144*; St. John, K.R., Ed.; American Society for Testing and Materials: Philadelphia, 1992; 90–108.
45.   Dicarlo, E.F.; Bullough, P.G. The biological response to orthopedic implants and their wear debris. Clin. Mater. 1992, *9*, 235–260.
46.   Kossovsky, N.; Mirra, J.M. Biocompatibility and bioreactivity of arthroplastic materials. In *Hip Arthroplasty*; Amstutz, H.C., Ed.; Churchill Livingstone: New York, 1991; 571–601.
47.   Mirra, J.M.; Amstutz, H.C.; Matos, M.; Gold, R. The pathology of the joint tissues and its clinical relevance in prosthesis failure. Clin. Orthop. 1976, *117*, 221–240.
48.   Pizzoferrato, A.; Ciapetti, G.; Stea, S.; Toni, A. Cellular events in the mechanisms of prosthesis loosening. Clin. Mater. 1991, *7*, 51–81.
49.   Spanier, S. Histology and pathology of total joint replacement. In *Total Joint Replacement* Petty, W., Ed.; W.B. Saunders: Philadelphia, 1991; 165–185.
50.   Willert, H.G.; Bertram, H.; Buchhorn, G.H. Osteolysis in alloarthroplasty of the hip. Clin. Orthop. 1990, *258*, 95–107.
51.   Nyquist, R.A.; Kagel, R.O. *Infrared Spectra of Inorganic Compounds*; Academic Press: New York, 1971; 171 pp.

52a. Willert, H.G. Failure modes of artificial joint implants due to particulate implant material. In *Implant Bone Interface*; Older, J., Ed.; Springer: New York, 1990; 67–75.

52b. Spector, M. Biomaterial failure. Orthop. Clin. North Am. April 1992, *23* (2), 211–217.

53. www.engin.umich.edu. Accessed October 2003.

54. Buckley, C.A.; Gilbert, J.J.; Urban, R.M.; Summer, D.R.; Jacobs, J.J.; Lautenschlager, E.P. Mechanically assisted corrosion of modular hip prosthesis components in mixed and similar metal combinations. Trans. Soc. Biomater. 1992, *15*, 58 (Implant Retrieval Symposium).

55. Collier, J.P.; Surprenant, V.A.; Jensen, R.E.; Mayor, M.B. Corrosion at the interface of cobalt-alloy heads on titanium-alloy stems. Clin. Orthop. 1991, *271*, 305–312.

56. Collier, J.P.; Surprenant, V.A.; Jensen, R.E.; Mayor, M.B.; Surprenant, H.P. Corrosion between the components of modular femoral hip prostheses. J. Bone Jt. Surg. 1992, *74-B* (4), 511–517.

57. Gilbert, J.L.; Buckley, C.A.; Urban, R.M.; Lautenschlager, E.P.; Galante, J.O. Mechanically assisted corrosive attack in the Morse taper of modular hip prostheses. Transactions of the 4th World Biomaterials Congress, Berlin, European Society for Biomaterials, 1992; 267 pp.

58. Mathiesen, E.B.; Lindgren, J.U.; Blomgren, G.G.A.; Reinholt, F.P. Corrosion of modular hip prostheses. J. Bone Jt. Surg. 1991, *73-B* (4), 569–575.

59. Urban, R.M.; Jacobs, J.J.; Gilbert, J.L.; Galante, J.O. Corrosion products of modular hip prostheses: micromechanical identification and histopathological significance. Trans. Orthop. Res. Soc. 1993, *18*, 81.

60. Gilbert, J.L.; Buckley, C.A.; Jacobs, J.J.; Bertin, K.C.; Zernich, M.R. Intergranular corrosion-fatigue failure of cobalt-alloy femoral stems. J. Bone Jt. Surg. 1994, *76-A* (1), 110–115.

61. Dumbleton, John H. *Tribology of Nature and Artificial Joints*; Elsevier Scientific Publishing: Amsterdam, 1981.

62. www.newbernortho.com/artificial_joints.html. Accessed October 2003.

63. http://arthritis.about.com. Accessed October 2003.

64. Ajaots, M.; Tamre, M. Biotribology General Approach, http://www.ttu.ee/mhk/art/sat/sat.html. Accessed October 2003.

65. Wang, A.; Essner, A.; Polineni, V.K.; Sun, D.C.; Stark, C.; Dumbleton, J.H. *New Directions in Tribology*; Hunchings I.M., Ed.; I Mechanical Engineering Publications: London, 1997; 443 pp. (London).

66. McKellop, H.A.; Campbell, P.; Park, S-H.; Schmalzried, T.P.; Grigoris, P.; Stutz, H.C.; Sarmiento, A. The origin of submicron polyethylene wear debris in total hip arthroplasty. Clin. Orthop. Relat. Res. 1995, *311*, 3–20.

67. Jasty, M.; James, S.; Bragdon, C.R.; Goetz, D.; Lee, K.R.; Hanson, A.E.; Harris, W.H. Patterns and mechanisms of wear in polyethylene acetabular components retrieved at revision surgery. 20th Annual Meeting of the Society for Biomaterials, Boston, MA, 1994; 103 pp.

68. Wang, A.; Sun, D.C.; Stark, C.; Dumbleton, J.H. Wear mechanisms of UHMWPE in total joint replacements. Wear 1995, *181–183*, 241–249.

69. Wang, A.; Stark, C.; Dumbleton, J.H. Role of cyclic plastic deformation in the wear of UHMWPE acetabular cups. J. Biomed. Mater. Res. 1995, *29*, 619–626.

70. Shi, W.; Dong, H.; Bell, T. Tribological behavior and microscopic wear mechanisms of UHMWPE sliding against thermal oxidation-treated Ti6A14V. Mater. Sci. Eng. 2000, *A 291*, 27–36.

71. Wang, A.; Essner, A.; Polineni, V.K.; Sun, D.C.; Stark, C.; Dumbleton, J.H. Lubrication and wear of ultra-high molecular weight polyethylene in total joint replacements. In *New Directions in tribology;* Hutchings, I.M., Ed.; 1st World Tribology Congress, Sept. 8–12, 1997.

72. Dumbleton, John H. Tribology of Natural and Artificial Joints. Tribology Series Vol. 3, Elsevier Scientific Publishing Company: Amsterdam, 1981.

73. Winter, D.A. *Biomechanics and Motor Control of Human Movement*; John Wiley & Sons Inc: New York, 1990.
74. Jost, H.P. *Lubrication (Tribology) Education and Research—A Report on the Present Position and Industry's Needs*; Her Majesty's Stationery Office: London, 1966.
75. Unworth, A. Tribology of human and artificial joints, Donald Julius Groen Prize lecture.
76. Morrison, J.B. Bioengineering analysis of force actions transmitted by the knee joints. Biomed. Eng. 1968, *3*, 164–170.
77. Ajaots, M.; Tamre, M. Design of rehabilitation devices for hip and knee joints. University of Oulu, Proceedings of OST-95 Symposium on Machine Design, Oulu, Finland May, 1995; pp. 106–111.

13

Biocompatible Metals and Alloys: Properties and Degradation Phenomena in Biological Environments

Alexia W. E. Hodgson, Sannakaisa Virtanen, and Heimo Wabusseg
Swiss Federal Institute of Technology, Zurich, Zurich, Switzerland

I. METALLURGY AND MECHANICAL ASPECTS OF IMPLANT MATERIALS

Metals are used as surgical implants in the human body for two main applications: for orthopedic purposes, they are chosen for the prosthetic replacement of natural joints such as hips, knees, elbows, shoulders, ankles, and fingers, as well as for internal fixation of bone fractures such as osteosyntheses plates, bone nails and screws, and bone rods or wires. Metallic biomaterials are the material of choice for angioplasty stents, pacemaker leads and cases, surgical clips, and staples, and are also often selected in dental applications.

Metallic implant materials have to fulfill a long list of various requirements. In terms of mechanical properties, the material must show a very good combination of elasticity, yield stress, ductility, toughness, ultimate strength, fatigue strength, creep behavior, hardness, and wear resistance. The biocompatibility of the metal should guarantee surface and structure compatibility as well as tissue protection from primary corrosion products and wear particles; its corrosion resistance has to avoid corrosive damage by choosing electrochemically stable materials.

Owing to the necessary compliance with these requirements, only a few metals and their alloys can currently be used for implant materials, and these are titanium and titanium alloys, cobalt-based alloys, stainless steel, and, more infrequently, tantalum and niobium.

A. Metallurgical Properties

1. Titanium and Titanium Alloys

Titanium is a transition metal with an incomplete shell in its electronic structure, which enables it to form solid solutions with most substitution elements having a size factor within $\pm 20\%$.

In its elemental form, titanium has a high melting point of $1678°C$, exhibiting a hexagonal close-packed (hcp) crystal structure α up to the β transus at $882.5°C$ and undergoing an allotropic transformation to a body-centered cubic (bcc) structure β above this temperature [1].

429

Figure 1 Microstructure of annealed cp titanium grade 4.

Commercially pure (cp) titanium is applied to unalloyed titanium and designates several grades containing minor amounts of impurity elements, such as carbon, iron, and oxygen. The microstructure of cp titanium is essentially all α titanium with relatively low strength and high ductility. The material may be slightly cold-worked for additional strength, but cannot be strengthened by heat treatment [2]. Therefore cp titanium does not possess sufficient strength for load-bearing applications and it is only used extensively for surface coatings and dental implants. Figure 1 shows the microstructure of cp Ti grade 4 in annealed condition whereas Table 1 illustrates the chemical compositions (in weight percent) of the four grades of cp titanium.

Titanium alloys are classified as α, near-α, α + β, metastable β, or stable β, depending upon their room temperature microstructure [1]. Alloying elements for titanium can be divided into three main categories: α-stabilizers, such as Al, O, N, and C; β-stabilizers, such as Mo, V, Nb, Ta (isomorphous), Fe, W, Cr, Si, Ni, Co, Mn, and H (eutectoid); and neutral elements such as Zr. Through process variations, the alloy microstructure can be controlled and the titanium alloys' properties can be optimized.

Although α and near-α titanium alloys show superior corrosion resistance, their utility as biomedical materials is limited by their low strength. α + β Alloys exhibit higher strength due to the presence of α and β phases with properties depending upon composition, the relative proportion of α and β phases, and the alloy's prior treatment and thermomechanical processing conditions. At this point in time, the most frequently used α + β titanium alloy is Ti–6Al–4V [4] or Ti–6Al–4V ELI [5], followed by Ti–6Al–7Nb [6] and Ti–3Al–2.5V [7]. More recent developments in the field of α + β titanium alloys are Ti–5Al–3Mo–4Zr [8] and Ti–15Zr/Sn–4Nb–2/4Ta–0.2Pd–N,O [9]. The combination of Al and V, Nb, Ta, Mo, or Fe enables both allotropic modifications to exist at room temperature. However, the alloying

Table 1 Chemical Composition (wt.%) of cp Titanium

	N	H	O	Fe	C	Ti
cp Ti grade 1	0.03	0.01	0.18	0.20	0.10	Balance
cp Ti grade 2	0.03	0.01	0.25	0.30	0.10	Balance
cp Ti grade 3	0.05	0.01	0.35	0.30	0.10	Balance
cp Ti grade 4	0.05	0.01	0.40	0.50	0.10	Balance

Source: Ref. 3.

Table 2 Chemical Composition (wt.%) of the Most Frequent Titanium $\alpha + \beta$ Alloys (Balance Ti)

	Al	V	Nb	Fe	C	H	N	O	Ta
Ti–6Al–4V	5.5–6.75	3.5–4.5	—	0.3	<0.1	<0.015	<0.05	0.2	—
Ti–6Al–4V ELI	5.5–6.5	3.5–4.5	—	<0.25	<0.08	<0.09	<0.05	<0.13	—
Ti–6Al–7Nb	5.5–6.5	<0.05	6.5–7.5	<0.25	<0.08	<0.009	<0.05	<0.2	<0.5
Ti–3Al–2.5V	3	2.5	—	<0.13	<0.05	<0.015	<0.01	<0.1	—

conditions also contribute to increased strength by solid solution hardening mechanisms. In the as-cast condition, the microstructure exhibits a lamellar duplex structure (α and β lamellas) after cooling, which has to be homogenized in an annealing process prior to any extensive deformation. In the hot-worked condition, $\alpha + \beta$ alloys have fine equiaxed α-grains, with a dispersion of β at primary α-grain boundaries. Equiaxed microstructures are characterized by small (< 10 mm), rounded grains. This class of microstructure is recommended for Ti–6Al–4V surgical implants [5].

The chemical composition (in weight percent) of some $\alpha + \beta$ alloys for surgical implant application is given in Table 2. Figure 2(a) and (b) illustrates the microstructure of Ti–6Al–4V in a smooth-turned condition and of Ti–6Al–7Nb in a hot-rolled condition consisting of a distribution of α (light) and β (dark) phases.

β Titanium alloys are defined as alloys that contain sufficient β stabilizers to retain 100% β upon quenching the above β transus. Metastable β alloys still have the potential of precipitating a second-phase α' upon aging, whereas in stable β alloys, no precipitation takes place during practical long-time thermal exposure. They show high strength, good formability, and high hardenability, and offer the unique possibility of combined low elastic modulus and superior corrosion resistance [10,11].

The biomedical β-type titanium alloys developed so far can be placed into the group of niobium-rich and molybdenum-rich metastable β alloys beside the β alloy Ti–30Ta [12] for dental applications. Ti–13Nb–13Zr [13], Ti–16Nb–10Hf [14], Ti–35Nb–5Ta–7Zr (TNZT) [15], Ti–29Nb–13Ta–4.6Zr [16], Ti–35Zr–10Nb [17], and Ti–45Nb [17] belong to the group of niobium-rich alloys originally developed as implant materials. Ti–13Nb–13Zr, for example, consists of hcp martensite α' in the water-quenched condition and with subsequent

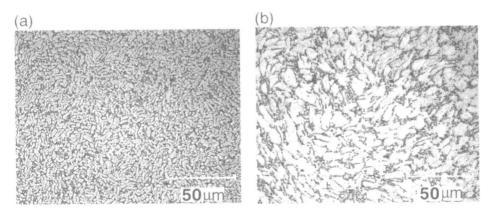

Figure 2 Microstructure of (a) Ti–6Al–4V in smooth-turned condition and (b) Ti–6Al–4V in hot-rolled condition.

Table 3 Chemical Composition (wt.%) of Metastable Nb-Rich β Titanium Alloys (Balance Ti)

	Nb	Ta	Zr	Hf	Fe	O	C	N	H
Ti–13Nb–13Zr	12.5–14	—	12.5–14	—	<0.25	<0.15	<0.08	<0.05	<0.012
Ti–16Nb–10Hf	15.5–16.5	—	—	9–10	—	—	—	—	—
Ti–35Nb–7Zr–5Ta	35.5	5.7	7.3	—	<0.05	<0.15	<0.05	<0.02	<0.015
Ti–35Zr–10Nb	10.5	—	35.2	—	<0.05	<0.1	<0.05	<0.002	<0.015
Ti–45Nb	45.5	—	—	—	<0.1	<0.1	<0.04	<0.01	<0.003

aging submicroscopical bcc β precipitates, thus strengthening and hardening the material. Ti–12Mo–6Zr–2Fe (TMZF™) [18], Ti–15Mo–5Zr–3Al [19], Ti–15Mo–3Nb–0.3O (21SRx) [20], and Ti–15Mo [21] can be classified as molybdenum-rich β alloys. Ti–15Mo, for instance, can retain a fine β-grain structure after solution annealing at 800°C with rapid quenching. The alloy exhibits excellent corrosion resistance and is being evaluated for orthopedic implant applications [21]. The chemical composition (in weight percent) of metastable β titanium alloys is given in Tables 3 and 4.

Titanium alloys are used in cardiovascular implants such as artificial heart valves, for spinal surgery applications, as well as in orthopedic implants such as total hip, finger joint, total knee, and total elbow replacements and fracture fixation appliances.

2. Cobalt-Based Alloys

Four cobalt–chromium alloys are currently used for surgical applications and of these, only the alloy F-75 [22] is fabricated by casting. The two most important additions to the cobalt base are chromium, to enhance corrosion resistance, and carbon, to improve the cast ability of the alloy. In the as-cast condition, F-75 consists of a highly cored dendritic matrix within which are dispersed chromium-rich carbides of the type $M_{23}C_6$, M_7C_3, and M_6C (with M as Cr, Mo, or Co) [23]. This coring causes the alloy composition to vary across the microstructure and this results in small differences in the properties between the cobalt-rich dendrites and the chromium-rich interdendritic regions. Modification of this microstructure by heat treatment may be conducted by homogenization treatment and by dissolution of the carbide particles in the matrix at temperatures between 1210°C and 1250°C [23].

Co-based alloys processed by thermomechanical treatment are the alloys F-75, F-90 [24], and F-562 [25], with the latter known as MP35N (multiphase alloy with 35% Ni). These materials are highly dependent on composition and thermomechanical processing history for their final microstructure. This is mostly due to the allotropic transformation from the hcp crystal structure to the face-centered cubic (fcc) structure, which occurs in cobalt on heating at about 430°C. However, because these two crystallographical forms are separated only by a very small energy difference, both may coexist at room temperature by alloying

Table 4 Chemical Composition (wt.%) of Metastable Mo-Rich β Titanium Alloys (Balance Ti)

	Mo	Zr	Nb	Al	Fe	O	C	N	H
Ti–12Mo–6Zr–2Fe	10–13	5–7	—	—	1.5–2.5	<0.28	<0.05	<0.05	<0.02
Ti–15Mo–5Zr–3Al	15	5	—	3	<0.2	<0.18	<0.02	<0.01	<0.02
Ti–15Mo	15	—	—	—	<0.1	<0.15	<0.05	<0.01	<0.015
Ti–15Mo–3Nb–0.3O	14–16	—	2.2–3.2	<0.05	<0.3	0.25–0.3	<0.05	<0.05	<0.015

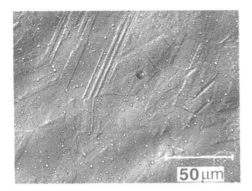

Figure 3 Microstructure of a rolled and annealed (1050°C/0.5 hr) F-75 alloy.

and mechanical working and, therefore, a substantial improvement in mechanical properties can be obtained. The hcp phase is promoted by addition of Cr, Mo, W, and Si, whereas the fcc phase is stabilized by C, Ni, Fe, and Mn [26].

Figure 3 shows the microstructure of a rolled and annealed (1050°C/0.5 hr) F-75 alloy. The material is strengthened by deformation and twinning and shows a fine dispersion of carbides. Table 5 illustrates the chemical composition (in weight percent) of surgical-grade cobalt–chromium alloys.

Surgical uses of cobalt–chromium alloys are prosthetic replacements of hips, knees, elbows, shoulders, ankles and fingers, bone plates and screws, bone rods and staples, heart valves, or wires.

3. Iron-Based Alloys

For implantation and particularly in those circumstances that include multicomponent weight-bearing or articulating devices, type 316L (L: extra low carbon) [27] stainless steel is most frequently selected because of its superior corrosion resistance in addition to good mechanical properties. For less rigorous environments and functions, such as some dental applications, austenitic stainless steels of lower corrosion resistance are also used. Stainless steels are relatively inexpensive and formable by common techniques; their mechanical properties are controllable over a wide range, providing optimum strength and ductility. However, stainless steels are not sufficiently corrosion-resistant for long-term devices such as hip implants and are better suited for temporary components in an orthopedic treatment such as bone screws, bone plates, or intramedullary rods.

Table 5 Chemical Composition (wt.%) of Surgical-Grade Co–Cr Alloys (Balance Co)

	Cr	Mo	Ni	Fe	C	Si	Mn	P	S	W	Ti
F-75 cast	27–30	5–7	<1	<0.75	<0.35	<1	<1	—	—	—	—
F-75 forged	26–28	5–7	<1	<0.75	<0.05	<1	<1	—	—	—	—
F-90	19–21	—	9–11	<3	<0.15	—	<2	<0.04	<0.03	14–16	—
F-562	19–21	9–10.5	33–37	<1	<0.025	<0.15	<0.15	<0.015	<0.01	—	<1

Figure 4 Microstructure of 316L stainless steel.

Type 316L stainless steel is fcc or austenitic down to room temperature. Due to the very low carbon content, precipitation of chromium carbides along the grain boundaries is suppressed and, therefore, the resistance against stress corrosion is improved.

The addition of bcc chromium to iron produces an alloy system, which is predominantly ferritic, restricting the γ (austenite) phase to a limited field, whereas nickel, the second main alloying element in 316L, stabilizes the austenitic phase.

During the last few years, further development was made toward nitrogen-rich austenitic stainless steels such as Rex734 (ISO 5832-9:E) [28] and nickel-free high-nitrogen austenitic steels such as PANACEA P558 [29] for medical devices. These alloys have a higher Mo and N content than 316L and are therefore more resistant against pitting and crevice corrosion. Figure 4 illustrates the microstructure of a 316L stainless steel whereas Table 6 shows the chemical composition (in weight percent) of stainless steels for medical devices.

B. Mechanical Properties

For surgical and dental implants, there are high mechanical properties required for their structural efficiency, but their volume and shape are restricted by anatomical realities. The mechanical properties of metals and their alloys are assessed by their microstructure, which are controlled by alloy design and thermomechanical processing. Static strength levels for implant materials such as yield strength, which fixes the forces above which implants lose their shape and will no longer fulfill a given function, are generally acceptable with adequate ductility. However, in the case of orthopedic implants, there is still concern over the high elastic modulus of the alloys compared to bone, which has an elastic modulus of between 17 and 28 GPa [30], which may eventually lead to prosthesis failure. This phenomenon, called "stress shielding," is related to the difference in flexibility or stiffness in part

Table 6 Chemical Composition (wt.%) of Stainless Steels for Medical Devices (Balance Fe)

	Cr	Ni	Mo	Mn	C	P	S	Si	N	Cu
316L	17–20	13–15.5	2–3	<2	<0.03	<0.025	<0.01	<0.75	—	—
Rex734	21	9	2.4	3.5	0.037	0.008	0.001	0.3	0.4	0.05
P558	17	0.08	3	10	0.2	0.012	0.001	0.4	0.45	0.04

dependent on different elastic moduli between the natural bone and the metallic implant material. It has been shown that the coupling of an implant with a previously load-bearing natural structure may result in tissue loss and bone resorption due to insufficient load transfer from the artificial implant to the adjacent remodelling bone. When the tension/compression load or bending moment to which the living bone is exposed to is reduced, decreased bone thickness, bone mass loss, and increased osteoporosis may ensue [31,32].

Orthopedic implants are subjected to cyclical loading during body motion as are dental implants when these devices are placed within the jaw bone and must withstand normal occlusal forces during chewing, grinding, and other masticatory actions, which could exceed the range of a million cycles per year.

It results in alternating plastic deformation of microscopically small zones of stress concentration produced by notches or microstructural inhomogeneities. The interdependency between factors such as implant shape, material processing parameters, and type of cyclical loading makes the determination of the fatigue resistance of a component an intricate, but very critical, task.

Testing an actual implant under simulated implantation and load conditions is a difficult and expensive process and, therefore, a variation of "standardized" fatigue tests including tension/compression, bending, torsion, and rotating bending fatigue (RBF) testing for the initial screening of orthopedic materials is selected. However, in the case of orthopedic implants, joint simulator trials are generally carried out in a later stage of the implant development process.

1. Titanium and Titanium Alloys

The mechanical properties of titanium implant materials developed so far are listed in Table 7 [17,19,33–37]. Titanium and titanium alloys are heat-treated to develop specific microstructures for particular applications and, therefore, the alloys show a very wide range of mechanical properties depending upon their heat treatment. To optimize the microstructures and properties of a component or device, heat treatments may be combined with special thermomechanical processing steps. The response of titanium alloys to heat treatment depends upon the chemical composition of the alloy, as well as on the effects of alloying elements on the α/β phase balance. Heat treatments, which are generally applicable to all titanium alloys, are stress relieving, mill annealing, and solution treatment plus artificial aging [38]. Stress relieving is a low-temperature treatment (480–700°C) followed by air or controlled cooling to relieve internal stresses that occur during forging, rolling, machining, etc. Mill annealing is conducted at about 650–820°C followed by air cooling and this treatment minimizes hardness and maximizes the ductility of the milled product. Solution treating plus aging is a high-temperature treatment with rapid cooling, followed by a low-temperature aging treatment to produce the highest tensile strength and moderate ductility combined with good fatigue strength, fracture toughness, and creep resistance [17].

The yield strengths of titanium and titanium alloys vary from 170 MPa for cp Ti grade 1 to 1284 MPa in the aged condition for Ti–15Mo–5Zr–3Al, with typical elongation values in the range between 10% and 25%. Ti–6Al–4V is still generally considered as a "standard material" when evaluating the mechanical properties of new orthopedic titanium alloys [39]. However, especially in terms of fatigue strength, the mechanical response of Ti–6Al–4V is extremely sensitive to prior thermomechanical processing history concerning prior β-grain size, the ratio of primary α to transformed β, α-grain size and morphology, and the density of the α/β interfaces. Microstructures with a small (< 20 μm) α-grain size, a well dispersed β-phase and a small α/β interface area, such as equiaxed microstructures, resist fatigue crack

Table 7 Mechanical Properties of Titanium Implant Materials

	Type	YS [MPa]	UTS [MPa]	Elongation [%]	EM [GPa]	FS [MPa]
cp Ti grade 1	α	170	240	24	102.7	—
cp Ti grade 2	α	275	345	20	102.7	—
cp Ti grade 3	α	380	450	18	103.4	—
cp Ti grade 4	α	485	550	15	104.1	430 (RBF)
Ti–6Al–4V	α/β	825–869	895–930	6–15	110–114	620–725 (UA)
						610 (RBF)
Ti–6Al–4V ELI	α/β	795–875	860–965	10–15	101–110	500–816 (UA)
Ti–6Al–7Nb	α/β	793–950	862–1050	8–15	114	580–710 (UA)
						500–600 (RBF)
Ti–3Al–2.5V	α/β	586	690	15	—	—
Ti–13Nb–13Zr	β	725–908	860–1037	8–16	79–84	500 (UA)
Ti–16Nb–10Hf	β	276–736	486–851	10–16	81	—
Ti–35Nb–7Zr–5Ta	β	530–976	590–1010	19–20	55–66	260 (UA)
						265–450 (RBF)
Ti–35Zr–10Nb	β	621	897	16	—	—
Ti–45Nb	β	448	483	12	—	—
Ti–29Nb–13Ta–5Zr	β	864	911	13	80	—
Ti–12Mo–6Zr–2Fe	β	965–1060	1000–1100	15–22	74–85	525 (RBF)
Ti–15Mo–5Zr–3Al	β	838–1284	852–1312	18–25	75–113	400 (UA)
Ti–15Mo	β	544–655	793–874	21–22	78	—
Ti–15Mo–3Nb–0.3O	β	945–987	979–1020	16–18	82–83	490 (RBF)

YS = yield strength; UTS = ultimate tensile strength; EM = elastic modulus; FS = fatigue strength; RBF = rotating bending fatigue tests at $R = -1$; UA = uniaxial fatigue tests at $R = -1$.

initiation best and have the highest fatigue strengths [40]. Lamellar microstructures, which have a greater α/β surface area and more oriented colonies, have lower fatigue strengths than equiaxed microstructures.

Titanium alloys show a high sensitivity in their fatigue properties to surface condition, which is associated with their high notch sensitivity. This makes surface finishing techniques and treatments a crucial point, for example, in the performance of porous coated implants for cementless prostheses, where the application of bead coating or wire coating produces preferential crack initiation sites at the porous coating/substrate interface [39].

2. Cobalt-Based Alloys

The mechanical properties of cobalt-based alloys for surgical applications are listed in Table 8 [26,41–45]. High ultimate and high fatigue strengths make cobalt–chromium alloys a prime alloy system used in joint reconstruction. However, the elastic modulus varies from 185 to 250 GPa, depending on the composition of the alloy, which is around twice as high as the elastic modulus of titanium α + β alloys. The F-75 alloys used in surgical devices are capable of exhibiting a wide range of mechanical properties and, generally, adequate properties are obtained in the as-cast condition to fulfil the requirements in most applications. However, the coarse grain size and interdendritic carbides usually of the M_7C_3-type and σ-phases present in cast F-75 limit the strength and ductility of the as-cast alloy [46]. Therefore, when a higher ductility is required, a solution treatment leads to increased elongation while improving tensile properties. Hot isostatic pressing of the F-75 alloy can

Table 8 Mechanical Properties of Surgical-Grade Cobalt-Based Alloys

	YS [MPa]	UTS [MPa]	Elongation [%]	FS [MPa]
F-75				
As-cast	430–490	716–890	5–8	300 (RBF)
Solution-treated	533	1143	15	280 (RBF)
Hot isostatically pressed	840	1275	16	765 (RBF)
ASTM minimum	450	655	8	—
Forged	962	1507	28	897 (RBF)
F-90				
Annealed	350–650	862–1220	37–60	345 (RBF)
Cold-worked	1180–1610	1350–1900	10–22	490–587 (RBF)
ASTM minimum	310	860	10	—
F-562				
Annealed	300	800	40	340 (RBF)
Cold-worked	650–1413	1000–1827	10–20	435–555 (RBF)
Cold-worked and aged	1586–1999	1793–2068	8–10	405–850 (RBF)

YS = yield strength; UTS = ultimate tensile strength; FS = fatigue strength; RBF = rotating bending fatigue tests at $R = -1$.

double the fatigue strength in comparison with the as-cast alloy. This processing technique points out the importance of fine grain size and carbide distribution to achieve optimum mechanical properties. In the same way, forged F-75 exhibits extremely high strength values because of a combination of fine grain size and the cold-worked-induced allotropic transformation. In this case, high ductility can be obtained even with a large degree of cold work because the carbon content is reduced to very low values [26].

The forging or worked alloys generally have better tensile properties arising, firstly, because some of the cobalt is replaced by nickel, imparting better ductility, and, secondly, because of the work hardening that is induced in the structure. In the annealed conditions, F-90 consists of a single-phase fcc microstructure and only relatively poor mechanical properties can be expected. With cold work, however, the allotropic transformation can be induced, and similar or superior properties to those of forged F-75 can be obtained. In the same way as for F-90, the alloy F-562 requires cold working to achieve its excellent mechanical properties. Due to the high nickel content, F-562 exhibits the poorest mechanical properties of all the cobalt-based alloys in the annealed condition. Nickel is very similar to cobalt and is therefore unable to provide solid solution hardening of the matrix. Cold working improves the properties to levels similar to those attained in other both forged or wrought Co–Cr alloys. In the case of F-562, on the other hand, artificial aging at temperatures around 550°C produces high strengths (highest value among surgical alloys) with concomitant elongation values of 8%, thanks to the precipitation of an intermetallic compound.

3. Iron-Based Alloys

The mechanical properties of iron-based alloys for medical devices are listed in Table 9 [29,47–49]. The mechanical properties of 316L stainless steel are primarily dependent on the degree of cold working. Because the yield and ultimate strengths of fully annealed stainless steel are relatively low, cold working strengthens these materials. Strengthening due to working will occur for deformations performed below the recrystallization and recovery temperature (for 316L below about 500°C) [50]. However, because 316L hardens fairly

Table 9 Mechanical Properties of Iron-Based Surgical Implants

	YS [MPa]	UTS [MPa]	Elongation [%]	FS [MPa]
316L				
Annealed	170–210	480–645	40–68	190–230 (RBF), 240 (UA)
Cold-worked	310–1160	655–1256	6–28	530–700 (RBF)
Rex734				
Annealed	440–584	810–898	39–48	420 (UA)
Cold-worked	912–1325	—	—	—
P558				
Annealed	610–720	980–1120	54–65	480 (UA)
Cold-worked	965–1419	1000–1630	12–35	540–600 (UA)

YS = yield strength; UTS = ultimate tensile strength; FS = fatigue strength; RBF = rotating bending fatigue tests at $R = -1$; UA = uniaxial fatigue tests at $R = -1$.

rapidly, numerous intermediate annealing treatments are necessary to avoid cracking on extensive cold working. The fatigue strength of 316L stainless steel follows a Hall–Petch relationship. Therefore, coarse grains weaken the structure and have been implicated in the failure of hip prostheses [51]. Because of their high nitrogen content in solid solution, Rex734 and especially P558 exhibit yield strengths and fatigue strengths in solution-annealed condition, which are two to three times higher than the yield strength and fatigue strength of 316L.

II. AN INTRODUCTION TO CORROSION PROCESSES OF PASSIVE METALS AND ALLOYS

Corrosion can be defined as the destruction or deterioration of a material due to its reaction with the environment. Corrosion—or material deterioration—concerns all material groups. This chapter, however, only discusses corrosion reactions of metallic materials, and more specifically electrochemical corrosion reactions, which can occur, for example, in aqueous solutions, in the atmosphere, and on the ground. The corrosion behavior is influenced by a wide variety of factors, including the materials, the environment, as well as the construction. In the following, the fundamentals of electrochemical corrosion processes—thermodynamic and kinetic principles including passivity—are shortly introduced. Thereafter, the special modes of corrosion encountered by passive metals and alloys are discussed, followed by a short description of the corrosion behavior of the most widely used biocompatible alloys. For further reading, a number of textbooks and handbooks on corrosion science and engineering are available (e.g., see Refs. 52–58).

A. Electrochemical Fundamentals of Corrosion Reactions

1. Thermodynamics

An electrochemical corrosion reaction involves electron transfer [59–63]. The actual corrosion reaction leading to material loss is anodic or oxidation reaction, whereby a metal dissolves while releasing electrons and ions (Eq. (1)):

$$Me \rightarrow Me^{n+} + ne^-$$

(1)

As mentioned later in the text, the fate of the oxidized metal ion depends on the metal/environment system. Possibilities include formation of a soluble ion (active dissolution) or an insoluble protective corrosion product.

Simultaneously, the released electrons are consumed in a cathodic or reduction reaction, which involves the reduction of species in the surrounding environment.

Depending on the electrolyte pH, one of the following reduction (cathodic) reactions will dominate:

(a) in acidic solutions:

$$2H^+(aq) + 2e^- \rightarrow H_2(g) \tag{2}$$

(b) in neutral and alkaline solutions:

$$2H_2O(l) + O_2(g) + 4e^- \rightarrow 4OH^-(aq) \tag{3}$$

Thermodynamic considerations allow the determination of whether a corrosion reaction of a metal in a certain environment can spontaneously take place or not. The driving force for the corrosion reaction is the difference of the electrode potentials of the anodic reaction (metal oxidation) and the cathodic reaction (reduction of redox partners in the electrolyte).

For the coupled red/ox cell, the emf (E) results as:

$$E = E_c - E_a \tag{4}$$

where E_c is the potential of the cathodic reaction, and E_a is the potential of the anodic reaction. The effective electrode potential can be determined from the standard electrode potentials (E^0) by using the Nernst equation; for instance, for the anodic iron oxidation reaction, this would be:

$$E_a = E^0_{Fe^{2+}/Fe} + \frac{RT}{nF}\ln[Fe^{2+}] \tag{5}$$

The Gibbs free energy (ΔG) of the reaction is:

$$\Delta G = -nFE \tag{6}$$

Thus, it can basically be predicted under which conditions the metal dissolution reaction (Me $\rightarrow$ Me^{n+}) will proceed thermodynamically. According to Eqs. (4) and (6), a metal or an alloy can be immune to corrosion due to thermodynamic stability, if its oxidation potential is more anodic than the reduction potential of species occurring in the surrounding environment—this is encountered with noble metals in most media. However, for most technologically important materials, this is not the case; hence, a reaction between the metallic material and its environment takes place.

From a practical point of view, the corrosion rate depends on the fate of the oxidized species (Me^{2+}): they can either be dissolved in the solution, or react with oxygen species in the solution to form a surface oxide layer. Such oxide layers can represent effective kinetic barriers against corrosion (see Sec. II.A.2).

Thermodynamic data can be presented in pH potential diagrams (so-called Pourbaix diagrams) [64]. As an example, Fig. 5 shows the Pourbaix diagram for iron in an aqueous environment. The diagram gives regions of stability for different species (dissolved: Fe^{2+}, Fe^{3+}, $HFeO_2^-$; or solid: Fe_3O_4, Fe_2O_3) as a function of pH and red/ox potential. The dotted lines represent the equilibria for the above cathodic reactions, (a) or (b), respectively. The solid lines indicate the stability ranges for Fe and its corrosion products. This diagram hence enables the prediction of whether iron will be inert (no reaction, region A: stability region of

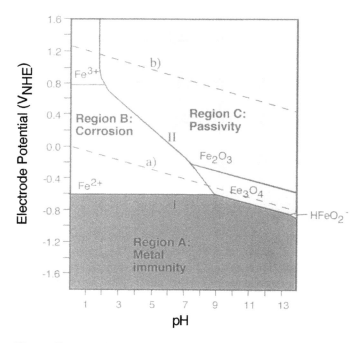

Figure 5   Potential pH (Pourbaix) diagram for Fe/H₂O at 25°C. (From Ref. 64.)

metallic iron), actively dissolve (region B: stability region of Fe^{2+} and Fe^{3+}), or form stable oxides (region C).

pH potential diagrams are available for many elements in aqueous environments and are often a valuable tool in the preliminary assessment of the (thermodynamic) stability of a system. However, it should be pointed out that these calculations are based purely on thermodynamic considerations; hence, this approach gives no information on the rate (kinetics) of the possible corrosion reactions. Even if a reaction is thermodynamically possible, its rate can be negligible to be of practical relevance. Moreover, in case of more complex practical systems (e.g., biological environment), local acidity changes, temperature variations, and the eventual presence of complexing agents can lead to corrosion behavior, which cannot be predicted by the simple pH potential calculations.

2. Kinetics

For most practically relevant material/environment combinations, thermodynamic stability is not provided because $E_a > E_c$ [60–62]. Hence, information on corrosion rates is needed. As for other electrochemical reactions, a variety of factors can influence the rate-determining step. In the most straightforward case, the reaction is activation energy-controlled (charge transfer-controlled; i.e., the ion transfer through the surface Helmholtz double layer involving migration and adjustment of the hydration sphere to electron uptake or donation is rate-determining). Active metal dissolution and hydrogen evolution reaction at low pH values are typical electrochemical reactions that are under activation control.

Alternatively, the mass transport properties in the solution can become rate-determining—the reaction is then diffusion-controlled. In this case, the charge transfer process is fast compared with the diffusion of the reacting anion to the surface, or dissolving cations from the surface. In aqueous solutions, diffusion control of uniform corrosion is frequently encountered when the cathodic reaction depends on the supply of $O_2(g)$, which is only sparingly soluble in water and therefore is present only in limited concentrations.

A specific case of kinetic control is passivity, which will be separately discussed in Sec. II.A.3.

In the corrosion process, the anodic and cathodic reactions occur simultaneously and with the same rate (due to the law of conservation of charge). The coupled situation of both redox equilibria is described by the so-called "mixed potential theory." The mixed oxidation (e.g., $Fe \rightarrow Fe^{2+}$) and reduction (e.g., $H^+ \rightarrow H_2$) systems will equilibrate to zero net current and the resulting potential that lies between $E_{Fe^{2+}/Fe}$ and $E_{H^+/H}$ is called the corrosion potential (E_{corr}).

The rate of corrosion is given by the current of metal ions leaving the metal surface in the anodic region. Thus, the corrosion current density i_{corr} is the anodic current of the coupled system. The corrosion current can be converted into material loss (M_{corr}) using Faraday's law according to Eq. (7):

$$M_{corr} = (M/nF)i_{corr} \tag{7}$$

where M is the molar mass of the metal, n is the number of electrons involved (= the charge number of the ion), F is the Faraday constant, and t is the time. This conversion assumes a 100% current efficiency regarding metal dissolution (i.e., no other competitive electrochemical reactions occur).

3. Passivity

The excellent corrosion behavior of today's high-resistant alloys is based on the spontaneous formation of a highly protective oxide film, the so-called passive film, on the metal surface, as it reacts with the environment (e.g., Refs. 65–71). These films, which in many cases show a thickness of only a few nanometers, act as a highly protective barrier between the metal surface and the aggressive environment and hence kinetically retard the reaction rate.

a. Passive Film Formation. Passivation of a surface represents a significant deviation from an ideal electrode behavior. As described above, for a metal immersed in an electrolyte, thermodynamic formation of a second-phase film—often a sparingly soluble oxide film—can become favored compared with the dissolution of the oxidized metal cation. Depending on the type of the corrosion product formed, such a surface layer can efficiently retard further dissolution—when dissolution is slowed down by many orders of magnitude, such surface layers are called passive films. The spontaneous formation of stable and highly protective passive films is the reason for the high corrosion resistance of many important construction materials such as aluminum or stainless steels.

The electrochemical behavior of a metal in a certain electrolyte can be characterized by the current/potential relationship. Figure 6 shows a schematic polarization curve for a metal showing passivation. At low anodic overpotentials, the current steeply increases as a function of the potential—in this region, the metal is actively dissolving (the current limiting factor is charge transfer). At a certain distinct potential—the so-called passivation potential E_p—the current density reaches a maximum (so-called critical current density i_{cr}). The lower the critical current density is, the better is the passivation ability of the material. The subsequent dramatical decrease in current, which can be on the order of several orders of magnitude, indicates passivation of the surface. In the passive region, the metal is dissolved with a very slow rate, corresponding to the passive current density (i_p). The passive current density hence is a measure of the protective quality of the passive film.

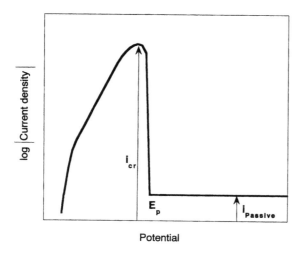

y-axis: log |Current density|

labels: i_{cr}, E_p, $i_{Passive}$

x-axis: Potential

Figure 6 Schematic anodic polarization curve for a metal showing an active/passive transition.

Often, the reaction scheme for passivation can be divided into active range, transition range, prepassive range, and passive layer formation. Models for the active/passive transition of Fe and Fe-based alloys describe that in the transition and prepassive range, the metal becomes increasingly covered by $M(OH)_x$ adsorbates [72–74]. These adsorbates increasingly block the active dissolution (apparent in polarization curves as a deviation from the active dissolution behavior). The passivation potential is reached when the surface is completely covered by adsorbates and deprotonating leads to the formation of a primary passivating film that mainly consists of MO_x (the valency of the metal cations depends on the metal and the passivating conditions).

    b. Properties of Passive Films. The protective quality of the passive film is determined by the ion transfer through the film as well as the stability of the film against dissolution. A variety of factors can influence ion transport through the film, such as the film's chemical composition, structure, stoichiometry and defectiveness, thickness, and compactness. The dissolution of passive oxide films can occur either chemically or electrochemically. The latter case takes place if an oxidized or reduced component of the passive film is more soluble in the electrolyte than the original component. An example of this is the oxidative dissolution of Cr_2O_3 films as CrO_4^{2-} [64,75,76], or the reductive dissolution of Fe_2O_3 films as Fe^{2+} (see Ref. 77 and references therein). Recently, a large number of investigations on the stability of passive films against chemical and electrochemical uniform dissolution have been carried out [76–78] (for local breakdown of passivity, see Sec. II.B).

    Exact information on the chemistry and structure of the passive film is necessary to clarify the mechanisms relevant to stability and protectiveness of passive films. Therefore, the nature of passive oxide films on many technologically important metals and alloys has been the subject of investigation for many years (for recent reviews on passivity, see Refs. 70 and 71). Ex situ surface analytical techniques such as x-ray photoelectron spectroscopy (XPS), Auger electron spectroscopy (AES), and secondary ion mass spectrometry (SIMS) have been widely used to investigate the chemical composition and thickness of films. Good agreement exists regarding a qualitative description of the chemistry of passive films on many metals; however, slightly different views can be found on the more detailed nature of the different films. This can be at least partially due to different experimental approaches or data analyses. Typical for many alloys is that the composition of the passive film can significantly differ from the composition of the underlying alloy. In the case of Fe–Cr alloys, for instance, a strong Cr enrichment in the passive film takes place. Furthermore, the passive film composition can be inhomogeneous in depth.

The structure of passive films has been the focus of interest of many research groups. For thick anodic oxide films or thick oxide films grown at elevated temperatures, the structure can be assessed by x-ray diffraction techniques. However, for thin passive films formed at low to moderate temperatures, the thickness of the films is usually less than 10 nm; hence, it is experimentally difficult to investigate the structure by traditional techniques. Additionally, the structure of a thin, mostly hydrated passive film formed under electrochemical conditions may change as it is removed from the condition under which it was formed. Lately, therefore, new in situ techniques [scanning tunneling microscopy (STM), x-ray scattering using synchrotron radiation (EXAFS)] for the study of the structure of thin oxide films have attracted considerable interest [79–81].

Electron transfer through the passive film can also be crucial for the corrosion behavior of a metal and, hence, interest has grown in studies of the electronic properties of passive films. Many passive films are of a semiconductive nature [70,81].

The thickness of native passive films on many highly corrosion-resistant metals and alloys is typically in the range of a few nanometers. Substantially thicker anodic films can be formed on so-called valve metals (e.g., Ti, Ta, Zr, etc.), which allow the application of high anodizing potentials (high electric fields) without dielectric breakdown.

Generally, it is important to note that passive films should not be considered as a rigid layer, but instead as a system in dynamic equilibrium between film dissolution and growth. In other words, the passive film can adjust its composition and thickness to changing environmental factors. Principally, the nature of electrochemically formed passive films depend (apart from the base metal) on the passivation potential, time, and electrolyte composition (i.e., on all passivation parameters), and hence a detailed treatment is beyond the scope of this chapter.

B. Corrosion Mechanisms of Passive Metals and Alloys

A metal or an alloy in a passive state dissolves with a very low corrosion rate (typical passive current densities are $< 10^{-6}$ A cm^{-2}). However, even in the case of stable passivity, some release of metal ions due to passive dissolution takes place. As mentioned previously, passive films can dissolve either chemically or electrochemically. If the dissolution is uniform electrochemical dissolution, the anodic and cathodic partial reactions occur statistically distributed over the complete surface.

Under certain conditions, localized breakdown of passivity can occur, leading to fast dissolution at the site of breakdown. The origin of the localization of corrosion phenomena can be the presence of inhomogeneities or heterogeneities either in the material or in the surrounding environment. Even though most of the surface is still covered by the intact passive film, and hence the total mass loss can be small, the locally activated sites can show very high corrosion rates and hence lead to deterioration of the whole system. Localized corrosion can lead to unexpected damage with disastrous consequences, especially because inspection of corrosion damage is, in many cases, difficult. Therefore, localized corrosion processes are more dangerous in nature and far less easily predictable than uniform corrosion.

Essential to localized corrosion phenomena is the separation of the anodic (active) and the cathodic (passive) sites of the surface. In all cases of localized corrosion, the ratio of cathodic to anodic area plays a major role in the localized dissolution rate. A large cathodic area provides high cathodic currents, and because of electroneutrality requirements, the small anodic area must provide an equally high anodic current. Hence, the local current density and therefore the local corrosion rate become higher with a larger cathode/anode ratio.

In the following, localized corrosion phenomena most relevant to the application of biomedical metallic materials are shortly introduced. Other modes of corrosion not discussed in this chapter include intergranular corrosion (corrosion damage due to enhanced dissolution in, or adjacent to, the grain boundaries of a metal), dealloying (selective corrosion), and different types of flow-induced (or flow-accelerated) corrosion phenomena, as well as stress-induced corrosion modes (stress–corrosion cracking, corrosion fatigue).

1. Pitting Corrosion

The passive state of a metal can, under certain circumstances, be prone to localized breakdown; in the case that local dissolution leads to the formation of cavities surrounded by an intact, passivated surface, this type of attack is called pitting corrosion [82–86]. Pitting occurs with many metals in halide-containing solutions (for most metals, chloride ion is the most aggressive halide species). Typical examples of metallic materials prone to pitting corrosion are Fe, stainless steels, and Al.

The actively corroding pits represent local anodic sites where the metal oxidation reaction (e.g., $Fe \rightarrow Fe^{2+} + 2e^-$) dominates and the surrounding passive surface represents a cathode where the reduction reaction takes place (e.g., O_2 reduction).* Due to the large cathodic area, as compared with the small anodic area, metal dissolution in the pit typically proceeds with a high dissolution rate. In addition, the pitting process is autocatalytic: After nucleation and initiation of pitting corrosion, the reactions taking place inside the pit establish conditions that further stimulate dissolution. Inside the pit, the metal dissolves. The M^{n+} species hydrolyze according to:

$$M^{n+} + mH_2O \rightarrow M(OH)_m^{(n-m)} + mH^+ \tag{8}$$

Because the exchange of solution is restricted in the occluded cell in the case of a small growing pit, reaction (8) leads to acidification of the pit electrolyte. Furthermore, to maintain charge neutrality, additional halide ions have to migrate inside the pit, thus increasing the local chloride concentration. Both of these factors can accelerate active metal dissolution and hinder repassivation.

Generally, the following two stages of pitting are distinguished: pit initiation (nucleation) and pit propagation. Different models have been suggested for the initiation of pits at distinct surface locations. Pit initiation has been ascribed to bulk metal inhomogeneities (inclusions, precipitates, grain boundaries, dislocations, etc.), or to properties of the passive film (mechanical film rupture, electrostriction, local composition, or structure variations generally often described as "weak sites" of the film). Initiation mechanisms that assign the key role to the passive film involve Cl^- penetration, local film thinning, and vacancy condensation, whereas mechanisms focussing on the bulk metal ascribe the key role to preferential dissolution at inhomogeneities.

---

*The passive surface can act as a cathode in cases where the passive film shows sufficient electron conductivity. In the case of insulating passive oxide films (e.g., the passive film on pure Al), both the anodic and cathodic reactions occur inside the active pit.

In an electrochemical polarization experiment of a passive system, the onset of localized dissolution can be detected by a steep current increase at a very distinct anodic potential (the pitting potential E_{pit}) (see Fig. 7). In the potential range anodic to E_{pit}, stable pit growth occurs. The value of E_{pit} is typically shifted to lower anodic potentials with increasing temperature, increasing Cl^- concentration, and decreasing pH, and is dependent on the presence of other anions in the electrolyte. In the potential range cathodic to E_{pit}, one frequently observes current transients, which indicate that the so-called metastable pitting takes place. In other words, local passivity breakdown events take place, but instead of pit propagation, the small pits do not remain active but repassivate relatively fast. Even though these type of events do not lead to a complete deterioration of the system, they nevertheless indicate that the metal is not completely stable in its environment.

The alloys typically used in biomedical applications are, in principle, all susceptible to pitting corrosion in chloride-containing environments, and hence the occurrence of pitting corrosion in the biological environment cannot be excluded. However, the relative susceptibility of the different materials employed to pitting corrosion strongly varies (see Sec. II.C).

2. Crevice Corrosion

Another type of localized corrosion closely related to pitting corrosion is crevice corrosion. This type of attack preferentially occurs in regions on the metal surface where mass transfer is limited (e.g., in narrow crevices or under deposits), and hence the concentration of aggressive species (halides), the lowering of the pH as discussed above, and depletion of oxygen can rapidly lead to activation of the surface in the crevice area. Metals, which are susceptible to pitting corrosion, also suffer from crevice corrosion. The presence of crevices on the surface often triggers localized corrosion already under conditions where stable pitting would not take place (e.g., with lower concentration of aggressive halides).

In biomedical devices, crevices are often present (e.g., between screws and plates, or between the head and stem of a hip implant). For such systems, clearly the eventual danger of crevice corrosion must be considered in the design and choice of materials.

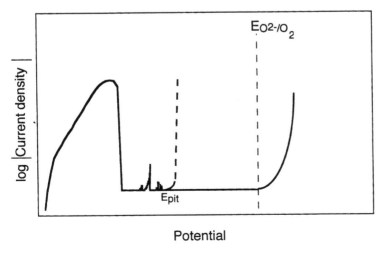

Figure 7 Schematic anodic polarization curve for a metal showing pitting corrosion (dashed line). The solid line shows the anodic polarization curve in the case of stable passivity until the region of oxygen evolution.

3. Galvanic Corrosion

This type of corrosive attack occurs when dissimilar metals (i.e., with different $E_{red/ox}$) are in direct electrical contact in corrosive solutions or atmospheres. Under such conditions, enhanced corrosion of the less noble part of the bimetallic couple takes place, whereas the corrosion rate of the more noble part of the couple is reduced or even completely suppressed (as in the case of corrosion suppression by cathodic protection). The difference in corrosion potential of the components of the couple provides the driving force for the corrosion reaction. With passive metals and alloys, the corrosion potential in an electrolyte is typically much more noble than the thermodynamic equilibrium value for a bare metal surface.

To determine the kinetics of galvanic corrosion, knowledge of the nature and kinetics of the cathodic reaction at the surface of the more noble metal as well as the nature and kinetics of the anodic reaction on the surface of the less noble metal are required. For passivated surfaces, therefore, both the ion transfer and electron transfer properties of the passive films become of relevance in determining the danger of galvanic corrosion. The nature and conductivity of the electrolyte solution determine the current and potential distribution: the larger the conductivity is, the further the coupling action is experienced from the contact site. An important factor in determining the danger of galvanic corrosion is the ratio of the cathode to anode areas: the higher the ratio is, the larger is the enhancement of the dissolution of the less noble metal due to coupling.

Galvanic corrosion can cause problems in biomedical applications if dissimilar materials are combined (e.g., in modular implant design).

4. Tribocorrosion and Fretting Corrosion

Tribocorrosion is defined as a conjoint action of mechanical wear and corrosive attack on a material surface [87]. A special mode of tribocorrosion, which is highly relevant to the field of biomedical implants, is fretting corrosion. Fretting corrosion is a form of damage that occurs at the interface of two closely fitting surfaces when they are subject to slight oscillatory slip and joint corrosion action. Almost all materials are subject to fretting, and hence its incidence in vibrating machinery is high. The damage is mostly of a localized form and any debris generated (mostly oxide) has some difficulty in escaping from the rubbing zone, and this can lead to an increase in stress. Apart from environmental factors, fretting corrosion is influenced by load, amplitude of slip, number of fretting cycles, frequency of oscillation, and lubrication. The hardness of the materials also plays a role, as in all tribocorrosion phenomena. An increase in hardness generally leads to a reduction of fretting wear. However, the hardness also controls the form and type of debris formed.

In most biomedical applications, the materials encounter mechanical stress. Fretting corrosion can occur, for instance, in hip implants due to constant minute movement between the shaft and the bone, or at the contact between the screw head and the bone plate.

C. An Introduction to the Corrosion Behavior of Some Biocompatible Alloys

1. Titanium and Titanium Alloys

Ti and its alloys typically used in biomedical applications (Ti–6Al–4V, Ti–6Al–7Nb) can be considered as the most corrosion-resistant of the alloy systems described here [88]. This is based on the very high stability of the TiO_2 passive film spontaneously formed on the alloy surface. Thermodynamically, TiO_2 is very stable in the pH range between 2 and 12, and only complexing species such as hydrofluoric acid (HF) or H_2O_2 lead to substantial dissolution

[64]. Localized breakdown of passivity—mainly crevice corrosion of Ti—only takes place at elevated temperatures (above 70°C in seawater and above 315°C in pure H_2O [89,90]. A review of studies on the active/passive transition of Ti can be found in Ref. 91. It is interesting to note that even though the passivation potential of Ti is more negative than the equilibrium potential of hydrogen evolution, and the critical current density is relatively low, Ti is not spontaneously passivated in deaerated strong acidic solutions. This is due to the relatively slow kinetics of the cathodic hydrogen evolution reaction on the Ti surface [92,93]. This also explains the observation of depassivation leading, to uniform corrosion of Ti in deoxygenated acidic media. In oxidizing acids, however, Ti is spontaneously passive. In neutral halide solutions, Ti–6Al–4V and Ti–6Al–7Nb alloys typically show a very similar corrosion behavior as pure Ti [94]. In polarization experiments (see Fig. 8), neither pure Ti nor Ti alloys are susceptible to passivity breakdown in Cl⁻-containing solutions in the potential region below the oxygen evolution region (corresponding to the conditions prevailing in biomedical applications).

Figure 8 shows a comparison of the pitting potentials measured for some stainless steels, CrCrMo alloys, and titanium alloys [94,95]. The susceptibility to pitting and crevice corrosion typically decreases in the order of stainless steels (e.g., type AISI 316L) > CoCr-Mo > titanium alloys [95,96]. However, a significantly higher wear corrosion rate has been reported for the Ti–6Al–7Nb alloy in comparison to Co–28Cr–6Mo, whereas Fe–22Cr–10NiN stainless steel showed the lowest wear corrosion rate [97]. Fretting corrosion, or fretting corrosion in combination with crevices, has been identified as one major factor in implant corrosion [98–100].

2. CoCrMo Alloys

Similar to stainless steels, alloying Co with Cr greatly enhances corrosion resistance [101]. The addition of Cr reduces the critical current density for passivity, thus widening the conditions under which the alloys will show spontaneous passivation. Therefore, Cr is the key alloying element in Co alloys with regard to corrosion resistance in oxidizing media.

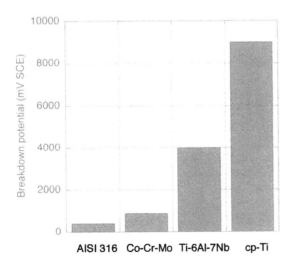

Figure 8 Breakdown potentials typically reported for AISI 316 stainless steel, Co–Cr–Mo alloy, Ti–6Al–7Nb alloy, and cp–Ti. Measurements were carried out in 0.17 M NaCl solution at 37°C.

Molybdenum has been found to be beneficial especially under active corrosion conditions (e.g., in hydrochloric acid). CoCrMo alloys can be susceptible to pitting and crevice corrosion in Cl^--containing environments. The pitting corrosion resistance also depends on the Cr and Mo content of the alloys [102]. Mechanisms of localized corrosion of CoCrMo alloys, however, have not been studied as systematically as is the case of stainless steels (see Sec. II.C.3).

3. Stainless Steel

Today, very many different grades of stainless steels exist [103]. Chromium is present in all stainless steels, whereas the content of other alloying elements such as nickel and molybdenum varies. All the alloying elements can affect the metallurgy of steels and, hence, the mechanical and chemical properties. Furthermore, the corrosion behavior is also influenced by the purity of the steels (e.g., content of C and S). The good corrosion resistance of stainless steels (Fe–Cr, Fe–Cr–Ni alloys) is based on the formation of a Cr-rich oxide layer on the alloy surface. Therefore, Cr is the alloying element mainly responsible for the high passivation ability of these alloys. The critical current density decreases with an increasing Cr content in the alloy. The addition of Mo further enhances the passivation ability. Austenitic Fe–Cr–Ni alloys show lower critical current densities than ferritic Fe–Cr alloys with a similar Cr content. Stainless steels, unfortunately, show a relatively high susceptibility to pitting and crevice corrosion in halide solutions. An increase of the Cr and Mo content greatly enhances the resistance against passivity breakdown. Another factor with a very significant influence on the pitting corrosion of stainless steels is the purity of steels, and especially the content of MnS inclusions in the steel. It has been demonstrated that pit initiation preferentially takes place at the inclusion/metal matrix interface.

III. DEGRADATION PHENOMENA IN A BIOLOGICAL ENVIRONMENT

In Sec. II, the mechanisms of metal degradation were introduced. In a biological environment, the foresight of the behavior of a metal implant is not straightforward because many known and unknown parameters need to be taken into consideration. Among these are the choice of materials and combination, the mechanical load under which the materials are placed, and the dynamic biological environment with all its associated chemical, biochemical, and biophysical equilibria. The simple fact that the material is placed in vivo renders the direct monitoring and prediction of reactions rather complex. Postoperative investigations and in vitro experiments enable an indirect understanding of possible processes occurring at the metal–biological interface and of the important parameters that can influence the acceptance and/or rejection of an implant. A great number of in vitro studies have been carried out in the biomedical field in the last 20 years, mostly focussing on specific aspects of the system under investigation, such as medical, biochemical, biological, or material aspects. From the metal implant viewpoint, surface science, corrosion, tribology, and mechanical investigations have been carried out in which biologically present parameters such as aggressive ions, proteins, and cells were taken into consideration. Although all of these parameters play a role in the degradation of the material, and hence on the lifespan of the implant in the body, no one aspect is solely responsible for the acceptance, rejection, or lifetime of an implant, and the interplay between each environment certainly needs consideration.

In Sec. III.A, an overview of the research efforts aimed at understanding the behavior of metal implants in biological systems will be given. Although examples will be taken from the main implant material families employed, namely Ti and its alloys, CoCr-based alloys, and iron-based alloys, particular focus will be placed on the first of these families, thus reflecting the current line of research interest.

A. Characterization of the Metal–Biological Interface

The implantation of each spontaneously passive metallic device is associated with the release of metal: on a small scale, due to the uniform passive dissolution resulting from the slow diffusion of metal ions through the passive film, and on a larger scale, due to the breakdown of passivity as a consequence of chemical (pitting and crevice corrosion) or mechanical (fretting corrosion) events [104–107].

Mechanical breakdown of passivity takes place, for example, when rubbing occurs between the passive film and a counterbody. Under these circumstances, local abrasion of the passive film leads to wear-accelerated corrosion due to the rapid dissolution of the depassivated metal surface, followed by repassivation. Wear-accelerated corrosion phenomena have been investigated and recently reviewed [108], in particular by combining electrochemical and wear test techniques. In addition, models relating wear-accelerated corrosion to passivation kinetics, or to mechanical, material, and operating conditions have been developed [109,110]. In the case of implanted prostheses, small amplitude sliding (or fretting) is known to occur between the metal implant and a counterbody such as bone or bone cement, leading to the release in the body of solid particles as well as dissolved metal ions [111]. The environment in which the implant finds itself is aggressive, comprising, among others, a saline electrolyte solution and proteins. These can play an important role in both corrosion and wear processes as the ions and polar organic molecules interact with the charged double layer established at the interface of the metal and the aqueous environment and act as lubricant toward the wear behavior. In the presence of wear and corrosion phenomena occurring simultaneously, a complex series of reactions at the metal/aqueous environment interface is to be expected [104–106,112,113].

It is therefore clear that the characterization of the properties of metals on exposure to a biological system is of primary importance in order to enable the prediction of reactions that ensue upon implantation. The monitoring of changes in the metal oxide upon exposure to simulated biological solutions and after wear and repassivation processes is an example because the oxide may be different in its composition and therefore influence whole sequences of reactions.

1. Surface and Electrochemical Characterization of Passive Films in Simulated Biological Environments

One of the most important features of an implant is that it will be in contact with the living tissues of the body, thus creating an interface between them. What happens at this interface is a matter of great interest, as it will largely determine the success or failure of the implant both in terms of the immediate reaction and the longer-term response. A wide variety of metals, polymers, and ceramics have found widespread application in medicine, but to be a successful implant material, its effect on the biological environment both at the local and systemic levels is of utmost importance [114]. Stainless steel and CoCr alloys have been used as implant materials over many decades, but the former is now gradually being replaced with either CoCr implants [115] or Ti and its alloys [116–118].

Ti was first introduced into the medical field in the early 1940s with the publication of an article by Bothe et al. [119] on the reaction of bone to multiple metallic implants and has since been employed as biomaterial gaining widespread interest in the last two decades, after a slow beginning due to the popularity of stainless steel and CoCr alloys. The superior corrosion resistance of titanium and its alloys [120,121] in comparison to stainless steel [122–125] and CoCr alloys [115] has been widely reported, and commercially pure titanium has shown impressive clinical results as dental, maxillofacial, and auditory implants. The oxide films are composed of TiO_2 or lower oxidation state oxides and have, in fact, been shown to be highly stable, fairly unreactive, and resistant to the most common corrosion mechanisms under body conditions [112,120,121,126–129]. Composition and structure are key factors in determining stability. Electrochemically formed oxide films on cp Ti can be of amorphous or crystalline nature, depending on the anodic potential applied and on the electrolyte into which it is immersed. Three stable forms of TiO_2 are known, namely, rutile, anatase, and brookite [130–135].

Alloys of Ti are widely used in the aircraft industry and have military applications because of their light weight, strength, and ability to withstand high temperatures. Ti_6Al_4V has been used for many years as implant material, mainly in the fabrication of orthopedic prostheses, thanks to the improved mechanical strength compared to the unalloyed form. In fact, Ti_6Al_4V and the Nb alternative Ti_6Al_7Nb are two-phase alloys, with a close-packed hexagonal α-phase and a bcc β-phase (see Sec. I). The former is stabilized by the presence of Al, which imparts higher mechanical strength, whereas V and Nb, respectively, are β-amorphous with bcc titanium. Ti–6Al–7Nb was specifically developed using the alloying element niobium to replace vanadium due to concerns over the biocompatibility of vanadium ions and over possible toxic effects [117].

The alloys, although also highly passive and fairly unreactive, are generally considered to be less corrosion-resistant than the metal itself [135–137]. This was also recently shown on a microscopical level in a study by Burstein and Souto, in which the authors described the use of microelectrodes to identify breakdown events in the passive films on cp Ti and TiAlV in Ringer's solution at pH 7.4 and 2.5. Local passive film instability, in the form of active–passive microtransients, was observed and found to be initiated by the presence of Cl^- ions. Although Ti showed a smaller number of breakdown events at lower pH, none was recorded at physiological pH [137,138]. The composition of the passive films on the alloys has been shown to contain oxides of the alloying elements, and the porosity and structure of TiO_2 have been reported as nonhomogeneous, reflecting the underlying α/β microstructure [132,139,140]. It is known that the oxide layer on TiAlV is predominantly TiO_2 accompanied by small amounts of the suboxides TiO and Ti_2O_3 closer to the metal/metal oxide interface [130]. Al and V have also been observed in the passive film—the former mostly in the outer layer of the oxide film in an enriched quantity as Al^{3+} (Al_2O_3), and the latter in reduced amounts and often undetectable, its presence strongly depending on the oxide film formation conditions. The oxidation states reported varied between V^{3+} and V^{5+}, thus forming the oxides V_2O_3 and V_2O_5, but they were also detected in the ionic form at interstitial and substitutional sites [139,140]. It is important, therefore, to bear in mind that the behavior and changes in the TiO_2 passive film on exposure to a biological environment cannot be expected to be identical for Ti (cp) as for its alloys.

Despite the thermodynamic stability of the TiO_2 passive films, Ti releases corrosion products into the surrounding tissues and fluids, and observations from revision surgery and results from in vivo experiments have indicated that a higher metal release and oxide layer growth occur than first envisaged from in vitro experiments set in simple saline solutions [112,141–144]. Accumulation of Ti as well as of its alloying elements in tissues adjacent to the implant has been reported, and the causes could not only be attributable to wear [128]. This triggered a series of investigations into the properties of the passive film on Ti and the specific changes resulting upon exposure to the aggressive conditions present in a biological environment. The importance of surface topography and, not the least, of the chemical composition of the surface oxide film with regards to tissue response is a well-known factor [145]. Biocompatibility, in fact, heavily relies on the properties of the surface and the interface phenomena that take place between the material and the biological environment in which it is placed.

A thorough understanding of the metals and alloys can be obtained through electro-chemical considerations of the kinetics and mechanisms of the surface processes that occur, and indeed electrochemical methods have provided insights into the properties of passive films whereas surface techniques have enabled the study of oxide film composition under varied physiological conditions and treatments, including electrochemically treated samples. The two methods are often used in combination, yielding complimentary information on the evolution of passive films. Milosev et al. [146], for example, recently utilized surface analytical techniques in combination with electrochemical oxidation to study the quasi in situ formation of oxide layers on TiAlV in physiological solutions, thus revealing the chemical composition, thickness, and structure of the passive layer in dependence on the oxidation potential. Electrochemical impedance spectroscopy was then used to characterize the electronic properties of the passive film [146]. The latter technique is a nondestructive technique that allows the passive film/electrolyte interface to be investigated in situ over periods of time. A further in situ technique employed for the characterization of passive films on biomaterials is photoelectrochemistry [147]. Studies on the electronic and semiconduc-tive properties of rutile and anatase TiO_2 (n-type semiconductor) film electrodes, important to the understanding of corrosion resistance, are found in the literature [148–151].

In the first stages of implantation, the implant surface is exposed to tissue fluids composed of biological macromolecules and inorganic or mineral components, salts, and ions. Because the latter are much smaller in size than macromolecules, they are expected to diffuse faster and adsorb first to the implant surface. In this way, they condition the adsorption of organic species and the further adhesion and reaction of biological tissues [152]. Calcium and phosphate ions, present in physiological media and fundamental constituents of the human bone, have been known to selectively interact with titanium oxide surfaces, causing a remodelling of the latter, which in turn have been shown to induce the nucleation of calcium phosphates of various stoichiometry on the surface. In this direction, the effects of mineral ions on the passive film remodelling and corrosion resistance of Ti (cp) have been studied via a number of surface and electrochemical methods. Tech-niques such as XPS, AES, SIMS, and SEM have been employed for the investigation of thickness, chemical composition, microstructure, and morphology of the surface oxide films, whereas a number of corrosion and electrochemical studies on Ti and, to a lesser extent, on titanium alloys have been carried out in saline solutions and physiological solutions.

Healy and Ducheyne investigated the changes in surface chemistry, stoichiometry, and adsorbed surface species on titanium oxide during exposure to model physiological environments and revealed that the oxide incorporates both Ca and P elements from the extracellular fluid and that it increases in thickness as a function of implantation time. The $H_2PO_4^-$ adsorption was thought to be an exchange reaction with the basic hydroxyl groups because the titanium surface, although amphoteric in character, at physiological pH is negatively charged. The authors also suggested a mechanism for the dissolution of Ti, in which they proposed that the dissolution rate decreases on incorporation of P in the passive film and on increase of the passive film thickness [129,153]. XPS studies conducted by Ellingsen [153], on the other hand, indicated that calcium is adsorbed by the negatively charged TiO_2 surface, which then further promotes the adsorption of calcium-binding proteins. Highly structured H$_2$O was shown to be physisorbed to the hydroxylated surface by hydrogen bonding, favoring inorganic ions with strong hydration tendencies, which preferentially adsorb to the surface.

Sousa and Barbosa [154] reported that in the presence of calcium and phosphate ions, a decrease in the corrosion resistance of Ti takes place both in the presence and absence of film breakdown. The authors reported that when film breakdown occurs, calcium phosphate decreases the film resistance to chloride attack. Energy-dispersive spectroscopy (EDS) analysis of corrosion products obtained from galvanostatic experiments showed the formation of different types of calcium phosphate on the titanium surface. Sundgren et al. [155,156] used AES to reveal that the titanium oxide layer grows and takes up ions during implantation time. This growth and uptake were demonstrated to occur even in the presence of an adsorbed layer of protein such as albumin, although fibronectin was reported as containing calcium-sensitive heparin-binding sites, which interfere with the biomineralization of the Ti substrates [155–157].

Hanawa and Ota [158] also reported that the titanium oxide layer on both Ti and Ti–6Al–4V reacts with phosphate and calcium ions upon immersion in artificial saliva. The authors investigated the incorporation of inorganic ions into the surface of titanium by XPS and Fourier transform infrared reflection–absorption spectroscopy (FTIR-RAS) and found that upon immersion of titanium in biofluids of different pH values, the surface adsorbed only oxygen, hydroxyl groups, water, calcium ions, and phosphate ions even though the solution contained other ions. The calcium phosphate formed on Ti at pH 7.4 was similar to apatite. The same authors investigated the repassivation of Ti after abrasion in bioliquids, and found that the kinetics was slower compared with that in saline or water solutions, leading to more Ti dissolution. The repassivation step was also found to selectively incorporate phosphate ions and, in the outer layer, both calcium and phosphate ions [159]. Frauchiger et al. also observed a spontaneously formed adsorbed layer of calcium phosphate on a titanium dioxide surface, and its thickness as well as the Ca/P ratio were found to increase with immersion time. After 71 days of immersion, the Ca/P ratio was reported to correspond to that of brushite or monetite, whereas after 6 months, it corresponded close to that of hydroxyapatite (HA). The presence of phosphate was seen to be a prerequisite to the adsorption of calcium ions [152], whereas according to Panjian and Ducheyne [160], the formation of TiOH groups in simulated body fluids appeared to be the prerequisite for the formation of a hydroxyapatite layer.

The deposition of calcium or phosphate ions onto the surface and the subsequent nucleation of a calcium phosphate layer were also found to depend on the surface pretreatment of the Ti metal and the resulting charge at the surface. Experiments in which Ti surfaces were previously alkali-treated were shown to yield alkali titania surfaces that hydrolyzed, giving a negative surface charge upon solution exposure. Calcium ions, in this case, were preferentially demonstrated to adsorb onto the surface followed by phosphate ions [161].

Surface roughness has also been shown to be a key parameter in the first surface response to a biological environment. Increased surface roughness is thought to cause an increase in P deposition after immersion in simulated physiological solutions, as well as to induce higher protein production and calcium uptake by osteoblast-like cells. This is in agreement with previous findings that porous or rough Ti implants have been suggested to cause microscopical tissue cell ingrowth, thus improving implant fixation. Blackwood et al. [120] and Meachim and Williams [162] reported that ultrastructural, microstructural, and macrostructural levels of surface topography influence the behavior of adjacent tissues. Ti surfaces containing Ca and/or P lead to osteoinduction of new bones and are hence regarded as being bioactive. Studies aimed at improving and controlling this biomineralization are also at the center of current research attention [163,164], as well as the study of the kinetics of the heterogeneous nucleation of calcium phosphates on anatase and rutile surfaces [165].

The nucleation of calcium phosphate in the presence of proteins has also been investigated, but at times yielding contradicting results. Lima et al. [166] reported that in Hank's solution, the presence of albumin prevented the formation of a calcium phosphate layer, which would otherwise form in the period of 1 or 2 weeks. Zeng et al. [167] found that 4 min were necessary to reach surface concentrations close to saturation value both on Ti and calcium phosphate films obtained by ion beam sputtering, in contrast to Klinger et al. [168], who suggested albumin adsorption on Ti surfaces to be much faster. Overall, it is known that electrostatic effects play a major role in albumin adsorption because at physiological pH, it is negatively charged, thus proving a suitable binding site for Ca ions. The incorporation of calcium ions in the albumin molecule favors its attraction toward negatively charged surfaces, such as TiO_2 (negatively charged at physiological pH), where calcium ions would have a bridging effect in the electrostatic adsorption of albumin on Ti. Calcium, rather than phosphate ions, would therefore favor the adsorption of albumin [169]. The application of an external electrode potential naturally can affect the surface charge of the Ti substrate and influence the electrostatic adsorption. The electrochemical properties and surface composition of Ti and the mutual effects of the presence of phosphate, calcium, and albumin on the deposition of a calcium phosphate layer were investigated in detail by Lima et al. [166], whereas the adsorption of human serum albumin onto nanocrystalline TiO_2 electrodes as a function of potential and pH was studied by Oliva et al. [170].

A further inorganic species known to influence the properties of titanium dioxide surfaces is H_2O_2. During an inflammatory response triggered by surgical operation, inflammatory cells release superoxide and hydrogen peroxide into extracellular space. Ti interacts with the hydrogen peroxide to form TiOOH, capable of soaking up OH radicals. These radicals are known to cause injury in biological systems, for example, by degradation of proteins in the extracellular fluids [171,172]. Pan et al. [173–176] monitored changes in the passive film with time due to the presence of H_2O_2 in phosphate-buffered solutions at room temperature as well as changes related to exposure to cell cultures after a H_2O_2 treatment using electrochemical impedance spectroscopy and XPS. Fonseca and Barbosa [177] also studied the effects of H_2O_2 on Ti via electrochemical impedance spectroscopy, but at 37°C. An increase of the corrosion rate of titanium was observed in the presence of H_2O_2. Ti was found to undergo chemical attack by the latter with the formation of a thick oxide layer on the metal surface with a consequent release of Ti ions to the solution. Although the interaction mechanism has not been fully described, it is known that the corrosion of the metal occurs with the dissolution of Ti, formation of a titanium oxide film on the surface, and a Ti-catalyzed decomposition of the peroxide. The thickness of the TiO_2 layer after exposure to H_2O_2 was observed as being far superior to that obtained by simple immersion in a saline solution. Both research groups [174–178] sustained a two-layer structure composed of an outer highly porous and hydroxylated layer and of an inner much thinner and insulating layer, and proposed similar representative electronic circuitry models. Hanawa et al. [159] reported a 40-nm layer after a 2-week exposure of a Ti sample to a saline solution in the presence of H_2O_2 in contrast to a mere 6-nm layer in the absence of peroxide. In the former case, more than 90% of the surface was represented by a hydroxylated layer. It has been suggested that a high concentration of anionic species must exist at the oxide/electrolyte interface, creating a strong electrostatic field in the oxide that in turn accelerates the titanium oxidation and the ionic migration toward the oxide/electrolyte interface [177]. The presence of a Ti-hydrated outer layer also finds confirmation in works by

Blackwood et al. [120], Blackwood and Peter [178], Blackwood [179], and Ohtsuka and Otsuki [180] in studies on the dielectric and semiconductive properties of titanium dioxide. The latter observed that the semiconductive properties of the anodic oxide film depended on the film growth rate, which could be controlled by the sweep rate. The dielectric constant, in fact, increased with increasing rate of oxide growth, whereas the donor density and the flat band potential decreased and shifted to more negative potentials, respectively. Moreover, when the oxide film was formed at relatively high potential scans, the oxide film was noted as being more hydrated and the dielectric constant was noted to be higher due to the high content of hydrating water or the presence of OH bridges in the film. The decrease in donor density of the n-type semiconductor as a function of growth rate was also explained by the increased hydration: The introduction of an OH bridge into the anhydrous TiO_2 film was thought to bring about electromigration of interstitial Ti^{3+} ions or ionized oxygen vacancies, both considered to be ionized donors in anhydrous TiO_2 [180].

2. Surface Modification: Coatings and Topography

As discussed earlier, in addition to environmental fluid conditions, the reactions that follow upon exposure of the implant surface to a body fluid environment are determined mainly by the material's surface characteristics, and among these are composition, structure, and surface energy, but also roughness and topography [163,181]. Many efforts have, therefore, been directed, on one hand, toward surface preparations and, on the other hand, toward tailored surface modifications both aimed at promoting bioactivity and obtaining early and high bone fixation. In fact, it is well known that Ti enables bones to grow close to the surface of the implant, in turn becoming firmly embedded into the bone [182,183]. This process is called osteointegration and was first observed in the 1950s in a work by Branemark [184], who observed that titanium chambers implanted into rabbit tibula could not be removed from the bone once it had healed because it had grown into the spaces occupied by the screw head. Surface topography is also a very important factor in influencing the biological response, in particular in relation to cells in their adhesion morphology, proliferation, and differentiation. The reader is referred elsewhere for an understanding of cell behavior [185–192].

    Titanium surface treatments employed in the biomedical field span from blast coating, heat treatment, anodic oxidation, hydrothermal synthesis, to HA plasma spraying, HA granule intrusion, alkali treatment, and ion implantation, to name a few [190]. The aim is to favor bone attachment to the implant surface by increasing the porosity and surface areas of the implant, to favor the adsorption of calcium and phosphate and the nucleation of natural HA, to apply hydroxyapatite coatings or favor the adherence of these coatings, to increase the wear resistance of the implant surface, or a combination of these [152,163,164,191]. Hydroxyapatite, thanks to its similarity to bone composition, has been shown to promote bone ingrowth and anchorage strength in porous coated implants [192–201], and a number of coating technologies designed for Ti surfaces have been described to produce adherent apatite with a thickness of a few micrometers [64,67,202–208]. The coatings are tailored to be porous because the interfacial bond between an implant and the bone can be improved by creating a rough or porous surface on the implant, thus inducing the bone to grow within the pores (100–300 µm). Surface treatments aimed at increasing the surface have included acid etching, flame-sprayed titanium powder, and Ti wire surface-bonded to the solid implant [183,209].

    Anodic oxidation of the implant materials to yield porous and thick oxide films is also often carried out. In fact, it has been shown that oxide films of different thickness on Ti and

Ti alloys, formed by anodic oxidation, can significantly influence the stability of the passive films in physiological solutions, and that a higher oxide thickness leads to much reduced passive currents and thus a much lower ion release or dissolution of Ti in the body [163,101,210–212]. Nonetheless, the oxide structure is also believed to play a role in lowering the ion release. Rutile, for example, is considered to be denser and has a closer-packed structure with fewer paths for ion diffusion compared to anatase, thus displaying improved dissolution resistance [213].

Modification of the metal surface by ion implantation has also often been described, the concept having been first introduced in the 1980s mostly as a means to improve the corrosive wear behavior of surgical Ti–6Al–4V for use in hip prostheses [141,214–219], but later also used, for example, to improve biocompatibility by implantation of Ca and P [216]. Implant relevant surface modifications, however, are known to lead to changes not only in the structure and morphology of the surface, but also in the oxide film composition, which is known to be as important with regard to corrosion resistance [217–220] and tissue response. The surface composition of the oxide film on cp Ti, Ti–6Al–4V, and Ti–6Al–7Nb having undergone surface finishing treatment has been shown to depend on both alloy composition and type of surface treatment and cleaning procedure. After particle blasting, for example, the surface composition is modified by the presence of particle residues [139,140], but differences between the surface oxides of cp Ti, Ti–6Al–4V, and Ti–6Al–7Nb are also encountered. The oxide layer is connected to the composition of the underlying metal surface and XPS measurements have shown that the composition of the oxide film is mainly TiO_2 with suboxides close to the metal/oxide interface for all three alloys mentioned. The presence of the alloying elements in the oxide layer has also been observed in the form of Al_2O_3 and Nb_2O_5 or VO_x, respectively, but the concentration has been found to strongly depend on surface finish. Al and, similarly, Nb, for example, are enriched in polished or passivated surfaces compared to the bulk metal composition, whereas V is impoverished. In the case of acid-etched surfaces, on the other hand, in Ti–6Al–7Nb, the oxide film is enriched in Nb and impoverished in Al, whereas in Ti–6Al–4V, it is enriched in Al, and V is higher in concentration than in the case above. It was demonstrated that the art of attack upon acid etching treatment, albeit rapid on both Ti and the alloys, differs between the metals: on cp Ti, the attack is regular; on Ti–6Al–7Nb, the α-phases are selectively attacked at a higher etch rate and the β-phases form ridges. The topography of Ti–6Al–4V was observed to closely reflect the microstructure probably because of more closely matching etch rates on the two phases [139–149].

3. Wear and Surface Roughness Analyses

Implanted materials conventionally consist of polymers, ceramics, and metals, which represent a wide range of mechanical and surface properties [221]. All implanted materials evoke a host tissue response, which is very much dependent on the disparate material properties, such as mechanical, surface chemistry, bulk chemistry, topography, shape, and degradation rate, all of which involve surface interactions [222,223]. The clinical success of these devices depends on the surgical technique, implant design, mechanical loading, and the favorable response of tissues to the implant materials. In addition to the chemical and biochemical reactions that occur at the implant–biological interfaces, micromotion between interfaces has also been pinpointed as an essential factor in the determination of the integration of an implant in the body [107,224–228]. Micromotion and wear particles are, in fact, thought to be involved in eliciting host-specific responses and in activating cell mechanisms such as the production of cytokines and the release of local inflammatory mediators and matrix metalloproteinases, which in turn have been reported to play a role in osteolysis [111,229–234].

It has been suggested that mechanical malpositioning of prostheses, or, in the case of cemented implants, deposition of an incomplete bone cement [poly(methyl methacrylate) (PMMA)] mantle around prostheses could lead to micromotion between the formed interfaces. This is thought to be the initial trigger responsible for inducing inflammation and a whole sequence of undesired reactions, leading to an increase in implant instability, which in turn provokes further wear of the prosthesis components and consequent tissue reactions [234,235].

Component wear also plays a critical role in artificial total hip and knee joint failure, and wear debris appears to be crucial in causing adverse tissue reactions [107,236–238]. Despite the limited information on metal component wear, metal femoral components also wear while articulating against ultra high molecular weight polyethylene (UHMWPE) components. Furthermore, second body and third body wear caused by metal, PMMA, and bone debris are documented damaging wear mechanisms known to increase the roughness of metal alloy surfaces [232,233,239,240].

A number of studies have been conducted to investigate the extent of damage by wear and fretting corrosion on prostheses [107,228,232,233,239,240]. However, devising of a methodology has proven somewhat challenging because meaningful evaluations of a metal component's surface may be complex because of uneven and inconsistent surface damage, as both the density and depth of the scratches generally change from one area to another. Moreover, the nature of the surface damage is not known a priori. The use of a simple, but purely subjective, grading scheme to describe damaged surfaces has been suggested but is limited in its applicability [241–243]. A number of recent investigations employed non-contact profilometry to measure the surface roughness of metal components. The surface area analyzed for each measurement represented, however, only a small fraction (usually less than 1 mm^2) of the total surface, and therefore the sampling methodology was found to be critical in quantifying the surface roughness of the total surface. Quantification of the surface roughness of the metal components has often been based on the average of only five or 10 surface roughness measurements [244–248]. Recently, Que et al. suggested more reliable and repeatable methods to quantify the roughness of articulating alloy surfaces. In their work, they carried out systematic measurements on CoCrMo alloy surfaces using noncontact white light profilometry to investigate the influence of measurement protocol on a representative measure of surface damage [231–233].

Much attention has been paid at characterizing the bone–bone cement and bone–prosthesis interfaces, but the bone cement–prosthesis interface has also been thought to contribute to the initiation of hip replacement failure, in particular with debonding phenomena between the bone cement and the prosthesis [227,249,250]. Recent studies have investigated this interface but, although surface topography of the prosthesis was proven to be an important factor affecting the surface contact, the cement–metal interfacial strength, and the kinetic friction process between the bone cement and the prosthesis, mechanisms for loosening were not clearly established [227,251–253], It was suggested that surface roughness alone was insufficient a parameter to describe a prosthesis surface, and that parameters such as correlation length, root mean square slope, and fluid retention also require consideration.

Degradation of implant materials by means of corrosive and/or tribology mechanisms is known to lead to the release of particles in the body. Quantification of particles released into the surrounding tissues, together with categorization according to species, size, and shape, has likewise been on the agenda of topical research because it can provide insights into the causal degradation mechanisms and elucidate host response mechanisms to their presence [229,230,255]. In the case of total hip arthroplasty, for example, wear particles may originate from wear of the prosthesis components such as metal from the femoral stem, PMMA from the bone cement, or polyethylene from the acetabular cups. Particular notice has been paid in the field of biotribology to UHMWPE used in acetabular cups of total hip prostheses due to the severe wear problem posed by the material. There is evidence that submicron debris released by UHMWPE acetabular cups is an important mediator of osteolysis and of loosening of prostheses [240,256]. The increased release of PE particles has been associated with the presence of abrasive particulate entrapped between articulating surfaces (adhesive wear mechanism), also causing an increase in the femoral head roughness in artificial hip joints [240], In fact, several retrieval studies correlated the roughening of the femoral head to the presence of hard particulate debris embedded in the polymeric component, which in turn leads to an increase of the femoral head roughness and a subsequent acceleration of polyethylene wear. This damaging mechanism was simulated in in vitro wear experiments, in which cross sections of a metallic scratch were shown to have two peaks of piled up material, which, during sliding contact, could act as rigid indenters, thus dramatically increasing the abrasion of PE [231,239,257–260]. However, damage on the prosthetic head has been reported only to be possible by a third-body particle entrapped in a CoCr/UHMPWE joint, if the particle is harder than the metal. Metallic debris cannot scratch the surface unless carbide, nitride, or other intermetallic particle or inclusion is released by gradual surface abrasion, or, alternatively, in the case of PMMA particles containing ZrO_2 or $BaSO_4$, which are known to be at least three times harder than CoCr alloy [231,240,159].

The high volumetrical wear of PE in many total hip replacement (THR) patients and the subsequent problems of osteolysis and prosthesis loosening have led to an increased interest in the development and use of more wear-resistant polymers and alternative material couplings such as metal/metal joint and ceramic/ceramic joint, as well as a renewed interest in total hip prostheses composed of metal-bearing surfaces only. In the latter case, in contrast to original conceptions, wear appears to be significantly reduced compared to metal on PE with a correspondingly lower volume of wear debris [117,118,230].

A number of techniques have been proposed for the investigation of particles in the tissues surrounding an implant [261–268]. In the case of metal particle identification, as with PE debris, the suggested methodologies can present limitations, such as those associated with contamination upon cutting tissue sections in addition to the problems of the measured size of the particles being limited by the thickness of the section and the orientation of the particles. In contrast, the potential problem with studying isolated metal particles can be the loss of particles during the isolation process and the formation of particle agglomerates [230,269,270]. Reviews on wear particle analysis and methodologies are to be found in Refs. 229, 230, and 268–270. Evidence of metal wear particles in synovial fluids, in acetabular cups, within the bone, and within interfacial membranes between the bone and the prosthesis, or the bone and the bone cement has been widely reported from a number of in vivo investigations. In general, the metal particles detected are considerably smaller than polyethylene particles. This may also be due to the fact that the former, in contrast to polyethylene, may corrode and eventually dissolve in the joint fluid and tissues, and subsequently spread throughout the body, thus leaving smaller and fewer metal particles in the periprosthetic tissue. Animal studies have, in fact, shown evidence of metal accumulation in the lymph nodes, spleen, liver, and bone marrow, as well as increased levels of metal in the urine and blood, which are then excreted [230].

The reaction of the host body to metal degradation products (dissolved ions from simple dissolution and/or corrosion reactions, and wear particles from fretting corrosion and/or micromotion) has been a cause of increasing concern over the years [271]. Obvious examples of this are the toxicities associated with nickel in stainless steel implants and, more recently, with the Ti alloying element, vanadium. In general, the tolerance toward the released metal is associated with the toxicity of the metal species formed and their concentration, and with the ease with which the metal can be excreted from the body [262,272–279].

B. Case Study—Total Joint Arthroplasty

The biological response at the interface between an implant and the host tissues is strongly reliant upon the site of implantation and the composition and surface properties of the implant. In the case of a hip prosthesis, the interface consists almost entirely of bone, whereas in the case of a permucosal dental implant, the material will be in contact with the bone, connective tissues, and epithelia. A common biological response to a foreign object such as an implant is the isolation of the latter from the immediate surroundings by formation of an encapsulating layer of fibrous tissues. However, when the implant is to perform as a load-bearing device, this type of response is not considered acceptable because it can have a destabilizing influence on the implant [117,280,281].

In this section, a short overview of the history of total joint replacement is given and of the problems that have emerged over the years. Another major field of research is that of revolving dental implants, and the reader is referred to book chapters and reviews elsewhere [118,282,283].

Total joint arthroplasty is a successful surgical operation, which has gained acceptance worldwide in the last few decades. Arthroplasty is a surgical technique, which replaces all articulating degenerated natural surfaces with artificial materials, with the aim of achieving pain relief and improved joint mobility. The idea was first introduced in 1949 with the use of a major bone and joint prosthesis to ease the suffering of patients with destructive lesions of the bone and joint, thereby avoiding the use of extensive external prostheses. Important advances ensued during the 1950s with G. K. McKee, who introduced metal on metal hip prostheses in which components were originally made of stainless steel. In order to mitigate the excessive friction and rapid loosening of the stainless steel pair, the components were quickly replaced by CoCrMo alloy (Vitallium™), which at this time constituted the most suitable material available [115]. Side plate fixation was required prior to the introduction of PMMA bone cement in the 1960s, a substitution that increased the short-term implantation success rate to >90%. PMMA bone cement, in fact, is used in total joint replacements to anchor implants to the underlying bone [117].

In the 1950s, the clinical acceptability of titanium became increasingly established and it was used for nails, plates and screws, and partial proximal femoral components. However, in prostheses that required a total joint bearing, cast CoCrMo alloy articulating surfaces were still employed. A combination of the two materials was initiated in the 1960s, but it soon became apparent that the fatigue properties of Ti proved to be inadequate for highly stressed intramedullary stems. It was further recognized that the use of identical metals in the tribological pair was necessary to avoid galvanic corrosion, but this unfolded concerns over tribological design [237,238,242,243]. In fact, a high rate of loosening was encountered with early metal on metal devices because of nonoptimum fit between the articulating surfaces. This led to the development of the concept of low-friction arthroplasty in the 1960s, first introduced by Sir J. Charnley with a new design consisting of a small-diameter metallic femoral head articulating with a polymeric acetabular cup—PTFE and later UHMWPE [104]. The latter material has prevailed for 30 years as a dominant orthopedic material in total joint replacement, despite that wear has consistently been observed when rubbing against metal femoral heads or femoral condyl components, and adverse tissue reaction and tissue reaction to the wear debris have been reported [112,127,233,238,240,243,255,256,284]. The choice of counterpart is known to be a critical factor, with Ti–6Al–4V imparting a more detrimental impact than CoCrMo and ceramic. In recent years, concomitant development of metal on metal technology through optimization of the prosthesis geometry and manufacturing practices has created the renaissance of metal on metal prostheses in Europe [230,269].

Although continued development of the technique, of the implant design, and of the biological aspects of the procedure has contributed to and increased the success of total joint replacements (275,000 hip and knee replacements were carried out in 1995 in the United States) with success rates of 95% and higher at 10 years and 85–90% at 15 years, every year, up to 20% of human patients require revision surgery due to complications, the major cause of implant failure being aseptic loosening [285].

Aseptic loosening has been associated with the formation of a synovial-like tissue at the interface between the bone and the implant, or between the bone and the cement mantle for cemented hip prostheses, in response to the interaction of mechanical, electrochemical, and biological processes [286–288]. Micromotion and wear particles are thought to be involved in eliciting host-specific responses, activating cell mechanisms such as the production of cytokines, and the release of local inflammatory mediators and matrix metalloproteinases, which in turn have been reported to play a role in osteolysis typical of aseptically loosened prostheses [285,289,290]. However, it is still discussed in the literature whether micromotion or wear particle formation is the initial trigger for the process of aseptic loosening [234]. It has been suggested that micromotion due to an incomplete cement mantle and/or mechanical malpositioning of the prosthesis may be the initial trigger inducing inflammation, leading to the formation of an interface membrane [291]. The latter, in turn, may release substances that may further accelerate bone resorption in the adjacent bone tissue. An increase in implant instability may then be the cause for the establishment of a vicious circle that will add to excessive wear of the prosthesis components. Debonding and failure of the prosthetic–PMMA interface have also been implicated as a primary cause of loosening in cemented femoral components [292,293]. Debonding has, in fact, been shown to result in substantial increases in stress levels in the PMMA bone cement mantle, leading to extensive cracking of the PMMA and the premature failure of the cemented systems. Saline solutions tend to enhance cracking between the interface [292]. Furthermore, the mechanical behavior is thought to be sensitive to formulation, preparation, and sterilization methods. Even though the material has been successfully employed in the last three decades, with over 60 different bone cements used in clinical practice, fear over its role in implant rejection has renewed interest in the field, with the appearance of innovative preparation techniques and modified formulations [292–294].

Because most studies are either based on retrieved material from clinical patients or on in vitro studies, it is almost impossible to determine the cascade of events occurring in clinical cases of aseptic loosening of hip prostheses. The establishment of experimental animal models has also been sought to enable the study of this clinical condition early in the process under controlled circumstances, thereby overcoming limitations posed by clinical investigations normally performed in human orthopedics. Various animal models for aseptic loosening have been described in the literature based on the addition of wear particles [134,295–297], or initial placement of loose implants [298]. In a recent study, an animal model for interface tissue formation in cemented CoCrMo hip replacements was exploited [299]. The aim of this work was to gain a clearer understanding of the processes that take place at the cement–tissue interface, and of the influences that each environment plays on the other. In contrast to many reported studies, in which the task on hand was approached from either a biological, clinical, or material angle, this work sought to investigate the interaction of the different processes by combining competence from medicine, biology, electrochemistry, corrosion, surface science, and biochemistry [300]. The investigation revealed that a generation of metal particles, or in other words alloy degradation mechanism, involving a mixture of wear and tribocorrosion took place in all of the implanted hip prostheses. This was also sustained by the detection, in interface tissues collected between the cortex and the PMMA bone cement, of the alloying elements. Furthermore, the concentrations of the metal detected and the quantity of interface tissue formation could be directly correlated to the stability and fixation of the hip prosthesis within the bone at postmortem. Although the surface degradation mechanisms of the metal could not be elucidated, the results indicated the modes of mass transport between the implant surface and the tissue to be key factors in the understanding of the cascade of events [300].

Although an understanding of the mechanisms that lead to aseptic loosening remains a priority to the medical and scientific world, longer human life expectancy and implantation in younger patients have driven bioengineers and material scientists to consider the long-term limitation of existing materials and the development of new materials and designs. Essential elements to ensure biocompatibility are suitable chemical composition to avoid adverse tissue reaction, high wear and degradation resistance (corrosion/fretting) to minimize debris generation, acceptable strength to sustain the cycle loading endured by the joint, and low modulus to minimize bone resorption. In recent years, the tendency to adopt existing materials for orthopedic application as exemplified by Ti–6Al–4V, developed for aerospace application, has been surpassed by a tailored research development of new materials and alloys. Ti alloys remain the metal of preference, thanks to their lower modulus, superior biocompatibility, and corrosion resistance [117].

REFERENCES

1. Collings, E.W. The physical metallurgy of titanium alloys. In *ASM Series in Metal Processing*. Gegel, H.L., Ed.; American Society for Metals: Cleveland/Metals Park, OH, 1984.
2. Bardos, D.I. Titanium and titanium alloys. In *Concise Encyclopedia of Medical and Dental Materials*. Williams, D., Ed.; Pergamon Press: Oxford, 1990; 360–365.
3. American Society for Testing and Materials. Annual Book of ASTM Standards 2000, ASTM F 67-95; Medical Devices and Services: West Conshohocken, PA, 2000; Vol. 13.01,1–3.
4. American Society for Testing and Materials. Annual Book of ASTM Standards 2000, ASTM F 1472-99; Medical Devices and Services: West Conshohocken, PA, 2000; Vol. 13.01,783–786.
5. American Society for Testing and Materials. Annual Book of ASTM Standards 2000, ASTM F 136-98; Medical Devices and Services: West Conshohocken, PA, 2000; Vol. 13.01, 19–21.
6. American Society for Testing and Materials. Annual Book of ASTM Standards 2000, ASTM F 1295-97a; Medical Devices and Services: West Conshohocken, PA, 2000; Vol. 13.01,638–640.
7. American Society for Testing and Materials. Annual Book of ASTM Standards 1999, ASTM B 348-98;. Nonferrous Metal Products: West Conshohocken, PA, 1999; Vol. 02.04,201–207.
8. Higo, Y.; Ouchi, C.; Tomita, Y.; Murota, K.; Sugiyama, H. Properties evaluation and its application to artificial hip joint of newly developed titanium alloy for biomaterials. In *Metallurgy and Technology of Practical Titanium Alloys*; Fujishiro, S., Eylon, D., Kishi, T., Eds.; The Minerals, Metals and Materials Society: Warrendale, PA, 1994; 421–425.

9. Okazaki, Y.; Ito, Y.; Ito, A.; Tateishi, T. Effect of alloying elements on mechanical properties of titanium alloys for medical implants. Mater Trans., JIM 1993, *34*, (12), 1217–1222.

10. Bania, P.J. Beta titanium alloys and their role in the titanium industry. In *Beta Titanium Alloys in the 1990's*; Eylon, D., Boyer, R.R., Koss, D.A., Eds.; The Minerals, Metals and Materials Society: Warrendale, PA, 1993; 3–14.

11. Schutz, R.W. An overview of beta titanium alloy environmental behavior. In *Beta Titanium Alloys in the 1990's*; Eylon, D., Boyer, R.R., Koss, D.A., Eds.; The Minerals, Metals and Materials Society: Warrendale, PA, 1999; 75–91.

12. Breme, J.; Biehl, V.; Schulte, W.; d'Hoedt, B.; Donath, K. Development and functionality of isoelastic dental implants of titanium alloys. Biomaterials 1993, *14* (12), 887–892.

13. American Society for Testing and Materials. Annual Book of ASTM Standards 2000, ASTM F 1713-96; Medical Devices and Services: West Conshohocken, PA, 2000; Vol. 13.01,1071–1074.

14. Teledyne Wah Chang Albany, Tiadyne 1610 Data Sheet, August 1993.

15. Ahmed, T.; Long, M.; Silvestri, J.; Ruiz, C.; Rack, H.J. A new low modulus, biocompatible titanium alloy. Proceedings of the Eighth World Conference on Titanium. Blenkinsop, P.A., Evans, W.J., Flower, H.M., Eds.; Titanium '95 Science and Technology: Birmingham, UK, 1999; 1760–1767.

16. Niimoni, M.; Kuroda, D.; Fukunaga, K.; Morinaga, M.; Kato, Y.; Yashiro, T.; Suzuki, A. Corrosion wear fracture of new β type biomedical titanium alloys. Mater. Sci. Eng. A 1999, *263*, 193–199.

17. Freese, H.L.; Volas, M.G.; Wood, J.R. Metallurgy and technological properties of titanium and titanium alloys. In *Titanium in Medicine: Material Science, Surface Science, Engineering, Biological Responses and Medical Applications*. Brunette, D.M., Tengvall, P., Textor, M., Thomsen, P., Eds.; Springer: Berlin, Heidelberg, 1993; 25–51.

18. American Society for Testing and Materials. Annual Book of ASTM Standards 2000, ASTM F 1813-97; Medical Devices and Services: West Conshohocken, PA, 2000; 1250–1252.

19. Steinemann, S.G.; Mäusli, P.-A.; Szmuler-Moncler, S.; Semlitsch, M.; Pohler, O.; Hintermann, H.-E.; Perren, S.M. Beta-titanium alloy for surgical implants. In *Titanium '92 Science and Technology*; Froes, F.H., Caplan, I., Eds.; The Minerals, Metals and Materials Society: Warrendale, PA, 1993; Vol. 13.01, 2689–2696.

20. Fanning, J.C. Properties and processing of a new metastable beta titanium alloy for surgical implant applications: TIMETAL® 21SRx. Proceedings of the Eighth World Conference on Titanium, Titanium '95 Science and Technology; Birmingham, UK. Blenkinsop, P.A., Evans, W.J., Flower, H.M., Eds.; 1995; 1800–1807.

21. Zardiackas, L.D.; Mitchell, D.W.; Disegi, J.A. Characterization of Ti–15Mo beta titanium alloy for orthopaedic implant applications. Proceedings of a Symposium on Medical Applications of Titanium and Its Alloys: The Material and Biological Issues; Phoenix, AZ. Brown, S.A., Lemons, J.E., Eds.; 1994;60–65.

22. American Society for Testing and Materials. Annual Book of ASTM Standards 2000, ASTM F 75-98; Medical Devices and Services: West Conshohocken, PA, 2000; Vol. 13.01; 4–6.

23. Clemow, A.J.T.; Daniell, B.L. Solution treatment behavior of Co–Cr–Mo Alloy. J. Biomed. Mater. Res. 1979, *13,* 265–279.

24. American Society for Testing and Materials. Annual Book of ASTM Standards 2000, ASTM F 90-97; Medical Devices and Services: West Conshohocken, PA, 2000; Vol. 13.01, 10–12.

25. American Society for Testing and Materials. Annual Book of ASTM Standards 2000, ASTM F 562-95; Medical Devices and Services: West Conshohocken, PA, 2000; Vol. 13.01, 87–89.

26. Weinstein, A.M.; Clemow, A.J.T. Cobalt-based alloys. In *Concise Encyclopedia of Medical and Dental Materials*. Williams, D., Ed.; Pergamon Press: Oxford, 2000; 106–112.

27. American Society for Testing and Materials. Annual Book of ASTM Standards 2000, ASTM F 138-97; Medical Devices and Services: West Conshohocken, PA, 2000; Vol. 13.01, 22–25.

28. Cigada, A.; Rondelli, G.; Vicentini, B.; Giacomazzi, M.; Roos, A. Duplex stainless steels for osteosynthesis devices. J. Biomed. Mater. Res. 1989, *23*, 1087–1095.

29. Thomann, U.I.; Uggowitzer, P.J. Wear–corrosion behavior of biocompatible austenitic stainless steels. Wear 2000, *239*, 48–58.

30. Davidson, J.A.; Gergette, F.S. State of the art materials for orthopaedic prosthetic devices. Proceedings of Implant Manufacturing and Material Technology; Society of Manufacturing Engineers, Em 87-122, 1986; 122–126.

31. Cheal, E.J.; Spector, M.; Hayes, W.C. Role of loads and prosthesis material properties on the mechanics of the proximal femur after total hip arthroplasty. J. Orthop. Res. 1992, *10*, 405–422.

32. Huiskes, R.; Weinans, H.; van Rietbergen, B. The relationship between stress shielding and bone resorption around total hip stems and the effects of flexible materials. Clin. Orthop. Relat. Res. 1992, *274*, 124–134.

33. Niimoni, M. Mechanical properties of biomedical titanium alloys. Mater. Sci. Eng., A 1998, *243*, 231–236.

34. Wang, K. The use of titanium for medical applications in the USA. Mater. Sci. Eng., A 1996, *213*, 134–137.

35. Semlitsch, M.F.; Weber, H.; Streicher, R.M.; Schön, R. Joint replacement components made of hot-forged and surface-treated Ti–6Al–7Nb alloy. Biomaterials 1992, *13* (11), 781–788.

36. Borowy, K.H.; Kramer, K.H. On the properties of a new titanium alloy $(TiAl_5Fe_{2.5})$ as implant material. Proceedings of the Fifth International Conference on Titanium Science and Technology, Munich, Germany. Lütjering, G., Zwicker, U., Bunk, W., Eds.; 1985; Vol. 2, 1381–1386.

37. Wang, K.; Gustavson, L.; Dumbleton, J. Low modulus, high strength, biocompatible alloy developed for surgical implants. *Beta Titanium in the 1990's*; The Minerals, Metals and Materials Society: Warrendale, PA, 1998; 49–60.

38. Wood, J.R.; Russo, P.A. Heat treatment of titanium alloys. Ind. Heat, April, 1997; 51–55.

39. Long, M.; Rack, H.J. Titanium alloys in total joint replacement—a materials science perspective. Biomaterials 1998, *19*, 1621–1639.

40. Stubbington, C.A.; Bowen, A.W. Improvements in the fatigue strength of Ti–6Al–4V through microstructure control. J. Mater. Sci. 1974, *9*, 941–947.

41. Devine, T.M.; Wulff, J. Cast vs. wrought cobalt–chromium surgical implant alloys. J. Biomed. Mater. Res. 1975, *9*, 151–167.

42. Bardos, D. High strength Co–Cr–Mo alloy by hot isostatic pressing of powder. Biomater. Med. Dev. Artif. Organs 1979, *7*, 73–80.

43. Hodge, F.G.; Lee III, T.S. Effect of processing on performance of cast prosthetic alloys. Corrosion 1975, *31* (3), 111–112.

44. Hughes, A.N.; Lane, R.A. The development of high-tensile properties in a wrought Co–Cr–Ni–W alloy of potential orthopaedic surgery use. Eng. Med. 1976, *5* (2), 38–40.

45. Lorenz, M.; Semlitsch, M.; Panic, B.; Weber, H.; Wilhert, H.G. Fatigue strength of cobalt-alloys with high corrosion resistance for artificial hip joints. Eng. Med. 1978, *7*(4), 241–250.

46. Pilliar, R.M.; Weatherly, G.C. Developments in implant alloys. Williams, D.F., Ed.; CRC Crit. Rev. Biocompat. 1986; *1*, 371–403.

47. Kohn, D.H. Materials for bone and joint replacement. In *Materials Science and Technology, A Comprehensive Treatment: Vol. 14. Medical and Dental Materials*. Williams, D.F., Ed.; VCH Verlag: Weinheim, 1992; 31–109.

48. Fraker, A.C.; Ruff, A.W. Metallic surgical implants: state of the art. J. Met. 1977, *29* (5), 22–28.

49. Speidel, M.O.; Uggowitzer, P.J. Biocompatible nickel-free stainless steels to avoid nickel allergy. *Materials Day, Materials in Medicine*. Speidel, M.O., Uggowitzer, P.J., Eds.; Department of Materials, ETH Zürich: Zürich, 1998; 191–207.

50. Pilliar, R.M. Manufacturing processes of metals: the processing and properties of metal implants. In *Metal and Ceramic Biomaterials 1*. Ducheyne, P., Hastings, G.W., Eds.; CRC Press: Boca Raton, 1984; 79–105.

51. Rostoker, W.; Chao, E.Y.S.; Galante, J.O. Defects in failed stems of hip prostheses. J. Biomed. Mater. Res. 1978, *12*, 635–641.

52. Uhlig, H.H.; Revie, R.W. *Corrosion and Corrosion Control: An Introduction to Corrosion Science and Engineering*; Wiley: New York, 1985.

53. Fontana, M.G. *Corrosion Engineering*; McGraw-Hill: New York, 1986.

54. Mansfeld, F., Ed.; *Corrosion Mechanisms*; Marcel Dekker: New York, 1987.

55. Kaesche, H. *Die Korrosion der Metalle*; Springer: Berlin, 1990.

56. Shreir, L.L., Jarman, R.A., Burstein, G.T., Eds.; *Corrosion*; Vols. 1 and 2; Butterworth-Heinemann: Oxford, 1994.

57. Marcus, P., Oudar, J., Eds.; *Corrosion Mechanisms in Theory and Practice*; Marcel Dekker: New York, 1995.

58. Revie, R.W., Ed.; *Uhlig's Corrosion Handbook*; Wiley: New York, 2000.

59. Atkins, P.W. *Physical Chemistry*; Oxford University Press: Oxford, 1986.

60. Vetter, K.J. *Elektrochemische Kinetik*; Springer: Berlin, 1961.

61. Bockris, J.O.; Reddy, A.K.N. *Modern Electrochemistry*; Plenum Press: New York, 1970.

62. Newman, J. *Electrochemical Systems*; Prentice-Hall: Englewood Cliffs, NJ, 1991.

63. Bard, A.J.; Parsons, R.; Jordan, J. *Standard Potentials in Aqueous Solutions*; Marcel Dekker: New York, 1985.

64. Pourbaix, M. *Atlas d'Equilibres Électrochimiques*; Gautier-Villars and Cie: Paris, 1963.

65. Frankenthal, R.P., Kruger, J., Eds.; Passivity of Metals; The Electrochemical Society: Princeton, NJ, 1978.

66. Froment, M., Ed.; *Passivity of Metals and Semiconductors*; Elsevier Science Publishers: Amsterdam, 1983.

67. Sato, N., Hashimoto, K., Eds.; *Passivation of Metals and Semiconductors*; Pergamon Press: Oxford, 1990.

68. Heusler, K., Ed.; *Passivity of Metals and Semiconductors*; TransTech Publications: Aedermannsdorf CH, 1995.

69. Ives, M.B., Luo, J.L., Rodda, J., Eds.; Passivity of Metals and Semiconductors, Proceedings of the 8th International Symposium; The Electrochemical Society: Pennington, NJ, 2000 (PV 99-42).

70. Schultze, J.W.; Lohrengel, M.M. Stability, reactivity and breakdown of passive films—problems of recent and future research. Electrochim. Acta 2000, *45*, 2499–2510.

71. Schmuki, P. From bacon to barriers: a review on the passivity of metals and alloys. J. Solid State Electrochem. 2002, *6*, 145–164.

72. Loorbeer, P.; Lorenz, W.J. *The Kinetics of Iron Dissolution and Passivation*; 1978; 607–613.

73. Epelboin, I.; Keddam, M.; Mattos, O.R.; Takenouti, H. The dissolution and passivation of Fe and Fe–Cr alloys in acidified sulphate medium: influence of pH and Cr content. Corros. Sci. 1979, *19*, 1105–1112.

74. Keddam, M.; Mattos, O.R.; Takenouti, H. Mechanism of anodic dissolution of iron–chromium alloys investigated by electrode impedance. Electrochim. Acta 1986, *31*, 1147–1165.

75. Heumann, T.; Rösener, W. Z. Elektrochem. 1955, *59*, 722–729.

76. Schmuki, P.; Virtanen, S.; Davenport, A.J.; Vitus, C.M. Transpassive dissolution of Cr and sputter-deposited Cr oxides studied by in situ x-ray near-edge spectroscopy. J. Electrochem. Soc. 1996, *143*, 3997–4005.

77. Schmuki, P.; Virtanen, S.; Davenport, A.J.; Vitus, C.M. In situ x-ray absorption near-edge spectroscopic study of the cathodic reduction of artificial iron oxide passive films. J. Electrochem. Soc. 1996, *143*, 574–582.

78. Schmuki, P.; Virtanen, S.; Isaacs, H.S.; Ryan, M.P.; Davenport, A.J.; Böhni, H.; Stenberg, T. Electrochemical behavior of Cr_2O_3/Fe_2O_3 artificial passive films studied by in situ XANES. J. Electrochem. Soc. 1998, *145*, 791–801.

79. Ryan, M.P.; Newman, R.C.; Thompson, G.E. An STM study of the passive film formed on iron in borate buffer solution. J. Electrochem. Soc. 1995, *142*, L177–L179.

80. Toney, M.F.; Davenport, A.J.; Oblonsky, L.J.; Ryan, M.P.; Vitus, C.M. Atomic structure of the passive oxide film formed on iron. Phys. Rev. Lett. 1997, *79*, 4282–4285.

81. Maurice, V.; Yang, W.P.; Marcus, P. X-ray photoelectron spectroscopy and scanning tunneling microscopy study of passive films formed on (100) Fe–18Cr–13Ni single-crystal surfaces. J. Electrochem. Soc. 1998, *145*, 909–920.

82. Zsklarska-Smialowska, Z. *Pitting Corrosion of Metals*; National Association of Corrosion Engineers: Houston, TX, 1986.

83. Boehni, H. Breakdown of passivity and localized corrosion processes. Langmuir 1987, *3*, 924–930.

84. Strehblow, H.H. Mechanisms of pitting corrosion. In *Corrosion Mechanisms in Theory and Practice*. Marcus, P., Oudar, J., Eds.; Marcel Dekker: New York, 1995; 201–238.

85. Baroux, B. Further insights on the pitting corrosion of stainless steels. In *Corrosion Mechanisms in Theory and Practice*. Marcus, P., Oudar, J., Eds.; Marcel Dekker: New York, 1995; 265–310.

86. Frankel, G.S. Pitting corrosion of metals: a review of the critical factors. J. Electrochem. Soc. 1998, *145*, 2186–2198.

87. Mischler, S.; Debaud, S.; Landolt, D. Wear-accelerated corrosion of passive metals in tribocorrosion systems. J. Electrochem. Soc. 1998, *145*, 750–758.

88. Godard, H.P.; Jepson, W.B.; Bothwell, M.R.; Kane, R.L. *The Corrosion of Light Metals*; John Wiley and Sons: New York, 1967.

89. Hughes, P.C.; Lamborn, I.R. J. Inst. Met. 1961, 89–95.

90. Schutz, R.W.; Thomas, D.E. Corrosion of titanium and titanium alloys. In *Metals Handbook, 9th Ed.* ASM International: Metals Park, OH, 1987; Vol. 13,669–706.

91. Kelly, E.J. Electrochemical behavior of titanium. Modern Aspects of Electrochemistry. Bockris, J.O.M., Conway, B.E., White, R.E., Eds.; Plenum Press: New York, 1959; 319–424 (No. 14, Chap. 5).

92. Stern, M.; Wissemberg, W. J. Electrochem. Soc. 1959, *106*, 756–764.

93. Straumanis, M.E.; Shih, S.T.; Schlechten, A.W. Hydrogen overvoltage (Cathodic potential) on Titanium in acidic and basic solutions. J. Phys. Chem. 1955, *59*, 317–321.

94. Schenk, R. The corrosion properties of titanium and titanium alloys. In *Titanium in Medicine*. Brunette, D.M., Tengvall, P., Textor, M., Thomsen, P., Eds.; Springer: Berlin, 1966;146–170.

95. Hoar, T.P.; Mears, D.C. Corrosion resistant alloys in chloride solutions. Proc. R. Soc., Ser. A 1966, *294*, 486–510.

96. Levine, D.L.; Staehle, R.W. Crevice corrosion in orthopedic implant metals. J. Biomed. Mater. Res. 1977, *11*, 553.

97. MacDougall, J.E.; Schenk, R.; Windler, M. Investigation into wear-induced corrosion of orthopedic implant materials. Guagliano, M., Aliabadi, M.H., Eds.; Advances in Fracture and Damage Mechanics II Hoggar: Geneva, 2001; 423–429.

98. Brown, S.A.; Merritt, K.; Payer, J.H.; Kraay, M.J. Fretting corrosion of orthopedic implants. In *Compatibility of Biomedical Implants*. Kovacs, P., Istephanous, M.S., Eds.; The Electro-chemical Society: Pennington NJ, 1993; 42–47, (PV 94-15).

99. Gilbert, J.L.; Buckley, C.A.; Jacobs, J.J. In vivo corrosion of modular hip prosthesis components in mixed and similar metal combinations. The effect of crevice, stress, motion, and alloy coupling. J. Biomed. Mater. Res. 1993, *27*, 1533–1544.

100. Khan, M.A.; Williams, R.L.; Williams, D.F. The corrosion behaviour of Ti–6Al–4V, Ti–6Al–7Nb and Ti–13Nb–13Zr in protein solutions. Biomaterials 1999, *20*, 631–637.

101. Crook, P.; Silence, W.L. Cobalt alloys. In *Uhlig's Corrosion Handbook*; Revie, R.W., Ed.; Wiley: New York, 2000; 717–728.

102. Rebak, R.B.; Crook, P. Improved pitting and crevice corrosion resistance of nickel and cobalt alloys. In *Critical Factors in Localized Corrosion III*; Kelly, R.G., Frankel, G.S., Natishan, P.M., Newman, R.C., Eds.; The Electrochemical Society: Pennington, NJ, 1996; 209–301 (PV 98-17).

103. Sedriks, A.J. *Corrosion of Stainless Steels*; Wiley: New York, 1996.

104. Bundy, K.J. Corrosion and other electrochemical aspects of biomaterials. Crit. Rev. Biomed. Eng. 1994, *22*, 139–251.

105. Zitter, H.; Plenk, J.H. The electrochemical behavior of metallic implant materials as an indicator of their biocompatibility. J. Biomed. Mater. Res. 1987, *21*, 881–896.

106. Kruger, J. Fundamental aspects of the corrosion of metallic implants. In *Corrosion and Degradation of Implant Materials, ASM STP 684*; Syrett, B.C., Acharya, A., Eds.; ASTM: Philadelphia, 1979; 107–127.

107. Hall, R.M.; Unsworth, A. Review: friction in hip prostheses. Biomaterials 1997, *18*, 1017–1026.

108. Watson, S.W.; Friedersdorf, F.J.; Madsen, B.W.; Cramer, S.D. Methods of measuring wear–corrosion synergism. Wear 1995, *181–183*, 476–484.

109. Jemmely, P.; Mischler, S.; Landolt, D. Electrochemical modelling of passivation phenomena in tribocorrosion. Wear 2000, *237*, 63–76.

110. Mischler, S.; Debaud, S.; Landolt, D. Wear accelerated corrosion of passive metals in tribocorrosion systems. J. Electrochem. Soc. 1998, *145*, 750–758.

111. Kovacs, P.; Davidson, J.A.; Daigle, K. Correlation between the metal ion concentration and the fretting wear volume of orthopaedic implant metals. Mech. Form. Biol. Conseq. 1992, *1144*, 160–176.

112. Khan, M.A.; Williams, R.L.; Williams, D.F. Conjoint corrosion and wear in titanium alloys. Biomaterials 1999, *20*, 765–772.

113. Bundy, K.J.; Luedemann, R. Factors which influence the accuracy of corrosion rate determination of implant materials. Ann. Biomed. Eng. 1989, *17*, 159–175.

114. Vallet-Regí, M. Introduction to the world of biomaterials. An. Quim., Int. Ed. 1997, *93*, S6–S14.

115. Williams, D.F. The properties and clinical uses of cobalt–chronium alloys. In *Biocompatibility of Clinical Implant Materials*; Williams, D.F., Ed; CRC Press: Boca Raton, 1998; 99–127.

116. Steinemann, S.G. Titanium—the material of choice. Periodontics 1998, *17*, 7–21.

117. Long, M.; Rack, H.J. Review: Titanium alloys in total joint replacement—a materials science perspective. Biomaterials 1998, *9*, 1621–1639.

118. Noort, R.V. Review titanium: the implant material of today. J. Mater. Sci. 1987, *22*, 3801–3811.

119. Bothe, R.T.; Beaton, K.E.; Davenport, H.A. The reaction of bone to multiple metallic implants. Surg. Gynecol. Obstet. 1940, 71–85.

120. Blackwood, D.J.; Peter, L.M.; Williams, D.E. Stability and open circuit breakdown of the passive oxide film on titanium. Electrochim. Acta 1988, *33*, 1143–1149.

121. Ducheyne, P. In vitro and in vivo modeling of the biocompatibility of titanium. 6th World Conference on Titanium, France, 1988; 551–556.

122. Sutow, E.J.; Pollack, S.R. The biocompatibility of certain stainless steels. In *The Biocompatibility of Clinical Implant Materials*. Williams, D.F., Ed.; CRC Press: Boca Raton, 1981; 45–98.

123. Thomann, U.I.; Uggowitzer, P.J. Wear–corrosion behavior of biocompatible austenitic stainless steels. Wear 2000, *239*, 48–58.

124. Pan, J.; Karlén, C.; Ulfvin, C. Electrochemical study of resistance to localised corrosion of stainless steels for biomaterial applications. J. Electrochem. Soc. 2000, *147*, 1021–1025.

125. Reclaru, L.; Lerf, R.; Eschler, P.Y.; Meyer, J.M. Corrosion behavior of a welded stainless-steel orthopedic implant. Biomaterials 2001, *22*, 269–279.

126. Clark, G.C.F.; Williams, D.F. The effects of proteins on metallic corrosion. J. Biomed. Mater. Res. 1982, *16*, 125–134.

127. Khan, M.A.; Willimas, R.L.; Williams, D.F. In vitro corrosion and wear of titanium alloys in the biological environment. Biomaterials 1996, *17*, 2117–2126.

128. Khan, M.A.; Williams, R.L.; Williams, D.F. The corrosion behaviour of Ti–6Al–4V, Ti–6Al–7Nb and Ti–13Nb–13Zr in protein solutions. Biomaterials 1998, *20*, 631–637.

129. Healy, K.E.; Ducheyne, P. The mechanism of passive dissolution of titanium in a model physiological environment. J. Biomed. Mater. Res. 1992, *26*, 319–322.

130. Roberts, M.W.; Tomellini, M. Mixed oxidation states of titanium at the metal–oxide interface. Catal. Today 1992, *12*, 443–452.

131. Lausmaa, J.; Mattsson, L.; Rolander, U.; Kasemo, B. Chemical composition and morphology of titanium surface oxides. Mater. Res. Soc. Symp. 1986, 55, 351–359.

132. Ask, M.; Rolander, U.; Lausmaa, J.; Kasemo, B. Microstructure and morphology of surface oxide films on Ti–6Al–4V. J. Mater. Res. 1990, 5, 1662–1667.

133. Lausmaa, J.; Kasemo, B.; Rolander, U.; Bjursten, L.M.; Ericson, L.E.; Rosander, L.; Thomsen, P. Preparation, surface spectroscopic and electron microscopic characterization of titanium implant materials. In Surface Characterization of Biomaterials; Ratner, B.D., Ed.; Elsevier Science Publishers: Amsterdam, 1988; 161–173.

134. Lausmaa, J.; Kasemo, B.; Mattsson, H. Surface spectroscopic characterisation of titanium implant materials. Appl. Surf. Sci. 1990, 44, 133–146.

135. Nowak, W.B.; Sun, E.X. Electrochemical characterisation of Ti–6Al–4V alloy in 0.2 N NaCl solution: II. Kinetic behaviors and electric field in passive film. Corros. Sci. 2001, 43, 1817–1838.

136. Sun, E.X.; Nowak, W.B. Electrochemical characteristics of Ti–6Al–4V alloy in 0.2 N NaCl solution: I. Tafel slopes in quasi-passive state. Corros. Sci. 2001, 43, 1801–1816.

137. Souto, R.M.; Burstein, G.T. A preliminary investigation into the microscopic depassivation of passive titanium implant materials in vitro. J. Mater. Sci., Mater. Med. 1996, 7, 337–343.

138. Burstein, G.T.; Souto, R.M. Observation of localised instability of passive titanium in chloride solution. Electrochim. Acta 1995, 40, 1881–1888.

139. Sittig, C.; Hahner, G.; Marti, A.; Textor, M.; Spencer, N.D.; Hauert, R. The implant material, Ti_6Al_7Nb: surface microstructure, composition and properties. J. Mater. Sci., Mater. Med. 1999, 10, 191–198.

140. Sittig, C.; Textor, M.; Spencer, N.D.; Wieland, M.; Vallotton, P.H. Surface characterization of implant materials cp Ti, Ti–6Al–7Nb and Ti–6Al–4V with different pretreatments. J. Mater. Sci., Mater. Med. 1999, 10, 35–46.

141. Buchanan, R.A.; Rigney, E.D.; Williams, J.M. Wear-accelerated corrosion of Ti–6Al–4V and nitrogen-ion-implanted Ti–6Al–4V: mechanisms and influence of fixed-stress magnitude. J. Biomed. Mater. Res. 1987, 21, 367–377.

142. Solar, R.J. Corrosion resistance of titanium surgical implant alloys: a review. In Corrosion and Degradation of Implant Materials, ASM STP 684. Syrett, B.C., Acharya, A., Eds.; ASTM: Philadelphia, 1979; 259–273.

143. Solar, R.J.; Pollack, S.R.; Korostoff, E. Titanium release from implants: a proposed mechanism. In Corrosion and Degradation of Implant Materials, ASM STP 684. Syrett, B.C., Acharya, A., Eds.; ASTM: Philadelphia, 1979; 161–172.

144. Solar, R.J.; Korostoff, E. In vitro corrosion testing of titanium surgical implant alloys: an approach to understanding titanium release from implants. J. Biomed. Mater. Res. 1979, 13, 217–250.

145. Sittig, C.; Wieland, M.; Marti, A.; Hauert, R.; Vallotton, P.H.; Textor, M.; Spencer, N.D. Surface characterisation of cp Ti, Ti_6Al_7Nb and Ti_6Al_4V. In ECASIA '97: 7th European Conference on Applications of Surface and Interface Analysis; Nyborg, L., Briggs, D., Olefjord, I., Eds.; John Wiley and Sons: Weinheim, 1997; 156–167.

146. Milosev, I.; Metikos-Hukovic, M.; Strehblow, H.-H. Passive film on orthopaedic TiAlV alloy formed in physiological solution investigated by x-ray photoelectron spectroscopy. Biomaterials 2000, 21, 2103–2113.

147. Silva, R.; Barbosa, M.A.; Rondot, B.; Belo, M.C. Impedance and photoelectochemical measurements on passive films formed on metallic biomaterials. Br. Corros. J. 1990, 25, 136–140.

148. Dolata, M.; Kedzierzawski, P.; Augustynski, J. Comparative impedance spectroscopy study of rutile and anatase TiO_2 film electrodes. Electrochim. Acta 1996, 41, 1287–1293.

149. Goossens, A. Intensity-modulated photocurrent spectroscopy of thin anodic films on titanium. Surf. Sci. 1996, 365, 662–671.

150. Thull, R. Semiconductive properties of passivated titanium and titanium based hard coatings for implants—an experimental approach. Med. Prog. Technol. 1990, 16, 225–234.

151. Aziz-Kerrzo, M.; Conroy, K.G.; Fenelon, A.M.; Farrel, S.T.; Breslin, C.B. Electrochemical

studies on the stability and corrosion resistance of titanium-based implant materials. Biomaterials 2001, *22*, 1531–1539.

152. Frauchiger, L.; Taborelli, M.; Aronsson, B.O.; Descouts, P. Ion adsorption on titanium surfaces exposed to a physiological solution. Appl. Surf. Sci. 1999, *143*, 57–77.
153. Ellingsen, J.E. A study on the mechanism of protein adsorption to TiO_2. Biomaterials 1991, *12*, 593–596.
154. Sousa, S.R.; Barbosa, M.A. Corrosion resistance of titanium cp in saline physiological solutions with calcium phosphate and proteins. Clin. Mater. 1993, *14*, 287–294.
155. Sundgren, J.E.; Bodo, P.; Ivarsson, B.; Lundstrom, I. Adsorption of fibrinogen on titanium and gold surfaces studies by ESCA and ellipsometry. J. Colloid Interface Sci. 1986, *113*, 530–543.
156. Sundgren, J.E.; Bodo, P.; Lundstrom, I. Auger electron spectroscopic studies of the interface between human-tissue and implants of titanium and stainless steel. J. Colloid Interface Sci. 1986, *110*, 9–20.
157. do-Serro, A.P.V.A.; Fermandes, A.C.; Jesus, B.D.; Saramago, V. Calcium phosphate deposition on titanium surfaces in the presence of fibronectin. J. Biomed. Mater. Res. 2000, *49*, 345–352.
158. Hanawa, T.; Ota, M. Calcium phosphate naturally formed on titanium in electrolyte solution. Biomaterials 1991, *12*, 767–774.
159. Hanawa, T.; Asami, K.; Asaoka, K. Repassivation of titanium and surface oxide film regenerated in simulated bioliquid. J. Biomed. Mater. Res. 1998, *40*, 530–538.
160. Panjian, L.I.; Ducheyne, P. Quasi-biological apatite film induced by titanium in a simulated body fluid. J. Biomed. Mater. Res. 1997, *41*, 341–348.
161. Yang, B.C.; Weng, J.; Li, X.D.; Zhang, X.D. The order of calcium and phosphate ion deposition on chemically treated titanium surfaces soaked in aqueous solution. J. Biomed. Mater. Res. 1999, *47*, 213–219.
162. Meachim, G.; Williams, D.F. Changes in non-osseous tissue adjacent implants. J. Biomed. Mater. Res. 1973, *7*, 555–572.
163. Zhu, X.; Kim, K.-H.; Jeong, Y. Anodic oxide films containing Ca and P of titanium biomaterial. Biomaterials 2001, *22*, 2199–2206.
164. Ishizawa, H.; Ogino, M. Hydrothermal prediction of hydroxyapatite on anodic titanium oxide films containing Ca and P. J. Mater. Sci. 1999, *34*, 5893–5898.
165. Wu, W.; Nancollas, G. Kinetics of heterogenous nucleation of calcium phosphate on anatase and rutile surfaces. J. Colloid Interface Sci. 1998, *199*, 206–211.
166. Lima, J.; Sousa, S.R.; Ferreira, A.; Barbosa, M.A. Interactions between calcium, phosphate and albumin on the surface of titanium. J. Biomed. Mater. Res. 2000, *55*, 45–53.
167. Zeng, H.; Chittur, K.K.; Lacefield, W.R. Analysis of bovine serum albumin adsorption on calcium phosphate and titanium surfaces. Biomaterials 1999, *20*, 337–384.
168. Klinger, A.; Steinberg, D.; Kohavi, D.; Sela, M.N. Mechanism of adsorption of human albumin to titanium in vitro. J. Biomed. Mater. Res. 1997, *36*, 387–392.
169. Wassell, D.T.H.; Hall, R.C.; Embery, G. Adsorption of bovine serum-albumin onto hydroxyapatite. Biomaterials 1995, *16*, 697–702.
170. Oliva, F.Y.; Avalle, L.B.; Macagno, V.A.; Pauli, C.P.D. Study of human serum albumin–TiO_2 nanocrystalline electrodes interaction by impedance electrochemical spectroscopy. Biophys. Chem. 2001, *91*, 141–155.
171. Tengvall, P.; Lundström, I.; Sjöqvist, L.; Elwing, H.; Bjursten, L.M. Titanium–hydrogen peroxide interaction: model studies of the influence of the inflammatory response on titanium implants. Biomaterials 1989, *10*, 166–175.
172. Tengvall, P.; Elwing, H.; Sjöqvist, L.; Lundström, I.; Burstein, L.M. Interaction between hydrogen peroxide and titanium: a possible role in the biocompatibility of titanium. Biomaterials 1989, *10*, 118–120.
173. Pan, J.; Thierry, D.; Leygraf, C. Electrochemical impedance spectroscopy study of the passive oxide film on titanium for implant application. Electrochim. Acta 1996, *41*, 1143–1153.

174. Pan, J.; Liao, H.; Leygraf, C.; Thierry, D.; Li, J. Variation of oxide films on titanium induced by osteoblast-like cell culture and the influence of an H_2O_2 pretreatment. J. Biomed. Mater. Res. 1998, *40*, 244–256.

175. Pan, J.; Thierry, D.; Leygraf, C. Electrochemical and XPS studies of titanium for biomaterial applications with respect to the effect of hydrogen peroxide. J. Biomed. Mater. Res. 1994, *28*, 113–122.

176. Pan, J.; Thierry, D.; Leygraf, C. Hydrogen peroxide toward enhanced oxide growth on titanium in PBS solution: blue coloration and clinical relevance. J. Biomed. Mater. Res. 1996, *30*, 393–402.

177. Fonseca, C.D.; Barbosa, M.A. Corrosion behaviour of titanium in biofluids containing H_2O_2 studies by electrochemical impedance spectroscopy. Corros. Sci. 2001, *43*, 547–559.

178. Blackwood, D.J.; Peter, L.M. The influence of growth-rate on the properties of anodic oxide-films on titanium. Electrochim. Acta 1989, *34*, 1505–1511.

179. Blackwood, D.J. Influence of the space-charge region on electrochemical impedance measurements on passive oxide films on titanium. Electrochim. Acta 2000, *46*, 563–569.

180. Ohtsuka, T.; Otsuki, T. The influence of the growth rate on the semiconductive properties of titanium anodic oxide films. Corros. Sci. 1998, *40*, 951–958.

181. Lee, T.M.; Chang, E.; Yang, C.Y. A comparison of the surface characteristics and ion release of Ti_6Al_4V. J. Biomed. Mater. Res. 2000, *50*, 113–119.

182. Groessner-Schreiber, B.; Tuan, R.S. Enhanced extracellular matrix production and mineralization by osteoblasts cultured on titanium surfaces in vitro. J. Cell Sci. 1992, *101*, 209–217.

183. Williams, D.F. Biocompatibility: performance in the surgical reconstruction of man. Interdiscip. Sci. Rev. 1990, *15*, 20–33.

184. Branemark, P.I. J. Prosthet. Dent. 1983, *50*, 399–407.

185. Biological performance. In *Titanium in Medicine*. Brunette, D.M., Tengvall, P., Textor, M., Thomsen, P., Eds.; Springer: Heidelberg, 2001.

186. Craighead, H.G.; James, C.D.; Turner, A.M.P. Chemical and topographical patterning for directed cell attachment. Curr. Opin. Solid State Mater. Sci. 2001, *5*, 177–184.

187. Perizzolo, D.; Lacefield, W.R.; Brunette, D.M. Interaction between topography and coating in the formation of bone nodules in culture for hydroxyapatite- and titanium-coated micromachined surfaces. J. Biomed. Mater. Res. 2001, *56*, 494–503.

188. Wieland, M.; Textor, M.; Spencer, N.D.; Brunette, D.M. Wavelength-dependent roughness: a quantitative approach to characterizing the topography of rough titanium surfaces. Int. J. Oral Maxillofac. Implants 2001, *16*, 163–181.

189. Brunette, D.M.; Chehroudi, B. The effects of the surface topography of micromachined titanium substrata on cell behavior in vitro and in vivo. J. Biomech. Eng. Trans. ASME 1999, *121*, 49–57.

190. Surface engineering. In *Titanium in Medicine*; Brunette, D.M., Tengvall, P., Textor, M., Thomsen, P., Eds.; Springer: Heidelberg, 2001.

191. Sohmura, T.; Tamasaki, H.; Ohara, T.; Takahashi, J. Calcium-phosphate surface coating by casting to improve bioactivity of titanium. J. Biomed. Mater. Res. (Appl. Biomater.) 2001, *58*, 478–485.

192. Sun, L.; Berndt, C.C.; Gross, K.A.; Kucuk, A. Material fundamentals and clinical performance of plasma-sprayed hydroxyapatite coatings: a review. J. Biomed. Mater. Res. (Appl. Biomater.) 2001, *58*, 570–592.

193. Ducheyne, P.; Healy, K.E. Surface spectroscopy of calcium phosphate ceramic and titanium implant materials. In *Surface Characterization of Biomaterials*; Ratner, B.D., Ed.; Elsevier Science Publishers: Amsterdam, 1988; 175–192.

194. Lange, G.L.D.; Donath, K. Interface between bone tissue and implants of solid hydroxyapatite or hydroxyapatite-coated titanium implants. Biomaterials 1989, *10*, 121–125.

195. Park, E.; Condrate, R.A.; Hoelzer, D.T. Interfacial characterization of plasma-spray coated calcium phosphate on Ti–6Al–4V. J. Mater. Sci., Mater. Med. 1998, *9*, 643–649.

196. Matsuura, T.; Hosokawa, R.; Okamoto, K.; Kimoto, T.; Akagawa, Y. Diverse mechanisms of osteoblast spreading on hydroxyapatite and titanium. Biomed. Mater. Eng. 2000, *21*, 1121–1127.

197. Overgaard, S.; Bromose, U.; Lind, M.; Bünger, C.; Søballe, K. The influence of crystallinity of the hydroxyapatite coating on the fixation of implants. J. Bone Jt. Surg., Br. 1999, *81-B*, 725–731.

198. Chou, L.; Marek, B.; Wagner, W.R. Effects of hydroxyapatite coating crystallinity on bio-solubility, cell attachment efficiency and proliferation in vivo. Biomaterials 1999, *20*, 977–985.

199. Kaciulis, S.; Mattagongo, G.; Napoli, A.; Bemporad, E.; Ferrari, F.; Montenero, A.; Gnappi, G. Surface analysis of biocompatible coatings on titanium. J. Electron. Spectrosc. Relat. Phenom. 1998, *95*, 61–69.

200. Kaciulis, S.; Mattogno, G.; Napoli, A.; Pandolfi, L.; Cavalli, M.; Montenero, A.; Gnappi, G. XPS study of apatite-based coating prepared by sol–gel technique. Appl. Surf. Sci. 1999, *151*, 1–5.

201. Fini, M.; Cigada, A.; Rondelli, G.; Chiesa, R.; Giardino, R.; Giavaresi, G.; Aldini, N.N.; Torricelli, P.; Vicentini, B. In vitro and in vivo behavior of Ca- and P-enriched anodized titanium. Biomaterials 1999, *20*, 1587–1594.

202. Wen, H.B.; Wijnm, J.R.D.; Blitterswijk, C.A.V.; Groot, K.D. Preparation of calcium phosphate coating on titanium implant materials by simple chemistry. Biomed. Mater. Res. 1998, *41*, 227–236.

203. Ban, S.; Maruno, S. Morphology and microstructure of electrochemically deposited calcium phosphates in a modified simulated body fluid. Biomaterials 1998, *19*, 1245–1253.

204. Feng, Q.L.; Wang, H.; Cui, F.Z.; Kim, T.N. Controlled crystal growth of calcium phosphate on titanium surface by NaOH-treatment. J. Cryst. Growth 1999, *200*, 550–557.

205. Kim, D.G.; Shin, M.J.; Kim, K.H.; Hanawa, T. Surface treatments of titanium in aqueous solutions containing calcium and phosphate ions. Biomed. Mater. Eng. 1999, *9*, 89–96.

206. Cleries, L.; Fernandez-Parads, J.M.; Morenza, J.L. Behavior in simulated body fluid of calcium phosphate coatings obtained by laser ablation. Biomaterials 2000, *21*, 1861–1865.

207. Cleries, L.; Fermandez-Paradas, J.M.; Morenza, J.L. Bone growth on and resorption of calcium phosphate coating obtained by pulsed laser deposition. J. Biomed. Mater. Res. 2000, *49*, 43–52.

208. Kim, H.-M.; Miyaji, F.; Kokubo, T.; Nakamura, T. Preparation of bioactive Ti and its alloys via simple chemical treatment. J. Biomed. Mater. Res. 1996, *32*, 409–417.

209. Vehof, J.W.M.; Spauwen, P.H.M.; Jansen, J.A. Bone formation in calcium-phosphate-coated titanium mesh. Biomaterials 2000, *21*, 2003–2009.

210. Sousa, S.R.; Barbosa, M.A. Effect of hydroxyapatite thickness on metal ion release from Ti_6Al_4V substrates. Biomaterials 1996, *17*, 397–404.

211. Ong, J.L.; Lucas, L.C.; Raikar, G.N.; Gregory, J.C. Electrochemical corrosion analysis and characterisation of surface-modified titanium. Appl. Surf. Sci. 1993, *72*, 7–13.

212. Ellingsen, J.E.; Videm, K. Effects of oxide thickness on the reaction between titanium implants and bone. Mater. Clin. Appl. 1995, 543–551.

213. Lee, W.H.; Park, J.W.; Puleo, D.A.; Kim, J.-Y. Surface characteristics of a porous-surfaced Ti_6Al_4V implant fabrication by electro-discharge-compaction. J. Mater. Sci. 2000, *35*, 593–598.

214. Buchanan, R.A.; Rigney, E.D.; Williams, J.M. Ion implantation of surgical Ti–6Al–4V for improved resistance to wear-accelerated corrosion. J. Biomed. Mater. Res. 1987, *21*, 355–366.

215. Sioshansi, P.; Oliver, R.W. Improvements in the hardness of surgical titanium alloys by ion implantation. Mater. Res. Soc. Symp. 1986, *55*, 237–241.

216. Wieser, E.; Tsyganov, I.; Matz, W.; Ruether, H.; Ostwald, S.; Pham, T.; Richter, E. Modification of titanium by ion implantation of calcium and/or phosphorus. Surf. Coat. Technol. 1999, *111*, 103–109.

217. Revie, R.W.; Greene, N.D. Corrosion behaviour of surgical implant materials: I. Effects of sterilization. Corros. Sci. 1969, *9*, 755–762.

218. Revie, R.W.; Greene, N.D. Corrosion behaviour of surgical implant materials: II. Effects of surface preparation. Corros. Sci. 1969, *9*, 763–770.

219. Browne, M.; Gregson, P.J. Surface modification of titanium alloy implants. Biomaterials 1994, 15, 894–898.
220. Wisbey, A.; Gregson, P.J.; Peter, L.M.; Tuke, M. Effect of surface treatment on the dissolution of titanium-based implant materials. Biomaterials 1991, 12, 470–473.
221. Ratner, B.D.; Johnston, A.B.; Lenk, T.J. Biomaterial surfaces. J. Biomed. Mater. Res. 1987, 21, 59–90.
222. Thomsen, P.; Gretzer, C..Macrophage interactions with modified material surfaces. Curr. Opin. Solid State Mater. Sci. 2001, 5, 163–176.
223. Werkmeister, J.A.; Tebb, T.A.; White, J.F.; Ramshaw, J.A.M. Collagenous tissue formation in association with medical implants. Curr. Opin. Solid State Mater. Sci. 2001, 5, 185–191.
224. Søballe, K.; Hansen, E.S.; Rasmussen, H.B.; Jørgensen, P.H.; Bünger, C. Tissue ingrowth into titanium and hydroxyapatite-coated implants during stable and unstable mechanical conditions. J. Orthop. Res. 1992, 10, 285–289.
225. Pilliar, R.; Deporter, D.; Watson, P. Tissue–implant interface: micromovement effects. Mater. Clin. Appl. 1995, 569–579.
226. Rabbe, L.M.; Rieu, J.; Lopez, A.; Combrade, P. Fretting deterioration of orthopaedic implant materials: search for solutions. Clin. Mater. 1994, 15, 221–226.
227. Chen, C.Q.L.; Scott, W.; Barker, T.M. Effect of metal surface topography on mechanical bonding at simulated total hip step–cement interfaces. J. Biomed. Mater. Res. (Appl. Biomater.) 1999, 48, 440–446.
228. Horswill, N.C.; Sridharan, K.; Conrad, J.R. A fretting wear study of a nitrogen implanted titanium alloy. J. Mater. Sci. Lett. 1995, 14, 1349–1351.
229. Savio, J.A.; Overcamp, L.M.; Black, J. Review: size and shape of biomaterial wear debris. Clin. Mater. 1993, 15, 101–147.
230. Doorn, P.F.; Campbell, P.A.; Worrall, J.; Benya, P.D.; McKellop, H.A.; Amstutz, H.C. Metal wear particle characterization from metal on metal total hip replacements: transmission electron microscopy study of periprosthetic tissues and isolated particles. J. Biomed. Mater. Res. 1998, 42, 103–111.
231. Que, L.; Topoleski, L.D.T. Surface roughness quantification of CoCrMo implant alloys. J. Biomed. Mater. Res. 1999, 48, 705–711.
232. Que, L.; Topoleski, L.D.T.; Parks, N.L. Surfaces roughness of retrieved CoCrMo alloy femoral components form PCA artificial total knee joints. J. Biomed. Mater. Res. 1999, 53, 111–118.
233. Que, L.; Topoleski, L.D.T. Third-body wear of cobalt–chromium–molybdenum implant alloys initiated by bone and poly(methyl methacrylate) particles. J. Biomed. Mater. Res. 2000, 50, 322–330.
234. Karrholm, J.; Borssen, B.; Lowenhielm, G.; Snorrason, F. Does early micromotion of femoral stem prosthesis matter? 4–7 Year stereoradiographic follow-up of 84 cemented prostheses. J. Bone Jt. Surg. Br. 1994, 76, 912–917.
235. Olmstead, M.L. Canine cemented total hip replacements: state of the art. J. Small Anim. Pract. 1995, 36, 395–399.
236. Goldberg, J.R.; Buckley, C.A.; Jacobs, J.J.; Gilbert, J.L. Corrosion testing of modular hip implants. Modul. Orthop. Implants, 1997, 157–176.
237. Brown, S.A.; Flemming, C.A.C.; Kawalec, J.S.; Placko, H.E.; Vassaux, C.; Merritt, K.; payer, J.H.; Kraay, M.J. Fretting corrosion accelerates crevice corrosion of modular hip tapers. J. Appl. Biomater. 1995, 6, 19–26.
238. Chan, F.W.; Bobyn, J.D.; Medley, J.B.; Krygier, J.J.; Yue, S.; Tanzer, M. Engineering issues and wear performance of metal on metal hip implants. Clin. Orthop. Relat. Res. 1996, 333, 96–107.
239. Raimondi, M.T.; Vena, P.; Pietrabissa, R. Quantitative evaluation of the prosthetic head damage induced by microscopic third body particles in total hip replacement. J. Biomed. Mater. Res. (Appl. Biomater.) 2001, 58, 436–448.
240. Saikko, V.; Ahlroos, T.; Calonius, O.; Keränen, J. Wear simulation of total hip prostheses with polyethylene against CoCr, alumina and diamond-like carbon. Biomaterials 2001, 22, 1507–1514.

241. Huo, M.H.; Salvati, E.A.; Lieberman, J.R.; Betts, F.; Bansal, M. Metallic debris in femoral endosteolysis in failed cemented total hip arthroplasties. Clin. Orthop. Relat. Res. 1992, *276*, 157–168.

242. Milliano, M.T.; Whiteside, L.A. Articular surface material effect on metal-backed patellar components—a microscopic evaluation. Clin. Orthop. Relat. Res. 1991, 204–214.

243. Milliano, M.T.; Whiteside, L.A.; Kaiser, A.D.; Zwirkoski, P.A. Evaluation of the effect of the femoral articular surface material on the wear of a metal-backed patellar component. Clin. Orthop. Relat. Res. 1993, 178–186.

244. McGovern, T.E.; Black, J.; Jacobs, J.J.; Graham, R.M.; LaBerge, M. In vivo wear of Ti_6Al_4V femoral heads: a retrieval study. J. Biomed. Mater. Res. 1996, *32*, 447–457.

245. Dowson, D.; Jobbins, B.; Seyedharraf, A. An evaluation of the penetration of ceramic femoral heads into polyethylene acetabular cups. Wear 1993, *162*, 880–889.

246. Cooper, J.R.; Dowson, D.; Fisher, J. Macroscopic and microscopic wear mechanisms in ultra-high-molecular-weight polyethylene. Wear 1993, *162*, 378–384.

247. Isaac, G.H.; Wroblewski, B.M.; Atkinson, J.R.; Dowson, D. A tribological study of retrieved hip prostheses. Clin. Orthop. Relat. Res. 1992, 115–125.

248. Nasser, S.; Campbell, P.A.; Kilgus, D.; Kossovsky, N.; Amstutz, H.C. Cementless total joint arthroplasty prostheses with titanium-alloy articular surfaces—a human retrieval analysis. Clin. Orthop. Relat. Res. 1990, 171–185.

249. Raab, S.; Ahmed, A.M.; Provan, J.W. The quasi-static and fatigue performance of the implant–bone cement interface. J. Biomed. Mater. Res. 1981, *15*, 159–182.

250. Jasty, M.; Maloney, W.J.; Bragdon, C.R.; O'Connor, D.O.; Haire, T.; Harris, W.H. The initiation of failure in cemented femoral components of hip arthroplasties. J. Bone Jt. Surg. 1991, *73B*, 551–558.

251. Stauffer, R.N. Ten-year follow-up study of total hip replacement. J. Bone Jt. Surg. 1982, *64A*, 983–990.

252. Mohler, C.G.; Callaghan, E.J.J.; Collis, D.K.; Oregon, E.; Johnston, R.C. Early loosening of the femoral component at the cement–prosthesis interface after total hip replacement. J. Bone Jt. Surg. 1995, *77A*, 1315–1322.

253. Stone, M.H.; Wilkinson, R.; Stother, I.G. Some factors affecting the strength of the cement–metal interface. J. Bone Jt. Surg. 1989, *71B*, 217–222.

254. Verdonschot, N.; Huiskes, R. Mechanical effects of stem cement interface characteristics in total hip replacement. Clin. Orthop. Relat. Res. 1996, 329–344.

255. Dallari, D.; Nigrisoli, M.; Catamo, L.; Gualtieri, G. Experimental wear in hip prostheses. Mater. Clin. Appl. 1995, 605–609.

256. Sochart, D.H. Relationship of acetabular wear to osteolysis and loosening in total hip arthroplasty. Clin. Orthop. 1999, *363*, 135–150.

257. Fisher, J.; Firkins, P.; Reeves, E.A.; Hailey, J.L.; Isaac, G.H. The influence of scratches to metallic counterfaces on the wear of ultra high molecular weight polyethylene. Proc. Inst. Mech. Eng., H 1995, *209*, 263–264.

258. Hampel, H.; Hector, L.G.; Nuhfer, N.T.; Piehler, H.R. Physical model of unit three-body abrasive wear of ultra high molecular weight polyethylene. Transactions of the 24th Annual Meeting of the Society of Biomaterials, San Diego, 1998; 221–235.

259. Wang, A.; Polineni, V.K.; Stark, C.; Dumbleton, J.H. Effect of femoral head surface roughness on the wear of ultra high molecular weight polyethylene acetabular cups. J. Arthroplast. 1998, *13*, 615–620.

260. McNie, C.; Barton, D.C.; Stone, M.H.; Fisher, J. Prediction of plastic strains in ultra high molecular weight polyethylene due to microscopic asperity interactions during sliding wear. Proc. Inst. Mech. Eng., H 1998, *212*, 49–56.

261. Yamac, T.; Kobayashi, A.; Bonfield, W.; Revell, P.A. The characterisation of polyethylene and metal wear particles, retrieved from tissue adjacent to failed total joint replacements using a novel extraction technique. New Biomed. Mater. 1998, *16*, 86–90.

262. Jacobs, J.J.; Skipor, A.K.; Doorn, P.F.; Campbell, P.; Schmalzried, T.P.; Black, J.; Amstutz,

H.C. Cobalt and chromium concentrations in patients with metal on metal total hip replacements. Clin. Orthop. Relat. Res. 1996, S256–S263.

263. McKellop, H.A.; Campbell, P.; Park, S.H.; Schmalzried, T.P.; Grigoris, P.; Amstutz, H.C.; Sarmiento, A. The origin of submicron polyethylene wear debris in total hip-arthroplasty. Clin. Orthop. Relat. Res. 1995, 3–20.

264. Campbell, P.; Ma, S.; Yeom, B.; McKellop, H.; Schmalzried, T.P.; Amstutz, H.C. Isolation of predominantly submicron-sized UHMWPE wear particles from periprosthetic tissues. J. Biomed. Mater. Res. 1995, 29, 127–131.

265. Margevicius, K.J.; Bauer, T.W.; McMahon, J.T.; Brown, S.A.; Merritt, K. Isolation and characterization of debris in membranes around total joint prostheses. J. Bone Jt. Surg. Am. 1994, 76A, 1664–1675.

266. Campbell, P.; Ma, S.; Schmalzried, T.; Amstutz, H.C. Tissue digestion for wear debris particle isolation. J. Biomed. Mater. Res. 1994, 28, 523–526.

267. Shanbhag, A.S.; Jacobs, J.J.; Black, J.; Galante, J.O.; Glant, T.T. Macrophage/particle interactions—effect of size, composition and surface-area. J. Biomed. Mater. Res. 1994, 28, 81–90.

268. Gwynn, I.A.P.; Wilson, C. Characterizing fretting by analysis of SEM images. Eur. Cells Mater. 2001, 1, 1–11.

269. Doorn, P.F.; Mirra, J.M.; Campbell, P.A.; Amstutz, H.C. Tissue reaction to metal on metal total hip prostheses. Clin. Orthop. Relat. Res. 1996, S187–S205.

270. Doorn, P.F.; Campbell, P.A.; Amstutz, H.C. Metal versus polyethylene wear particles in total hip replacements. Clin. Orthop. Relat. Res. 1996, S206–S216.

271. Black, J. Systemic effects of biomaterials. Biomaterials 1984, 5, 11–18.

272. Bianco, P.D.; Ducheyne, P.; Cuckler, J.M. Titanium serum and urine levels in rabbits with a titanium implant in the absence of wear. Biomaterials 1996, 17, 1937–1942.

273. Black, J.; Maitin, E.C.; Gleman, H.; Morris, D.M. Serum concentration of chromium, cobalt and nickel after total hip replacement: a six months study. Biomaterials 1982, 4, 160–164.

274. Woodman, J.L.; Black, J.; Nunamaker, D.M. Release of cobalt and nickel from new total finger joint prosthesis of Vitallium. J. Biomed. Mater. Res. 1983, 17, 655–668.

275. Koegel, A.; Black, J. Release of corrosion products by F-75 cobalt base alloy in rat: I. Acute serum elevations. J. Biomed. Mater. Res. 1984, 18, 513–522.

276. Hallab, N.J.; Mikecz, K.; Jacobs, J.J. A triple assay technique for evaluation of metal-induced, delayed-type hypersensitivity responses in patients with or receiving total joint arthroplasty. J. Biomed. Mater. Res. 2000, 480–489.

277. Merritt, K.; Brown, S.A. Release of hexavalent chromium form corrosion of stainless steel and cobalt–chromium alloys. J. Biomed. Mater. Res. 1995, 29, 627–633.

278. Tomás, H.; Carvalho, G.S.; Fernandes, M.H.; Friere, A.P.; Abrantes, L.M. Effects of Co–Cr corrosion products and corresponding separate metal ions on human osteoblast-like cell cultures. J. Mater. Sci., Mater. Med. 1996, 7, 291–296.

279. Yang, J.; Merritt, K. Detection of antibodies against products in patients after Co–Cr total joint replacements. J. Biomed. Mater. Res. 1994, 28, 1249–1258.

280. Wilson, J. Biocompatibility and tissue response to implants. An. Quim., Int. Ed. 1997, 93, 15–16.

281. Hunt, J.A.; Stoichet, M. Biomaterials: surface interactions. Curr. Opin. Solid State Mater. Sci. 2001, 5, 161–162.

282. Buser, D. Titanium for dental implants (II): implants with roughened surfaces. In *Titanium in Medicine*. Brunette, D.M., Tengvall, P., Textor, M., Thomsen, P., Eds.; Springer: Heidelberg, 1983; 875–888.

283. Esposito, M. Titanium for dental applications (I). In *Titanium in Medicine*. Brunette, D.M., Tengvall, P., Textor, M., Thomsen, P., Eds.; Springer: Heidelberg, 1983; 827–874.

284. Kadoya, Y.; Kobayashi, A.; Ohashi, H. Wear and osteolysis in total joint replacements. Acta Orthop. Scand. Suppl. 1998, 278, 1–16.

285. El-Warrak, A.O.; Olmstead, M.L.; von-Rechenberg, B.; Auer, J.A. A review of aseptic loosening in total hip arthroplasty. Vet. Comp. Orthop. 2001, 14, 115–124.

286. Goldring, S.R.; Schiller, A.L.; Roelke, M.; Rourke, C.M.; O'Neil, D.A.; Harris, W.H. The synovial-like membrane at the bone–cement interface in loose total hip replacements and its proposed role in bone lysis. J. Bone Jt. Surg. 1983, *65*, 575–584.

287. Johanson, N.A.; Bullough, P.G.; Wison, P.D.; Salvati, E.A.; Ranawat, C.S. The microscopic anatomy of the bone–cement interface in failed total hip arthroplasties. Clin. Orthop. 1987, *218*, 643–648.

288. Xu, J.W.; Kontinen, Y.T.; Lassus, J.; Natah, S.; Ceponis, A.; Solovieva, S. Tumor necrosis factor-alpha (TNF-alpha) in loosening of total hip replacement (THR). Clin. Exp. Rheumatol. 1996, *14*, 643–648.

289. Horowitz, S.M.; Purdon, M.A. Mechanisms of cellular recruitment in aseptic loosening of prosthetic joint implants. Calcif. Tissue Int. 1995, *57*, 301–305.

290. Jiranek, W.A.; Machando, M.; Jasty, M.; Jevsevar, D.; Wolfe, H.J.; Goldring, S.R. Production of cytokines around loosened cemented acetabular components; analysis with immunohistochemical techniques and in situ hybridisation. J. Bone Jt. Surg. Am. 1993, *75*, 863–879.

291. Olmstead, M.L. Canine cemented total hip replacements: state of the art. J. Small Anim. Pract. 1995, *36*, 395–399.

292. Ohashi, K.L.; Dauskardt, R.H. Effects of fatigue loading and PMMA precoating on the adhesion and subcritical debonding of prosthetic–PMMA interfaces. J. Biomed. Mater. Res. 2000, *51*, 172–183.

293. Lewis, G.; Mdlasi, S. Correlation between impact strength and fracture toughness of PMMA-based cements. Biomaterials 2000, *21*, 775–781.

294. Giddins, V.L.; Kurtz, S.M.; Jewett, C.W.; Foulds, J.R.; Edidin, A.A. A small punch test technique for characterising the elastic modulus and fracture behaviour of PMMA bone cement used in total joint replacement. Biomaterials 2001, *22*, 1875–1881.

295. Goodman, A.; Aspenberg, P. The effects of particulate cobalt chrome alloy and high density polyethylene on tissue ingrowth into bone harvest chamber rabbits. 40th Annual Meeting of Orthopaedic Research Society, New Orleans, 1994; 840–865.

296. Thornhill, T.S.; Ozuna, R.M.; Shortkroff, S.; Keller, K.; Sledge, C.B.; Spector, M. Biochemical and histological evaluation of the synovial-like tissue around failed (loose) total joint replacement prostheses in human subjects and a canine model. Biomaterials 1990, *11*, 69–72.

297. Dowd, J.E.; Schwendeman, L.J.; Macaulay, W.; Doyle, J.S.; Shanbhag, A.S.; Wilson, S. Aseptic loosening in uncemented total hip arthroplasty in a canine model. Clin. Orthop. 1995, *319*, 106–121.

298. Turner, R.M.; Urban, R.M.; Sumner, D.R.; Galante, J.O. Revision, without cement, of aseptically loose, cemented total hip prostheses; quantitative comparison on the effects of four types of medullary treatment on bone ingrowth in a canine model. J. Bone Jt. Surg. Am. 1993, 75–97.

299. El-Warrak, A.O.; Olmstead, M.L.; Noetzli, H.; Akens, M.; von-Rechenberg, B. A new model for interface tissue formation in cemented hip replacements. *in press.*

300. Hodgson, A.W.E.; Mischler, S.; Virtanen, S. Degradation of CoCrMo implants—a corrosion and wear. *submitted for publication*, 2003.

14
Epilamization/Barrier Films

Zygmunt Rymuza
Warsaw University of Technology, Warsaw, Poland

I. INTRODUCTION

The modification of the surface to prevent oil from spreading or creeping from the rubbing area of elements is very important task particularly in watch industry. The lubrication of miniature and micromechanisms differs from the lubrication of machines because they are usually lubricated for life during the assembly process. A small amount of oil (one drop) must ensure efficient lubrication over a long period of time. The amount of oil is usually 0.5–100 mg. Watch bearings need only 0.01 mg of oil for several years of operation. Barriers films to prevent oil spreading or creeping are applied in bearings of fine instruments, such as above-mentioned watches, clocks, fuse mechanisms, or in miniaturized equipment (electrical and electronic equipment to reduce surface electrical leakage) where lubricants have to be confined carefully to avoid deterioration of nearby components or materials. Sometimes during their normal use, such instruments need to be subjected to abrupt temperature changes from subzero to high temperature in presence of high humidities. In spacecraft, the control of oil migration is especially important for planned missions of 7–10 years duration. In addition to bearing reliability, the possible contamination of critical sensor surfaces and thermal shield materials due to oil migration is also of concern.

The oil is placed between two elements, usually with high surface free energies (e.g., metals and minerals), and because tension of oil is 30 or more times less than the surface free energy of metal or mineral, the oil spreads out or creeps from the bearing [1,2]. The surface tension of oil decreases with time because of the aging process [2]. The specific loads in the contact area of the rubbing elements can reach 1 GPa or more which, combined with the dynamic conditions of operation and high temperatures, are conductive to migration of the oil from the microbearing or other microtribosystem. This migration from the rubbing region can be prevented, for example, by special coating techniques called epilamization. Sometimes, these techniques refer to so-called barrier films surrounding the oil's drop and preventing it from spreading out.

There are three ways to prevent oil from spreading away from the microbearing: the oil can be inherently nonspreading, it can be made nonspreading by the careful use of additives, or the solid surface can be specially treated or modified. Sometimes, the magnetic field is applied to keep the oil in the rubbing region. This method is, however, rarely used because it needs the application of additional magnets and magnetic fluids which demonstrate poor lubricating properties particularly when applied as one drop during for-life lubrication. The

475

first aforesaid solution is the simplest, but the chemical stability of nonspreading oils, obtained mainly by various refining procedures from porpoise jaw, blackfish, olive, neat's-foot, and bone oils, is very poor. These oils age and their lubrication properties change with time. The second and third solutions will be discussed below. The latter, the most common solution, will be described first.

II. FUNDAMENTALS—CAUSES OF OIL MIGRATION

The dynamics of liquid spreading on a solid surface can be described by the change in the contact angle θ (Fig. 1). Full spreading occurs when $\theta \rightarrow 0$. The variation in the contact angle of a liquid laid on a solid surface is conditioned by the following forces [3]: 1) cohesive forces on the liquid; 2) interaction forces between molecules of solid and liquid (adhesive forces); 3) attractive (molecule) forces of solid at the perimeter of the drop; and 4) gravity forces. The first two kinds of forces resist the liquid spreading and the last two encourage it. The equilibrium (without taking gravity forces into consideration) is described by Young's equation $\gamma_l \cos\theta = \gamma_s - \gamma_{sl}$ where θ is the contact angle, γ_l is the surface tension of liquid, γ_s is the surface free energy of solid, and γ_{sl} is the interfacial tension. The spreading of oil drop is therefore the result of the acting force $\Delta\gamma$ related to the perimeter unit of the liquid–solid interphase circle $\Delta\gamma = \gamma_s - \gamma_{sl} - \gamma_l \cos\theta_m$, where θ_m is the momentary value of the contact angle. Because the value of γ_{sl} is relatively small, the spreading dynamics depends on the difference between γ_s and γ_l.

It is possible to achieve relatively small differences between γ_s and γ_l when the oil drop (with small surface tension 12–36 mN/m) is placed on the polymer surface (the surface free energy of polymers is lower than 100 mJ/m^2) [4]. The surface free energy of polymers can be described with the value of the critical surface tension of wetting γ_c introduced by Zisman [5]. He found that the plot of $\cos\theta$ vs. γ_l for a homologous series of liquids on a given low-energy solid is generally a straight line. The empirical value of γ_c is defined as the value of γ_l at the

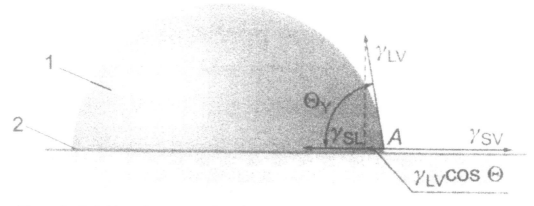

Figure 1 Definition of contact angle and geometrical interpretation of Young's equation. 1—liquid, 2—substrate material.

intercept of the plot $\cos \theta$ vs. γ_l with the horizontal line, $\cos \theta = 1$. Liquids with γ_l less than γ_c would be expected to spread on the solid surface.

The important requirement for the oil is strong adhesion to the solid surface. The adhesion between the surface is described by the classic Dupre equation $W_a = \gamma_l(1 + \cos \theta)$ where W_a is the specific work (energy) of adhesion. The work of adhesion decreases as γ_l decreases and θ increases, i.e., when the value of γ_s (at a given γ_l) decreases. The use of low surface free energy solids to prevent the spreading of the oil drop, in the oil-on-solid combination, is not a satisfactory solution to this problem. Because the work of cohesion W_k of the oil is $W_k = 2\gamma_l$, the nonwetting of the solid surface by the oil will occur when $W_a < (1/2)W_k$, limited wetting ($0° < \theta < 180°$) at $W_a > (1/2)W_k$, and full wetting at $W_a > W_k$. It is possible to describe therefore the optimum lubrication of the bearing in terms of non-spreading and maximum adhesion as $W_a > (1/2)W_k$ and $W = \mathrm{MAX}(W_a)$. Practical experience with instrument oils shows that the optimum relationship between adhesion of the oil to a solid surface and the tendency to spread occurs at the contact angle $\theta = 25–40°$.

Natural migration of oil is caused by the difference of the surface free energy of materials (steels, brasses, bronzes, and minerals) used in the construction of small mechanisms and the surface tension of used instrument oils. The surface free energy of these materials is usually higher than $1000\ \mathrm{mJ/m^2}$ and the surface free energy of typical instrument oils is $12–36\ \mathrm{mN/m}$. Spreading of oils is possible in practice on polymeric elements because of inhomogeneity of the material on the surface, in particular, when the polymeric material contains some additives. The migration of oil occurs mainly during operation of a mechanism, and also such phenomenon is observed during static shelf storage. Oil creeping is caused because of aging of oil which effects in the decrease of the surface tension of oil. Such phenomenon is characteristic for the first phase of aging (oxidation) when the surface tension of oil decreases. The rate of spreading out depends on the roughness of the surface. The migration is more intensive on the rough surfaces. The gravity forces also play a very important role when the layer of oil is thick enough. At higher viscosity of oil, the rate of spreading out is smaller. The spreading out of the oil can be described by the function $A = C \exp[w \ln(t)]$, where A is the area of oil contact with the substrate material, C and w are constants for the defined substrate–oil interface, and t is time.

The migration of oil can be caused by the movement of elements. It is observed, for example, in rolling bearings. The observation of oil creeping on the surface of the inner ring of the bearing revealed that the oil migrated because of centrifugal forces. The other serious cause of oil migration observed particularly in space mechanisms is creeping the oil because of temperature's gradient [6,7]. A small temperature gradient leads to rapid and complete migration of thin oil films to the colder regions. This phenomenon is so-called Marangoni effect. The velocity v of the thermally induced migration is governed by the equation

$$v = \frac{h}{2\eta} \frac{d\gamma_l}{dT} \frac{dT}{dx}$$

where h is the film thickness, η is the viscosity of the oil, γ_l is the surface tension of the oil, T is the temperature, and x is the distance.

Another mechanism for oil migration is capillary flow which describes the tendency of oil to flow along surface scratches and corners. This flow is driven by pressure gradients in the oil caused by variations in the radius of curvature of the oil–vapour interface. It does not occur on flat polished surfaces. In some spacecraft mechanisms, oil-filled reservoirs are usually incorporated to replenish the lubricant that has been lost through evaporation. This replenishment is effective only on roughest surfaces (e.g., by sand-blasted technique). On smooth surfaces, the presence of the reservoir creates an instability in any existing oil film such that dewetting occurs [8]. In this situation, no oil leaves the reservoir.

III. PREVENTING OIL MIGRATION BY SURFACE MODIFICATION

The prevention of oil migration can be achieved by the geometrical or chemical modification of the solid surface. An example of the geometrical modification of the surface of the mineral cover bearing of a balance in a watch is shown in Fig. 2, where the oil can only reach and fill the barrier circle.

The other geometrical approach to nonspreading is based on the edge effect. It is a well-known fact expressed by Laplace's equation that the pressure P inside a liquid is related to the curvature R and the surface tension γ_l of the liquid–vapour interface by $P = P_o + \gamma_l/R$ where P_o is the pressure outside the liquid and R is taken to be positive when the center of curvature is on the liquid side of the interface [3,7]. If an edge was uniformly coated with oil, the curvature of the oil–air interface would have to mimic that of the solid surface (Fig. 3a). This would produce pressure gradients in the oil given by

$$\frac{dP}{dx} = \frac{\gamma_l}{R^2}\frac{dR}{dx}$$

These pressure gradients drive the oil away from the edge and cause dewetting there (Fig. 3b). The dewetted edge acts as an effective barrier to subsequent migration (Fig. 3c).

The situation is reversed in the case of a corner where the pressure gradients cause the oil to be attracted and held. The process generally results in dewetting on either side of the corner (Fig. 4). However, since corners tend to act as sinks for large quantities of oil, they would appear to be less desirable than edges for use in blocking oil migration.

The effectiveness of a geometrical barrier against oil migration depends on the quantity of oil involved. When the film thickness exceeds, a critical value flow over a given barrier

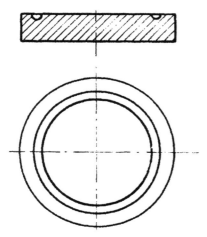

Figure 2 Geometrical modification of the surface of a cover balance bearing to prevent oil migration.

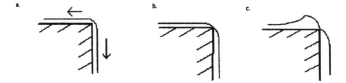

Figure 3 Dewetting of an edge by capillary flow.

would occur. The geometrical barriers do not prevent migration when the surface carried a rough finish. Rough edges do not prevent oil migration.

The standard method of preventing oil migration is through the use of chemical agents. However, this is sometimes considered to be undesirable because of the extra expense involved in their application as well as the fear that the chemical may interact with and somehow degrade the lubricant.

The following chemical methods of solid surface modification are used:

1. The surface is completely coated by monolayers of low surface energy compounds on which oil drops with higher surface tension will not spread.
2. The oil drop is surrounded with a narrow ring coating (epilame/barrier film) of a compound with low surface energy, or the entire surface of the element is coated and the coating is then removed from the area where the oil drop will be laid (Fig. 5). This is the so-called "Stop-Oil" method [9].
3. The solid surface is chemically treated to lower the surface free energy.

The first method was invented by Woog in France in the 1920s, the second method was developed in U.S. Navy laboratories after the last war, and the chemical treatment method was invented by Osowiecki and applied in watch industry in Switzerland in 1964.

The coating (epilame/barrier film) is laid on the solid surface from liquid or gas solution (or dispersion). The thickness of the layer usually does not exceed 500 nm. The active polar groups of the epilame macromolecules are physically adsorbed or chemisorbed on the solid surface. Osowiecki's method is very effective in the case of mineral solid surfaces. The effect of the chemical treatment of the surface is very stable and, for example, ultrasonic cleaning or chemical interactions between the oil and the material of the surface layer do not change the energetic properties of the surface. Only strong alkalis or acids can spoil the treated surface layer of ruby or sapphire jewels. Osowiecki's method has been applied by Reno S.A. and by KIF Parechoc in Switzerland to prevent oil migrating from the rubbing region of the cover bearing in the watch balance, especially in the Incabloc system.

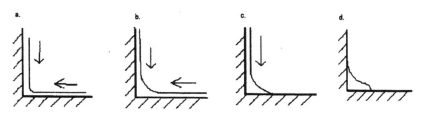

Figure 4 Accumulation of oil in a corner.

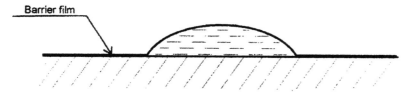

Figure 5   Stop-Oil method for preventing oil migration.

IV. EPILAMES/BARRIER FILMS

The requirements for the special coatings (epilames/barrier films) to prevent oil from spreading or creeping are as follows [1,2]:

1. Low surface free energy.
2. Strong adhesion to the solid surface.
3. High chemical stability.
4. The coating compound must be inert to the oil and the coated materials.
5. Broad temperature range of use.
6. Suitable hardness and shear strength.
7. Homogeneity of the coating layer on the whole solid surface (thickness, mechanical and physicochemical properties).
8. The smallest possible thickness.
9. The compound for the epilame should be soluble in common solvents.
10. The techniques for depositing coatings must not be too sophisticated.

The surface free energy of the epilame depends on the chemical structure of the macromolecules. The surface free energy of the epilame decreases for the macromolecules with side chains of the following structures [5]:

$$-CH_3 > -CF_2- > -CF_2H > -CF_3$$

The epilames with $-CF_3$ groups are characterized by their low surface free energy (ca. 6 mJ/m^2) (see Table 1 according to data taken from Ref. 5).

The compound for the coating is usually deposited on the solid surface from a liquid of gas solution (or dispersion). In the case of gas solution, an inert gas, e.g., argon, flowing through the chamber with the epilame's compound, takes its evaporated (at high temperature) molecules and brings them to the container with the elements to be coated. The polytetrafluoroethylene (PTFE) coatings are normally applied from dispersion in water, but it is difficult to meet the requirement of a brief heating period at high temperatures, which is necessary to sinter the PTFE particles, obtain good adhesion, and eliminate the water and additives. This treatment is often impractical for the manufacture of instruments. A more generally applicable and simpler method, which does not require any pretreatment, is the use of soluble compounds. The low surface energy films can be readily laid down from a solution in an appropriate fast-evaporating fluorosolvent, which, after evaporation, leaves a suitable coating.

The compounds actually used on the solid surface have the important advantage of not being readily dissolved by any of the cleaning solvents commonly used in the instrument field. Three groups of compounds are applied for the epilames: organic materials such as oleic or stearic acids, polysiloxanes, and fluorinated polymers. First two groups of com-

Table 1 Critical Surface Tension of Wetting γ_c of Layers with Different Chemical Structures of Surface Groups

Surface constitution	γ_c at 20°C (mN/m)
A. Fluorocarbon surfaces	
$-CF_3$	6
$-CF_2H$	15
$-CF_3$ and $-CF_2-$	17
$-CF_2$	18
$-CH_2-CF_3$	20
$-CF_2-CFH-$	22
$-CF_2-CH_2$	25
$-CFH-CH_2-$	28
B. Hydrocarbon surfaces	
$-CH_3$ (crystal)	20–22
$-CH_3$ (monolayer)	22–24
$-CH_2$	31
$=CH-$ (phenyl ring edge)	35
C. Chlorocarbon surfaces	
$-CClH-CH_2-$	39
$-CCl_2-CH_2-$	40
$=CCl_2$	43
D. Nitrated hydrocarbon surfaces	
$-CH_2ONO_2$ (crystal)	40
$-C(NO_2)_3$ (monolayer)	42
$-CH_2NHNO_2$ (crystal)	44

pounds have, nowadays, rather historical meaning. P. Woog in France in 1930 first patented a process in which a dilute solution of the fatty acids in a volatile solvent was used to coat the bearing surfaces with a thin film (called epilame) of fatty acid. Epilames based on polysiloxanes have been elaborated by Cetehor in Besancon in France in 1970s. The critical surface tension of wetting γ_c of these two kinds of epilames are very high (ca. 30 and 17–25 mN/m, respectively). Effective epilames have been developed in the Naval Research Laboratory in Washington, DC [9,10]. They are based on commercially available fluorinated polymers whose low γ_c values render them especially attractive for such application. The best compounds are fluorinated esters of poly(acrylic acid) or poly(methacrylic acid) with the values of γ_c 11.1 and 10.6, respectively. The lowest value of critical tension of wetting was found on the surface with an adsorbed fluorinated fatty acid monolayer. Modern industrially manufactured antimigration barrier films (epilames) are based on 0.1–2% solution of fluoropolymer in Freon-free perfluorinated solvent (Table 2). Such coatings can be deposited by dipping, spraying, or brushing on metals, minerals, glasses, and plastics. Recent studies [11] of the physicochemical interactions in metal coating of epilame system have shown that in such systems, the course of processes of change of a phase condition of materials, catalysis, and inhibition of oxidation of metal surfaces is possible. The products of contact reactions such as salts of greasy acids have been found.

Quite new antimigration barrier film formulation was elaborated recently by NASA [12,13]. This formulation consists of a fluoroalkyl silane having a low surface energy part, a liquid crystal silane operable for enhancing the orientation of the molecules of the

Table 2 Industrially Manufactured Antimigration Coatings (Epilames)

Description and producer properties	NyeBar Q Nye Lubricants Inc., Fairhaven, MA, U.S.A.	Antispread F2/50FK60 Dr. Tillwich-Werner Stehr, Horb-Ahldorf, Germany	Fixodrop FK/BS-10 Moebius et Fils, Allschwil (Basel), Switzerland
Material	Fluoropolymer	Fluoropolymer	Fluoropolymer
Solvent	Fluoroalkane (0.1–2% solution)	Fluorinated hydrocarbon (0.5–2% solution)	Fluorinated hydrocarbon
Critical surface tension of wetting γ_c of dried film (mN/m)	11	Test silicone (50 mm² sec⁻¹ at 20°C) oil forms a drop with contact angle 5–45°	Silicon oil form a drop with contact angle about 45°
Density (mg mm⁻³)	1.7 (25°C)	1.7 (20°C)	1.54
Boiling point (°C)	Below 60	30–60°C	47.6
Film thickness (nm)		40 (F2/50)	0.01–005%
Thermal stability of dried film (°C)	Up to 200	−75 to +200	−75 to +200
Flash point of solution	Nonflammable	Nonflammable	Nonflammable
Toxicity	Vapor or liquid contact with intense heat sources should be avoided, since pyrolysis products can be highly toxic	At 250°, fluorine is secreted	At 250°C, fluorine is secreted
Appearance	Clear and colorless liquid	Clear and colorless	Clear and colorless
Chemical stability	Can be used for coatings deposited on polymers	Can be used for coatings deposited on polymers	Can be used for coatings deposited on polymers
Technique for depositing films	Dipping, brushing, spraying. Air drying for 5 min. Baking the applied film for 15 min at 100°C produces a more durable film	Dipping, brushing, printing, spraying. Dipping: 5–10 s at 20°C Dries at 20°C in 10 s	Dipping, brushing, spraying. Dries at 50–80°C in 2–5 min
Productivity		ca. 100 g/m²	
Storage	At room temperature	12 months in closed container	12 months in closed container

fluoroalkyl silane and for cross-linking with the fluoroalkyl silane, and a transport medium (which can be an alcohol such as methanol or ethanol) for applying the fluoroalkyl silane and the liquid crystal silane to the surface of a substrate. The surface free energy of the film was found to be 11.5 mJ/m^2. The coating resisted water, oils, greases, various solvents, and moderate levels of abrasion such as rubbing with cloths.

Investigations carried out on metal-on-polymer and polymer-on-polymer miniature systems have shown that lubricating such combinations of materials are very effective way to significantly reduce their wear rates [1]. There is still a problem, however, with the oil drop migration from the rubbing region. The cause of the oil migration in polymeric systems is not quite clear; pure polymers have relatively very low surface free energy $<60 \text{ mJ/m}^2$ (see the Chapter 13). Because of the low work of adhesion between polymer and oil, the drop can be easily displaced on the polymeric surface.

The spreading of liquids on low-energy solid surfaces does not depend on the polar and nonpolar combinations of polymers and liquids [5,14]. The adhesion of hydrocarbon surfaces to liquids in aliphatic hydrocarbons was better than in liquids containing oxygen or fluorine but not as good as in aromatic hydrocarbons. The adsorption energy of hydrocarbon liquids on low-energy surfaces was largely due to dispersion interactions; the observed heats of adsorption set a maximum value for the potential in these systems [15].

The irregularities and roughness of realistic polymer elements affect the spreading of oils. The formation of an adhesive bond between a liquid and a solid surface depends on the development of a maximum area of molecular contact and the displacement of air from the microirregularities on the surface. The rate of penetration of liquids into capillaries and slits on the surface effects the dynamics of flow. The migration of the oil from polymeric miniature bearings is probably also affected by the application of fillers, plasticizers, solid lubricants, etc. to the polymeric material (composite). Fillers such as glass or carbon rise the surface free energy of polymers. The inhomogeneity of the surfaces can be very high. The coating (epilame) can be used to make the polymer surface uniform or as an antimigration barrier film.

V. SELF-COATING (AUTOEPILAMIZATION)

The idea of self-coating is very simple. The oil (specially prepared or with special additives) does not spread after being laid on the solid surface, but a layer with low surface energy is formed. This layer is formed as the result of chemical action or the selective adsorption of molecules, e.g., very active molecules of special additives in the lubricant.

Attempts have been made to synthesize a self-coating oil [9,16]. Three classes of nonspreading liquids have been distinguished:

1. Autophobic liquids—exemplified by molten stearic acid, octyl, alcohol, tricresyl phosphate, and trichlorodiphenyl.
2. Numerous esters able to spread completely on metal surfaces but unable to spread on glass, silica, or sapphire. The ester hydrolyzes immediately upon adsorbing onto these hydrated solid surfaces; of the two products of the hydrolytic reaction, the one with the greater average lifetime of adsorption remains to coat the surface with a close-packed monolayer, thereby blocking further progress of the hydrolysis reaction. If this protective monolayer has a critical surface tension of wetting, which is less than the surface tension of the liquid ester, nonspreading results.

3. Liquids whose surface tension is so high and adhesion energy is so low that the energy of adhesion is lower than that of cohesion and spreading is thus thermodynamically impossible. If such liquids existed, they would differ from autophobic liquids in not leaving a film behind them when rolled over a horizontal polished solid surface.

Pure nonspreading instrument oil with good lubrication properties and chemical stability has unfortunately never been found.

The best results have been obtained with the esters. The oils can be prevented from spreading by the addition of selected solutes which act in one of three ways. The first way is based on the ability of the solute to adsorb onto a high-energy surface and form a monolyaer with a critical surface tension of wetting less than the surface tension of the original oil. The second way is based on the gradient at the edge of an oil drop that opposes the spontaneous spreading of the oil. The third way is based on the ability of the solute to react chemically with the solid surface, forming a low-energy surface.

The five-ring polyphenyl ether isomeric mixture was inherently autophobic [i.e., was not spreading on its own surface film until its surface tension was less than the critical surface energy of spreading (γ_c) on steel surface in a dry nitrogen atmosphere] [17]. Two different techniques were used to determine the critical surface energy of spreading: a tilting-plate apparatus and a sessile drop apparatus. Surface tension was measured by a differential maximum bubble pressure technique over the range of 23–220°C. It was found that the critical surface energy/tension of spreading was 30–31 mJ/m^2, while the surface tension of oil was linearly decreasing vs. the temperature from 46 mN/m (at 20°C) to 28 mN/m (at 220°C). These values indicate that the phenyl ether does not spread on its own surface film until its surface tension falls below 30–31 mN/m. This surface tension corresponds to a fluid temperature of approximately 190–200°C.

The problem of autoepilamizing instrument oils is very complex and the suggested solutions are far from satisfactory. The chemical structure and concentration of the addition used to obtain nonspreading properties must be very carefully selected. The compound must be easily soluble in the oil at application temperature, adsorb quickly from the spreading edge of the drop, and form a surface layer with $\gamma_c < \gamma_l$. The surface tension of the base oil must not be lowered more than 5–6 mN/m. Nonvolatility and chemical stability on the solid surface are also very important requirements which are difficult to satisfy.

VI. COATING (EPILAME/BARRIER FILM) TECHNOLOGY

The selection of the correct technology for the deposition of the epilame/barrier film is extremely important. By "technology" we understand the procedure for preparing the solid surface to receive the coating (epilame), the process of depositing, and checking the quality of the coating.

The solid surface should be very clean before the deposition of a coating to ensure a strong adhesive bond between the coating and the solid surface. There are many techniques to choose [18–20]. Elements which are assembled on a production line to keep costs low are usually cleaned by chemical methods. Cleaning techniques using a solvent are effective enough for preparing a solid surface for the deposition of an epilame [1,21]. The elements are usually cleaned by using several solvents successively, for example, nonpolar (e.g., benzine) followed by polar (such as water or alcohol). If water is used, the last step in the cleaning procedure is the removal of its residue, for example, with acetone, or by using a centrifuge or drying with hot air. Microelements such as watch components are usually cleaned once in benzine, twice in soap solutions (at temperature of 60–70°C), and twice in acetone. Before each washing procedure, the elements are dried using a centrifuge. Ultrasonics is normally used to circulate the cleaning liquids in the baths. Often, the ultrasonic cleaning is completed with an ordinary washing process.

The evaluation of the cleanliness of the solid surface is difficult. There are many methods, but their efficiency depends on the cleanliness requirements. For industrial applications, the optical methods are very handy and cheap. Wettability is also commonly used to monitor the cleanliness of a surface. The other more sophisticated methods are ellipsometry, microfluorescence, evaporative rate analysis (ERA), and spectroscopic methods (AES, ESCA, ISS, and SIMS) [22,23].

A coating (epilame) can be deposited on the solid surface by dipping, brushing, or spraying. Dip coating is usually the most convenient and produces a uniform film. Dipping can last anywhere between a few seconds and over half a minute for the modern fluoropolymer epilames. A fine camel-hair brush can be used to paint bands of any desired width (between 1 and 3 mm for example). This method is usually applied with the Stop-Oil technique. Spraying is also very efficient but does not produce a very uniform film and handling problems arise (e.g., the need for adequate ventilation).

The fresh coating (epilame) is usually dried in air at room temperature or hotter (see Table 2). The film can be also cured at 50°C in vacuo for 3 to 4 hr [1,10]. The type of solvent and the mode of film drying affect the profiles of the film. An example of the industrial procedure for depositing Fixodrop epilame films (Table 2) follows with the cleaning and drying of the elements to be coated; the elements are placed then in a basket and are dipped into the epilame bath for 30 to 60 sec with slow rotation. After the treatment, the basket is lifted out of the liquid and the excess epilame is expelled by high-speed rotation for a few seconds. The elements then have to be dried for 2 to 5 min at 50–80°C. It is essential to dry the elements immediately after dipping because as a consequence of the high volatility of solvent in high atmospheric humidity, condensation can prevent the proper formation of the epilame film.

The quality of the epilame film depends on its surface free energy, mechanical properties, adhesion to the solid surface, topography (on the solid surface), thickness, and oil resistance. The surface free energy can be approximated by the value of the critical surface energy (tension) of wetting γ_c. Measuring the contact angle using a homologous series of liquids is a simple way to find this value. Rotating the drop allows more precise measurements to be made of the surface and also of the interfacial tension of the epilame–liquid systems [24]. The method proposed, for example, by Owens and Wendt [3,25] gives the values of bonding component γ_h (polar component) and dispersion force component γ_d of the surface free energy of the solid γ_s following Fowkes that $\gamma_s = \gamma_h + \gamma_d$ [14].

Using two liquids (e.g., water and methylene iodide) with known two above-mentioned components of the surface tension pf, the liquid γ_{ld} and γ_{lh}, and measuring the contact angle θ, the values of the polar and dispersion components of the epilame film can be estimated from the system of two equations based on the following modification of Young's equation:

$$\gamma_l(1 + \cos\theta) = 2\gamma_d^{1/2}\gamma_{ld} + 2\gamma_h^{1/2}\gamma_{lh}$$

More precise values of the polar and dispersion components of the free surface energy of the epilame film can be evaluated using the formula for calculation of interfacial tension (for polymer–liquid or polymer–polymer systems) γ_{sl} proposed by Wu [26]

$$\gamma_{sl} = \gamma_s + \gamma_l - \frac{4\gamma_d\gamma_{ld}}{\gamma_d + \gamma_{ld}} - \frac{4\gamma_h\gamma_{lh}}{\gamma_h + \gamma_{lh}}$$

When γ_{sl} is known, by using Young's equation and determining the values of θ for two liquids with known dispersion and polar components of surface tension, it is possible, as with Owens and Wendt method, to evaluate the dispersion and polar components of the epilame film.

The simple method used in practice to evaluate the surface energetics of an epilame film is to test it by using test liquids with a known surface tension. The series of test liquids may consist of, for example, polysiloxanes (e.g., with surface tension as low as 19–13 mN/m). A drop (as small as 1 mm in diameter) of such low surface tension liquid cannot spread on the epilame film giving the contact angle say 5–45°.

To determine the surface energetics of strongly curved solids surfaces is a rather difficult task. In this case, the analysis of the spreading dynamics of the test liquid can be proposed as a method. The author has shown [27] that this method can be satisfactorily used to estimate the surface free energy of polymer surfaces with a curvature radius of about 1 mm.

The methods useful to test the mechanical properties and the adhesion of the ultrathin (e.g., polymeric) films to solid surfaces can be applied to the investigation of epilame films. Hardness and elasticity modulus can be successfully estimated using nanoindentation technique [28]. The simple peel test, where the force required to peel the surface apart or the angle at which the surfaces spontaneously readhere can be determined [29], rolling methods, where a rigid cylinder or sphere rolls down an inclined polymeric plane, or the method of Schallamach waves or curves of detachment moving through the polymeric interface can also be applied.

The thin epilame films are transparent and are difficult to see with the naked eye. The local topography (on nanoscale, in normal direction to the substrate, and in micrometer scale, in lateral direction) of the films can be estimated effectively by the use of atomic force microscopy (AFM) [28]. Since the film is transparent, this difficulty seriously hampers inspection of the films for location and continuity. The incorporation in the fluoropolymer solution of a fluorescent indicator so that the resulting solidified coating might fluoresce sufficiently under ultraviolet radiation (UV) to be readily detectable was proposed [30]. The major drawback of fluorescent barrier films is the local variation in fluorescent intensity which restricts its use for monitoring the continuity of the coatings. The loss of fluorescence due to humidity is also a characteristic of fluorescent barrier films.

Another method for the inspection of a barrier film proposed by Borja [31] seems a more effective way to verify the proper application of barrier films. The barrier film can be observed and its thickness can be measured by means of a polarization interferometer. Interference fringes are superimposed on the barrier film surfaces. The interference fringes, the lines of constant phase difference, are like contour lines of the barrier film surface which follow each other at a level spacing equal to half a wavelength λ of the transmitted light. The use of a sodium vapor lamp ($\lambda = 589$ nm) with monochromatic light of excellent quality enables accurate measurements to be made because the interference bands appear in particularly good contrast. Small local irregularities on the substrate, such as scratches, are clearly shown up by displacements of the interference bands. A single scratch across the observed area of the barrier film surface without scoring enables the metal substrate to be observed by fringe displacement.

A qualitative scaling utilizing optical microscopy (with crossed polars and coaxial lighting) has been established relating thickness to color (that is, the thickness can be directly related to the color of the film). A film of optimum thickness (less than 250 nm) is invisible (transparent) but has a slight brownish tinge. Thicker, less desirable films (e.g., deposited on the steel surface of the ball bearing) have Newtonian fringes (rainbow colors). Even thicker, unacceptable films are gray and are very susceptible to mechanical abrasion and flaking. The optimum thickness of the film less than 250 nm ensures maximum resistance to abrasion while maintaining nonwettability and selective uniformity (as shown by experiments [10]).

Testing the oil resistance of any epilame film is usually reduced to checking the initial wetting of the film and the reaction to it. The film samples are immersed in the oil for certain time intervals at controlled temperatures, usually elevated. Then they are washed with detergent and water to make them free of oil. The measurement of contact angles with the test fluids has been shown to be a sensitive detector of surface changes [10,30]. The advancing contact angle with these liquids should be measured before and after oil immersion.

Scanning electron microscopy (SEM) or AFM analysis of the film surface is also an efficacious tool to reveal the surface changes after oil immersion. The surface pitting and roughing can be observed. Small patches and blisters appear where the film does not adhere firmly to the substrate. Spectroscopic methods can be used to inspect the effects of chemical reactions in the oil–epilame film system. The ESCA is a spectroscopic tool par excellence for studying in considerable detail aspects of structure and bonding in the surface regions of polymers. The typical sampling depth is 10 nm, and single ESCA experiment provides information which makes it possible to study the finest details of the surface regions of inhomogeneous samples.

VII. EFFICIENCY OF EPILAMIZATION AND FINAL REMARKS

The efficiency of epilamization is expressed by non-spreading of the oil from the bearing during long time of operation or storage. The epilamization of the elements of bearings of fine instruments, such as watches, clocks, fuse mechanisms, and ball bearings, increases reliability and lifetime of such components and various devices. On electrical and electronic equipment barriers, films reduce surface electrical leakage; they demonstrate low adhesion for dust, oil, grease, and water. The invisible barrier film can increase the life of oil-lubricated instrument ball bearings tenfold when the film is properly applied to the bearing ring faces [32]. The epilamization of the steel shaft in miniature porous bearings reduces effectively the creep of oil on the shaft's surface and its rapid evaporation. The deposition of the surface of miniature polymeric elements of fine mechanisms reduces the tendency of oil to spreading out and increases the lifetime in particular of polymer-on-polymer journal bearings [1,33]. The deposition of epilame film around the friction area of a bearing (Stop-Oil method) is more effective than the epilamization of the whole surface of the rubbing elements. The deposition of epilame in some area to compose a "net" of epilame islands is more effective than the use of continuous epilame film covering completely the surface of rubbing element(s) [34].

The efficiency of epilamization results mainly from the ability of epilame to keep the oil in the rubbing area. Some trials to use only epilame films on rubbing area to perform the lubrication functions have shown that the epilame film can reduce the friction force [34,35].

The strong adhesion of the film to the substrate is essential. It is particularly important under higher loads. The epilames produced on the basis of fluoric acid R_FCOOH are effective under light loads, but the epilame films produced from NN-dihydrazimine of fluoric acid $R_FCOOHN(CH_3)_2$ are effective under high loads [34]. The epilame film reduces the contact pressures in the rubbing area that reduces the structural changes in the material (e.g., steel) during tribological processes, so the presence of the epilame film can reduce the wear by this way. The high efficiency of epilames in reducing friction and wear in many metal, ceramic, and polymeric tribosystems including cutting tools have been proven [35]. An epilame film has slight influence on the reduction of the stick-slip phenomena (e.g., in the guides of machine tools or positioning systems) [36]. Modern epilames can be applied on any surface of metals, ceramic materials, minerals, glasses, and polymers.

The migration of oil is induced by difference in the surface free energy of the solid surface and the low tension of oil applied by capillary forces, temperature gradients, and gravity. The prevention of oil from spreading or creeping can also be realized by geometrical modifications of the surfaces of critical elements around the rubbing area. The smooth sharp edges are sometimes as effective as chemical agents (epilames) in preventing oil migration [7]. Rough edges do not exhibit this effect, however. Another possibility is to use special oils with autoepilamization properties. The synthesis of such oil is a very difficult task, they are very expensive, and their efficiency is lower as compared with the traditional epilamization.

REFERENCES

1. Rymuza, Z. *Tribology of Miniature Systems*; Elsevier: Amsterdam, 1989.
2. Tillwich, M.; Glaser, G. Tribologie der Uhren und Feinwerktechnik. *Handbuch der Chronometrie und Uhrentechnik Band II, Kapitel 8.4*; Ebner Verlag: Ulm/Donau, 1987.
3. Neumann, A.W., Spelt, J.K., Eds.; *Applied Surface Thermodynamics*; Marcel Dekker: New York, 1996.
4. van Krevelen, D.W. *Properties of Polymers*; Elsevier: Amsterdam, 1990.
5. Zisman, W.A. Relation of equilibrium contact angle to liquid and solid constitution. Adv. Chem. Ser. 1964, *43*, 1.
6. Fote, A.A.; Slade, R.A.; Feuerstein, S. Thermally induced migration of hydrocarbon oil. Trans. ASME—J. Lubr. Technol. 1977, *99*, 158–162.
7. Fote, A.A.; Slade, R.A.; Feuerstein, S. The prevention of lubricant migration in spacecraft. Wear 1978, *51*, 67–75.
8. Fote, A.A.; Slade, R.A.; Feuerstein, S. The behaviour of thin oil films in the presence of porous lubricant reservoirs. Wear 1978, *46*, 377–385.
9. Bernett, M.K.; Zisman, W.A. Prevention of liquid spreading and creeping. Adv. Chem. Ser. 1964, *43*, 332–340.
10. Kinzig, B.J.; Ravner, H. Factors contributing to the properties of fluoropolymer barrier films. ASLE Trans. 1978, *21*, 291–298.
11. Potecka, V.L.; Napreev, I.S. Investigation of features of physical–chemical processes in the system metal-coating of epilame. Proc. Natl. Acad. Sci. Belarus, Ser. Phys. Eng. Sci. 1, 21–24.
12. Durable low surface energy treatments. NASA Tech Brief July, 1992, *16* (7), 64.
13. Durable low surface-energy surfaces. US Patent 5,266,222, 1992.
14. Fowkes, F.M. Dispersion force contributions to surface and interfacial tension, contact angles, and heats of immersion. Adv. Chem. Ser. 1964, *43*, 99–111.
15. Whalen, J.W. Adsorption on low-energy surfaces. J. Colloid Interf. Sci. 1968, *28*, 443.
16. Bascom, W.D.; Cottington, R.L.; Singleterry, C.R. Dynamic surface phenomena in the spontaneous spreading of oils on solids. Adv. Chem. Ser. 1964, *43*, 355–379.

17. Jones, W.J. Contact angle and surface tension measurements of a five-ring polyphenyl ether. ASLE Trans. 1986, *29*, 276–282.
18. Mittal, K.L., Ed.; *Treatise on Clean Surface Technology*; Plenum Press: New York, 1987; Vol. 1.
19. Hudson, J.B. *Surface Science: An Introduction*; J. Wiley: Chichester, 1998.
20. Steigerwald, J.M.; Murarka, S.P.; Gutman, R.J. *Chemical Mechanical Planarization of Microelectronic Materials*; J. Wiley: Chichester, 1997.
21. Handelsman, Y.M. *Jewel Bearings*; Mashinostroyenie: Moscow, 1973. *in Russian.*
22. Clark, R.J.H., Hester, R.E., Eds.; Spectroscopy for Surface Science; J. Wiley: Chichester, 1998.
23. Hummel, D.O. *Handbook of Surfactant Analysis: Chemicals, Physico-Chemical and Physical Methods*; J. Wiley: Chichester, 2000.
24. Patterson, H.P., et al. Measurement of interfacial and surface tension in polymer systems. J. Polym. Sci., C 1971, *34*,31.
25. Owens, D.K.; Wendt, R.C. Estimation of the surface free energy of polymers. J. Appl. Polym. Sci. 1969, *13*, 1741.
26. Wu, S. Calculation of interfacial tension in polymer systems. J. Polym. Sci., C 1971, *34*, 19.
27. Rymuza, Z. Spreading of oil drops on highly curved polymer surfaces. J. Colloid Interface Sci. 1986, *112*, 221–228.
28. Bhushan, B., Ed.; *Handbook of Micro/Nanotribology*; 2nd Ed. CRC Press: Boca Raton, 1999.
29. Mittal, K.L., Ed.; *Adhesion Measurements of Films and Coatings*; VSP: Utrecht, 1995.
30. Fitz Simmons, V.G.; Shafrin, E.G. The detection, wettability and durability of fluorescent barrier films. ASLE Trans. 1974, *17*, 135–140.
31. Borja, P.C. Invisible barrier film. Lubr. Eng. 1981, *37*, 446–450.
32. Fitz Simmons, V.G.; Murphy, C.M.; Romans, J.B.; Singleterry, C.R. Barrier films increase service lives of prelubricated miniature ball bearings. Lubr. Eng. 1968, *24*, 35–42.
33. Rymuza, Z. Effektivität der Epilamisierung. Tribol. Schmier.tech. 1993, *40* (4), 221–224.
34. Garbar, I.I.; Kisiel, A.S.; Riabinin, N.A.; Sapgir, E.V. Nature and activity mechanisms of epilames in friction. Part 1 and 2. Trenie Iznos 1990, *11* (5 and 6), 292–800, 987–995. *in Russian.*
35. Poteha, V.L. Tribological efficiency of epilamization of cutting tools and machine components. Trenie Iznos 1992, *13* (6), 1070–1076. *in Russian.*
36. Lapidus, A.S.; Gitis, N.V.; Chizhov, B.N. Influence of epilamization on anti stick-slip effects in guides. Stanki Instrum. 1998. 26–28. *in Russian.*

Index

Milton Keynes UK
Ingram Content Group UK Ltd.
UKHW052024071024
449327UK00027B/2416